Cultural Understanding of Soils

Nikola Patzel • Sabine Grunwald
Eric C. Brevik • Christian Feller
Editors

Cultural Understanding of Soils

The Importance of Cultural Diversity and of the Inner World

Springer

Editors
Nikola Patzel
Freelance Scientist
Office for Soil Communications
Überlingen, Germany

Sabine Grunwald
Soil, Water, and Ecosystem Sciences
Department
University of Florida
Gainesville, FL, USA

Eric C. Brevik
College of Agricultural, Life, and Physical
Sciences
Southern Illinois University
Carbondale, IL, USA

Christian Feller
UMR Eco & Sols
Institut de Recherche pour le
Développement
Montpellier, France

ISBN 978-3-031-13168-4 ISBN 978-3-031-13169-1 (eBook)
https://doi.org/10.1007/978-3-031-13169-1

Cover credit: *A window into soil dug by school classes and Nikola Patzel in southern Germany. The soil is formed in glacial deposits. For more than 800 years it supported a common pasture.*

This Springer imprint is published by the registered company Springer Nature Switzerland AG
The registered company address is: Gewerbestrasse 11, 6330 Cham, Switzerland

Paper in this product is recyclable.

Visual introduction to cultural understanding of soils, in the form of a rose window (rosette) that creates a mandala. Credits: For the images see (clockwise starting with the lowest image) Chap. 4 for the soil microbe, Chap. 9 for the Native American Kiva, Chap. 3 for Adam, Chap. 18 for Darwin, Chap. 12 for Buddha, Chap. 2 for Ceres, Chap. 15 for Banerjee, Chap. 8 for the soil squares, and the soil cover image forms the centre. Concept M. Dold, composition Nikola und Silvan Patzel.

Preface and Acknowledgments

This book project began in 2016 in Rio de Janeiro, Brazil, at the inter-congress meeting of the International Union of Soil Sciences (IUSS). Such meetings take place every 4 years between World Congresses of Soil Science at the future venue of those who do the structural work for this scientific union and preparatory work for the congress.

Christian Feller from France had just been elected Chair of IUSS Division 4: "The Role of Soils in Sustaining Society and the Environment." His priority for this term of office was to strengthen the importance of the cultural human-soil relationship within the IUSS. Based on a collaboration on chapters for the book *Soil and Culture* 7 years earlier, Nikola Patzel was invited to come to Rio and co-found an official working group of the IUSS on "Cultural Patterns of Soil Understanding." More than 35 interested colleagues quickly volunteered to be founding members, which was a condition for the official establishment of the working group within the IUSS. On the spot, Christian Feller proposed a book project as the group's first major goal.

Two years later, at the 2018 World Congress of Soil Science in Rio, there were more soil culture presentations than ever before in this setting. These broadly international presentations laid a partial foundation for this book. From the outset, this publication was planned to be more than a collection of chapters, but all in all a steppingstone towards a new scientific paradigm in relation with nature and soil, as well as for soil communication. It is simply about "the importance of cultural diversity and the inner world"—as our subtitle reads—but to work with these is not easy at all. To implement the concept of this book, intense cooperation between the participants was necessary. In addition to years of writing, hundreds of hours of videoconferencing and email correspondence to work on the chapter texts as well as profound content debates and reviews were the result, both within the editorial team and with the authors.

We, the editors, would like to sincerely thank all the chapter authors for their great commitment and patience in the process of creating this work. The authors of all the chapters dealt here with topics that were not in their respective job descriptions or financed by project funds. The conditions under which this book came into being also confirm its necessity.

We thank Christian Feller's successors Damien Field and Claudio Zaccone as chairs of Division 4 and the IUSS Board of Directors for their support throughout the long process of the project. And we thank Springer for welcoming this project and promoting it with their publishing program.

Nikola Patzel thanks first and foremost Sabine Mayer-Patzel and his family for bearing the significant restrictions that increasingly arose over the years in various areas of life due to the workload of this project. He warmly thanks his teachers and collegial advisors in the fields of soil science and psychology for projects and competence building in the 25 years prior to this project, which created a potential for it; Theodor Abt shall be mentioned as his first mentor—since 1992—on the topics of this book. He is very grateful for the inter-cultural learning community that emerged from this IUSS working group in its pretty personal scientific exchange.

Sabine Grunwald thanks her many friends and colleagues from the transpersonal psychology and mindfulness community who were inspirational in intersecting soil science and pedometrics with psycho-spiritual and cultural themes for chapters of this book.

Eric C. Brevik thanks his wife Corinne and son Theodore for their patience and understanding as he devoted large amounts of time to this project. He also thanks mentors who showed him early in his career that the study of soils could extend to areas beyond just the science of soils, including Thomas Fenton, Dan Yaalon, Roy Simonson, John Tandarich, and Ed Landa. Finally, he thanks his co-editors for many thoughtful, interesting, enlightening, and sometimes animated discussions during the creation of this book.

Christian Feller thanks his three fellow editors for the great journey made together with this book. But he wants to say a special thanks to Nikola Patzel, who invested a lot of time in this task as moderator of the whole. This is especially true since Nikola works as a consultant and had to significantly reduce his other professional activities in order to make this project move forward. He also thanks the co-authors of Chaps. 17 and 18, J.-P. Aeschlimann, L. Feller (his wife), and N. Patzel. Finally, he turns to Ed Landa, a great friend, to say "thank you Ed" because the first and wonderful adventure we lived together with the co-edition of the book *Soil and Culture* (2010), which is at the origin of this new book.

We wish and hope that this book will be received both within and outside academic fields. May it be rediscovered from time to time in the future and may the insights and questions of its authors become fruitful for soil relations, soil management, and soil sciences.

Überlingen, Germany — Nikola Patzel
Gainesville, FL, USA — Sabine Grunwald
Carbondale, IL, USA — Eric C. Brevik
Montpellier, France — Christian Feller

Contents

About the Editors

Dr. Nikola Patzel received his degrees in Environmental Science and Depth Psychology and received his doctorate in 2002 at the ETH Zurich (Switzerland) for research on “Soil Science and the Unconscious”. After 10 years of research on that, he published a book, *Symbols in Agriculture*, in 2015. He works in research and consulting on nature relationships and soil communication, and gives courses on soil for farmers, which integrate scientific, cultural, and psychological aspects. Patzel chairs the working group on “Cultural Patterns of Soil Understanding” in the International Union of Soil Sciences (IUSS) and is Head of Commission VIII “Soil in Education and Society” at the German Soil Science Society (DBG).

Dr. Sabine Grunwald earned a Ph.D. in Environmental Sciences from Giessen University, Germany, and is Professor in the Soil, Water and Ecosystem Science Department at the University of Florida (UF), Gainesville, Florida, USA. She has research expertise in soil-ecosystem modeling, AI machine learning and deep learning, soil carbon quantification and modeling, pedometrics, digital soil mapping, soil health, soil proximal sensing and remote sensing, environmental quality assessment, and geospatial analysis. She has published over 213 peer-reviewed publications, is a highly cited researcher, and is actively engaged in the national and global soil science community. She also earned a Ph.D. in Integral and Transpersonal Psychology from the California Institute of Integral Studies, San Francisco, CA. Dr. Grunwald has served as the Director of the University of Florida Mindfulness Program since 2015 and teaches mindfulness meditation sessions and workshops. She is also a trained Embodied Life Coach providing services in ontological coaching to foster wellness, health and wellbeing and connect people to nature and the environment.

Dr. Eric C. Brevik is the Dean of the College of Agricultural, Life, and Physical Sciences at Southern Illinois University Carbondale. He has also taught courses in geology and soil science at Dickinson State University (DSU) and Valdosta State University, coordinated the DSU Environmental Science degree program, and advised student research. Dr. Brevik's research interests include combining information from soil science and geology, soil genesis, and the impact of humans on soil properties and processes, as well as soil science history, education, and links between soil science and culture. He is an active member of the European Geoscience Union, International Union of Soil Sciences, and Soil Science Society of America.

Dr. Christian Feller is an Emeritus Soil Scientist and the former Director of Research at the "Institut de Recherche pour le Développement" (IRD) in Montpellier, France. He earned his M.S. (1969) and Ph.D. degrees in Organic Chemistry (1972) from the Sorbonne University (Faculty of Sciences) in Paris, and his Doctorate of Science (1994) in Soil Science from the Louis Pasteur University in Strasbourg. His research focuses on soil organic matter studies applied to soil fertility and environmental services—in particular, the impact of agroecological practices on soil-plant carbon sequestration in tropical and subtropical areas; he has worked extensively in Senegal, French West Indies (Martinique), Brazil, and most recently, Madagascar. Christian is a member of the French Academy of Agriculture, and was the first recipient of the Soil Science Society of America's Nyle C. Brady Frontiers of Soil Science Lectureship in 2006. He served as Chair (2014–2018) of the Division 4 on The Role of Soils in Sustaining Society and the Environment of the International Union of Soil Sciences.

Part I
Introduction to Cultural Soil Dimensions

Chapter 1
Introduction: What Do We Know About Soil and Culture?

Nikola Patzel, Sabine Grunwald, Eric C. Brevik, and Christian Feller

1.1 What on Earth Is Soil?

"Soil" is a word amongst many, and it is plurivalent. In French, if you say that you work "on soil", it is understood that you are not in the sky like an airplane pilot, but that you are actually employed on the ground such as at an airport. Other variations of "soil" include the "sol de nos ancètres" (land of the ancestors) or the "droit du sol" (law of the land).

In the earth sciences, the soil of the archaeologist, the geotechnician, or even the soil of the agronomist is not the same as the soil of the soil scientist. Within the discipline of pedology, or soil science, one can evoke the microbial and rhizospheric, organic, and mineral soil. One may speak of the living soil, health of the soil, the fatigue or degradation of the soil, etc. Some of these qualifiers show quite

N. Patzel (✉)
Freelance Scientist, Office for Soil Communications, Überlingen, Germany
e-mail: nikola.patzel@soilcom.com

S. Grunwald
Soil, Water, and Ecosystem Sciences Department, University of Florida, Gainesville, FL, USA
e-mail: sabgru@ufl.edu

E. C. Brevik
College of Agricultural, Life, and Physical Sciences, Southern Illinois University, Carbondale, IL, USA
e-mail: eric.brevik@siu.edu

C. Feller
UMR Eco & Sols, Institut de Recherche pour le Développement, Montpellier, France
e-mail: christian.feller@ird.fr

N. Patzel et al. (eds.), *Cultural Understanding of Soils*,
https://doi.org/10.1007/978-3-031-13169-1_1

organic concepts of soil, maybe organicism and elements of anthropomorphism. Are they scientific? Are we still in the realm of outer observables, or to some extent in that of the human interior world of imagination and a vast context of associations, feelings, and meanings? In any case, it is clear that scientific definitions of soil have changed continuously, sometimes due to non-scientific circumstances. Today's definitions are no longer the same as those, for example, from the time of the recognized founder of pedology, Vasilii Vasil'evich Dokuchaev (1883).

Analyzing how scientists talk about soil and "define" it shows how scientific approaches towards soil depend on the scientists' self-image as researchers in their scientific and social context (Patzel et al., 2000), whether a scientific author claims the power to define a term or not. An example of a normative definition is: "Soil fertility should not only be understood as representing yield opportunities, but also as the integral of various pedologic and socio-economic quality indicators." Such a normative definition claims that the reader should use the term in the proposed manner, basically implying that the author's definition is the correct one, or at least "better" than others. But often, authors do not stipulate a definition, but just report (or promote) what can be observed in the literature as an apparently common definition, for example, "Soil fertility is generally understood as the potential of the soil to bring forth yield."

The second differentiation in how soils are scientifically seen can be made by the kind of definition itself, whether it says what a soil *is* or what it *is for.* A mainly descriptive definition is: "Soil is the material in the top layer of the surface of the earth in which plants grow, and it is formed by living organisms, organic matter as well as water, air and mineral particles" (quotation from EU Commission 2021, see below). This type of definition fits common scientific practice well because it allows the definition of the term to be given as a set of generally measurable features and properties. Such descriptive definitions of soil allow one to discern characteristics. As a result, descriptive definitions are frequently used to make a scientific statement without explicitly committing to personal or cultural values.

Natural scientists are used – or tempted – to say what *is*. At its extreme, this can lead to an essentialistic claim, which wants to say something about how things really are in their essence. The old, almost magical, belief that the name given to something may grasp its being can also occur in modern science, for example: "This is an Oxisol!" "No, this is a Ferralsol!" In the case when *no* such supposedly "objective claim of truth" (ontologic claim) is made or intended, terms of soil are rather of operational significance, helpful for communication and open to be changed depending on scientific and also broader cultural context and purpose.

Another way to address soil is to tell the audience not what soil *is*, but what a soil *is for*. It is about the significance of soil, often carrying the author's (or his/her donor's) idea about its function, purpose, and benefits derived from soils. For example, the definition "a fertile soil brings forth yield" is a functional definition; the purpose is the yield, and the way that yield is achieved matters less. This goal-oriented definition type is also very handy for an "interdisciplinary" or "integrated" approach that tries to incorporate different desired qualities.

Functional views of soil are particularly connected to scientific and societal zeitgeist, and often enter into legislation, production guidelines, and policy papers on sustainability or "development goals". For example, the European Commission framed soil in its public consultation on its new "Soil Strategy" in April 2021: "Soils are essential ecosystems that deliver valuable services such as the provision of food, energy and raw materials, carbon sequestration, water purification and infiltration, nutrient regulation, pest control and recreation. Therefore, soil is crucial for fighting climate change, protecting human health, safeguarding biodiversity and ecosystems and ensuring food security." This functional type of definition usually results from a negotiation process in either the scientific discourse or public arena among all those who manage to get a (more or less loud) voice in the discourse as stakeholders. It can therefore be particularly useful for incorporating socially expressed features, including the cultural dimension of soil.

Soil scientists typically agree that soils form at Earth's surface due to physical weathering and biogeochemical transformation processes, and soils are capable of supporting life. Most definitions of "soil" include these concepts (Brevik & Arnold, 2015), but concrete, scientific definitions of soil differ widely. Some call to change or refine the definition of soil such that humans are recognized as the assumed dominant soil forming factor (Sobocká, 2005), particularly as the concept of the Anthropocene has gained acceptance (Richter, 2020). Some think that reliance on biogeochemical transformation processes should be removed from the definition because there is no evidence of such processes on extraterrestrial bodies like our moon, and the surface regolith of these bodies should be included in the definition of soil (Ibáñez & Montanarella, 2013). Others go in the opposite direction, and argue that soils should be defined as a society or community that is "governed" by soil organisms (Morozov, 1998). The definitions as provided by soil scientists have changed over time on multiple occasions (Hartemink, 2016) or have been based on the specific interests driving the definition (Mizuta et al., 2018, 2021). The Soil Science Society of America recently sought to create a new definition of soil that was inclusive of diverse environments (including urban, hydric, desert, extraterrestrial, etc.), broadened the concepts of horizons and particle arrangements, was less theoretical, and included greater flexibility (van Es, 2017). There seem to be as many different proposed definitions for soil as there are perspectives on and uses of soil. *Soil itself does not care, it has its preconceptual existence regardless of what people call it and what they include or do not include in its definition. However, whether people care about the soil is likely to depend a lot on their cultural understanding and image of soil.*

Even beyond the definition of soil itself, it may be good to acknowledge all the words that can address soil. Just in the English language, words like "earth," "nature," "sod," "dirt," "clay," "turf," and "land" can evoke images that include or serve as synonyms for "soil." The word "earth" also refers to many cultural interpretations, such as "mother earth," "promised land," "land of the ancestors," "earth below our feet," or "planet earth." The story of Adam and Eve is a common mythologem of the three Abrahamic religions. In the Hebrew Bible, man was created from "earth," "soil," "clay," or "dust," depending on the translation and book being read

(Chap. 3, compare to Chap. 6). Other cultures have similar creation connections to soil. For example, the narratives of human creation from various Native American tribes include clay, earth, mud, and soil as materials that humans were made from (Chap. 9), and the shamanistic cultures of Siberia and the Far East believe that humans were made of soil (Chap. 5).

1.2 What About Culture?

Like the word "soil," the word "culture" has many meanings and definitions. Among these are formally narrow or semantically very broad ones. The word culture "derives from the Latin root *colere* and is associated with activities of preservation, of tending to and caring for" (Benhabib, 2002, p. 2). The United Nations Educational, Scientific and Cultural Organization (UNESCO) World Conference on Cultural Policies in Mexico City stated (Final Report 1982): "The cultural heritage of a people … includes both tangible and intangible works through which the creativity of that people finds expression [p. 43, § 23]. … Every people therefore has the right and a duty to defend and preserve its cultural heritage … in which they find a source of creative inspiration [§ 24]. … Science and technology teaching should be conceived above all as a cultural process which develops critical awareness" [p. 44, § 35]. The UNESCO Universal Declaration on Cultural Diversity states (2001, p. 3): "Reaffirming that culture should be regarded as the set of distinctive spiritual, material, intellectual and emotional features of a society or social group, and it encompasses, in addition to the art and literature, lifestyles, ways of living together, value systems, traditions and beliefs."

By this international and extensive definition, culture refers, for a given society or social group, to practically all the traits that characterize humanity and its ways of being, whether it concerns a materialistic and rationalist (such as science and technology) or refers to the spiritual-religious (religions, myths, beliefs) or ideological (dogmatic, partisan) approach. Tales and narratives, artistic works, and any experiences of relation to nature that trigger emotions and create meaning: all of that can inform cultural patterns of soil understanding too. Cultural patterns give people orientation; cultural understanding provides links between distinct areas of life experience, including what can be called the outer and inner world, the outer and inner nature, the outer and inner soil of humans on earth. The popular and also scientific tendency in the Global West to place the "outer" with "objectivity" and the "inner" with "subjectivity" may be questioned, and the distinction between "inner" and "outer" cannot be taken absolutely either. The opposite may also be true for both or none of them.

The cultural dimension is important in the context of communication about soil with a general non-scientific public (Brevik et al., 2019). Chapter 24, provides multiple educational lenses of soil curricula across the globe emphasizing the diversity and plurality of understanding soils. The German philosopher of culture Gottfried Herder (1744–1803) saw culture as a product or a "blossoming" out of the natural and historical situation of a people, and at the same time being a factor of formation of each individual: "Just as a man who comes into the world knows nothing: he must learn what he wants to know"; people would also learn in this way (Herder, 1813, p. 204). This was essentially done through the social and natural environment. The more the culture of a state is "a human wisdom learned from nature," "the lower" its center of gravity is, the more stable it can remain (Herder, 1813, p. 207). Herder exclaimed in this context: "Great Mother Nature, on what trifles you have tied the fate of our species" (1813, p. 52, transl. by eds.). In many cultures the images of mother earth are central in human nature and soil relation, and often appear in the shape of a goddess.

Depending on the approach or context, many definitions have been developed for the concept of culture, including more than 150 in the work by Kroeber and Kluckhohn (1952). For example:

- In philosophy, the word culture can designate that which is different from nature. But there is nothing universal about the opposition between nature and culture; it essentially characterizes Western thought and its scientific approach. This was recently widely debated by the anthropologist Philippe Descola in his work *Par-delà nature et culture* (Beyond nature and culture, Descola, 2005).
- Classically, in social and cultural anthropology, culture is used to distinguish human groups occupying a certain geographical space during a given period and their associated rituals, beliefs, values, etc. Here "a culture" can mean, for example, the Indus Valley Civilization in the Bronze Age and its tilling practices with water buffalos; or the art of cheese-making in 18th century Central Switzerland with its mountain pasture and hay farming techniques and beliefs; or the present traditional way of living and doing agriculture with square plots practiced by the Dogon people in West Africa.

On the influence of the cultural environment on scientific choices, current examples are the concepts of "living soil" and "soil health." These concepts are part of an idea of a nature that is soaked by vitalist currents. Such theories have been developed since Antiquity and were re-expressed in the 20th century with the writings of Lady Eve Balfour (1948, *The Living Soil*), Sir Albert Howard (1947, *The Soil and Health*), and many others. Currently, these notions are the subject of research and reflection within and at the highest level of soil science, with numerous articles, international declarations of the IUSS, and congresses addressing them. The question is not to pass judgment on these approaches, but simply to be aware of the fact that scientists in their theories and practices are not cut off from historical, or any kind of, socio-cultural influences (Brevik & Hartemink, 2010; Patzel, 2015).

From all of this, we will note that at the most general level, culture appears as a social phenomenon that is brought forth by people. Social includes all aspects of a society as viewed through a variety of lenses. For example, shamanistic cultures in Siberia (Chaps. 5 and 16), Buddhist cultures (Chaps. 11 and 12), Judeo-Christian rooted cultures (Chap. 3), Islamic culture (Chaps. 6 and 7), and urban cultures (Chap. 14) have influenced how people relate to and care for soils. Through time, cultures have adapted or co-created different views, beliefs, and uses of soils as demonstrated in North America (Chap. 9) ranging from creation myths of Native Americans to Judeo-Christian concepts of domination over the land to control and rule it (on "God's behalf"), agricultural commodification of soils, to industrialization and capitalist culture competing with environmentalism. Isolated cultures are the exception and cultures that communicate with each other are the rule, as has been proven many times for Europe, the Middle and Far East as well as for the Americas since the Stone Age, among other places and times.

The terms *culture* and *civilization* are understood differently in various linguistic and scientific areas that have adapted these words from Latin: sometimes showing significant, even ideological opposites, sometimes as quasi-synonyms. We find it interesting that towards the end of this introduction it becomes clear that books on soil have tended to speak of civilizations rather than cultures, which could possibly mean a certain emphasis on material culture (see Sect. 1.3).

A culture can entail intimate relations between various aspects of a society with soil, for example, as sacred, divine, or to be cared for. There can also be rather indifferent relations between culture and soil, a distancing, even a disconnect between or disregard of people for soils and the land. Another cultural distinction is among rather demarcated cultures with strong notions of cultural identity in contrast to the globalization of the so-called "Western culture," as well as the mingling or merging of individual elements from different cultural areas in a kind of global cultural tissue. In many places across time cultural separatism and identity politics have stood in contrast to global integration fused into a diverse amalgam of cultures, a global village. In many and very different societies, traditionalists and conservatives have argued to preserve cultures to keep them separate and "pure," while progressives embrace the hybridization of cultures into a larger whole asserting unity-in-diversity of cultures. According to Pieterse (2015), globalization involves cultural transformation with modernization, osmosis, reform, revivalism (e.g., reconnecting with indigenous cultures), naive romanticism, antagonism, and more. This has fostered hybridization of cultures that are blending and blurring, and sometimes clashing literally in the form of culture wars. The globalization of cultures has impacted individual cultures, such as Native American (Chap. 9), Dogon (Chap. 8), Chinese (Chap. 13), and Bengali (Chap. 15) and their relations to soils and nature. Cosmopolitanization and global mélange have contributed to create power structures that led to the marginalization, destruction, suppression, and oppression of specific cultures (e.g., Indigenous cultures) through time. Specifically, colonialism and postcolonialism have contributed to the expression of different eco-identities that refer to how people relate to soil ecosystems and nature and the land in general (Chap. 20).

In this sense, cultures are usually not as homogeneous as is often assumed (Benhabib, 2002). Thus, although each of the chapters in this book is written through (a) specific cultural lens(es) – religious and spiritual, social and philosophical, educational and psychological – we invite readers to hold these views lightly. It is worthwhile to assume that specific cultures do not only have some specific characteristics that may sharply distinguish them from others, but that there is also a great deal of fluidity, blurring and blending of diverse cultures, and amorphous traits that reside within cultures. Each reader of this book may be situated in multiple cultural perspectives. Our understanding of soils is viewed through a plurality of cultural contexts where one is not *a priori* better or worse than another, but simply different. A research soil physicist or engineer may look at soil through the physical-pedological cultural lens, while an urban gardener may relate to soils as a living breathing space within the city to provide beauty. A shaman in a local community views soil as sacred, a farmer in rural America considers soils to be cropped for food and livelihood, while a pedometrician views soil as something to be placed in a digital representation, an abstract image of a soil with associated properties represented through machine generated code. One may pendulate from an external relating to soils (outer soil) to an "inner soil" due to whatever cultural perspective, lens, and personal experience comes to the fore in a given moment. Such dynamics of culturally informed relations to earth and nature, land and soils remind us that we all hold the capacity to slip into different psychic contexts and cultural perspectives and immerse with awareness in what they have to offer. Whatever dominant cultural view(s) one may hold, the chapters in this book are an invitation to learn from a plurality of diverse cultural understandings of soils. Familiarity with the spectrum of cultural understandings of soils hold potential to individually embody and to socially co-create deeper meaning and purpose to view, use, care, and relate to soils in new ways.

To the view that culture dwells mainly in collective memory and shared convictions, there is a rather exotic view of what culture is: a deeply embodied pattern of identity of how things are and should be related to, that is, paradigms and codes of behavior, even involving the subconscious and unconscious of a member of a social group or nation, or this in whole. In this understanding, while culture is not necessarily "repressed" like parts of our inner nature (instincts, drives and dreams) may possibly be, culture is viewed as becoming a kind of tacit knowledge, a set of implicit basic assumptions, resonating with deep layers of the psyche or emerging from it, and even remaining there when cultural consciousness is lost. This perspective led to the statement of French writer Emile Henriot (1889–1961): "Culture is what remains when we have forgotten everything."

Having shown glimpses into the open horizon of various meanings of "culture", here follows a more emotive perspective that speaks to many chapters in this book: *cultural understanding of soils can be seen tentatively to be one that is in touch with the intangible.* A world of inner images and traditional symbols of meaning has been given expression and shape in the arts, sacred buildings, and landscapes that are filled with religious ideas or their substitutes. Cultural development may also happen after experiencing visions and other numinous inner experiences if they are perceived as collectively significant beyond individual scope, and become part of the corpus of imagination and ideas of a cultural tradition. This has occurred throughout the history of humankind in many cultures and settings.

Case studies in this book show how people's relations to soil have been shaped by natural and/or cultural patterns including the central values of their culture. Or they show that this is often not the case in the contemporary world, that many people have turned their backs to the inner orientation that is provided by the great myths, revelations and tales of their cultural substrate. The resulting cultural and societal landscape of soil understanding may be either overly homogenized or highly fragmented. A re-connection and new connections, as well as diversity sprouting anew, may then lead one to experience again the touch and resonance or dissonance between the inner and the outer dimension of living from soil, dwelling in the land, or being part of the earth or one with earth. Feelings of coherence in diversity might help to overcome fragmentation of perceptions and understandings within and between cultural spheres. Poetic expressions and storytelling, feelings and emotions, deep sense and "purposeless purpose" of humans on earth may show their effects or acausal coincidences with how soil is tilled or not, cared for or not, revered or not. This co-creates landscapes that record human actions and respond to them. The outer landscapes of humans may mirror the inner landscape of those acting on it.

1.3 Previous Works on the Cultural Dimension of Soils

Scientific works on the cultural dimension of soils are relatively limited compared to other subfields of soil science. However, the works that are available reach out to several of the cultural aspects mentioned above.

An early influential example of linking soil and culture was *Farmers of Forty Centuries* by Franklin Hiram King in 1911. Farmers in eastern Asia were practicing many of the ideas that have become the core of organic agriculture long before those ideas caught on in the West. The organic-based agricultural management practices developed in eastern Asia created a system that was sustainable for some 4,000 years (King, 1911). These practices included the use of green manure, composts, ashes, animal and human manure, and growing legumes in crop rotations to increase soil fertility. Basically, a culture had developed that did not let anything organic go to waste. Organics were eaten, burned for fuel, worn as clothing, or used as fertilizers. In the interest of full disclosure, King also identified favorable geography, the ability to grow multiple crops per year, intercropping, good crop selection, appreciation for the value of water in crop production, and the use of multiple nutrient sources for fertilization as other factors that contributed to long-term sustainability. King's publication became highly influential in the Western organic movement. It was also an early example of work by a Western scientist that recognized value in non-Western practices.

Another early work on the subject was *Soils and Civilization* by Hyams Edward (1952, 3rd ed. 1976). It is an analysis of the decline of great civilizations or empires, and that decline was linked to poor knowledge about or management of soils. The notion of great civilizations or empires, according to the author, means the existence of great cities and a strong administration. Civilization decline was explained, among other things, as a result of poor knowledge or management of soils or by a low level of interest in this resource leading to land degradation, with a strong emphasis on soil erosion. In this same perspective, one can also cite later works such as *Out of the Earth: Civilization and the Life of the Soil* (Hillel, 1991) and *Dirt, The Erosion of Civilizations* (Montgomery, 2007) as works that particularly focused on soil erosion as a fundamental factor in the disappearance of civilizations. Diamond (2011) also focused on soil degradation in a number of his examples of civilization decline in *Collapse: How Societies Choose to Fail or Succeed* (2011). A recent work, according to its title, seems to be part of this same cultural dimension. It is *Earth Matters. How Soil underlies Civilization* (Bardget, 2016). In fact, it is a personal reflection of the author on the need to make the non-scientific public aware of the immense services that soils render to humanity. It focuses on agriculture, biodiversity, climate change, and urban soils. It is an interesting book but its cultural dimension is reduced to a historical approach to soil science and an original chapter on soil and war. All these books use the key word "civilization" in their titles and not "culture," by that rather focusing on the material existence of large and complex societies, less (but yet also) on their immaterial foundations and structures.

Another approach is to conduct a historical analysis of how different societies around the world perform the management of their land. This is the case of *Soils and Societies. Perspectives from Environmental History* (McNeill & Winiwarter, 2006). The issue here is more about land management than land perception or cultural representation. The scientific and technical view of the soil remains the same as in the previous works.

The ethnopedology approach has been very successful. One of the characteristic publications is a special issue of the journal Geoderma entitled *Ethnopedology* (Winkler-Prins & Sandor, 2003). Ethnopedology focuses on "local knowledge of soils possessed by populations living in a particular environment for a certain period of time." This is also called traditional agriculture, with a focus on the diversity of perceptions by practitioners of soils and their properties, their place in the landscape, and possibly their classifications. This perception is often put into perspective by comparing it to international soil classifications and knowledge systems.

Soil as seen from a cultural point of view can refer to what is known in English as the "humanities" including arts and letters. This is the case of the IUSS work *Soil and Culture* (Landa & Feller, 2010), which has various chapters on soil in literature, poetry, visual arts, cinema, comics, color, war, etc. This book illustrates how the soil also participates in the sensitive world of each person.

Finally, another cultural approach can be cited, one that looks at the way in which the soil and earth is inscribed in the great founding texts of creation in mythologies and religions. There is a work in French that is not widely distributed but is specifically on this theme: *Sols et Sociétés. Regards pluriculturels* [Soils and Societies. Pluricultural Perspectives] (Lahmar & Ribaud, 2001). It uses short testimonies from a large number of specialists from various religions and spiritual traditions from societies around the world. It includes religions such as Baha'ism, Buddhism, Christianity, Confucianism, Islam, Judaism, and Shintoism, and also gives particular case studies from societies such as the Andean countries, Buryats (Siberia), Dogons (Mali), Eastern Amazonia, Romania, and the animism in Central Africa. In the same perspective, one should also cite the first chapter (pp. 3–20), "For Soil thou Art," of Hillel's above-mentioned work *Out of the*

Earth: Civilization and the Life of the Soil (1991) in which Hillel gives a "soil" reading of Genesis in the Hebrew Bible. This includes, among other things, the two accounts of creation in Genesis (compare also Hillel 2006, *The Natural History of the Bible)*, the origin of the word Adam, man's relationship with the "fertility" of the earth, plowing, the question of caring for the soil and the question of spirituality and reconnection with nature. This last approach is thus in the field of the sensitive dimension of the perception of soil. Churchman and Landa (2014) in *The Soil Underfoot: Infinite Possibilities for a Finite Resource*, presented multiple historical and contemporary cultural perspectives of soils and values that cultures hold for soil. In multiple chapters of this book sacred religious, economic, power structures, ownership, technological, and other cultural lenses were contrasted to explain soil awareness and blindness. The authors provided multi-perspectival views of soil that invite the reader to ponder how historic wisdom can be used for wise contemporary management of soils perceived as a finite resource. How might different attitudes towards soils fare in the face of enormous global challenges, such as global climate change and soil degradation?

In this book, the cultural dimension of humanity's relationship with soil is explored across its broad interpretations in the project in hand, including societal and individual, philosophical, religious, spiritual, psychological, political, and educational aspects. This approach addresses cultural understanding of soils by appealing to the cultural diversity and the importance of the inner world to inform individuals and the collective by means of case studies and conceptual chapters.

The book is, between the editors' introduction and conclusions, organized in five parts:

1. World Cultures: Religious, Spiritual and Other Worldviews on Soils (13 chapters)
2. Politics on Soil (2 chapters)
3. Case Studies on the Inner Soil of Individual Scientists (2 chapters)
4. Soil Relation and Care (4 chapters)
5. Soil Education and Cultural Language of the Soil (3 chapters)

1.4 Who Are the Audiences for This Work?

As a primary audience, we want to address people that are bridge-builders, who care about soil and utilize soil information routinely but are not necessarily soil scientists themselves. These bridge-builders include socio-culturally active farmers and extension workers, people from environmental and agricultural non-governmental organizations (NGOs), from politics and governmental agencies, from institutions supporting environmental education or agricultural development; and last but not least soil scientists who are open to knowing that a purely natural scientific approach is not the only valuable and effective approach to soil, and therefore want to enlarge their perception and communication skills. Thus, the targeted primary audience is at an intermediate level between "solely soil scientists" and "the general public". These bridge builders are people who can act as messengers, networkers, and knowledge brokers between different realities of life and different types of knowledge, and can communicate this in different scientific and agricultural communities and to the general public.

As an additional audience, the editors would like to reach the next generation of scientists and other academics with an interest in soils. The book can serve to broaden their horizons regarding what is relevant about soil that sustains humans by facilitating access to the human part of the soil relationship, and enabling multi-perspectivity on soils for those who are currently learning about them.

NASA sent a Mars vehicle called "Curiosity" to explore extraterrestrial soils. We may now send out our own curiosity vehicles all over the world to explore the richness of the cultural soils of our home planet. Not demarcation and fear, but happiness to learn from the "other" outside and inside of ourselves.

References

Balfour, Lady E. (rev. ed. 1948, 1st ed. 1943): *The living soil: Evidence of the importance to human health of soil vitality, with special reference to national planning.* Faber & Faber.

Bardget, R. (2016). *Earth Matters: How soil underlies civilization.* Oxford University Press.

Benhabib, S. (2002). *The claims of culture: Equality and diversity in the global era.* Princeton University Press.

Brevik, E. C., & Arnold, R. W. (2015). Is the traditional pedologic definition of soil meaningful in the modern context? *Soil Horizons, 56*(3). https://doi.org/10.2136/sh15-01-0002

Brevik, E. C., & Hartemink, A. E. (2010). History, philosophy, and sociology of soil science. In W. Verheye (Ed.), *Soils, Plant Growth and Crop Production.* EOLSS Publishers. http://www.eolss.net

Brevik, E. C., Steffan, J. J., Rodrigo-Comino, J., Neubert, D., Burgess, L. C., & Cerdà, A. (2019). Connecting the public with soil to improve human health. *European Journal of Soil Science, 70,* 898–910.

Churchman, G. J., & Landa, E. R. (Eds.) (2014). *The soil underfoot: Infinite possibilities for a finite resource.* CRC Press.

Descola, P. H. (2005). *Beyond nature and culture.* Par-delà nature et culture. Paris: Gallimard.

Diamond, J. (rev. ed. 2011). *Collapse, how societies choose to fail or succeed.* Penguin Publishing Group.

Hartemink, A. E. (2016). The definition of soil since the early 1800s. *Advances in Agronomy, 137,* 73–126.

Hillel, D. (1991). *Out of the Earth: Civilization and the Life of the Soil.* University of California Press.

Hillel, D. (2006). *The Natural History of the Bible: An Environmental Exploration of the Hebrew Scriptures.* Columbia University Press.

Hyams, E. (1976, 1st. ed. 1952). *Soils and Civilization.* Harper and Row (NY).

Howard, Sir A. (1947). *The soil and health: Farming and gardening for health or disease.* Rodale Press.

Herder, J. G. von (1813). *Collected works on philosophy and history, vol. 5.* Werke zur Philosophie und Geschichte, fünfter Theil: Ideen zur Geschichte der Menschheit dritter Theil (first published in 1787).

Ibáñez, J. J., & Montanarella, L. (2013). *Magic numbers: A meta-analysis for enlarging the scope of a universal soil classification system* (JRC technical rep. EUR 25849 EN). Joint Research Centre of the European Commission.

King, F. H. (1911). *Farmers of Forty Centuries*. Organic Gardening Press.
Kroeber, A., Kluckhohn, C. (1952). *Culture: A critical review of concepts and definitions*. Harvard University Press.
Lahmar, R. and Ribaut, J.-P. (2001). *Soils and Societies. Multicultural Perspectives.* Sols et sociétés. Regards pluriculturels. Éditions Charles Léopold Mayer.
Landa, E. R., and Feller, Ch. (2010). *Soil and Culture.* Springer.
Mizuta, K., Grunwald, S., & Phillips, M. A. (2018). New Soil Index Development and Integration with Econometric Theory. *Soil Science Society of America Journal, 82*, 1017–1032.
Mizuta, K., Grunwald, S., Cropper, W. P., Jr., & Bacon, A. R. (2021). Developmental history of soil concepts from a scientific perspective. *Applied Sciences Journal, 11*(4275). https://doi.org/10.3390/app11094275
Morozov A. I. (1998). *Soil as a polis of fungi. On the problem of global simulation of soil.* 16th International Congress of Soil Science, Montpellier, France.
McNeill, J., Winiwarter, V. (2006). *Soils and Societies. Perspectives from Environmental History.* The White Horse Press.
Montgomery, D. (2007). *Dirt: The Erosion of Civilizations.* University of California Press.
Patzel, N., Sticher, H., & Karlen, D. (2000). Soil Fertility—Phenomenon and Concept. *Journal of Plant Nutrition and Soil Sciences, 163*, 129–142.
Pieterse, J. N. (2015). *Globalization and culture: Global mélange* (3rd ed.). Rowan & Littlefield.
Patzel, N. (2015 [1st. ed. 2003]). *Soil science and the unconscious. A contribution to depth psychology of natural sciences.* Bodenwissenschaften und das Unbewusste. Ein Beitrag zur Tiefenpsychologie der Naturwissenschaften. Oekom.
Richter, D. (2020). Game changer in soil science: the Anthropocene in soil science and pedology. *Journal of Plant Nutrition and Soil Science, 183*, 5–11.
Sobocká, J. (2005). A contribution to the anthropogenic soils understanding. *Vedecké Práce, 27*, 111–118.
van Es, H. (2017). A new definition of soil. *CSA News, 62*, 20–21. https://doi.org/10.2134/csa2017.62.1016
Winkler-Prins, A., Sandor, J. (2003): "Ethnopedology". Special Issue of *Geoderma 111 (3-4),* p. 165–538.

Part II
World Cultures: Religious, Spiritual and Other Worldviews on Soils

Chapter 2
Cultural Patterns of Soil Cultivation in Europe 1: Polytheistic Context

Nikola Patzel

In Europe, soil has been understood and treated by farmers and other stakeholders in ways that trace back to the *religious and scientific history* of this multifarious cultural area and sub-continent. Its historical background of soil understanding will be explored in this chapter on human-soil relation in polytheistic context, and the following two on the monotheistic and scientific cultural strands in European cultural history.

The oldest fairly known cultural roots of human-nature relations in Europe are in the Celtic and Greco-Roman, Balto-Slavic, Germanic, and Finno-Ugric religions, all except the last one sharing Indo-European commonalities with more or less surviving lore and record to date. Later came the European developments of the Abrahamic religions of Judaism, Christianity and Islam, which mainly originated in the Near (Middle) East, thought influenced by Hellenism. All these religions have led to more or less secularized reverberations in European societies to this day.

Religion as well as science, religious mission as well as science-based agricultural and military technology, both have been effective in Europe's strong and multi-layered expansion with colonizing powers and ideologies, which have effectuated almost global influences and vibes. However, Europe's "inner soil"[1] is much more manifold than what has been exported globally. Europe still contains structures of worldviews relevant for nature and soil relations that have been the object of subjugation first inside and then outside this cultural area. That is important because the general ideas and values of societies about soils are assumed to be impregnated with

[1] "Inner soil" is about the inner world of humans that co-shapes the cultural patterns linked with human nature perception, conceptualization, and relations. For more, see Chap. 22.

N. Patzel (✉)
Freelance Scientist, Office for Soil Communications, Überlingen, Germany
e-mail: nikola.patzel@soilcom.com

N. Patzel et al. (eds.), *Cultural Understanding of Soils*,
https://doi.org/10.1007/978-3-031-13169-1_2

the inner soils of their members including soil scientists – and do influence scientific communities in turn.[2]

2.1 A Prelude: Cultural Embedding of Agriculture in a Shamanic Hunter-Gatherer Culture

People in some European subarctic areas "practiced a mixed economy (…) which seek[s] to minimize the risks inherent in their environment," combining "hunting, fishing, keeping small herds, and trade, as well as slash-and-burn agriculture" (Pentikäinen & Poom, 1989, p. 162). The extensive mythological compilation Kalevala that represents the Karelian Finno-Ugric culture shows the symbolic cultural embedding of agriculture in a former hunter-gatherer society. The divinity and magic bard Väinämöinen sang (the following quotes are from Lönnrot & Crawford, 1891): "*I the seeds of life am sowing /… In this soil enriched with ashes / In this soil to sprout and flourish*" (p. 23 f.). Whilst in some verses, this god places tree seedlings according to different soil qualities, in others Väinämöinen becomes the prototype of the farmer, and invokes the goddess (p. 24): "*Mother of the fields and forests / Bring the rich soil to producing / Bring the seed-grains to the sprouting / That the barley well may flourish. /… Rise, O earth, from out thy slumber /… Let the barley-grains be sprouting /… From my plowing and my sowing / From my skilled and honest labor.*"

The Rune songs of the Kalevala know a Mother of the Earth, a Mother Earth and a Forest Mother. The latter was called to protect the herds, and to spawn the metals from the underground for human culture, which can be understood physically or spiritually (p. 186): "*Forest-mother, formed in beauty / Let thy gold flow out abundant / Let thy silver onward wander / For the hero that is seeking.*" For metalworking as a prerequisite of easier tilling, in the Kalevala a heavenly blacksmith came to earth, talking to the iron (note the animistic way of speaking, p. 109): "*Then the blacksmith, Ilmarinen / Came to earth to work the metal /… Thus addressed the sleeping iron …/ Should I place thee in the furnace / Thus to make thee free and useful?*" But for the required plowing of the "snake fields," iron alone was not enough, and the divine bride advised her cultural hero (p. 283): "*Forge thyself a golden plowshare / Forge the beam of shining silver / And of copper forge the handles.*" The use of noble metals indicated plowing as a magical act made possible with the help of spiritual and symbolic forces from the Earth Mother's domain.

[2] European societies are complexly structured, communicatively intertwined with each other and globally to varying, partly very high degrees. Soil scientists and their professional organisations are part of media and social networks influenced by all kinds of individual and institutional actors. These networks can be involved with communicative and power asymmetries, e.g. in science communication or in the ways how research programmes are set with selective participation of social groups and opinions.

Equipped in this way with noble plow and iron armor, Ilmarinen talked to the earth's serpents (p. 283 f.): "*Ye whose wisdom comes from Ukko / And whose venom comes from Hisi / Ukko is your greater Master /... Get ye hence before my plowing /... Crawl ye to the nearest thicket / Keep your heads beneath the heather ... / Thus the blacksmith, Ilmarinen / Safely plows the field of serpents /... Harmless all against his magic.*" The snakes symbolically understood as earth spirits and concretely as embodiment of wild and dangerous nature, had to subordinate themselves to the new cultural impulse of tilling the soil; not by brute force, but by the magic of words and skills.[3]

2.2 A Paradigmatic Greek Case: Demeter

In exploring the religious dimension of cultural patterns of soil cultivation in Europe the most extensive example will be the splendid myth of the Greek goddess Demeter: she who initiated humans to the secrets of life and resurrection, and she who taught them to practice agriculture in fertile soil. Our main source describing the myth was discovered in 1780 CE in Moscow (Deichgräber, 1950, p. 503/3).[4] The quoted English translation is by Gregory Nagy from Harvard 2018 (italics set by auth.).

2.2.1 The Myth

"I begin to sing of Demeter [Δημήτηρ], the holy goddess (…), *she who glories in the harvest*." Her myth tells that Demeter gave benefits to the mortals and immortals. Demeter and her divine Brother Zeus (Ζεύς) united to create a daughter named Kore (Κόρη = the maiden) who was immersed in beautiful wild nature. "She was picking flowers: roses, crocus, and beautiful violets" in a kind of paradise state where different seasons were flourishing simultaneously.

Then Kore was seized by Hades (Ἅιδης), the Lord of the underworld, with his brother Zeus' acceptance. Kore was displaced to the lower world, the underground kingdom of the souls. The description in the myth of how the divine maiden, screaming, was carried away in a golden chariot to another realm, corresponds in detail to the ritual "rape" which was in ancient Greece part of the wedding ritual (Kledt, 2004, p. 47 f.). So the virgin dies; the woman rises and will take a new role and rule (Fig. 2.1). Now Kore bears the name Persephone, "she who beats the sheaves

[3] In Northern and Eastern Europe, Finno-Ugric mythologies and related ideas are with peoples of Siberian ("Uralic") origin. Larger groups are the Hungarians, the Finns and Estonians, Sámi and Karelians as well as the Mordvins (Erzyans and Moksha). Although quite well documented especially in the north, they could be addressed here with the short examples from the Karelians only, whose area stretches from the Gulf of Finland to the White Sea.

[4] For an overview on other primary sources of this myth, see Allen et al. (1936/1980), pp. 108–114.

Fig. 2.1 Divine rulers of the underground: Persephone and Hades seated together on a throne, plants in their hands, a cock underneath, 5th century BCE. Terracotta unearthed from the Greek Sanctuary of Persephone in Locri, Southern Italy. (Credit: Museo Nazionale Archeologico Reggio Di Calabria, Italy; picture by AlMare/Wikimedia commons)

[threshes the corn]." Hades made Persephone queen of life, after she agreed to come back to the lower world every year for 4 months (Homer 359–369). The name Hades literally means "the invisible maker," and he is the *nourisher of plant life* (Abenstein, 2016, p. 85).

The death and resurrection part of this mythical story can mean that the goddess, who first appeared as the sweet maiden Kore, had to assimilate something from the behavioral pattern of forcing death from Hades – but in contrast to human beings,[5]

[5]The idea of reincarnation existed in ancient Greece too, but rather in philosophical schools (Pythagoras, Plato, and others) than as part of religious myths or folk beliefs.

she also won resurrection from Hades for herself and for the life cycle of the plants: to be observed with the harvester in the concrete symbolism of the "grim reaper" as well as with the "death and resurrection" pattern of the burrowed seedlings.

Demeter was a cultural heroine, known in the Greek world for about 1000 years to be the one who gave humans crops and the art of agriculture as well as the initiation to the mysteries of life.[6] But she was surprised by the transient death of her daughter aspect Kore. When Hades seized the maiden, another goddess heard Kore screaming, that is Hekate (Ἑκάτη), who is intimately related to night and the moon, knowing all hidden paths and the thresholds between the different realities. Hekate accompanied Demeter when she asked Helios (the god shining as the sun) "about *the girl born to me, a sweet young seedling*" (Homer 66). In the myth, the first time that Persephone was forced to Hades, her reappearance on earth appeared not to be granted at all. Thus, the mourning great goddess Demeter became a *restless wanderer over barren soil* until coming to the court of the human city of Eleusis. After some time there she revealed her divinity by saying:

> "I am Demeter (…). Let a great temple, with a great altar at its base, be built by the entire *demos* [people] (…). And *I will myself instruct you in the sacred rites* so that, in the future, you may perform the rituals in the proper way and thus be pleasing to my *noos* [spirit]" (Homer 268…274). The king and people obeyed. "Demeter sat down and stayed there [in the temple] (…). *The Earth did not send up any seed. (…) Many a curved plough was dragged along the fields by many an ox – all in vain. Many a bright grain of wheat fell into the earth – all for naught*" (Homer 306–309).

Still, none of the other gods could persuade the crucial goddess to allow life to germinate in the soils. So Zeus sent Hermes (the world-crossing messenger god) to tell Hades that Demeter is "*to destroy the tribes of earth-born humans (…) by hiding the seed underground*" (Homer 353). Thus, Hades "smiled" in a knowing way (357) and allowed Persephone to head back above to the surface. Then Zeus "assented that her [Demeter's] daughter, every time the season came round, would spend a third portion of the year in the realms of dark mist underneath, and the other two thirds in the company of her mother and the other immortals" (445–7). When this decision was conveyed to Demeter by her and Zeus' mother Rhea, all *grains which "had stayed hidden underneath, through the mental power of Demeter (…), began straightaway to flourish with long ears of grain (…). On the field, the fertile furrows began to overflow* with cut-down ears of grain lying on the ground." And "Straightaway she [Demeter] sent up the harvest from the land with its *rich clods of earth*" (452–471). And she revealed to the leaders of the people, "the way to perform the sacred rites, and she pointed out the ritual to all of them (…). Blessed is he among earthbound mortals who has seen these things" (476).

We see here the "mental power" of the goddess as an effective reason for plant growth from the soil. Later on, such a vital force was assumed by philosophers and scientists in Europe until about the mid-20th century (see Chap. 4). The mythopoietic tendency was to symbolize the sprouting force and spirit out of the goddess with

[6] See Isokrates, 4th century BCE, as well as Himeros, 4th century CE (Kledt, 2004, p. 11).

male attributes, such as the phalloi, symbolizing Dionysus and plant life, growing out of Demeter's garden (Fig. 2.2); her son the god Dionysus was also called "endendros," that is, the divine life "in the tree" (Neumann, 1997, p. 240). The life-bursting phallic god Dionysus was not only seen to be a child of Persephone with Hades, but also seen in identity with the underworld god himself, as Heraclitus stated in the 4th century BCE: "But Dionysos in whose honor they rave and hold revels, is the same as Hades."[7]

With that the story has not yet ended. Demeter sent out one of her first initiates, the "sound-minded Triptolemos" (Homer 153), on a chariot pulled by *winged snakes* (a union of spirits from "below" and "above"), to spread seeds for human nourishment and teach humans how to perform better agriculture (Fig. 2.3). Later he became king of Eleusis (symbolically: the new ruling principle). A new wave of knowledge about how to cultivate soils reached the people. Moreover, from

Fig. 2.2 Ancient Greek ritual as part of a festival celebrating the birth of Dionysus (Thesmophoria): A female figure spreads seeds, phallic figures made of clay were stuck into soil (Deubner, 1956, p. 65 f.; Cook, 1914, pp. 684–686). The phalloi were appearances of the god Dionysus who was viewed to be the son of Demeter/Persephone or identified with Hades himself. (Credit: The Trustees of the British Museum, Object no. 1865,1118.49 from 440-430 BCE, image license CC BY-NC-SA 4.0)

[7] There are mythic images where Persephone united with Hades, or with Zeus in shape of a chthonic snake: This gave rise to the phallic god Dionysus, who made wine, milk, and honey flowing out of soil. And even some of Dionysus-Zagreus' transformed pieces became part of the loam from which finally humans were created (various Orphic and other antique sources). – The quote of Heraclitus is his fragment no. DK 15, translation quoted from Harris, 1994, p. 77.

Fig. 2.3 Triptolemus, foster-son and initiate of Demeter, was given by the goddess Kore/Demeter seeds and knowledge to bestow them upon the land to be cultivated. His chariot with snakes and winged wheels can symbolize a connection between the chthonic and heavenly spirits with their sustaining and elating qualities. Attic cup from about 470–460 BCE. (Credit: Photograph by Marie-Lan Nguyen, published with Wikimedia Commons. Cup from the Louvre museum (Paris), accession number G 452)

Demeter's union with the young man or god Iasion in the furrow of a three times ploughed field, there came the children Bootes, that is "he who plows with oxes," and Plutos, "the rich one," attributed with the cornucopia.

2.2.2 What Did the Demeter Myth Signify for Agriculture and Soil Understanding?

Some Propositions:

1. The whole of soil fertility and plant life were dependent on, and under control of a transcendent and immanent feminine spiritual power. All success of agriculture was dependent on the grace of the mother-daughter goddess in the threefold appearance Demeter-Kore-Persephone (moreover including Demeter's mother

Rhea, daughter of the Earth Gaia/Gaea …). In this domain the goddesses were insuperable by the principal god Zeus, that is, by the male heavenly part of the god image.
2. The Greek "underworld" had a double meaning of the underground beneath human feet and the spiritual realm of those who are dead from the visible world. Hades – signifying the god and his kingdom at the same time – has a forceful yet finally fruitful role. The story shows a life goddess (Kore) and with her the seedlings for food, the basis of human sustenance, in a tension and then circular sequence between flourishing and fruiting, dying and resurrecting. Hades of the lower world represents a "dark knowledge" and concealed action in that. His fruitful aspect is shown by the fact that Hades was also sometimes depicted with the cornucopia of fertility, like the goddesses related to him, as well as identified with Dionysus.
3. It seems that agriculture already existed way before these mythic events, following the pattern of the story and historical evidence. This may hint – when we interpret the myth as symbolic expression that goes along with cultural change – at an existential agricultural crisis and how it was overcome.
4. The contribution of humans to reconcile Demeter was to build her a temple to perform initiation rituals in, and seasonal festivities in the whole country. That means it was deemed necessary to cultivate a religious dimension, acknowledged to be effective in one's own life and performing agriculture.
5. The story is also one of initiation: of women entering into marriage, and for all humans to become more sustainable farmers (Kledt, 2004; for the latter compare Lincoln, 1981, pp. 71–90).

The fact that the Demeter cult and the mysteries of Eleusis were sustained for about 1000 years in Greece that was, together with (later) Rome, the culturally leading civilization of European antiquity, hints to their enormous mental strength and persistence as oriental knowledge. The Demeter myth and ritual practices probably exerted a compensatory function to show borders to the masculine spirit and to strengthen the feminine spirit, striving to obtain a certain balance. Demeter was reported to have been a "rural people's goddess" to whom farmers and especially women prayed, and "who smiled to them with corn ears in her hand" (Abenstein, 2016, p. 50).

What we miss as elements to base an interpretation on is knowledge of what happened in the temple of Eleusis, Demeter's main sanctum, built about 600 BCE and rebuilt several times. What we know is that every initiated had to swear not to talk about their experience near the *sancta sanctorum* (holy of holies) of Eleusis. "Anyone could be initiated into the mysteries of Eleusis who spoke Greek and was not guilty of the shedding of blood, men and women alike" (Kerényi, 1949/1993, p. 138). The emphasis on initiation signifies that the *personal experience was considered to be most essential* and way too supreme of a teaching to be spread on the market. The hidden teaching was the experience *inside* (the temple and the soul), whereas knowledge about agriculture was said to be the teaching given by the initiate then *outside* at the courts of human society (the teaching of concrete knowledge was the role of Triptolemus, following the myth quoted above). But then the link between mystery cult and agriculture broke off increasingly, and a crucial deathblow occurred in 395 CE, when the temple of Demeter in Eleusis was finally completely destroyed by raging Christian barbarians under the Visigothic warlord Alarich I.

2.2.3 Soil in the Time of the Greco-Roman Religion

In Greek mythology, humans descended from creatures of Gaia, the primordial Mother Earth. In one version, the mediating deity Prometheus formed humans (with Zeus' consent) from clay and water, their animation coming in some versions from the goddess Athena (Abenstein, 2016). Prometheus also brought fire and pottery to humans. Moreover, soil was not only part of the mythical, but also of the scientific mind of ancient Greeks. In the year 360 BCE, Plato gave a clear description of soil degradation in Attica, which is the region of Athens, in his dialogue Critias (bold by the author):

> *... there has never been any considerable accumulation of the* ***soil*** *coming down from the mountains, as in other places, but the* ***earth*** *has fallen away all round and sunk out of sight. The consequence is, that in comparison of what then was, there are remaining only the* ***bones*** *of the wasted body, as they may be called, as in the case of small islands, all the* ***richer and softer parts of the soil having fallen away,*** *and the mere skeleton of the land being left. But in the primitive state of the country, its mountains were high hills* ***covered with soil,*** *and the plains (...) were full of* ***rich earth,*** *and there was abundance of* ***wood*** *in the mountains.*[8] *Of this last the traces still remain, for although some of the mountains now only afford sustenance to bees, not so very long ago there were still to be seen roofs of timber cut from trees growing there, which were of a size sufficient to cover the largest houses; and there were many other* ***high trees, cultivated by man*** *and bearing abundance of food for cattle. Moreover, the land reaped the benefit of the* ***annual rainfall,*** *not as now* ***losing the water which flows off the bare earth*** *into the sea, but, having an abundant supply in all places, and receiving it into herself and* ***treasuring it up in the close clay soil,*** *it let off into the hollows the streams which it absorbed from the heights, providing everywhere abundant fountains and rivers, of which there may still be observed sacred memorials in places where fountains once existed; and this proves the truth of what I am saying.*
>
> *Such was the natural state of the country, which was cultivated, as we may well believe, by true husbandmen, who made husbandry their business, and were lovers of honour, and of a noble nature, and* ***had a soil the best in the world, and abundance of water,*** *and in the heaven above an excellently attempered [temperate] climate.* (Plato, 360 BCE/1875 CE, p. 692 f.)

We can note that there existed a clear-sighted observation of scientific soil facts, completely understandable even 2500 years later! And that this view obviously coexisted with the mythological view and the Demeter myth that was strong in this time. Following Tietz (2015), the soil erosion in Attica was mainly due to deforestation in war- and post-war times, probably mainly for the construction of war fleets.

---

[8] Indeed before the Bronze Age, Greece was apparently full of oak forests (*Quercus pubescens*), then replaced by rather sparse forests with the smaller oak *Quercus ilex* together with *Pinus* and shrubs. Large trees like cedars were much (over-)used for shipbuilding (Thommen, 2009, p. 41).

But the remote regions far from the major cities must have provided sufficient food to nourish the Greek people for many centuries. However, agriculture also contributed its share to soil damage, as a somewhat critical remark of Sophocles (442 BCE) about permanent tilling may hint at in his tragedy "Antigone," in a paragraph describing human dominion over nature (lines 376–379): "*Earth, of all Gods, from ancient days, the first, mightiest and undecayed, he [man], with his circling plough, wears ever year by year.*"

There is evidence for at least a regional soil-environmental crisis in ancient *Greece*, and a certain awareness of the problem of soil exhaustion by permanent tilling without compensation in the early historical time of Demeter worship. Concerning what today would be called scientific knowledge, "Grecian understanding of the soil reached its zenith in the 5th–4th century B.C." by not only generalizing their own "rich experience," but also by integrating knowledge from Mesopotamia, Egypt, "as well as those collected at the time of military expeditions and travels" (Krupenikov, 1993, p. 24). In the 4th century BCE, the region of Athens imported about 2/3 of its cereals, mostly from the region of the Black Sea (Thommen, 2009, p. 37). On the spiritual-religious side, a compensation for experiences of deep agricultural crisis was given in the Demeter myth that even contains reference to new agricultural knowledge, but without spelling out this new knowledge in the documents we have today.

It is documented by contemporary literature and modern pollen analysis that the woodland and soil preservation in Antiquity in the **Roman Empire** was done much more thoroughly and quite well in the Italian mainland as compared to the captured lands in the North and South, where forests were cut brutally (North) and salinization was not considered (South) (Tietz, 2015, pp. 328–331). There is also a description of soil degradation mixed with a lamentation over cultural decline in a symbolic language, by Roman writer Lucretius from the 1st century BCE: "*It is happening already;* ***our poor earth, worn out, exhausted,*** *brings to birth no more great eons, titans, huge majestic beasts (…), but the same earth who nourishes them [humans] now once brought them forth, and brought them, to their joy, vineyards and shining harvests (…) we wear down our strength, whether in oxen or in men, we dull the edges of our plowshares, and in return* ***our fields turn mean and stingy, underfed.*** *And so today the farmer shakes his head, more and more often sighing that his work, the labor of his hands, has come to naught. When he compares the present to the past, the past was better, infinitely so. (…) all things, little by little, waste away as* ***time's erosion*** *crumbles them to doom*" (excepts from the lines 1150–1174 of De Rerum Natura, vol. 2; p. 85 in Lucretius, 1969).

Remarkable, in contrast, is the documented oath of Titus Victor, a Roman procurator for the Alpine space in the 2nd century CE, to plant 1000 big trees to honor the god Silvanus, when this divinity helped him in the cultivation of his Italian estates after his return home.[9] We see here a god of wild nature that is considered to be relevant for agriculture too; and we see actions of (probably) reforestation having a symbolic and religious meaning that was coupled with the hope for agricultural suc-

[9] Tietz, 2015, p. 327, citing the Corpus Inscriptiones Latinarum Berlin 1862 ff., vol. XII, p. 103.

cess. Also in remarkable contrast to the supremacy of Jupiter in the Roman pantheon, this invocation of the earth from the time of the Emperor Augustus may be perceived: "Holy Earth, Goddess and Queen, great Mother of the Gods," may she send her powers in the medicinal herbs (Franz, 1909: I, p. 395 f.).

2.2.4 *Demeter's Life in Rome and Until the 19th Century in Europe*

The Etruscan[10]-Roman mother goddess Ceres, associated with the earth and underworld god Tellus, together with her daughter Proserpina assimilated most of the myth of Demeter-Persephone early in Roman history. Therefore, both goddesses were equated with each other and they had the same attributes (snake, cornucopia etc.; Wissowa, 1899). This was facilitated by the long-lasting Latin and Greek peoples' interface in southern Italy. The cult of Ceres also spread north of the Alps with the Roman Empire and appeared, for example, in French folk religion until the 19th century CE, known as "Cérès, la mêre du blé" (Mannhardt, 1884, p. 318 f.), that is: the Corn Mother (compare Fig. 22.2 of Chap. 22).

As symbols for something longed for to reconnect, or as metaphors for agriculture and agricultural knowledge, the goddess images of Demeter and Ceres became revitalized in European iconography from the Renaissance age to signify a reconnection with symbolic imagination from Antiquity seen as a cultural stimulus. Their images were used emblematically for agricultural scientific and educational societies of the 19th century (see Fig. 2.4 from the 19th century Berne). This happened not only in intellectual middle-class historicism, it also regained fascination in the context of romanticism and vitalism. The German researcher Wilhelm Mannhardt (from the brothers Grimm school), who explored agricultural customs and rural beliefs in the 1860s in Germany and neighbouring countries by means of huge polls, identified a "spirit" and "energy" of plant growth (Mannhardt, 1875, p. 217, 303) associated with the ancient Demeter, calling it a "divine being," and "divinity of growth" (1875, p. 315 f.). Specifically, Demeter was called (Mannhardt, 1884, p. 243) "the principle of life, the causa efficiens [efficient cause] of plant life (…), the immanent psyche of the halm people" (corn), which brings forth the fruits, "giving birth to the child, the new corn." Similar to the ancient Greek custom of bringing seeds into Demeter's temple and returning some of them to the fields afterwards (Kledt, 2004, p. 139 f.; comp. Mannhardt, 1884, p. 223), ears of corn were brought into churches at the Maria assumption day (15 August) and partly mixed with the seeds for the fields afterwards, as documented, for example, in 19th century Switzerland (Swiss dialect dictionary/Idiotikon II, 1885, column 1303; comp. Wuttke, 1925, p. 423).

---

[10]The Etruscan advanced civilisation of central and northern Italy was probably of indigenous pre-Indo-European origin, was in close exchange with the Greeks, later assimilated by the Romans, and it unfortunately cannot be discussed here with regard to its relationship to nature.

Fig. 2.4 Allegorical Ceres on the letterhead of a Swiss agricultural society in 1814. The tip of the ploughshare is in the soil. The woman holds a cornucopia and wears ear wreaths. (Credits: Haupt publishers, Bern; repro from Stuber et al., 2009)

One of the first modern soil scientists, Friedrich Albert Fallou, solemnly called farmers "priests of Isis and Ceres" (1857, p. 3). Later, a group of biodynamic organic farmers in 1932 patented the goddess' name "Demeter" for agricultural practices that follow the instructions of their founder Rudolf Steiner.

We also see symbolism of female death and resurrection documented by Mannhardt in the Alps in the 1860s: a bunch of corn ears was called the "grandmother" and buried in the soil. From the strength of the germination of her seedlings from soil, which is symbolically her power to resurrect, the next harvest volume was concluded; "for the religious-minded man, the *grain* expressed an inexpressible divine reality rather than that the goddess, Demeter's daughter, expressed the grain. (…) According to the myth, the ripened wheat grains would have to fall to earth, the scene of their death and resurrection" (Kerényi, 1949/1993, p. 117 and 119). Not herself being the corn, but the source of it was called in this letter to Mannhardt from the north alpine Lake of Constance (1860s): "*Whether these are legends or only invented stories, I do not know. Many times, one tells the little children about the so-called 'Lady in Heaven'. To good people she gives plenty of corn as their harvest, while she causes harm to bad people*" (Patzel, 2015a, p. 250, transl. by auth.). We will come back to the "Lady in Heaven and Earth" who gives soil fertility when coming to Slavic, Germanic, and Christian religions below.

In a poem of Michèle Robert (1985) on "Persephone," the author did assimilate parts of the Demeter myth with the symbolic proximity of sexual union, death wedding and an underworld journey in symbolic soil from a female perspective:

> *this spring my lover – came to me again; this – time my mother did not – hold me back; (…) my lover was a dark man – we embrace in the garden, in the grave; his – twisting root is clotted with my black earth – as I break open, and take him in (…) – this time when we go – underground we go together (…).*

2.3 Baltic and Slavic Religion

In the family of European peoples with autochthonic Indo-European languages, the Slavs and related Balts are the large groups in today's Eastern Europe; and Baltic culture forms a substrate in large areas that are Slavic today.[11] A basic trait of the mythology of the Slavs as well as Balts before their Christianisation was a God of the above, a God of the below (in the elder religious layer it was a Goddess) – and an ambivalent or mediating Goddess who is at the same time heavenly and earthly: She reigns over human and plant life, over the sources of the earth and literally *the "gates" of the soil surface* (Katičić, 2003; Reiter, 2009), that is the pores as well as the springs. This goddess is the "Land Lady," in Latvian *Laime,* in Russian *хозяйка or пани* (that is: Lady, owner, host), who "lets it flow." She was the mistress of *the sources from earth,* whose water contains "silver" and "gold", and the flow of life in general: "You Laime, who creates everything, who determines the course of life" (Katičić, 2003, p. 42), you are the "mother of life (Mūža māte)" (Biezais, 1975, p. 363).

2.3.1 Guardian of the Soil's Doors

In Latvian (Lettish) ancient belief god, that is the goddess Laime's husband, and Laime often struggle over wealth for humans. For example, god wants to *concede land for bread to nourish humans,* but Laime withheld it and then they negotiate. "Dievs said this is a bread place, [but] Laima – bitter tears," or "Laima refused the ploughman" (Biezais, 1975, p. 361 f., transl. by auth.). This goddess herself was viewed to have two sides, named Laime and Nelaime, to give and to take or withhold life and livelihood, so that people prayed: "Throw, Laime, a stone in the back of the Nelaime." With her benevolent side, *Laime opens the earth in spring* with her golden key, or it is her son to whom she gives the key to open the soil; she opens the doors of the underworld, that is, at the same time of the beyond and of the earth. She lets the water flow, the rose flourish, and enables the sacral wedding of her young divine siblings, so that the summer may be fertile again (Katičić, 2003, pp. 79–84, as a Baltic and pan-Slavic idea). She could sow the wheat and bless the crops, wearing a dress that matched each one (Biezais, 1975, p. 364). We may see a common Indo-European layer shining through here, of what is called "spring" in English: water bursting forth from below, germinating seeds, birth, that is: spring time! It was the goddess who lets it flow.

The door guarded by the goddess is also that to the realm of the souls below and called "door of the dead ones" and of "their children." The netherworld's master,

[11] While the Baltic as a geographical region on the Baltic Sea today includes Estonia, in terms of origin and language this is only the case for the Lithuanians, Latvians and related smaller Baltic peoples; the Estonians, in contrast, are close relatives of the Finns (Finno-Ugric family).

"Velesъ" (or as mistress Vela, Veļu māte), *reigns over warmth and moisture in soil to bring forth plants*. These were ideas of who reigns in the soil and what is important. Compare them to the Greek author of about 400 BC who asserts that "with regard to the plant 'embryo' the role of the mother organism is played by the soil, more precisely, by its 'warmth' and 'moisture content', because 'every growing organisms on the earth lives by the moisture of the earth (…)'" (Krupenikov, 1993, p. 26).

This described goddess was named the *"mother of the earth" who rules it: "she who has enough sons to plough the field, she who has enough daughters to mill the corn"* (Katičić, 2003, p. 87). We see here the analogy of agriculture with processes in the netherworld. The master of the underworld to whom the goddess is related, reigns about the meadows for the souls (comparable to Hades and the Greek Ἐλύσιον πεδίον = the Elysian *soil*!), and there in the concrete mythical underground he gives a place for the deceased souls as well as shelter and warmth to wild animals and even birds during wintertime (compare Fig. 2.5). This signifies the underground to be a convivial and life-sustaining place. It is the goddess who is at the same time

Fig. 2.5 Goddess figure with symbols of death and life or resurrection. Pictured is one of the plates which make up the silver "Cauldron of Gundestrup" that was possibly made on Celtic behalf by Thracian-Scythian craftsmen. Dating to the Latène culture a few centuries BCE, the mythical picture gallery of this vessel shows motifs from East and West as a testimony to a lively cultural exchange in ancient Europe. One image also shows a "cauldron of rebirth" into which people are immersed (Hillgruber, 2021). (Credit: Object of the National Museum of Denmark (Copenhagen); photo by Erich Lessing/akg-images)

consort of the heavenly father and closely related to the master of the underworld; thus, a queen of and a *mediatrix* mediating between both realms (Katičić, 2003, pp. 89–108). Laime later mingled with or was identified with the Christian Mother of God, he holy Mary (Biezais, 1975, p. 364 f.). And the Lord of below became identified with the Christian Devil, and his formerly friendly realm below turned to a nasty, punishing hell (Katičić, 2003, pp. 90–94, 98).

2.3.2 *The Knowing Powers in the Underground – and Human Struggles*

In east Slavian, the earth goddess was named Mokošь: This name means "the wet one" or "the wetting one." This had the double meaning *to signify the aroused (and receptive) feminine and the fruitful soil.* She could act beneficially for farmers as well as for women in childbirth (Reiter, 2009, pp. 69–74). She was associated with the underworld, whose Lord was Velesъ. Regionally, the function of Mokošь was fulfilled by local goddesses named the vila, a kind of fairies, who are local interface phenomena between this world and the other, related to springs and to fertile land (Reiter, 2009, p. 146 f.). Many tales of the Vila/Vilen, White Ladies and similar, have been noted up to the 19th and even 20th century in the eastern Alps, designating them as mistresses and knowledge carriers of weather, animals, and plant growth.

In more north-eastern regions of Europe, powerful "white ladies" from the woodlands, following popular legends, taught people in their dreams how to practice agriculture, and helped to protect the fields (Veckenstedt, 1883, p. 186 f.). Or they appeared like a vision during rest at noon, then testing women for their agricultural knowledge – with death threats if they failed (Patzel, 2015a, pp. 332–6). It is not clear whether this cruel attitude was deeply rooted in historical imagination or not. We may consider that (1) with Christianization pagan gods were treated as demons, then (2) with the onset of witch hunts following the late medieval agrarian crisis there was a collective paranoia to see "the evil" in many women (a dissociative parallel with the rise of Marian devotion); and (3) then with the main wave of rational-scientistic "enlightenment" of the rural population in Central and Western Europe in the 19th century, all spiritual content in nature relation became an adversary. All these cultural blows inhered the tendency to reduce the goddess, divinities and spirits to their "dark" sides, then to make witches, demons and child frighteners out of them, and then to reduce the common legends rationally to be caused for example by "chest fear" or "panic attacks," and finally to: nothing (comp. Patzel, 2015a, p. 353 f.).

The Lithuanian people worshipped *snakes*, as said by a source of about 1400 CE: "Every household kept a snake lying mutely in the corner of the house, which he fed and to which he sacrificed"; and another source from 1587: "They consider snakes and all forms of ugly creeping things as their gods" (quoted from Bojtár, 1999, p. 292 f.). These reporting Christian missionaries then ordered beating to death and burning these animals so intimately related to soil.

In Slavic belief, the Lord of the underworld Velesъ was also the "king of snakes." In this shape he was able to teach people the *languages of plants and animals* (Reiter, 2009, pp. 198–200, 213), that is, knowledge and meaning of nature. By spitting or blowing into the mouth of a human, this chthonic god could initiate him or her to a deeper perception of nature's utterances. Symbolically, sharing saliva or breath can mean to communicate a spirit from nature, or psychologically seen a "spirit from the unconscious" that inspires knowledge and creates or reinforces a personal bond between a human and the wider psycho-spiritual world (compare it with god's snake saliva coming to man in the Dogon mythical perception, Chap. 8). But Velesъ could also cause droughts by blocking water sources: then nothing flows (Reiter, 2009, p. 199). There are strong structural similarities between Laime and Velesъ and the Greek Demeter and Hades (including the symbol of the snake), whilst the Slavic versions might be closer to their common Indo-European religious ground.

There existed almost *shamanic varieties* of spiritual fighting for the coming harvest of the new year, which is concealed below ground, documented from the Balkans to the Baltic (Ginzburg, 1966). For example, in the year 1691 CE, in the Livonian and today Estonian city of Tartu/Dorpat, an 85-year-old farmhand was accused of paganism (Anonymus, 1692). Asked without torture, he reported that three times per year he went underground in the shape of a werewolf, which he translated as "gods gundog," *to rescue together with comrades the flowers, fruits and animals and generally the "growth"* from being trapped underground by demonic sorcerers and their master devil. Missing this fight would result in bad harvests. The farmhand said that already at Christmas, *the green leaves and corn were present underground, but have to be freed* and brought up through the gate and thrown into the air and over the whole land, as a blessing for poor and rich ones. These actions were a service to god and no sin at all, reported this Baltic farm worker concerning his shamanic enterprise below ground.

In the Slavonic area, this conflictual part of the mythical and spiritual events, that the wealth stays concealed below ground and has to be freed, also developed into a classical dragon fight motive; and in the course of history, the Christian Saint George with his sword and expertise in dragon fighting took over the role of the heavenly God or of the Goddess' son (or of the shaman), "to open up the earth" (Katičić, 2003, p. 83). This has been common practice also in many Catholic regions of the world, to cover or replace so-called pagan deities with saints, or to give saints attributes of former deities (compare Chap. 10). The Archangel Saint Michael, in functional unity with Saint George, absorbed some characteristics of the archetypal pattern of the mythical hero that overcomes the snake (or the worms), the "giants," or other adversary forces, like former respectively non-Abrahamic gods did. Christ himself – "I have not come to bring peace, but a sword," Matthew 10.34 – was sometimes attributed this role of "the stronger god," especially in the context of missionary efforts.

More generally seen, we have here the mythical structure of the harvest concealed underground being addressed by human action. In contrast to the behavioural pattern of veneration and prayer (for the earth goddess and so forth), or of initiation to mysteries of life (like in the Demeter temple in Eleusis), there is the pattern of heroic human spiritual actions, that is, fights of their souls, to overcome adverse spiritual forces and stagnation in the flow of life. This reported special pattern of spiritual underground fighting for blessing and harvest was not a unique case in Europe.[12] It concedes to humans a spiritual role in the sequence of seasons and circles of life, and suggests that *the success of agriculture is not merely dependent on physical, but also on spiritual processes*. And it promotes the meaning that "inside hidden matter," that is, below the soil surface, there is the realm where the renewed life is prepared before it will become visible again. Today, some structurally similar ideas can be found among others in esoteric, geomantic, and anthroposophist communities with regard to the success of the agricultural year that is considered to be influenced by spiritual powers.

2.4 Celtic Beliefs About the Underground of Life

The Celtic culture was widespread in Europe for millennia, with an initial core area in the northern Alps and Alpine foothills. Celtic culture had a maximum extension from Anatolia to Ireland, and from Northern Italy to Spain, thereby integrating many peoples that all adopted Common Celtic (which then fanned out into related languages). The Celtic cultural area emerged autochthonously and spread its religious and knowledge system widely.[13] A characteristic of the Celtic relationship with nature was the intriguing closeness, and very permeable separation, between the "here and now" and "the netherworld." The underground appeared to be especially inhabited to the Celts, vivified by many species of chthonian spirits (Kuckenberg, 2019, p. 130).

Much of Celtic art was rich in ornamentation that cultivates a double entendre of floral life and spiritual wickerwork that "may mirror the Celt's most dynamic ideas of never ending circular movements and renewing life circles in nature," including a connection and entanglement of the spheres of here and now and netherworld (quot. Kuckenberg, 2019, p. 133, and paraphrase from p. 79, transl. by auth.).

Early Celtic farming included quite a number of (mixed) crops: early wheat species like emmer and einkorn, other cereals like rye and oat, legumes like pea, lentils and field beans,[14] and multi-use crops like flax, linseed, linseed dodder and dyer's

[12] There are documented parallels from the Balkans as well as from the north Italian province of Friuli, documented by Ginzburg (1966). These phenomena have been somewhat interpreted by Patzel (2015b, pp. 93–112).

[13] A pre-Indo-European people of Europe that was not culturally assimilated by either the Celtic or the Romanic and Germanic dominations, the only that has preserved its language until today are the Basques near the western edge of the Pyrenees. Their mythology could not be addressed here.

[14] Stone age agriculture in the later Celtic homeland shows peas to be the only legume; in the late Bronze age, field beans and lentils became very important crops that preserve soil fertility (Jacomet & Maier, 2016).

woad. The domesticated animals were pigs, sheep, goats, horses, dogs, and a bit later also chickens (Kuckenberg, 2019, p. 24); they brought organic fertilizers.[15]

2.4.1 *Vessels of Fecundity*

A pan-European symbol of fecundity and plenty is the cornucopia, that is, the large horn of an animal (or patterned after it), filled with plant life and food. Perhaps the first known representation in stone of a symbol close to that of the cornucopia is the carved horn in the hand of the female figure "Venus of Laussel" from about 25,000 years before present, in Palaeolithic Europe, that is, a long time before the invention and adoption of agriculture (Fig. 2.6).

Fig. 2.6 25,000 years old European representation of a "fertility goddess." One hand holds a horn like a vessel with 12 nicks, the other is on her belly. Palaeolithic limestone relief from Laussel in the Dordogne (France). (Credit: Musée d'Aquitaine, Bordeaux. Photo L. Gauthier)

[15] Nitrogen isotope relations ($\delta^{15}N$ values) prove the use of fertilizers in agriculture in the northern alpine region in the 40th century BCE, combined with intense hook tilling (Dieckmann et al., 2016; Jacomet & Maier, 2016).

“Vessels of Fecundity” or “Cauldrons of Plenty” were common symbols in Celtic religious representations (Mac Mathúna, 2014, p. 68, 75). This source was held by the god Dagda (Nagy, 1974), an Indo-European omniscient progenitor of all living beings, having the bottomless cauldron overflowing with food, and cosmogonic and chthonic properties (Shaw, 2018; Maier, 2003, p. 94). Some of his characteristics were later adopted by legendary figures (Mac Mathúna, 2014, p. 75). Vessels of all types and sizes played an extremely important social and symbolic role in Celtic culture (an example is Fig. 2.7).

The marvellously nourishing Holy Grail – that is a vessel – of medieval literature might also stand in this tradition (merged with the symbolism of the blood of Christ), as well as inexhaustible vessels in fairy tales that can be found all over Europe. On

Fig. 2.7 Celtic bronce cauldron wagon from the grave probably of a priestess (Strettweg/Austria), Hallstatt culture, 600 BCE. A superhumanly large figure, interpreted as a goddess, carries a pot. The human figures on the chariot hold stags with oversized antlers (Fürhacker & Modl, 2021). (Credit: Exponate no. 2000 from the Johanneum Museum UMJ, Graz (Austria); photo by N. Lackner)

the continent, the Celtic god Cerunnos was typically represented with an overflowing vessel named cornucopia, following the form of a large horn, from which the fruits of the earth emerged to humans. Cerunnos' deer antler on the head may hint to his wilderness as well as to his annual renewal; his snake can be a symbol of his connection to the underground (Petersmann, 2016, p. 171, 174).

The richness of soil and the land was also represented in the *cornucopiae* and fruit baskets of the *matronae* or *matres*, that is, of female divine trinities. They held flowers, ears, fruits, and new-born children on their womb. "Attributes like baskets of fruits, the cornucopia or infants indicated that the Matres were closely associated with ideas of abundance and fertility" (Egeler, 2013, p. 31; comp. Garman, 2008; see Fig. 2.8). In these figures, Celtic, Germanic, and Roman attributes merged (Egeler, 2013). Single figures with cornucopiae were, for example, Rosmerta, "she who allocates the food," and Epona on her throne with the fruits in her hands and vessels (Petersmann, 2016, pp. 186–188; Egeler, 2013, p. 31).[16]

Fig. 2.8 Female trinity of "girl, mother and old woman" from the Romano-Celtic area of southwest Germany, 2–3 century CE. Two goddesses hold fruit baskets, one holds a bunch of ears. (Credit: Exponate from the Landesmuseum Baden-Württemberg, found near Stuttgart; Photo: Nikola Patzel)

[16] In Greece, the cornucopia is called the "Horn of Amaltheia" to this day. Amaltheia was a mythical "divine goat," the nurse of the newborn god Zeus who lost one horn in a fight to protect the divine child against its jealous father Kronos. In human shape, she nourished him and others by means of this horn of plenty. The poet Phokylides of the 6th century BCE then identified the fertile arable field with the cornucopia. It became an attribute of the underworld god Hades in the 5th century BCE, and of Zeus in the 4th century BCE (Bemmann, 1994, p. 16, 19–22). Also Dionysus, god in wilderness and ecstasy and phallic states, the cultural hero Herakles, and the Egyptian god Osiris Serapis (influenced by Greece), and also Kore and Nymphs were often shown with filled horns (Bemmann, 1994).

The Celtic goddess Brixta ("the bright one") developed near the Alps to the mythical figure "Perchta" (merging with a Germanic goddess) who was documented in legend collections right up to the 20th century. She was a simultaneously earthly and heavenly goddess, and queen of the "White Ladies" (fairies). The Perchta could mend the plough of the peasant (its shavings turned to gold), brought fertility to the lands and could take it with her when she left (Grimm, 1875, p. 228 f., 232; Waschnitius, 1913, p. 27, 98; Timm, 2003, p. 44, 168, 318 f.).

2.5 Germanic Peoples' Religion

The Germanic tribes, their languages being related with the Celtic and other European ones by their common Indo-European origin, settled in northern Central and Northern Europe during early Antiquity. They expanded to Southern Germany and the Alps, to England and Scotland, France, Spain, Northern Italy, the Balkan, Ukraine, Tunisia, and Sicily in the migration period of late Antiquity.

2.5.1 *The Benevolent Divine Couple*

Germanic religion included divine couples who brought love, life, and soil fertility. In an older layer this couple was Nerthus (f) and Njördr (m), they were later Freyja (f) and Freyr (m), or their variants Frigg and Fricco (Näsström, 1995, p. 54, 70, 104 f.). They were the female and male sides of bringing forth life. Nerthus was functionally translated by Roman scholar Tacitus as "Mother Earth" (Näsström, 1995, p. 48). Nerthus' name may have come from the Anglo-Saxon "Erce, mother of the earth (Erce, eorthan modor)," who was invoked "when the first furrow was cut" (Näsström, 1995, p. 51). This prayer continues: You are "growing and sprouting fields, bellying and strongly shooting. … *Hail to you, land of soil [Erdflur], the human's mother, be greening in God's embrace, full of nourishment,* availing men" (Ninck, 1937, p. 114, transl. by auth.).

In the younger divine generation, Freyr/Fricco ("the master") was represented as ityphallic, that is, with a phallus of large size, symbolizing super-human fecundity, and attributed with stories of a "big lover" (Näsström, 1995, p. 70, 78). His sister-consort Freya/Frigg ("the Lady", ibid. p. 104) received divine lovers as well as the souls of the honorable deceased ones. Freya "supports fertility for beast and soil; but her special sphere was human love" and sexuality (Näsström, 1995, p. 77 f.), similar to the Greek goddesses Demeter and Aphrodite.

Freya/Frigg was the wife (that is, the female counterpart) of the north Germanic Odin or the south Germanic Wotan, who was at this stage in the history of Germanic religion the highest male god; and at the same time she had a "Hades" quality like the Valkyries, by receiving women and men who had an honorable death. She was thus both giving life and welcoming death. She shared the qualities of a death goddess with Urdr (Näsström, 1995, p. 81, 88 f., 148). Hence, Freya's characteristics "compose a special pattern which we recognise as characteristic of the Great Goddess of the Indo-European myth" (Näsström, 1995, p. 90; comp. Egeler, 2013, p. 21).

In the 11th century, the capitular clergy Adam von Bremen (Northern Germany) described Fricco: "They imagined his picture supplied with an enormous virile member" (1888, p. 223, transl. by auth.). After Christianization, the ithyphallic god Freyr/Fricco might have lived on in the Germanic harvest custom with "the old one," that was a big sheaf puppy with large penis on the harvest fields, documented in northeast Germany through the 19th century (Patzel, 2015a, p. 197 f.; Fig. 2.9a).

Freya, the life-giving goddess, lived on after Christianization under her by-name and later cover-name "Holle" (or Holda), meaning "the benevolent one," from the 7th century until the 19th century CE (Timm, 2003). In the German script of a monk, written in the 13th century, we read: *"On the night of Christmas some women lay the table for the Queen of Heavens – which is called 'Lady Holle' by the common folks, in order that she may help them"* (quoted from Timm, 2003, p. 21). According to the conception of the peasant beliefs and to legend, Lady (Frau) Holle governed within the earth, in deep waters and wells. She drove across the country in a chariot pulled by animals, and rode floating on a white horse (for Freya, the chariot was previously pulled by cats, and she rode on a boar). Holle brought children and she could take them back again. She made the weather and "*every year, she goes around in the land and bestows fertility upon the fields*" (Timm, 2003, pp. 132–45, 239, 242–48, 260–63; quote p. 248). According to the Grimm fairy tale Mother Holle, she gave gold to those who respected nature's needs, and tar pitch (meaning bad luck) to those who wanted to profit without service in return (KHM 24 in Grimm & Hunt, 1884, pp. 104–107).

Other names for the remnants of Freya/Holle were "the Corn Mother" and "she the Old One." In 19th century western Germany, ears of corn were left on the field for her, while saying: "We give it to the Old One / She shall keep it / Be she to us next year / As merciful as she was this year" (Mannhardt, 1884, p. 337 f., transl. by auth.; see Fig. 2.9b).

Today, there is still certain veneration: In 2012 a Swiss mountain farmer woman told me how important Holle was for her. In tales and legends about Holle, she has found the right attitude for farming: Knowing that one has to talk to soil, to ask and to give. And to be there working with it at the right moment, not wanting to impose her own will or laziness on the garden or the pastures.

In the following Chap. 3 on European monotheism, we will see how the female qualities of this religious substrate partly survived and changed in the context of Christianity. But, to round off the presentation of the older dimension of European religiosity with regard to the soil, we will have a glance at another underground gen(i)us: the dwarfs.

Fig. 2.9 (**a**) and (**b**): Drawings of figures made of ears of grain in the context of harvest customs from northeastern Germany in the 19th century. These are functional and motivic correspondences with Germanic deities. (Credits: German State Library/Prussian Cultural Heritage. Repros from Patzel 2015a)

2.5.2 The Psychophysic Beings Under Foot

We start with records from 19th century Switzerland (Patzel, 2015a, p. 278, transl. by auth.):

> "Some stories are told from the so-called earth manikins, little men that dwelled in the neighbouring forest. As elderly people tell, it was the dwarfs that brought meals and drinks in vessels of shining silver for the working people in the fields. But one day a farmer was so impudent as not to render the vessels but took them home. From that day on the little beings disappeared." Similar legends gave more interesting details: The dwarfs would appear when people slept during their midday rest. They would come out of a rock crevice or other openings to the underground. They not only gave "manna" to the field workers, *they helped during night preparing the soil of the arable fields.*

In the oldest known Nordic images of dwarfs, they emerged from the flesh of a cosmogonic giant. They showed typical trickster images but were very knowledgeable. Simek saw a cultural continuity in the Germanic area, and summarized the dwarfs' middle age characteristics as "the creative forces of the underworld" (Simek, 2003, p. 15). They taught new cultural techniques like the making of cheese and agricultural knowledge, for example, when to seed. They themselves stayed away from culture: "I am of wild nature, never my essence was looked at," says a dwarf in a tale from about 1300 CE (Anonymous).

In Christian context, the existence of dwarfs was not negated, but it seemed hardly possible to see them originating from below. Thus, the biblical myth of the fallen angels was adapted to them. In Luke (10.18) it was Lucifer, together with the angels following him, who fell down from heaven (apparently from the sky) into the earth (symbolically: into matter).[17] Further developing this mythologeme, from the 6th until the 12th century CE, the ideas developed that all semi-physical, psycho-physic beings in the soil and in wild nature were in their essence angels that stayed neutral in the struggle between God and Satan; or, alternatively, angels that fell less deep, not as far as hell, and therefore stayed close to humans and their soil. The interesting point of this aetiology is that the spiritual side within nature was valued to be beyond good or evil and not, or not fully, part of the dichotomy of the Christian god image (Patzel, 2015a, pp. 281–290). One can say these spiritual or psycho-physical beings were seen as "just nature," not following moral instances.

Simultaneously to this semi-Christian adaptation of folksy ideas of spirits in nature, there remained the pagan interpretation: dwarfs and their consorts were spirits or demons of unclear origin that – upon the arrival of Christ in the country – moved back to hideaways in the rocks, woods and waters (Lecouteux, 1995). Of course, these were spiritual places of Celtic and Germanic times, and this story can also be read as a typical cultural phenomenon that an ancient religious worldview, if not integrated into the newly dominating one, retires and rumbles underground, that is, in the unconscious and in unfathomable nature (compare Fig. 3.2 in Chap. 3).

Today, there are still beliefs to be found with farmers which reckon with "elementary beings" dwelling in the soil and mother rock. Dwarfs, water and fire beings, so-called elemental beings, are acknowledged with symbolic acts, or called on directly in plots of land and asked for collaboration. Offerings of mash or singing and drumming were done to please them. I have observed this in some bio-dynamic farming context (see Chap. 4) as well as rather "geomantic" esoteric systems. These actions and belief systems express, besides others, a wish for a personal spiritual relation to nature, to land and soil.

2.6 Concluding Remarks

What characterises the European subcontinent as a cultural area and mental space? There is a pronounced cultural diversity based on several layers of common ground. Most of the described different pre-Christian European religions have developed from a common Indo-European origin of the peoples, possibly integrating ideas and symbols from older European populations. They all share a strong religious and spiritual involvement with soil and agriculture, and especially the common pattern

---

[17] Maybe a forerunner of this image is shown in Isaiah 14 where the king of Babylon was brought down by God Yahweh, as if the morning star fell down from heaven in an earth pit filled with maggots, he then being covered with worms. The motive of "stars" falling down is also presented in the Bible in Isaiah 34.4, Revelation 8.10 and 9.1.

of a goddess as a ruling gatekeeper and essential driver for the cyclic renewal of life and of soil fertility.

There were male associates in their roles of consort, son – or counterpart. "Consort" hints to a coherent duality of the god-image, "son" means that the male spirit came out of the goddess; "counterpart" means that there were functional correspondences between god and goddess, which could lead to a certain competition of them in the human worldview, historically often leading to a transfer of qualities and functions from the goddess to the god. So the male associate with the goddess – as long as her existence was not negated – could be seen as a kind of spirit of life preparing in the underworld, sprouting from soil, as well as fecundating her and it from above.

The archetypal mythical structure shows "Mother Earth and Father in Heaven," but can, following our evidence, certainly not be reduced to such an attribution only. Indeed, the religious world views of the spiritual forces in the context of nature clearly show both the presence of a masculine spirit in the underworld and in the plant world that springs from the ground; and the presence of a feminine spirit in life with a celestial, cosmic origin, dwelling on earth and below ground, too. This finding may still stand today in opposition to simplistic projections of gender roles in religious and natural contexts.

The polytheistic approach with regard to nature relations offers vivid possibilities to involve divine drama in natural and human actions. In the polytheistic cosmos, "deep earth" and soil have mainly positive connotations, as they shelter and prepare the wealth of the future. But soil fertility can be concealed or blocked by adverse forces, or by decision of the divine lady. Then, events between the deities may occur, and also human effort may be required to try to convince the goddess (or the divine couple) to let fertility unfold again; or it may be necessary to engage in spiritual fighting, or magic to overcome forces that are adverse towards life including plant growth.

We can see here the psychological and historical observation that psychic and cultural renewal may come after decline, and that life may come back after events of hindering or destruction. That is inherent in religious myths like that of Demeter (or later that of Christ), having been understood as inextricably involved with what happens in nature, soils and also in agriculture. In this view, it seems to matter how people relate to the inner world, to the psycho-spiritual cosmos inherent to the world and to human microcosms. Humans were understood as a necessary factor in the course of events not only by their physical actions, but also by their mental and symbolic behaviour.

In Polytheism, spiritual forces in nature are multiple; and there are multiple ways of religious observation. In essence, findings suggest similarity of the typical fundamental patterns and dynamism within and between humans on the one hand, and in nature and between its forces on the other.

These assumptions can be a gateway to far-reaching questions and different interpretations for human nature relations that include religious motives and spiritual experiences, and scientific research.

In some mythical stories, the beyond[18] or its cultural messenger-heroes bring knowledge for the here and now, that is, orientational knowledge, force and technical skills for agriculture and nature relation. From today's perspective, this can be seen as a possibly complementary counterhypothesis to the prevailing quite positivist and emphasised ratio-oriented natural science's paradigm of outer observation, knowledge acquisition and theory-building in soil science, agronomy and nature protection. It may signify that looking more for and at "the other" inside us and within nature, in pursuit of symbols and whatever touches mind and heart from near and far there, might ultimately be good for human wholeness and sustainable relationships with nature; who knows? Human cultural and knowledge history has always opened up new chapters and more than one story at a time.

References

Abenstein, R. (2016, 4th rev. ed.). *(Greek mythology.)* Griechische Mythologie. Ferdinand Schöningh publisher.

Adam von Bremen (1888, first edition about 1070 CE). *(Church history of Hamburg.)* Hamburgische Kirchengeschichte. Nach der Ausgabe der Monumenta Germaniae übersetzt von Johann Laurent (2nd ed.). Original: Gesta Hammaburgensis ecclesiae pontificum. Publisher: Dyk'sche Buchhandlung.

Allen, T., Halliday, W., & Sikes, E. (1936/1980). *The Homeric Hyms* (pp. 108–114). Clarendon Press / Hakkert publisher.

Anonymus (1692). *(Memorandum of the manorial court of Dorpat: Arrived 31 of October 1692.)* Produktionsvermerk des Hofgerichtes zu Dorpat: Eingekommen d. 31. October ao 1692. Published in the annex of Bruiningk (1924): *(The werewolf in Livonia and the last therefore hold tribunal at the district court of Wende and the manorial court of Dörpt in 1692.)* Der Werwolf in Livland und das letzte im Wendeschen Landgericht und Dörptschen Hofgericht i. J. 1692 deshalb stattgehabte Strafverfahren. Mitteilungen aus der Livländischen Geschichte 22: 163–220.

Anonymus (about 1300 CE). Reinfried von Braunschweig, lines 18387-9, quoted from Bartsch (1871, p. 536): *Reinfrid von Braunschweig. Bibliothek des litterarischen Vereins in Stuttgart* (vol. CIX).

Bemmann, K. (1994). *(The cornucopiae in the classic and hellenistic period.)* Füllhörner in klassischer und hellenistischer Zeit. Peter Lang publisher.

Biezais, H. (1975). *(Baltic religion.)* Baltische religion. In Ström, Å., & Biezais, H. (Eds.), *Germanische und Baltische Religion* (pp. 311–384). Kohlhammer publisher.

Bojtár, E. (1999). *Foreword to the past. A cultural history of the Baltic people.* Central European University Press.

Cook, A. (1914). *Zeus. A study in ancient religion.* Cambridge University Press.

Deichgräber, K. (1950). *Eleusinian piety and Homeric world of imagination: the homeric myth of Demeter.* Eleusinische Frömmigkeit und homerische Vorstellungswelt im Homerischen

[18] "Beyond" is a traditional term for what is beyond the threshold of death, one may also say beyond conscious accessibility. Being ungraspable by definition, there are plenty of old or more recent terms for approaching it in various ways and with different attitudes: like "netherworld," "heaven/underworld," "hereafter," "transcendence," "spiritual world," "collective unconscious," "objective psyche" – or "god(s)"; and of course there are many different and contradictory or paradoxical opinions about it. There is no intention in this text to define or hammer anything down there.

Demeterhymnus. Akademie der Wissenschaften und der Literatur, Abhandlungen der Geistes- und Sozialwissenschaftlichen Klasse, Nr. 6. Franz Steiner Publisher.
Dieckmann, B., Warwath, A., Heumüller, M., et al. (2016). *(A brief village history.)* Eine kurze Dorfgeschichte. Hornstaad-Hörnle IA am Bodensee. In Archäologisches Landesmuseum Baden-Württemberg, Landesamt für Denkmalpflege: 4000 Jahre Pfahlbauten (pp. 80–92). Thorbecke Publisher.
Deubner, L. (1956). *(Attic festivals.)* Attische Feste. Publisher: Akademie Verlag.
Egeler, M. (2013). *Celtic influences in Germanic religion. A survey* (Münchener Nordistische Studien, Vol. 15). Publisher: Utz Verlag.
Franz, A. (1909). *(The ecclesiastical benedictions in the Middle Age.)* Die kirchlichen Benediktionen im Mittelalter (2 vols). Herder publisher.
Fürhacker, R., Modl, D., & Universalmuseum Johanneum Graz (2021). *(The Cult Wagon of Strettweg – An object biography. Restoration and reception of an archaeological icon.)* Der Kultwagen von Strettweg – eine Objektbiographie. Restaurierung und Rezeption einer archäologischen Ikone. Schild von Steier Beiheft 11/2021.
Garman, A. (2008). *The cult of the matronae in the Roman Rhineland. An historical evaluation of the archaeological evidence*. Edwin Mellen Press.
Ginzburg, C. (1966). *(The benandanti. Investigations on witchcraft and agrarian cults in the 16th and 17th century.)* I benandanti: Ricerce sulla stregoneria e sui culti agrari tra cinquecento e seicento. Einaudi publisher.
Grimm, J. (1875). *(German mythology, vol. 1.)* Deutsche Mythologie. 1. Band. Vierte Ausgabe, besorgt von Elard Hugo Meyer. Publisher: Ferdinand Dümmlers Verlagsbuchhandlung.
Grimm, J., & Grimm, W. (1884). *Grimm's household tales, vol. I.* (M. Hunt from the second German edition, 1812, Trans.). Publishers: Realschulbuchhandlung / George Bell and Sons (transl.).
Harris, W. (1994). *Heraclitus: The complete fragments: Translation and commentary and the Greek text.* Internet resource from Middlebury College.
Hillgruber, F. (2021). *(A Cauldron full of Gods. In: Magical land. Cult of the Celts in Baden-Württemberg.)* Ein Kessel voller Götter. In: Magisches Land. Kult der Kelten in Baden-Württemberg. Achäologisches Landesmuseum und Landesamt für Denkmalpflege (pp. 89–119). Publisher: wbg Theiss.
Jacomet, S., & Maier, U. (2016). *(Wide-ranging. Food plants between field and forest.)* Breit gefächert. Nahrungspflanzen zwischen Feld und Wald. In: Archäologisches Landesmuseum Baden-Württemberg, Landesamt für Denkmalpflege: 4000 Jahre Pfahlbauten (pp. 320–328). Thorbecke publisher.
Katičić, R. (2003). (*The landlady at the gate. On the track of the Great Goddess in fragments of Slavic and Baltic sacral poetic work.*) Die Hauswirtin am Tor. Auf den Spuren der großen Göttin in Fragmenten slawischer und baltischer sakraler Dichtung. Schriften über Sprachen und Texte 6. Peter Lang Publisher.
Kerényi, K. (1993, 1st ed. 1949). Kore. In Jung, C. G., & Kerényi, K. (Eds.), *Essays on a science of mythology. The myth of the divine child and the mysteries of Eleusis* (pp. 101–155) (R. Hull, transl.). Princeton University Press.
Kledt, A. (2004). *(The abduction of Kore. Studies on the athenic-eleusinic Demeter religion.)* Die Entführung Kores. Studien zur athenisch-eleusinischen Demeterreligion. Palingenesia Schriftenreihe für Klassische Altertumswissenschaft (Vol. 84). Franz Steiner Publisher.
Krupenikov, I. (1993). *History of soil science. From its inception to the present* (Russian translations series no. 98). Publisher: Balkema.
Kuckenberg, M. (2019). *(The Celts.)* Die Kelten. wbg Theiss publisher.
Lecouteux, C. (1995). *(Demons and spirits of nature in the land in the Middle Ages.)* Démons et génies du terroir au Moyen Age. Publisher: Imago.
Lincoln, B. (1981). *Emerging from the Chrysalis. Studies in rituals of women's initiation.* Harvard University Press.

Lönnrot, E. (1891). *Kalevala. The epic poem of Finland* (Crawford, J. (original Finnish edition 1835), trans).

Lucretius (Titus Lucretius Carus; 1st cent. BCE). De Rerum Natura (Published translated 1969 by Rolfe Humphries) *The way things are. The de rerum natura*. Indiana University Press.

Mac Mathúna, S. (2014). The relationship of the Chthonic World in early Ireland to Chaos and Cosmos. In J. Borsje, A. Dooley, S. M. Mathúna, & G. Toner (Eds.), *Celtic cosmology. Perspectives from Ireland and Scotland* (Papers of mediaeval studies 26). Pontificial Institute of Mediaeval Studies.

Maier, B. (2003). *(The religion of Germanic peoples: Gods, myths, worldview.)* Die Religion der Germanen: Götter, Mythen, Weltbild. C.H. Beck publisher.

Mannhardt, W. (1875). *(Agrarian and forest cults. First volume: The tree cult of the Germanic and neighbouring tribes. Mythological investigations.)* Wald- und Feldkulte. Erster Band: Der Baumkultus der Germanen und ihrer Nachbarstämme. Mythologische Untersuchungen. Repro WBG 1963.

Mannhardt, W. (1884). *(Mythologic research from the legacy of Wilhelm Mannhardt.)* Mythologische Forschungen aus dem Nachlasse von Wilhelm Mannhardt. Hgg. von Herrmann Patzig. Band LI der Reihe Quellen und Forschungen zur Sprach- und Culturgeschichte der german. Völker. Repro Olms 1998.

Nagy, J. (1974). *The Dagda: The giver and taker of food*. Harvard University Press.

Nagy, G. (2018, December 12). *Homeric Hymn to Demeter*. Center for Hellenic Studies in Harvard. Published online on https://chs.harvard.edu/CHS/article/display/5292

Näsström, B. (1995). *Freyja – The great goddess of the North* (University of Lund studies in history of religions, vol. 5). Department of History of Religions, University of Lund.

Neumann, E. (1997). *(The great mother. A phenomenology of the female forms of the unconscious, 11th edition [1st ed. 1956]).* Die Große Mutter. Eine Phänomenologie der weiblichen Gestaltungen des Unbewußen. 11. Auflage. Walter publisher.

Ninck, M. (1937). *(Germanic gods and belief in the beyond.)* Götter und Jenseitsglauben der Germanen. Diederichs publisher.

Patzel, N. (2015a). *(Symbols in agriculture. On the spiritual side of nature relation in Swiss agrarian culture.)* Symbole im Landbau. Zur spirituellen Naturbeziehung in der Schweizer Agrarkultur. Oekom Publisher.

Patzel, N. (2015b, 1st ed. 2003). *(Soil science and the unconscious. A contribution to depth psychology of natural sciences.)* Bodenwissenschaften und das Unbewusste. Ein Beitrag zur Tiefenpsychologie der Naturwissenschaften. Oekom Publisher.

Pentikäinen, J. (1989). *Kalevala mythology* (R. Poom, Trans. and Ed.). Indiana University Press.

Petersmann, A. (2016). *(The Celts. An introduction to Celtic studies from the perspectives of archaeological history, linguistics and history of religion.)* Die Kelten. Eine Einführung in die Keltologie aus archäologisch-historischer, sprachkundlicher und religionsgeschichtlicher Sicht. Publisher: Universitätsverlag Winter.

Plato (360 BCE/1875). Critias. In *Dialogues of Platon* (B. Jowett, 2nd revised ed., Vol. III, pp. 687–703, Trans.). Clarendon Press.

Reiter, N. (2009). *(The deposit of faith of the Slavs in its European connections.)* Das Glaubensgut der Slawen im europäischen Verbund (Slawistische Studien 21). Harrassowitz publisher.

Robert, M. (1985). The mirror of the mother. Quoted from H. Foley (1993) *The Homeric Hymn to Demeter. Translation, Commentary, and Interpretative Essays* (p. 164). Princeton University Press.

Shaw, J. (2018). Fashioner gods in Ireland and India: The Dagda and Tvaṣṭṛ. In E. Lyle (Ed.), *Celtic myth in the 21st century. The gods and their stories in a global perspective* (pp. 149–162). University of Wales Press.

Simek, R. (2003). *(Germanic religion and mythology.)* Religion und Mythologie der Germanen. wbg Theiss publisher.

Sophocles (442 BCE). *Antigone* (E. H. Plumptre, Harvard classics (Vol. VIII, Part 6, pp. 1909–1914), transl.). Collier & Son publisher.

Stuber, M., Moser, P., Gerber-Visser, G., & Pfister, C. (2009). *(Potatoes, clover and clever minds. The economic and charitable society of the Canton of Bern.)* Kartoffeln, Klee und kluge Köpfe. Die Oekonomische und Gemeinnützige Gesellschaft des Kantons Bern OGG (1759–2009). Haupt publisher.

(Swiss dialect dictionary.) Schweizerisches Idiotikon. Wörterbuch der schweizerdeutschen Sprache (Vol. 2), 1885. Published by Antiquarische Gesellschaft in Zürich.

Tietz, W. (2015). *(Herdsmen, farmers, Gods. A history of Roman agriculture.)* Hirten – Bauern – Götter. Eine Geschichte der römischen Landwirtschaft. Beck publisher.

Timm, E. (2003). *(Lady Holle, Lady Percht and similar figures, considered from a Germanic linguistic perspective 160 years after Jacob Grimm.)* Frau Holle, Frau Percht und verwandte Gestalten 160 Jahre nach Jacob Grimm aus germanistischer Sicht betrachtet. Hirzel publisher.

Thommen, L. (2009). *(Environmental history of antiquity.)* Umweltgeschichte der Antike. Beck publisher.

Veckenstedt, E. (1883). *(Myths, tales and legends oft the Lithuanians.)* Die Mythen, Sagen und Legenden der Zamaiten (Litauer) (Vol. I). Winter publisher.

Waschnitius, V. (1913). *(Perht, Holda and related figures. A contribution to the German history of religion.)* Perht, Holda und verwandte Gestalten. Ein Beitrag zur deutschen Religionsgeschichte. Sitzungsberichte der Kais. Akademie der Wissenschaften in Wien, Philos.-Historische. Klasse, 174 Band, 2. Abhandlung. Published by K.K. Hof- und Staatsdruckerei.

Wissowa, G. (1899). *(Ceres. In: Encyclopaedia of classical antiquity)* Ceres. In Paulys Realencyclopädie der classischen Altertumswissenschaft (vol. III(2), columns 1970–1979). Metzler publisher.

Wuttke, A. (1925, first ed. 1860). *(German popular superstitions of present age.)* Der deutsche Volksaberglaube der Gegenwart (4th ed). Wiegand & Grieben publisher.

Chapter 3
Cultural Patterns of Soil Cultivation in Europe 2: Monotheistic Context

Nikola Patzel

3.1 Emerging Monotheism and the Human-Nature Relationship

In the Greco-Roman cultural area of antiquity, there were early philosophical trends towards monotheism. The first known such statement was from Xenophanes in the 6th century BCE:[1] There was "*One single God, greatest among gods and men, not at all like mortals in shape or thought*" (fragment 23). The same Xenophanes wrote: "*For all things are from earth and all becomes earth at the end*" (fr. 27).[2] All water in clouds, rain and rivers had come from the sea (fr. 28), thus "*Earth and water is everything that becomes and grows*" (fr. 29) and "*we are all come into being out of earth and water*" (fr. 33). Xenophanes had a transcendent philosophy of monotheism (or henotheism) in combination with a concrete philosophy of nature, something that strongly recurred in Europe some 2000 years later.

In Ancient Greece, some Pythagoreans (Pythagoras was a contemporary of Xenophanes) and later in the 4th century BCE Plato and Aristotle spoke of an almighty, all-seeing, single one and permanent God (Staudt, 2012, pp. 28–36). From the philosophical school of the Stoa (founded around 300 BCE) came the statement that "the one Zeus" formed "all men from one single loam" (Staudt,

[1] All quotes are taken from Diels, 1906, p. 50 f., and translated by the author partly using the English edition by McKirahan, 2010, pp. 60–64.

[2] Compare from the Jewish tradition Sirach 40.11: *"All things that are from the earth turn back to the earth, and what is from the waters returns to the sea,"* written about 190–180 BCE in Jerusalem in a Hellenistic cultural context.

N. Patzel (✉)
Freelance Scientist, Office for Soil Communications, Überlingen, Germany
e-mail: nikola.patzel@soilcom.com

N. Patzel et al. (eds.), *Cultural Understanding of Soils*,
https://doi.org/10.1007/978-3-031-13169-1_3

p. 42), a conception in which the monotheistic tendency combined with the classical Greek myths of the creation of humans from earth (see also Chap. 2).

Another Greek monotheist approach saw the god Pan to be the one cosmic god of nature. As part of this development, Paon (Πάων = the shepherd) became Pan (Πάν = all), as said in a hymn: "*Heaven, Sea, All-Queen Earth, and Immortal Fire, for all are members of Pan*": whose spirit-seed would be the vital force of plants and in every being, creating and sustaining everything (Wassmann, 2003, p. 111, 114). The god Dionysus, who is similar in nature to Pan, also became the "one": son of Zeus and born by Demeter as well as by Persephone, being, among other appearances, the "spirit of vegetation" (Daraki, 1994, p. 29).

In Antiquity there emerged also a kind of feminine monotheism such as with Apuleios of Madaura (2nd century CE) who said: "*Mother of all is nature, ... the single face of all Gods and Goddesses, ... the one and single Goddess, revered all over the world under different names.*" She was at the same historical time identified with: "*I Isis am the single ruler of the world ... nothing ever comes into being without me.*" And there is even one document from the same period where Demeter's daughter Kore is "*One God [*sic*!], ruler of all, great is the invincible Kore*" (Staudt, p. 55, 57, 59, transl. by auth.). We may see in the feminine monotheism a kind of recurrence of the "great Goddess," their world described by Neumann (1997) as the female matrix and vessel that contains, bears and brings forth everything.

In the Roman Empire, it was the *Sol invictus* (the undefeated/invincible sun) who in its later period became for his adherents the one God. "As I see it, pagan monotheism is a possible interpretation of change in the religion of the Roman Empire from the first century onwards, (…) change was gradual and took many different forms and tendencies" (van Nuffelen, 2010, p. 32 f.). Finally, Judeo-Christian monotheism showed the most enduring persuasion on the peoples of the Roman Empire (at the time including the Near-East and North African coastal regions) and then spread, often by force, northwards all over Europe. The classic pre-Christian Greek formula εἷς (ὁ) θεός = (the) one God then became the standard Greek formula to express monotheism in the Christian New Testament (Staudt, 2012, pp. 318–20).

3.2 Judaism and the Old Testament

Judaism was a historical breakthrough of Monotheism in the Levant, in the southern ancient Near East. In this region, the Neolithic turn with prehistoric people becoming farmers had happened much earlier than in Europe: Wild cereals were deliberately cultivated and stored in sophisticated granaries in the Jordan Valley more than 11,000 years before present (Kuijt & Finlayson, 2009). In this ancient cultural land, the Jewish religion developed. In the Book of Joshua (Tanach, and Old Testament: 17.15 RSV) that was composed about the 7th–6th century BCE, it is said about an expansion of the cultivated land: "*If you are a numerous people, go up to the forest, and there clear ground for yourselves.*" Landscape-shaping cultural techniques such as terrace building were practised even before the first Jewish state-formation, and

"Either through experience or by borrowing the agricultural skills of the indigenous population, the Israelites gradually mastered the cultivation of the soil. The Talmud describes their predecessors as 'well versed in the cultivation of the land'."[3]

Especially since the Babylonian exile (597–539 BCE), there have been large Jewish communities in the Orient and throughout the Mediterranean, including Alexandria and Rome. Hellenistic Judaism developed prominently. Then in the year 6 CE, Judea itself became part of the Roman province of Syria. The great Jewish war against the Roman occupation ending with the destruction of the Temple of Jerusalem in the year 70, later insurgences and especially the Bar Kochba revolt in the 130s CE caused large-scale migration or abduction of Jews to Rome and other places in Europe. For these reasons, the Jews were no longer a regionally situated people, but settled (in addition to Persia, Iraq etc.) to a large extent within the Roman Empire, including Iberia, France, the Rhineland and England, later also moving to Eastern Central Europe. Therefore, beginning with Jews expelled and displaced, many demographically and culturally strongly developing Jewish communities emerged in Europe since the Antiquity. Jewish culture became an important factor in European cultural history (Fig. 3.1).

But since about the 9th century CE, persecution, banishment or pogroms against Jews increased in European countries, ordered by feudal authorities or done by mobs from local Christian (or in a few cases Muslim) majorities. This was accelerated in the 14th century with the devastating persecution of Jewish communities in many important cities during the Black Death. Then in Spain after about one century of regional pogroms and banishments, the Alhambra Decree of the joined kings of Castile and Aragon after the Reconquista of Spain was completed in 1492, forced Jews to convert or to emigrate and largely destroyed the Jewish culture of Iberia (Hottinger, 1995, pp. 332–338). The Holocaust in the 20th century was not based on Christian anti-Semitism, but the German Third Reich exploited the still virulent anti-Jewish cultural substrates when enacting its horribly cruel and incomparably systematic extermination policy in most parts of Europe.

Almost every time, Jews were forbidden to buy agricultural land (in the German state of Bavaria, for example, until 1816; the gradual attainment of full civil rights throughout Germany lasted until 1871), and they were forced to live only on special streets or in ghettos in the cities.[4] However, the Kibbuz movement of Jewish agricultural egalitarian communities started already in Eastern Europe from 1810 on, especially in Belarus and Galicia, and indeed prepared the later Kibbuz movement in Palestine and then the new state of Israel (Naor, 1996/98).

It can be hypothesized that the most effective historical contributions of Judaism to the European image of soil came through the integration of large parts of its holy scriptures into the Christian Bible, that is, into the Old Testament. Therefore, the

---

[3] https://www.jewishvirtuallibrary.org/history-and-overview-of-agriculture-in-israel (Jan. 2022).

[4] Also at present, "There are hardly any Jewish pastoralists today and not very much farmers. (Even in modern Israel, their numbers are few.) Most of the Jews in the world now live, and have lived for many centuries, in environments far removed from those of their distant progenitors" (Hillel, 2006, p. 224).

Fig. 3.1 Adam who had to work the hard soil by the sweat of his brow. French Illustration from the 15th century. (Credit: Musée Condé, Inventory number Ms139-folio4recto. Photograph: RMN, René-Gabriel Ojéda, Creative Commons License)

historical effect of Judaism on European soil understanding would have mainly been its impact on Christianity, or more precisely, the cultural continuity of Judaism in Christianity.[5]

In Mosaic Genesis in the Old Testament, agriculture was presented in the context of a punishment on humans and a spell on soil brought about by the original sin of Adam and Eve, who ate from the tree of knowledge in collaboration with a mysterious snake. Thus, Yahweh's verdict was (Genesis 1: 24–29 Revised Standard Version

[5] I am not referring here to modern Israeli soil science, which was and is internationally highly influential in pedology through scientists such as Dan Yaalon (e.g., 2000).

RSV from 1971):[6] "*... cursed is the ground because of you; in toil you shall eat of it all the days of your life; thorns and thistles it shall bring forth for you; and you shall eat the plants of the field. By the sweat of your face you shall eat bread until you return to the ground, for out of it you were taken; you are dust, and to dust you shall return.*" This framing might be seen difficult for the cultural image of agriculture and tilling. On the other side, the necessity of agricultural soil cultivation can be seen as a correlation to a major change in the human mental and cultural situation: Humans were no longer part of nature, they were now confronted with nature (Fig. 3.1).[7]

In other parts of the Old Testament, farming was affirmatively addressed: "*Noah was the first tiller of the soil*" (Genesis 9.20); "*but he will say, 'I am no prophet, I am a tiller of the soil; for the land has been my possession since my youth*"' (Zechariah 13.5); "*and he had farmers and vinedressers in the hills and in the fertile lands, for he loved the soil*" (Chronicles 26.10). There are about 800 occurrences of the words "*earth*" and "*soil*" in the English Revised Standard Edition (RSV 1971) of the Bible. Only the smallest part of these mentions refers to the natural soil. Aware that these English translation words are not synonymous with the greater variety of Hebrew words and their meanings in the original scriptures, the review of these passages formed the basis for the following description.

3.2.1 Men of Dust

The mythologeme that God made man out of clay or similar is widespread (Genesis 2.7): "*Then the Lord God formed man from the dust of the ground, and breathed into his nostrils the breath of life; and the man became a living being.*" Abraham refers to this statement later (Gen 18.27): "*Let me take it upon myself to speak to the Lord, I who am but dust and ashes.*"

The definition of humans and all beings on earth to be animated matter in the sense of dead matter animated by God only is stated in Psalm 104.29 f.: "*... when you take away their breath, they die and return to the dust. When you send your Spirit, they are created, and you renew the face of the ground.*" Death is equal for all humans, as Ecclesiastes 3.20 states: "*All go to the same place; all come from dust, and to dust all return*" – and Job 1.26 added precisely: "*and the worms cover them.*"

[6] In this text, the Bible is normally quoted following the Revised Standard Version in its 1971 edition. This text still has much language in the tradition of the King James Bible, but has also integrated most modern findings of biblical studies. Only in some exceptions where it seemed to be necessary for unfolding the content, more versions of the bible are quoted, in other English versions, or in the Hebrew, Greek, German or French languages.

[7] Compare the Kalevala quotes in Chap. 2 and the introduction of Chap. 4.

The association of soil with death was reinforced by the formulation "the dust of death" (Ps 22.15).

Ecclesiastes questioned whether or not there is a difference between humans and animals (3.21): *"Who knows if the human spirit rises upward and if the spirit of the animal goes down into the earth?"* A response is given in Ecclesiastes 12.7: *"And the dust returns to the earth as it was, and the spirit*[8] *returns to God who gave it."*

The motherly, caring and autonomous aspect of soil and earth was apparently avoided by speaking of mere *"dust."* However, there is one different quotation (Job 1.21): *"Naked I came from my mother's womb, and naked shall I return"* (traditional translations say "return there").[9] However, even for the "man of and in dust," the central Christian promise of resurrection was prefigured in the Old Testament, also the Christian dichotomy of heaven and hell and the "final judgement" (Daniel 12.2, which is one of the youngest books of the Old Testament): *"Multitudes who sleep in the dust of the earth will awake: some to everlasting life, others to shame and everlasting contempt."*

3.2.2 God Watering the Land and Letting Plants Grow

God being facilitator and maker of plant growth is mentioned in the Psalms of the Old Testament: *"He covers the heavens with clouds, he prepares rain for the earth, he makes grass grow upon the hills"* (Psalm 147.8). *"From thy lofty abode thou waterest the mountains; the earth is satisfied with the fruit of thy work"* (Ps 104.13). *"Thou dost cause the grass to grow for the cattle, and plants for man to cultivate, that he may bring forth food from the earth"* (Ps 104.14). Job varies it saying, *"he gives rain upon the earth and sends waters upon the fields"* (5.10). We can see in these quotations the importance of watershed (mountains), of direct rainfall and water for maybe flooding and irrigation of the fields and meadows.

By God's action, humans can experience prodigies of wealth: "*... and he ate the produce of the field, and He made him suck honey out of the rock, and oil out of the*

[8] In this Biblical book of Kohelet, the Hebrew word that is translated as *"spirit"* is *Ruach* (רוּחַ), in the Greek version it is *Pneuma* (πνεῦμα): meaning about "God's breath" or "divine spirit" as all-pervading and inherent spirit of life. This is what became in the New Testament the "Holy Spirit." The *vis vitalis* (Lat. "force of life") of living beings in vitalism is a more or less secular analogous conception, which at times was also part of the scientific world of ideas and still remains so in niches (see Chap. 4).

[9] The New Revised Standard and New American Standard say "return there," King James says "return thither," the New International Version says "naked I will depart." The French Louis Second Bible (1910) says most explicitly: "et nu je retournerai dans le sein de la terre" (will I return in the breast of the earth); and most German bibles, including the present standard version (Einheitsübersetzung) say: "Nackt kam ich hervor aus dem Schoß meiner Mutter; nackt kehre ich dahin zurück" (naked I appeared from my mother's womb, naked will I return there).

flinty rock" (Deuteronomy 32.13). Thus, human wealth was interpreted as a divine blessing: *"May God give you of the dew of heaven, and of the fatness of the earth, and plenty of grain and wine"* (Genesis 27.28). By grace, even nature may speak to humans: *"... plants of the earth, and they will teach you, and the fish of the sea will declare to you"* (Job 12.8); and Yahweh promised Israel to *"make a covenant"* to its advantage with *"the creatures that move along the ground,"* so that *"the earth shall answer the grain"* (Hosea 2.18, 22). Daniel Hillel interpreted "references to animals with the power to speak" and generally mythic animals to be "remnants of ancient myths preserved in the Bible" (2006, p. 207).

3.2.3 Destruction of Soils and Swallowing Earth

The presence of the "one and only God" was seen as a precondition for the vitality of his creation. But Yahweh has not been a good God only, but also showed destructive potential, not alone in the deluge (Genesis). An example is given by Job's bitter lament (14.19): "*... the torrents wash away the soil of the earth; so thou destroyest the hope of man*" (meaning that either God causes the erosion, or he acts against humans like erosion does). Ezekiel (26.4) reported God's statement: "*They shall destroy the walls of Tyre, and break down her towers; and I will scrape her soil from her, and make her a bare rock.*" Destruction and natural hazards were also interpreted as punishment for sins: "*The earth lies polluted under its inhabitants; for they have transgressed the laws, violated the statutes, broken the everlasting covenant*" (Isaiah 24.5). Therefore, one can say (Hillel, 2006, p. 215): "The land became a sort of seismograph, an indicator of the nation's collective behaviour. Its manifestations were to be watched at all times for tell-tale signs of the return of the desert. (...) To the Israelites, it was not merely the land that might revert, but – in a symbolic way – they themselves. (...) The labile nature of the land [given by God to the people led by Moses] amplified itself in the consciousness of the people living on it."

The Bible reports the idea that the earth annihilates enemies: "*... and the earth opened its mouth and swallowed them up, with their households and all the men that belonged to Korah and all their goods*" (Numbers 16.32); "*... the earth opened and swallowed up Dathan, and covered the company of Abi'ram*" (Psalm 106.17). And finally: *"But those who seek to destroy my life shall go down into the depths of the earth"* (Ps 63.9), which was in that time the realm of the dead.

But a kind of resurrection was also hoped for, similar to that from dust quoted above: "*... let the earth open, that salvation may sprout forth*" (Isaiah 45.8). And as humans may wish: *"Thou who hast made me see many sore troubles wilt revive me again; from the depths of the earth thou wilt bring me up again"* (Ps 71.20).

3.2.4 Adam and Eve

Soil scientist Daniel Hillel introduced to his scientific community the meaning of *Adam* (Hebrew אָדָם), which came from *Adamah* (אֲדָמָה) that means besides other "soil"; and the meaning of *Eve*, which came from חַוָּה = *ḥavāh* or *cḥavah*, that means "the living one" or "life giving"; as well as *homo* (the human) came from the Latin *humus* = "soil" (Hillel, 1992, p. 14; Gesenius, 2013, pp. 217–226). The meaning of Adamah covers the broad semantic field of "loose earth" and "clay", "fertile soil" (modern Hebrew קרקע/*karka*) and "arable land", "land" and "home," "earth" and "world" (Bührer, 2014, p. 1). The Hebrew noun Adam is used in the Old Testament above all appellatively as a collective singular with the meaning "humans" or "humanity" of both sexes (Pfeiffer, 2006). The biblical word עָפָר (*'āfār*) mainly translated as "dust" could also mean "world", "underworld" and "realm of the dead ones", whilst the related word אֵפֶר (*'efær*) for "dust" mainly means "ashes". In addition there are about six more other Hebrew words that are part of the semantic fields for underworld, soil, clumps, dust, and territory (Bührer, 2014, p. 1 f.). The etymologies of *Adam* and *homo* are often quoted and have been part of many popular discourses between soil scientists since Hillel first published his observations in the soil science community. When taking such statements for more than rhetoric, they can be good doors to enter the deeper cultural realm of human nature relations.

The name "Eve" (חַוָּה *ḥavāh*) is arranged in Genesis 3.20 with the root "life" (חיה *hjh*), and Eve is supposed to mean the "mother of all" living (also in Sirach 40.1). The image of Eve being made of Adam's rib could have been issued from a Sumerian word meaning "rib" and "life" at the same time (Pfeiffer, 2006).[10]

3.2.5 Jewish Theocentrism, Stewardship, and the Earth

The "original blessing" of the earth, being God's creation and "very good" (following Genesis I), contributed to the foundations for Jewish nature ethics. Moses Maimonides (1135–1204 CE), "uniquely in all of Jewish thought, challenged the primacy of humanity in the order of Creation," viewing animals and humans as equals and saying "that this whole of being is one individual and nothing else," because "the One [God] has created one being" (Seidenberg, 2017, p. 21). In the Jewish mystic tradition of the Kabbalah, "there is no place empty of God," so that "the presence of God can be found in every creation and being" (Seidenberg, 2017, p. 25) (Fig. 3.2). Therefore, "the dirt of the Earth itself includes the seal, structure,

[10] Whether the connection of חַוָּה *ḥawwāh* "Eve" with חיה *ḥjh* "life" is based on historical etymology is disputed. Reference is made, for example, to *ḥwt* "living" in Ugaritian. Among the rather vague alternatives to that, especially the relation to Aramaic *ḥwj'* "snake" has already been discussed by medieval rabbis (Pfeiffer, 2006). – Hardly any of this etymology is undisputed. In the comparative and etymologic analysis of Phoenician-Punic languages Hebrew is part of, also referring to their roots in Mesopotamian languages such as Akkadian, there would still be a lot to question and add.

Fig. 3.2 Yahweh appears in a tree and talks to Moses who sits on the "holy earth", as the text above the image says. Late 14th century Bible of the Roman-German and Bohemian King Wenceslas IV in Prague. (Credit: Bible of Wenceslas (Wenzelsbibel) vol. 1. Austrian National Library, Vienna, Cod. 2759, fol. 55v)

and image of God, which became part of Adam," so that in Hasidism, following rabbi Shneur Zalman (1745–1813), "the very substance of the Earth manifested the greatest revelation of divinity. (…) He described the growth from year to year of plants from the soil as the completion of Adam Kadmon" [who can be interpreted being the original and complete human being], the love that gives birth to Creation" (Seidenberg, 2017, p. 28).

The modern "Jewish environmental movement emerged in the early 1970s as an apologetic defence against the charge of Lynn White Jr. (1967) that the Judeo-Christian tradition, rooted in the Bible, was responsible for the current ecological crisis" (Tirosh-Samuelson, 2013). Stewardship ethics was developed on the basis that "Since the Creator is the true owner of the Earth, … God's tenant-farmers" are to "care for the Earth rather than exploit it for one's needs" (Tirosh-Samuelson, 2017, p. 182). The Jewish Austrian and then Israeli existential philosopher Martin Buber (1878–1965) provided a philosophical basis for a more relational approach towards nature: "In treating Nature as a 'Thou' rather than an 'It', Buber personified natural phenomena and recognized not only the need of humans to communicate with natural objects, but also the inherent rights of Nature. … What makes possible I-Thou relations with non-conscious Nature is God, the Eternal Thou, who is present everywhere and in everything" (Tirosh-Samuelson, 2017, p. 188). This is a pantheistic argument, "not anthropocentric but theocentric." More simply said, Buber understood the human relationship to God as a necessarily dialogical encounter, and through God's presence in his creation humans gained access to a dialogue with nature as well.

There are interesting questions regarding how this connects to the old ideas that when human beings conduct themselves in accordance with God's will the earth is fertile; when they sin against each other and against God, the earth loses its fecundity. Tirosh-Samuelson takes the ethical stance that the "human task is not understood as managerial 'control and command', but rather as attentive 'stewardship' or 'caregiving'" (2017, p. 190), but it may remain a question what in that is then "God's will". Israeli soil scientist Daniel Hillel concluded (2006, p. 220): "Restoring the precious balance between the spiritual and the earthly is now an essential task for Judaism, as it is for other religions."[11]

3.3 Christian Religion and Soil

Christianity began as a Jewish reform and prophetic revivalist movement in Palestine/Israel. The Christianization of almost all European and neighbouring peoples took about 1500 years after the crucifixion of Jesus Christ. It started with little communities of apostolic followers in Greece, Asia Minor and the Levant, spread very slowly in the Roman Empire, later convinced Celtic Ireland and the Frankish Emperor, spread eastwards from Byzantium and Bulgaria, and came by mission, imperious order or brute force to more and more of the remaining European people. There were of course many processes of acculturation between the new religious system and different European socio-cultural substrates, that is, the history and state of people Christianity was adopted by or superimposed to; there happened also splits into diverse denominations and theological traditions. Here we look at some characteristics that refer to soil and nature relations in the Christian context.

Part of standard Christian missionary rules was the strong interdiction to observe rocks and sources, trees and dreams.[12] This gave a heavy blow to people's relation to nature and caused a deep split at some level in the collective psyche (Fig. 3.3). But some features have pressed for a comeback to cultural awareness within a Christian framework, getting partly integrated into and by that in fact broadening the Christian god-image. Some symbols that are well known from cosmic as well as

[11] I am grateful to Danny Itkin for reviewing the first draft of this part on Judaism and the Old Testament, especially his comments on history and his admonition to be careful with etymology. The author is aware that this sub-chapter deals with its subject only very incompletely, and also that quite different approaches to presentation and interpretation on this subject would be well possible and insightful.

[12] See, for example, from the 8th century CE the mission brevier of Pirmin in Switzerland (edited by Hauswald, 2006, p. 79, 87).

Fig. 3.3 Arrival of the missionary and monk Pirmin, who founded monasteries in the 8th century on an island in Lake Constance (Southern Germany). According to legend, all the "worms" fled from the island. Depiction from about 1624 in a local church. (Credit: Minster of Mittelzell in Reichenau)

earthly goddesses have become attributes of Mary. Because of their special relevance for soil, such elements from folk tradition and official representations will precede the biblical quotations of the New Testament on soil here.

3.3.1 The Non-goddess and Her Divine Properties

At the Council of Ephesus in 431 Mary, mother of Jesus, was officially attributed the name God-bearer (Gr. *Θεοτόκος*, Lat. *Deipara*) in the meaning "Mother of God"; this name had already been in use for about 200 years. As Mary was regarded as an earthly human, with her an intimate connection between Jesus and the earth was initiated. Complementary to that, the name Mary Queen of Heavens emerged from the 6th century CE on. This notwithstanding the fact that in the Old Testament Jeremiah (6th century BCE) had complained about women who bake offering cakes for the "Queen of Heavens" (Jer 7.18). Jeremiah may have been referring to the Assyrian goddess Ishtar or the Syrian-Palestinian Astarte that were also called with that name and who were both cosmic goddesses and goddesses of fertility (Keel & Uehlinger, 1998). Both goddesses were depicted with a crown of stars and a crescent moon, which later were attributed to the Christian Mary (and merging her with the cosmic Lady of Revelation 12.1 that appears at the end of the New Testament). Concerning agriculture, Mannhardt (1875, p. 398) reported from Southern Germany "riders trough the seeds," walking around the fields of the city, chanting "be glad and happy, Maria, queen of Heavens".

In some writings of the 14th century, the heavenly Mary was recognized in the star constellation Virgin (Lat. *virgo*) and depicted with *ears of corn* in her hand, corresponding to the name of the brightest star of this star constellation, called *spica,* which means "corn ear" (Thomas, 1968, p. 84). Before in Mesopotamia, this star constellation was known as field furrow and as the divine ear of corn, later it was also named Divine Virgin of the Ear of Corn (Burrows, 1927). The Romans believed they saw the goddess Ceres in this constellation. In the 15th century, the German priest Heinrich von Laufenberg named Mary *Regina celi, terre et maris*: that is "queen of heavens, of earth, and of the sea(s)"; and the same author called her "*fertile soil*" (Wackernagel, 1867, p. 551).

The soil aspect of Mary was also highlighted in her image "Mary in the Garment of Ears," that is, being covered by a cloth that shows corn ears, an imagery that emerged in the 15th century (Fig. 3.4). "*Mary, you are the blessed earth, from which becomes the grain of wheat.* You have born it, let its strength come into myself" (von Laufenberg 1429 in Wackernagel, 1867, transl. by author). During this time in history, plant paintings came back into the churches, and literature was flooded with plant names, metaphors and symbols directed at Mary (de Gubernatis, 1878; Gerhard & Schnell, 2002). "*Oh Maria, you healing herb, you Rose and Cinnamon, you Myrrh and Incense, you are the Bride of God*" (von Laufenberg, 1415 in Wackernagel, 1867, p. 552). Already in the early 12th century, the influential poet and first cantor of the Notre Dame cathedral in Paris, Adam of St. Victor, named Mary the *lampas soli,* that is, the "light of the soil … brighter than the sun." Adam sang of Mary's independent splendour: "Holy root, living root, flower, vine and olive tree, not needing a grafted scion for bringing fruit"[13] (quoted from Becht-Jörgens, 1993, p. 47 and 49, transl. by auth.).

Early apocryphal texts show the belief in the Assumption of Mary. The first liturgies for the celebration of her bodily assumption into heaven have been known since the 9th century (Fournée, 1970, p. 277). About 500 years later, in the catholic dogma text of the *Assumptio Mariae*,[14] it was stated that Maria enters heaven in a bridal chamber. Her bodily rising to heaven might be interpreted as a tendency of the "earthly" coming somewhat back into the Christian God-image, by that Assumption of Mary to heaven re-gaining a cosmic dimension that diverse "mother earth" goddesses, such as, for example, the Germanic Freya/Holle, Baltic Laime, or Egyptian/Roman Isis previously had too (see Chap. 2). One element of the reception history of this dogma change was promoted by the psychologist C.G. Jung in a letter to the clergyman Victor White in 1950 (Jung Letters II, p. 203): "It [the assumptio mariae]

[13] *«Radix sancta, radix uiua / flos et uitis et oliula / quam nulla uis insitiua / iuuit, ut fructificiet»* (this Medieval Latin used the letter "u" for "v" and "u," both).

[14] *declaratio assumptio mariae*, done in the year 1950 by pope Pius XII, making it an official dogma of the catholic church that the virgin Mary was received into heavens with her body and her soul: "Immaculatam Deiparam semper Virginem Mariam, expleto terestris vitae cursu, fuisse corpore et anima ad caelestem gloriam assumptam." http://www.kathpedia.com/index.php/Munificentissimus_Deus_

Fig. 3.4 Mary in the garment of ears, standing on a soil with herbs, surrounded by ornamental plants and two angels. About 1450, Bavarian woodcarving after an Italian model. (Credit: Repro from Glaser, 1923, tables)

means no more and no less than a spiritual fact which can be paraphrased as the integration of the female principle into the Christian image of God" (Fig. 3.5).

Of course there is a strict taboo in Christian theologies of all denominations that no Goddess can ever really exist, and the dogma does not say that Mary is sensu stricto divine, although Pope Pius XII wrote in 1954 in his encyclical *Ad Caeli reginam* (on the Queen of Heaven, § 39), that "from her union with Christ she receives the royal right to dispose of the treasures of the Divine Redeemer's Kingdom."[15] However, the iconography of Mary in the catholic church and partly also in the orthodox east is not distinguishable from that of a goddess, reproducing well-known cosmic and earthly symbols of goddesses from Europe, the Near East and Egypt, at least in some of her aspects. Theology aside, this indicates a strong inner need of people.

[15] https://www.vatican.va/content/pius-xii/en/encyclicals/documents/hf_p-xii_enc_11101954_ad-caeli-reginam.html

Fig. 3.5 The heavenly Mary with her divine child as vision and patroness of the tilling farmer (19th century), Stará Boleslav/ Altbünzlau (Bohemia). (Credit: Reverse painting on glass from Josef Riederer's private collection)

3.3.2 Christ and Soil

The enormous breadth of the Christian imagination and theologies about the nature and actions of the man and God Jesus Christ also made possible a great breadth of mystical and folk-religious developments around this figure. It is interesting in regard to nature and soil that a symbolic or mystic identity was seen between wheat and Christ, both born out of Mary, the Mother of God that was in some respect identified with the Earth. This motif linked back with Isaiah 45.8 (Old Testament), whose translation in the first Latin standard Bible, the *vulgata*, reads: *aperiatur terra et germinet salvatorem*, in analogy to Genesis 1.11 *germinet terra herbam*. That means literally: "The earth may open wide and bring forth the saviour," in analogy to "earth let sprout the herb." In most of the later Bible editions, it was said instead that the earth brings forth salvation (not the saviour himself). But the image took its course. A popular German chant first printed in 1622 says (transl. by auth.): "*Oh earth, come out, come out, oh earth / that hills and dales they all will green / oh earth, this little flower bring / oh saviour, from the earth do spring.*"[16]

The idea of Christ being born from Mary who is here the earth links to Jesus' saying in John (12.24): *"I say to you, unless a grain of wheat falls into the earth and dies, it remains alone; but if it dies, it bears much fruit."* Here, death, resurrection and birth are together and one, a pattern similar to the Demeter-Persephone myth. In another image, Mary was called "a magnificent tree in paradise, its fruit is nourishment to all the world"; this chant explains: "the tree that is your Queen Maria, its fruit is the holy Christ" (Wackernagel, 1867, citing 13th or 14th century). So the tree of life was identified with Mary, Christ being symbolically a fruit of it and her – and then he was nailed down to his place of origin, was nailed to a cross made of wood from a tree that is identified with Mary (Fig. 3.6).[17] This mythopoietic development with its mystic connections to nature was not mainstream theology. But it symbolically expresses that God – or the new god-image of this time – was born out of and later reconnected forcefully to the living matter of the Earth for some time.

This close symbolic connection of Jesus Christ to the living soil and the tree of life is contrasted by severe problems that Christianity had with the earthly underground, and with life in soil. These problems at first relate to the notion of "hell", that the fallen angels and especially Satan have their seat underground and their kingdom in "this world": that is, inside and on earth, and with the worldly affairs down there. And second, Christians had of course problems with the pagan spirits living in soils and waters, woods, and rocks. Practically, both negative associations with earth led to practices of "exorcism of nature" (Franz, 1909: I, p. 197, 298). Not only were there pictures showing all earthly animals escaping when the missionary arrived *(see picture 3)*,[18] there were also rituals by priests to chase all "worms" (snakes, earthworms, caterpillars) away to wilderness, to leave a pure soil behind that is receptive for the Holy Ghost and Christian farmers – documented in 19th century Bavaria, for example (Patzel, 2015, p. 112).[19]

Intriguingly, there exist transitional forms from spiritual to chemical cleaning practices in agriculture: Holy (blessed) water and copper sulphate ($CuSO_4$) were combined with prayers and rituals to prevent fungal infections of crops; later, the chemical was used alone: that is, a transition happened between exorcism and pesticides with a similar pattern of behaviour, but changing to "technical" means for

[16] German Text: "O Erd', schlag aus, schlag aus, o Erd' / Dass Berg und Tal grün alles werd' / O Erd', herfür dies Blümlein bring / O Heiland, aus der Erden spring."

[17] «*Arbor pulcherrima est virgo Maria, … Lignum densarum frondium est crux Christi*» (de la Haye, 1653: Cap. XXI, p. 404). This means that the cross of Christ is made of the "wood of Mary."

[18] One of the early Christian cultural and missionary centers of the post-Roman period north of the Alps was the monastery of Reichenau on an island in Lake Constance. The legend is illustrated on a picture in the minster that when Saint Pirmin arrived to establish this monastery in 724, he first touched the soil with his crosier, causing all kinds of "grisly worms," some serpent-like or amphibian-like, to leave the island in flight. In the imagination of the time, these animals were equated with evil or pagan demons.

[19] Sources: Lexikon für Theologie und Kirche III: 1125 f.: *(Exorcism.)* "Exorzismus"; Franz, 1909: II, pp. 140–3, 152, 160–3; von Amira, 1891, pp. 17, 21–23, 28 f.; compare Ohrt (1941).

Fig. 3.6 Christ nailed to the tree of life and death, around which coils a snake with a human head. Fresco from Giovanni di Pietro Falloppi (Giovanni da Modena), Basilica di San Petronio of Bologna (Italy), about 1420. (Credit: Repro from Dufour-Kowalska, 1985, p. 91)

realizing it (John, 1900, p. 36; Patzel, 2015, p. 106). This raises the question of whether nature is seen to be potentially evil, threatening, and to be overcome by control, or less.

More on soil and earth in the New Testament: For Christ, being swallowed by the underground was part of a dramatic transition (Matthew 12.40): "*For as Jonah was three days and three nights in the belly of the whale, so will the Son of man [Jesus] be three days and three nights in the heart of the earth.*" And the earth remained a source of threat, as Revelation 13.11 says: "*Then I saw another beast which rose out of the earth; it had two horns like a lamb and it spoke like a dragon.*" – Less metaphysically, the earth was addressed in Mark 4.28: "*The earth produces of itself, first the blade, then the ear, then the full grain in the ear.*" In James 5.18 the close connection of God with earth was stated: *"Then he prayed again and the heaven gave rain, and the earth brought forth its fruit"*; and the Apostle Paulus confirmed (1 Corinthians 10.26) that *"the earth is the Lord's, and everything in it."* However, Paulus also stressed the "ashes to ashes" metaphoric field by saying (1 Corinthians 15.47): *"The first man was from the earth, a man of dust; the second man is from heaven."*

And finally the most known soil metaphor from the Bible is the parable of the sower (Matthew 13.5–6, 8) "*Other seeds fell on rocky ground, where they had not much soil, and immediately they sprang up, since they had no depth of soil, but when the sun rose they were scorched; and since they had no root they withered away. ... Other seeds fell on good soil and brought forth grain, some a hundredfold, some sixty, some thirty*" (other versions are given in Mark 4.8 and Luke 8.8).

The quotations from the New Testament as well as non-canonical folk beliefs show that the earth was seen ambiguously: The deeper underground was feared as a source of dangers, and the beasts in soil associated with that, but Christ or the Holy Spirit were expected to overcome these dangers, at least splitting them off from a site and chasing them away from the cultivated land, soil and soul. Closer to the surface, soil as a receptacle and matrix for seeds and corn was seen rather positively, identified with Mary and compared with the human soul that receives and breeds the seeds of God.

3.3.3 Recent Changes in Christian Approaches Towards Nature and Soil

For most of history Christian ethics was social ethics and an ethics of personal conscience. Confronted with huge destruction of nature by Christian people, especially in the colonies, but also in Europe and the USA, a discourse started – one may say after White (1967) – regarding whether this was based on statements in Genesis that humans were to "have dominion over" the creation, or not (Fig. 3.8).

Beforehand, there was not much discussion of Christian ethics regarding nature relations, not to mention soil. Some mystics and monks differed, talking emphatically of nature, like Saint Hildegard von Bingen (1089–1178) and Saint Francis from Assisi (1181–1226 CE). The latter said in his "Canticle of the Creatures" (1999/13th century CE): "Praise be to you, my Lord, through our Sister, *Mother Earth, who sustains and governs us,* and who produces various fruit with coloured flowers and herbs." This was quoted in the first paragraph of Pope Francis' encyclical *Laudato si'* from 2015. The pope comments subsequently (§ 29; italics by auth.): "This sister [Mother Earth] now cries out to us because of the harm we have inflicted on her by our irresponsible use and abuse of the goods with which God has endowed her. *We have come to see ourselves as her lords and masters,* entitled to plunder her at will. The violence present in our hearts, wounded by sin, is *also reflected in the symptoms of sickness evident in the soil,* in the water, in the air and in all forms of

Fig. 3.7 People in the countryside watch angels falling to earth. The black shape probably shows the devil's hole of impact in the soil, which is reported in Swiss legends. Mural of a Swiss house, around 1700. (Credit: Historical Society of Wädenswil, Canton Zurich, and Hauser, 1992, p. 153)

life." This is a clear turning away from the long-lasting interpretation of Genesis that men are the landlords of the earth.[20]

And it follows (§ 53): "*Never have we so hurt and mistreated our common home [the earth] as we have in the last two hundred years*. Yet we are called to be

[20] However, the author tries to keep on the Genesis report in the Bible, just turns from here to the classical position of stewardship ethics (Laudato si', § 116): "An inadequate presentation of Christian anthropology gave rise to a wrong understanding of the relationship between human beings and the world. Often, what was handed on was a Promethean vision of mastery over the world, which gave the impression that the protection of nature was something that only the faint-hearted cared about. Instead, our 'dominion' over the universe should be understood more properly in the sense of responsible stewardship." And explicitly (§ 67): "Although it is true that we Christians have at times incorrectly interpreted the Scriptures, nowadays we must forcefully reject the notion that our being created in God's image and given dominion over the earth justifies absolute domination over other creatures. The biblical texts are to be read in their context, with an appropriate hermeneutic, recognizing that they tell us to 'till and keep' the garden of the world (cf. Gen 2.15). 'Tilling' refers to cultivating, ploughing or working, while 'keeping' means caring, protecting, overseeing and preserving."

Fig. 3.8 Picture "The blessing of the wheat in Artois" by Jules Breton, 1857. Field processions including Mary with crown and sceptre were already adopted in early Christianity even in detail akin to the customs of Roman religion (Felbecker, 1995, p. 194) and similar to Celtic and Germanic customs, too (Franz, 1909: II, p. 8). – Jules Breton painted rural scenes with warm romantically inspired color moods, combined with sober realism in a time of cultural wars over religion and science as well as social issues. (Credit: b p k Bildagentur für Kunst, Kultur und Geschichte, Berlin)

instruments of God our Father, so that our planet might be what he desired when he created it and correspond with his plan for peace, beauty and fullness. *The problem is that we still lack the culture needed to confront this crisis.*" "Everything is related, and we human beings are united as brothers and sisters on a wonderful pilgrimage, woven together by the love God has for each of his creatures and which also unites us in fond affection with *brother sun, sister moon, brother river and mother earth" (§ 92).* Francis becomes spiritually most precise concerning soil, stating that *"we can feel the desertification of the soil almost as a physical ailment*, and the extinction of a species as a painful disfigurement" (Laudato si', § 89).[21]

Here we see a large shift in a most important branch of Christian ethics of nature. To say harming soil is hurting ourselves is revolutionary for catholic ethics. And directly addressing soil in an unprecedented manner for a Pope, Francis wrote about "substances which contribute to the *acidification of soil and water, fertilizers, insecticides, fungicides, herbicides and agrotoxins in general,*" mentioning the "increase in changed uses of the *soil,* principally deforestation for agricultural purposes" (§ 23), and summarizing: "*Soil, water, mountains:* everything is, as it were, a caress of God. The history of our friendship with God is always linked to particular places which take on an intensely personal meaning" (§ 84). The task were to see the "harmonious ensemble of organisms existing in a defined space and functioning as a system." Looking for a cure, "We need only recall how ecosystems interact in dispersing carbon dioxide, purifying water, controlling illnesses and epidemics, *forming soil*" (§ 104).

[21] The quote in paragraph 89 comes from the Apostolic Exhortation Evangelii Gaudium (24 November 2013), 215: AAS 105 (2013), 1109, of the same author.

But "*To seek only a technical remedy to each environmental problem which comes up is to separate what is in reality interconnected and to mask the true and deepest problems of the global system*" (§ 111). In contrast, the Pope asks for an "*integral ecology, which by definition does not exclude human beings,*" and quotes from the Old Testament (Ecclesiastes 38.4): "*The Lord created medicines out of the earth, and a sensible man will not despise them*" (§ 124). To learn that, "*All Christian communities have an important role to play in ecological education*" (§ 214); and "*Here, I would like to offer Christians a few suggestions for an ecological spirituality*" that "*cannot be sustained by doctrine alone, without a spirituality capable of inspiring us, without an 'interior impulse which encourages, motivates, nourishes and gives meaning to our individual and communal activity'*" (§ 216). Francis tries to establish a new normative requirement by saying: "*Living our vocation to be protectors of God's handiwork is essential to a life of virtue; it is not an optional or a secondary aspect of our Christian experience!*" (§ 217).

In the present theological discourse about this shift in Catholic Christian ethics of nature relation, Protestant theologian Altner (2019) said that Christianity had "forgotten about creation," being concerned with the salvation of the human soul only. This "salvation egoism" of Christians has prepared the "secular utopias and hopes in 19th and 20th century, where men take the place of God … and it seems that modern man suffers a 'god complex' in how he/she deals with nature." Catholic theologians Vogt and Frankenreiter commented that creation theology now may shift from "anthropocentrism" to "anthroporelationalism," where "adam" belongs to the arable soil "adamah" (2019).[22] Furthermore, "It was overdue that Francis broke the self-referential refusal of dialogue on the part of theology" (Vogt, 2018, p. 249, transl. by auth.). The tasks were, referring to White (1967, p. 1206): "rethinking our own religion," to overcome the "mental structures of forgetfulness of nature" and an "imperial relation to nature" (Vogt, 2018, p. 251).

Miller (2017) commented that "Integral ecology is the central theme of *Laudato Si',*" being "the essence of reality" (p. 11). And exemplified (p. 15): "Consider soil. We know that it is more than dirt, of course. … which gives us life and is a community to which we give back through care as well as our own death and decay. … Integral ecology inspires us to gaze with the patient openness to learn these connections."

In the *Orthodox Church* (that is the eastern main branch of Christianity), a "narcissistic worldview of the church and the broad failure of Christians to properly relate to nature" was complained about, for example, by Chryssavgis (2017). He quoted Symeon of Constantinople (10th century): "What is more painful than to be separated from life?" (p. 279). But in contrast, "orthodox theology emphasizes precisely the significance of Creation as the incarnation of God in the widest and deepest sense" (Chryssavgis, 2017, p. 283). The dichotomy of the Greek Orthodox Church between early ecological awareness and discourse on the one hand, and problems with pluralistic societies and ecological debates on the other, is presented by the Metropolitan of

[22] Further reading: Drummond et al., (2017) and Kärmer, Vellguth (2018).

Pergamon (Zizioulas, 2016) and Orthodox theologians (Asproulis, 2016); in the Russian Orthodox Church, this problem may not be smaller.

In the diverse Protestant Christian churches, discussions of ecology started earlier in the 20th century preceding developments in the Catholic Church. Similar to the debates concerning Jewish religious environmental ethics, most of these discussions circumambulate the question of replacing an ethics of absolute human dominion over nature by an ethics of stewardship, whatever that means (compare Chap. 21).

3.4 Islam in Europe

Islam was a major religion in Southern Europe for centuries, was then pushed back to small countries, and has become relevant again in most countries of Western and Southern Europe in the last 60 years or thereabouts due to immigration and growing religious communities.

The prophet Mohammed saw himself to found Islam following God's intention to establish true monotheism by renewing the religion of Abraham on the basis of gotten divine revelations. This movement in a short time developed an enormous mental and military force, resulting in rapid Islamic expansion. The Iberian Peninsula, where Christianised Visigoth rule had succeeded the Romans, was the first part of Europe extensively converted during an early stage of the Islamic expansion. This area became Al-Andalus until its complete end after 781 years of Islamic rule and culture in 1492 CE. Through the transfer of knowledge and goods between the widespread Islamic dominions, citrus fruits, rice, eggplants and other cultivated fruits reached southern Europe via Spain, together with their cultivation methods. The Spanish words for irrigation ditch, cistern, draw wells and irrigation wheel and even for the "hollow round the base of a tree to hold water" are of Arabic origin (Watt, 1972, p. 22). Quite parallel to Spain, Sicily was dominated by Muslims for about 250 years in the Medieval. "Thus trough trade contacts and though political presence in Spain and Sicily the superior culture of the Arabs gradually made its way into western Europe" (Watt, 1972, p. 29).

After the fall of Cordoba in 1492 which was the last Muslim domain in Iberia, tolerance accords were short-lived. Islamic peasants were bullied, forcibly converted and finally expelled, and the entire Arab and Islamic culture was bitterly fought by the new Christian rulers and the so-called Holy Inquisition (Hottinger, 1995).[23]

[23] On the east, powers developed differently. In the early 13th century, the Mongols gained dominance over the Russian principalities. The "Great State" of the Golden Horde in today Southern Russia mainly converted to Islam one century later, causing a partly lasting establishment of this religion in some southern regions of later Russian empire that came into being and became independent with Ivan the Great (ruling 1462–1505). Simultaneously the Osman expansion – conquest of the Byzantine empire and finally Constantinople in 1453 – brought more than 500 years of Islamic rule in Greece, the Balkans and other parts of south-east Europe, some areas have remained mainly Islamic until the present.

The Andalusian Sufi philosopher and mystic Muhyī d-Dīn Ibn ʿArabī (1165–1240) stated that creation is female and like a beautiful mirror in which Allah can regard himself (Schimmel, 1995, p. 21). Similarly, in the Shiite realm, the Persian mystical poet Jalāl al-Dīn Muhammad Rūmī sounded: "The heaven is male and the earth is female; what he pitches in, that brings fruit" (Schimmel, 1995, p. 108, transl. by auth.). The extra-Koranian tradition on Muhammad (collections of Hadiths) contains the statement: "Take care of the earth for she is your mother."[24] There was a "tendency found in both classical Sufism and medieval Christianity" to equate "'the world' with a woman"; however, in Islam "Eve was never made responsible for the fall of Adam" (Schimmel, 1975, p. 428 f.).

An important indirect influence on the understanding of soil in Europe may have come from Arabic alchemy that strongly influenced European alchemy (by Ibn Umail/Senior, and others) in its symbolic as well as its physical branch, and thereby building some roots of chemistry and general nature study. See this exemplary quotation from Muḥammad Ibn Umail, who lived in the early tenth century (Abt & Madelung, 2003, § 149.2): "*The water existed before the creation of the earth. Then God created the earth. Then he created Adam from it [the earth]. And the earth does not live without water. And Hermes said: 'The secret of all things is the water and the earth because all things get cultivated in the two of them.*[25]*' 'And like that our earth becomes alive by the water of our heaven and our water is from our earth. From it (the earth) it rises up and it returns'* (§ 157.12 ff.). *So, the earth holds what the water contains, even after its evaporation*" (§ 23.2–6).

The connection of Islam to soil following the Quran is presented in the Chap. 6 by Makki & Juilleret, and the mysticism of Rumi regarding soil is introduced in the Chap. 7 by Alavipanah et al.

3.5 Concluding Remarks

Monotheism has the challenge to integrate different aspects or appearances of the divine into one God-image – or to stay incomplete because of splitting off some non-accepted parts of the numinous from being part of the "one and only God." Therefore, monotheism apparently had more difficulty than polytheism to deal with the spiritual dimension of deep nature, especially of the soil. The Judeo-Christian religion shows strong ambiguities towards soil. Following a myth of Genesis, mankind fell from the state of paradise to the soils of the fields; and following the New

[24] Hadith book Majma'uz Zawaid from the 14th century CE including the collection by Al-Tabarani "Al-Mu'jamul Kabir," there Hadith no. 4596. This reference was searched for by the author after having seen the quote in Makki & Safaei-Shahverdi, 2015, p. 200.

[25] Compare Xenophanes' fragment 33 quoted in the first paragraph of this chapter. This may hint to an early interreligious exchange and some continuity of ideas in the Eastern Mediterranean, which is best documented with Aristotle's writings in Arabic translations.

Testament, the devil, which was thrown out of heaven, fell underneath (Fig. 3.7). The serpent as well as the other beings crawling on soil dwelling inside soil and earth became dubious like the pagan spirits that co-inhabit the underground. (The association of the subsoil with dead bodies waiting for physical resurrection after the final judgement was not discussed here.) Moreover, there is in the Judeo-Christian (and Islamic) tradition a certain loss of soul and devaluation of matter as "mere dust" – but a dust that is perceived as vivified and animated by the divine spirit, the creator god.

In Christianity, however, there have been strong movements from the beginning until the present to reconcile with nature in regard to creation, and to re-integrate some of the feminine that lacks a place in the doctrinal god-image; these are: (1) An intimate connection of Christ with the earth, especially by his birth from Mary, his death at the wooden cross, and his dive to and resurrection from the depths; (2) the adoption of some religious needs that earlier had been linked with Goddesses by the "Mother of God" Mary, including her identification with earth and soil and a certain cosmic spiritual side; (3) recently the move especially in the catholic realm to change its ethical position of extreme anthropocentrism in relation to the physical world, towards a more relational approach of "fellowship" between all living beings, explicitly including soil. Concurrently, Judaism developed a stronger move towards ecological stewardship ethics, and similar happens in protestant Christianity.

The Abrahamic religions of Judaism, Christianity and Islam include the prominent position of human beings as representatives and therefore stewards of God in creation. At the same time they all show the conception of a brother- or sisterhood within creation and the unity of the whole of creation. A second common issue of Abrahamic monotheism is that on the one side there is a certain devaluation of nature as "mere dust," "just matter" or even inhabited by the evil – and on the other side they say that God is mirrored, effective or inherent in his creation, or even incarnated in it, and that human beings were made of soil animated by God. These different religious ideas lead to the question whether we are relatives or rulers of soil, and if soil is for humans a counterpart, that is, an autonomous subject of its action with some finality inside, or a mere object of human action.

In addition to the sometimes wide-open ambivalences of human relations with nature in connection with various scriptures and traditions of monotheistic religions, there is the classic behavioural gap. This means that the structurally and also concretely lived handling of the soil all too often, almost regularly, does not correspond to some respectful and caring core orientations as they appear in the holy scriptures or religious traditions. Why is that? Can this be changed and if so, in which ways? We will come back to this problem in the overall conclusions of this book (Chap. 26).

Common to all described religious elements is the essential function of religion: to (re-)connect with and carefully relate to a spiritual dimension that is more than human,[26] that is, to acknowledge and venerate what is essentially beyond the scope of human consciousness and grasp, and to find deeper purpose and meaning in it. Symbolic ideas and actions are typical ways to fulfil such a mentally orientating and relating function, often in the frame of a myth, a "revelation," or other kind of religious tale and literature.

Religiously coloured and spiritually asking ways of inner orientation, of thinking and conceptualizing reality can challenge or maybe relativize the current scientific notion of causality, where a soil scientist habitually would ask the question: "What are the effects of all that on soil, and why?" It is hard enough to track causality in a given soil profile. But here the questions involve humans and more as parts of the system. *Causally* modelling soil systems including the human factor may all too often produce shortcomings or failures because of uncertainty and indeterminism – and creativity! – that transgress the possibilities of modelling. Then it might be an interesting venture to look at qualitative mental patterns effective in human behaviour towards soil: with their historical substrate and roots, their present fruits and future promises, being receptive whilst considering their ambiguous potentials. It is about functions and meaning of religious contents and spiritual experiences in human nature relations and especially human (humus) soil relation – at the collective socio-cultural levels as well as maybe for a personal quest for orientation, meaning and embeddedness.

References

Abt, Th., & Madelung, W. (Eds.). (2003). *Book of the explanation of the symbols*. Kitāb Ḥall ar-Rumūz by Muhammad Ibn Umail. Living Human Heritage Publications.

Altner, G. (2019). (The human being between nature and culture.) Der Mensch zwischen Natur und Kultur. *Politische Ökologie, 37*, pp. 56–57.

Asproulis, N. (2016). Creation, history and the Church. Towards an orthodox eco-theology and social ethics. In *Dietrich Werner and Eisabeth Jeglitzka: Eco-theology, climate justice and food security. Theological education and Christian leadership development*, pp. 187–208. Published by Globalethics.net.

Becht-Jörgens, G. (1993). *(Aspects of Latin hymnody.)* Aspekte lateinische Hymnendichtung (pp. 39–65). In Theo Stemmler: An die Gottheit. Bittgedichte aus zwei Jahrtausenden. Narr publisher. Becht-Jörgens took the Adam of St. Victor quotes from F. Wellner (1955): Adam von St. Victor. Sämtliche Sequenzen lateinisch-deutsch; no. 38. Kösel publisher.

Bible, the Holy (set down in writing from the 2nd cent. CE till about 400; the part issued from the Tanach recorded from about the 9th cent. BCE on): Several editions in different languages are quoted.

Bührer, W. (2014). *(Soil)* Boden, in: Das Wissenschaftliche Bibellexikon im Internet (www.wibilex.de). Deutsche Bibelgesellschaft.

[26] The Latin word *religio* can be traced back to the verb *religere*, which means "to take into careful account," as well as to the verb *religare*, which means "to tie oneself back to some prior state of existence, to an earlier source of being"; the first mentioned etymology apparently is older (Edinger, 2018, p. 35).

Burrows, E. (1927). Ḫurrian Sala(s). In *The journal of the Royal Asiatic Society of Great Britain and Ireland no. 2* (pp. 318–320). Cambridge University Press.

Chryssavgis, J. (2017). The face of God in the World. Insights from the Orthodox Christian Tradition. In J. Hart (Eds.), *The Wiley Blackwell companion to religion and Ecology* (pp. 273–285). Wiley.

Daraki, M. (1994 [1st ed. 1985]). *(Dionysos and the Goddess Earth.)* Dionysos et la déesse terre. Flammarion.

de Gubernatis, A. (1878). *(Plant mythology or the legends of the plant kingdom.)* Mythologie des plantes ou les légendes du règne végétal (Vol. 1). Reinwald publisher.

de la Haye, I. (1653). *(Collected works of the saints Francis of Assisi and Anthony of Padua.)* Sancti Francisci assisiatis, minorum patriarchae, nec non S. Antonii Paduani, eiusdem ordinis, opera omnia. Veith publisher.

Diels, H. (1906). *(Fragments of the pre-socratic philosophers.)* Fragmente der Vorsokratiker, griechisch und deutsch (2nd ed., Vol. 2). Weidmann publisher.

Drummond, C., Bermann, S., & Vogt, M. (2017). *Religion in the Anthropocene*. Wipf and Stock publishers.

Dufour-Kowalska, G. (1985). *(The tree of life and the cross. An essay on visionary imagination.)* L'arbre de vie et la croix. Essai sur l'imagination visionnaire. Editions du Tricorne.

Edinger, E. (2018). *The new God image: A study of Jung's key letters concerning the evolution of the western God image*. Chiron Publications.

Felbecker, S. (1995). *(The procession. Historical and systematic studies on a liturgical expressive action.)* Die Prozession. Historische und systematische Untersuchungen zu einer liturgischen Ausdruckshandlung. Münsteraner theologische Abhandlungen (Vol. 39). Oros publisher.

Fournée, J. (1970). (Assumption of Mary.) "Himmelfahrt Mariens". In *Lexikon der christlichen Ikonographie* (Vol. 2). Herder publisher.

Francis of Assisi. (1999/13th cent. CE). Canticle of the creatures. In R. Armstrong, J. Hellmann & W. Short (eds.): *Francis of Assisi, Early Documents* (Vol. 1, pp. 113–114). New City Press.

Franz, A. (1909). *(The ecclesiastical benedictions in the Middle Ages.)* Die kirchlichen Benediktionen im Mittelalter (Vol. 2). Herder publisher.

Gerhard, C., & Schnell, B. (2002). (*In words, herbs and stones is God. On the understanding of nature in German-language illustrated herbal books of the Middle Ages.*) In verbis in herbis et in lapidibus est deus. Zum Naturverständnis in den deutschsprachigen illustrierten Kräuterbüchern des Mittelalters. Paulinus publisher.

Gesenius, W. (2013, 18th ed.; 1st ed. 1810). *(Hebrew and Aramaic hand dictionary of the Old Testament.)* Hebräisches und Aramäisches Handwörterbuch über das Alte Testament. Edited by H. Donner. Springer publisher.

Glaser, C. (1923). (*Gothic woodcuts.*) Gotische Holzschnitte. Propyläen publisher.

Hauser, A. (1992). (*Old folk art on Lake Zurich.*) Alte Volkskunst am Zürichsee. Neue Zürcher Zeitung publisher.

Hauswald, E. (2006). (*Pirmin's excerpts* [catechism and mission breviary in early medieval Latin.]) Pirmins Scarapsus. Einleitung und Edition. Private edition (PhD thesis).

Hillel, D. (1992). *Out of the Earth*. Aurum Press.

Hillel, D. (2006). *The natural history of the Bible. An environmental exploration of the Hebrew Scriptures*. Columbia University Press.

Hottinger, A. (1995). *(The Moors. Arab culture in Spain.)* Die Mauren. Arabische Kultur in Spanien. Wilhelm Fink publisher.

John, A. (1900). *(Customs in farming of Egerland.)* Egerländer Ackerbräuche. Unser Egerland. Blätter für Egerländer Volkskunde 4(3/4), p. 36 f.

Jung, C. G. (1950). Letter to Victor White, 25th of December. In *(Letters)* A. Jaffé (Ed.), Briefe (Vol. 2), 1972–1973. Walter publisher.

Kärmer, K., & Vellguth, K. (2018). *Creation. Living together in our common house*. One world theology publisher, OWT vol. 11.

Keel, O., & Uehlinger, C. (1998). *Gods, Goddesses, and images of God in Ancient Israel*. Fortress publisher.

Kuijt, I., & Finlayson, B. (2009). Evidence for food storage and predomestication granaries 11,000 years ago in the Jordan Valley. *PNAS, 106*(27), 10966–10970.

Lexikon für Theologie und Kirche III: 1125 f.: *(Exorcism.)* "Exorzismus".

Makki, M., & Safaei-Shahverdi, M. (2015). (*Islam and soil protection.*) Islam und Bodenschutz. In G. Wessolek (Ed.), *(From the very below. Why we need to better protect our soils.)* Von ganz unten. Warum wir unsere Böden besser schützen müssen (pp. 195–210). Oekom publisher.

Mannhardt, W. (1875). (*Forest and field cults. Vol. 1: The tree cult of the Germanic people and their neighbouring tribes. Mythological studies.*) Wald und Feldkulte. Erster Band: Der Baumkultus der Germanen und ihrer Nachbarstämme. Mythologische Untersuchungen. Borntraeger publisher.

McKirahan, R. (2010). *Philosophy before Socrates. An introduction with texts and commentary* (2nd ed.). Hackett Publishing Company.

Miller, V. (2017). Integral ecology: Francis's spiritual and moral vision of interconnectedness. In *Miller: The theological and ecological vision of Laudato Si': Everything is connected* (pp. 11–28). Bloomsbury publisher.

Naor, M. (1996/98). *The Twentieth Century in Eretz Israel – A pictorial history* [סֵפֶר הַמֵּאָה – הִיסְטוֹרְיָה מְצֻלֶּמֶת שֶׁל אֶרֶץ־יִשְׂרָאֵל, Tel Aviv]. (J. Krausz, Trans.). Könemann publisher.

Neumann, E. (1974/1997). *(The Great Mother. A phenomenology of the female forms of the unconscious.)* Die Große Mutter. Eine Phänomenologie der weiblichen Gestaltungen des Unbewussten. Walter publisher.

Ohrt, F. (1941). (Blessings against worms.) Wurmsegen. In Hanns Bächtold-Stäubli (1927–1941): *(Hand dictionary of German superstition.)* Handwörterbuch des deutschen Aberglaubens (HdA). de Gruyter publisher.

Patzel, N. (2015). *(Symbols in agriculture. On the spiritual side of nature relation in Swiss agrarian culture.)* Symbole im Landbau. Zur spirituellen Naturbeziehung in der Schweizer Agrarkultur. Oekom publisher.

Pfeiffer, H. (2006). (*Adam and Eve.*) Adam und Eva. In *Das Wissenschaftliche Bibellexikon im Internet* (www.wibilex.de). Deutsche Bibelgesellschaft.

Pope Francis (2015). *Encyclical letter Laudato si' of the Holy Father Pope Francis on Care for our Common Home*. Vatican.

Schimmel, A. (1975). *Mystical dimensions of Islam*. University of North Carolina Press.

Schimmel, A. (1995). *(My soul is a woman. The feminine in Islam.)* Meine Seele ist eine Frau. Das Weibliche im Islam. Kösel publisher.

Seidenberg, D. (2017). Eco-Kabbalah. Holism and mysticism in Earth-Centred Judaism. In J. Hart (Ed.), *The Wiley Blackwell companion to religion and ecology* (pp. 20–36). Wiley publisher.

Staudt, D. (2012). *(The one and single God. Monotheistic formulas in early Christianity and their prehistory among Greeks and Jews.)* Der eine und einzige Gott. Monotheistische Formeln im Urchristentum und ihre Vorgeschichte bei Griechen und Juden. Vandenhoeck & Ruprecht publisher.

Thomas, A. (1968). (*The Madonna in the garment of ears.*) Ährenkleidmadonna. In *Lexikon der christlichen Ikonographie* (Vol. 1, pp. 82–85). Herder publisher.

Tirosh-Samuelson, H. (2013). Ecology in Judaism. In A. L. C. Runehov & L. Oviedo (Eds.), *Encyclopedia of sciences and religions*. Springer publisher.

Tirosh-Samuelson, H. (2017). Jewish environmental ethics. The imperative of responsibility. In J. Hart (Ed.), *The Wiley Blackwell companion to religion and ecology* (pp. 179–194). Wiley publisher.

van Nuffelen, P. (2010). Pagan monotheism as a religious phenomenon. In *Stephen Mitchell and Peter van Nuffelen: One God. Pagan monotheism in the Roman Empire* (pp. 16–33). Cambridge University Press.

Vogt, M. (2018). (*God perspectives in the sustainability discourse.*) Gottesperspektiven im Nachhaltigkeitsdiskurs. In *Leben im Anthropozän. Christliche Perspektiven für eine Kultur der Nachhaltigkeit* (pp. 247–258). Oekom publisher.

Vogt, M., & Frankenreiter, I. (2019). *In their internal report on Laudato si' for the Club of Rome*.

von Amira, K. (1891). (*Lawsuists and verdicts against animals.*) Thierstrafen und Thierprocesse. *Separatdruck aus den Mitteilungen des Instituts für österreichische Geschichtsforschung, XII*(4), 545–601.

von Laufenberg, H. (1415/1429). *Quoted after Wackernagel 1867.*

Wackernagel, P. (1867). *(The German choral from the eldest times till the early 17th century.)* Das deutsche Kirchenlied von der ältesten Zeit bis zu Anfang des 17. Jahrhunderts … . Vol. II. Teubner publisher.

Wassmann, X. (2003). (*The death of the great Pan. On the downfall of the god of nature in antiquity.*) Der Tod des grossen Pan. Zum Untergang des Naturgottes in der Antike. Jungiana publisher.

Watt, M. (1972). *The Influence of Islam on Medieval Europe*. Islamic Surveys, vol. 9. Edinburgh University Press.

White, L., Jr. (1967). The historical roots of our ecological crisis. *Science, 155*(3767), pp. 1203–1207.

Yaalon, D. (2000). Down to earth. Why soil – and soil science – matters. *Nature, 407*(6802), p. 301.

Zizioulas, Metropolitan John of Pergamon (2016). Pope Francis' encyclical Laudato Si'. In *Dietrich Werner and Eisabeth Jeglitzka: Eco-theology, climate justice and food security. Theological education and Christian leadership development* (pp. 187–208). Published by Globalethics.net.

Chapter 4
Cultural Patterns of Soil Cultivation in Europe 3: Scientific Context

Nikola Patzel

Science is just as much an "independently flying object" in European societies as are their religions; both realms have developed to become integral parts of European societies' identities. Modern soil science is rooted in physicochemical approaches on one side, and in bio-ecological approaches on the other. This chapter deals with the history of cultural patterns of scientific soil perception and understanding focused on the development since the nineteenth century: an era that has been marked by a very strong development of science and technology. These developments were a counternarrative and practical alternative to religious world views and rituals in agriculture; on the other hand, they themselves also carry mythical and (meant in a non-judgemental way) basically irrational patterns of imagination and ideas inside.

This chapter also includes the subcultural agricultural movements of organic farming that emerged in Europe. They form an interesting special case: small disciplines, often rather niche subjects of academic science combined with alternative research approaches as well as with deep-rooted non-scientific traditions and ideas, forming original cultural developments in nature and soil relations.

But first, let us step back to the cultural developments that led to modern science:

4.1 Origins and Development of Scientific Mindset

The neolithization of Europe was a complex cultural change that included people becoming farmers. This process took many millennia for spatial expansion and regional adaptations, along with changes of the mental and physical landscapes of

N. Patzel (✉)
Freelance Scientist, Office for Soil Communications, Überlingen, Germany
e-mail: nikola.patzel@soilcom.com

N. Patzel et al. (eds.), *Cultural Understanding of Soils*,
https://doi.org/10.1007/978-3-031-13169-1_4

almost the whole subcontinent. Soil cultivation in Europe started with immigrating farmers who first converted the most fertile valleys and basins, before taking in more and more land in sloping areas, this then led to Bronze Age erosion and alluvions (Seidel, 2016). An extreme and paradigmatic case of induced soil erosion was the North Germanic settlement in Iceland. Predominantly Norwegian farmers brought around the year 900 CE their farming techniques from heavy soils of post-glacial origin and transferred them to the extremely easily erodible andosols of the volcanic island of Iceland, with disastrous results and very lengthy remediation and healing requirements (Arnalds & Aradóttir, 2015).

Gatherers and hunters more and more becoming farmers coincided with a mental change in the human-nature relationship. New mythical images came into the relationship with nature and changed human self-understanding.[1] A distinction was necessary between the "cultivated," protected or even fenced agricultural land and the outer space, then becoming "wilderness."[2]

Let us look at a site of birth: The sanctuaries of the "belly hill" Göbekli Tepe in Anatolia (today Turkey) are the site where it is highly probable the Neolithic turn of the broad western Eurasian cultural area started 12,000 years ago. Stone Age people that practised a hunter and gatherer culture started to build circles of stone pillars that were up to 7 metres high, *differentiating between an inner space and an outer space* (Abt, 2014). These circular structures were rebuilt again and again over several thousands of years. The discoverer of Göbekli Tepe, archaeologist Klaus Schmidt, who spent decades working at the site, suggested to "not underestimate the significance of these spiritual driving forces for the entire cultural process underwent in that age" (Schmidt, 2012, p. 231). The "circle symbolism [served] as an expression of demarcation, protection and a new unity of development" (Abt, 2023, p. 48). These circles of pillars have *two large pillars in their center* (Fig. 4.1). "When a centre is manifested so clearly as a duality, it mirrors a dualistic worldview emerging from the unconscious. ... It signifies a potential for clear differentiation, reflection, ambivalence" (Abt, 2023, p. 52). This would reflect a cultural achievement of the Neolithic revolution, the transition from hunter-gatherer societies to those that practice plant production and animal husbandry, combined with a settled existence (Abt, 2023, p. 58).

On a religious level, another change from unity to duality may also have taken place with the Neolithic turn. While in the Palaeolithic the regeneration of life was likely perceived more as a *rebirth* from the maternal earth which naturally contained the male, too, in the Neolithic the symbolism of renewal from a (sexual) *union of opposites* became strong. This can be inferred from observed changes in religious art – rock paintings and statuettes – from the predominantly represented female

[1] Compare the quotes from the Karelian (Finnish) Kalevala myth at the beginning of Chap. 2 on European polytheism and soil.

[2] For a sophisticated critical review of human ecology of neolithization in Northern Europe, and theories on it, see Johansson, 2003. For many case studies on the diversity of this process in Southern Europe, see Manen et al., 2014.

Fig. 4.1 Stone pillar circles with twin pillars in the center: symbols whose formation coincided with the Neolithic revolution towards agriculture and permanent settlements. This building of sacral artefacts started about 10,000 BCE. (Credit: Deutsches Archäologisches Institut)

form to the rather equivalent representation of both sexes, including the emergence of phallic symbols (Vierzig, 2009, e.g., p. 127, 177 f.).

In the sanctuaries of Göbekli Tepe and the remnants of settlements found around it, a distinction happened between "inner" and "outer" areas. This might have gone along with developing distinctions between cultivated versus wild land, wild animals versus domestic animals, and maybe an all-over strengthening of the human capacity for conscious discernment and reflection, supported by mental patterns of duality. These achievements coincided with people becoming farmers, and they have been premises of *scientific thinking* to the present day, later leading to questions like: "Is this alluvial soil *or* an in-situ pedogenesis?" What happened with the emergence of agrarian culture and its knowledge systems were steps on a path that led to a scientific-technical civilization in Europe and other parts of the world – Neolithization not assessed as a sufficient condition, but perhaps as a necessary one.

The fundamental innovations for metal processing came to Europe from the southeast.[3] Thus ploughs could be shod with bronze and later iron (look at the set of tools in Fig. 4.2). Advanced agrarian cultures such as the Celtic one developed autochthonous innovations and corpora of knowledge and skills as described by the Roman author Pliny in the 70s CE: "*Not long ago in Rhaetian Gaul the invention was made …. the width of the ploughshare turns the turf.*" Then Pliny quoted Cato's golden rule of agriculture: "*What does it mean to cultivate the field well? To plough well. What is the second? Ploughing. What comes third? Fertilising.*" But after

Fig. 4.2 Early metallurgy was more than making weapons and jewelry. Equipment of a Celtic craftman's workshop (found in a moor): Pliers, hammer heads, chisels … everything you need to make hub rings for wheels, barrel hoops, and bronce vessels (second-first century BCE). (Credit: Objects of Federseeuseum Bad Buchau (Southern Germany), Presentation at an Exhibition of Archäologisches Landesmuseum in Konstanz 2021. Photo: Nikola Patzel)

[3] Plough blades of bronze and then iron were used in the Middle East some 4000, 2000 years, respectively, before these iron plough shares became common in Europe (Kaser, 2011).

reporting this and some examples of different ploughing depths adapted to the soil, Pliny added a warning concerning regional sustainability: *"In every field of agriculture, but in this one [ploughing] in particular, the well-known principle applies: 'As each area can stand it'"* (Plinius, 1995, § 173 f., transl. by auth.). Erosion control has been an issue from the beginning of agriculture: "The Celts in Britain cultivated fields across the slope to slow erosion, bench terraces were used in modern day France that possibly date back to the Phoenicians" (Brevik & Hartemink, 2010, p 24).

Scientific thinking about nature was enormously sprouting in Greek Antiquity. This was more than philosophy and rhetoric, it led not only to the hoisting crane for temple building (Coulton, 1974), but probably also to the Archimedean screw that allowed pumping water upwards into irrigation canals since about 500 BCE (for Archimedes' ground-breaking inventions see Hasan, 2006).

Much later, in the retrospectively so called Renaissance period, old knowledge was rediscovered, and new techniques invented. The printing press – invented in Germany and using paper technology that the Arabs learned from the Chinese and brought to Europe – came in use to spread texts, this coinciding with the new assimilation of knowledge and ideas from Arabic and Byzantine corpora across Europe (starting from Italy) that draw back to Antiquity. One could get the impression that it was "only in the Renaissance (…) that the scientific study of natural resources, including the soil, started in the western world" (Brevik & Hartemink, 2010, p. 26). Technical innovations that were rather playful and inconsequent in the Renaissance (see the work of artist and inventor Leonardo da Vinci 1452–1519) later developed to become the most relevant structural elements of the scientific-technological complex of the so-called Industrial Revolution that started in the United Kingdom in the eighteenth century. Self-energizing interconnected feedback-loops of innovations occurred and accelerated over time, based on a pragmatic and technophile scientific rationalism as well as an increasingly ample goods trade, including food, supported by new means of transport, all powered by fossil fuels. Fossil fuels and industrial technologies have also powered the Haber-Bosch synthesis of reactive nitrogen (N) (invented in Germany during the First World War for ammunition production) that subsequently strongly changed the global N-cycles and many soil properties with advantages and disadvantages both, contributing to strong augmentation of harvests over many decades in many regions of the world.

In this chapter the focus is on Western science of the modern age, and its consequences for human-soil relationships. A special attention is given to the cultural roots of organic agriculture because there we see an especially clear effectiveness of some deep cultural "groundwaters" regarding human-soil relations. Minor parts of the following are reused or redrafted from the book chapter *The Soil Scientist's Hidden Beloved* (Patzel, 2010).

As a kind of disclaimer, I want to emphasize that the global development of agricultural practices can by no means be reduced to European (and the US) technological development, its history of ideas, and its regional prehistory. Agriculture was invented in several places around the world, one of them described above, not in Europe or North America. When looking for agri-cultures that were at least regionally sustainable over millennia, good examples can be found in Mesopotamia,

Egypt, China, and South America (Brevik et al., 2018). These authors concluded (p. 17): "To prevent negative anthropogenic impacts on the soil resource, it is important that we carefully study the experiences of past civilizations, both successes and failures, and learn from what they did."

4.2 Philosophical Ground of Modern Science

An important precondition to develop *modern* European mainstream natural sciences has been to remove God and any deities or spirits from nature.[4] Creation was reduced to an initial event in the past only, or nature was no longer seen as a divine creation at all. *This fundamental change in worldview has not just been a "disenchantment" of This World, but basically a disconnecting of the inner and the outer world in the experience of humans.*

The act to exclude purposefulness from processes in nature was explicitly done by the very influential philosopher of science Francis Bacon (1623, vol. I, part 5): "*nam causarum finalum inquisitio sterilis est,*" which means, "thus, the investigation of final causes [finality] is sterile." This was extremely important because by that Bacon aborted the idea of purposeful development and intentional processes, that is, finality (do something "for the sake of …") by reducing them to deterministic effects of determined causes, that is, to a natural philosophy of "determinism." The idea of determinism ongoing everywhere in nature promoted a collective mental change that showed a culmination about 200 years later in Darwin's theory of natural selection. Darwin replaced purposeful creation by a theory of "*Struggle for Life*" (quote from the subtitle of his *Origin of Species*, 1859), and the evolution of nature by selection of non-purposefully appearing variations of organisms in function of their environmentally adaptive and especially reproductive value (compare Chap. 18). Applied to soil, this attitude facilitated the development of theories in soil science that are based on soil forming factors that determine the processes that develop soils and change them from one type to another – instead of God's intention to create all species of soils and to place them as he saw fit.

Bacon's equally influential French colleague René Descartes de-animated life in his worldview in a way that was summarized with the famous dictum *animalia sunt automata*: "animals are machines" (see Descartes, 1637, p. 185). He argued that "it seems consistent with the reason that nature also produces its automatons," animals being such ones (Descartes, 1649.[5]) Descartes accepted "reason" and "thinking" as occurring with humans only, nobody and nothing else in nature. But at the same time, he believed that it was reasonable that nature behaves like man's reason and his machines; the mechanistic "clockwork" metaphor regarding nature was popular

[4] See also Chap. 17 on Bernard Palissy.

[5] "*il semble conforme à la raison que la nature produise aussi ses automates*" (Descartes, 1649).

in early modernity. *Descartes reduced reason to humans and expanded its structures to the whole of nature simultaneously.* One may ask whether some present global modelling approaches are more sophisticated expressions of a similar paradigm.

In early modern science, strongly in the nineteenth century and party until present it has been an important activity to remove religion, spirits, and all kinds of "superstitions" from the images of nature, of soil and agriculture. This was intended to enable and purify science and to establish progress in modern land use. This could go hand in hand with the idea that nature behaves according to the theories and models of rational humans. There were times, for example, when soil physics strongly relied on artificial laboratory models using glass beads; or where soil was conceptualized to be a mere vessel and matrix for chemical plant nutrient elements; or people think soil organic matter content was mainly controlled and limited by the laws of thermodynamics and kinetics. Of course, there are not only simple but highly complex ways of modelling any soil "system" on earth. For the assumption of the modelability of nature, the idea of "general systems theory" as a "logico-mathematical field" was important, with "logical homology" of different system levels allowing "the use of systems which are simpler or better known as models for more complicated and less manageable ones" (Bertalanffy, 1950, p. 139, 142).

Mechanistic models, including the stochastic methods associated with them, dominate today's soil and environmental sciences worldwide. However the societal acceptance of scientific methods and results varies. Where fundamentalist movements or very conservative positions in social majority religions are strong, as for example in the USA or some Eastern European countries, the cultural questioning of the scientific worldview is much more socially relevant than in other countries. In Christian countries of the Northern Hemisphere, the classic *struggle between creationists and evolutionists* has existed since the nineteenth century, and still is a big struggle in the US "Bible belt" and other regions with fundamentalist Christian milieus that have strong influence (or dominance) on whole society. It is about whether all natural forms and life on Earth are God's creation once and forever, or the result of natural becoming, selection and evolution. This *kulturkampf* (culture clash) began primarily with Darwin's publications, but was foreshadowed by preceding debates, for example, about geological research results on orogeny, plutonism and marine sediments, glaciations, and fossils of extinct species.[6]

A variant of the idea of the creationism opposed by evolutionism has been the assumption of a *continuing divine creativity in nature and within evolution,* which already existed before Bacon and Darwin, and may be named "theistic

[6] See for the geology (and partly soil forming factors) debate between biblical and scientific models the works of Niklaus Steno (1638–1687), Jakob Lehmann (1719–1767), Giovanni Arduino (1714–1795), James Hutton (1726–1797), Abraham Werner (1749–1817), Georges Cuvier (1769–1832) and Charles Lyell (1809–1882), the last being a teacher of Darwin. Names and dates taken from Weissert and Stössel's historical outline of geology (2015).

evolutionism". The classic term for that is *creatio continua*. The idea was introduced in Christian theology by the Church's Father Augustine from Hippo around 400 CE. In a commentary on Psalm 113 of the Old Testament, Augustine said that whilst God remains eternally the same ("*in sua aeternitate p[er]sistit*"), he creates whatever he wants on heaven and earth ("*omnia quaecumque voluit fecit in coelis et in terra*"; Augustine, 1497, p. 565). By that, Christian theology went a step further than the Judaic basement where God's presence in creation was rather understood as a sustaining and preserving one,[7] not as ongoing new creations. This concept was further developed by Thomas Aquinas in the thirteenth century (Regenbogen & Meyer, 2005). The idea of *creatio continua* was also used by psychologist Carl Gustav Jung in his theory on creativity and meaningful coincidences in human and natural context that are not covered by models of causality, which he developed together with physicist Wolfgang Pauli in the 1950s.[8]

4.3 The Nutrient Elements Paradigm

The nutrient paradigm was conceptually prepared by Bernard Palissy (see Chap. 17) and probably first formulated by Swedish chemist and mineralogist Johan Wallerius in 1765 in his *Chemical Principles of Arable Farming* (p. 92 f. transl. by auth.): "*A fertile soil is the one that contains the right amount and proportion of nourishing elements, corresponding to the nature of all plants. A barren soil, on the other hand, is one to which only a few or no nourishing particles are added. A barren earth can therefore be made fertile if it is mixed with those elements which will feed the plants.*"[9] This differed from the previously influential idea of Van Helmont (1577–1644) who "believed that water was the sole nutrient of plants" (Tisdale et al., 1985, p. 9).

Hence, the soil became a rich yielding domain of chemistry. It was Swiss researcher Nicolas Théodore de Saussure who in 1804 first proved experimentally that plants use their roots to assimilate salts from the soil solution and their leaves to take up carbon from the air (not from humus; see Fig. 4.3). With this, de Saussure opened up the field of soil-plant material flow analyses. Then, German scientist Karl Sprengel wrote in 1828 (p. 424, transl. by author): "Chemical analysis and chemistry in general, provides the farmer with the best and cheapest method, how he can

---

[7] See for that Gen. 8.22, Jer. 31.35 f., Ps. 104.30–32, Ps. 145.15 f., and exemplarily Job 38.37 f. (New King James Bible): "*Or who can pour out the bottles of heaven, When the dust [soil] hardens in clumps, and the clods cling together?*"

[8] Jung & Pauli (1957). Cited from Jung's collected works 8, § 957 f.

[9] "*Eine fruchtbare Erde ist überhaupt diejenige, welche das beschiedene, und der natur aller Pflanzen zuständige maaß und verhältniß der nährenden grundteile enthält. Eine unfruchtbare hingegen diejenige, welcher nur gar wenige, oder gar keine nährende theilchen beigemischt sind. Es kan demnach eine unfruchtbare Erde fruchtbar gemacht werden, wenn sie mit jenen theilchen vermischt wird, welche den Pflanzen nahrung geben.*"

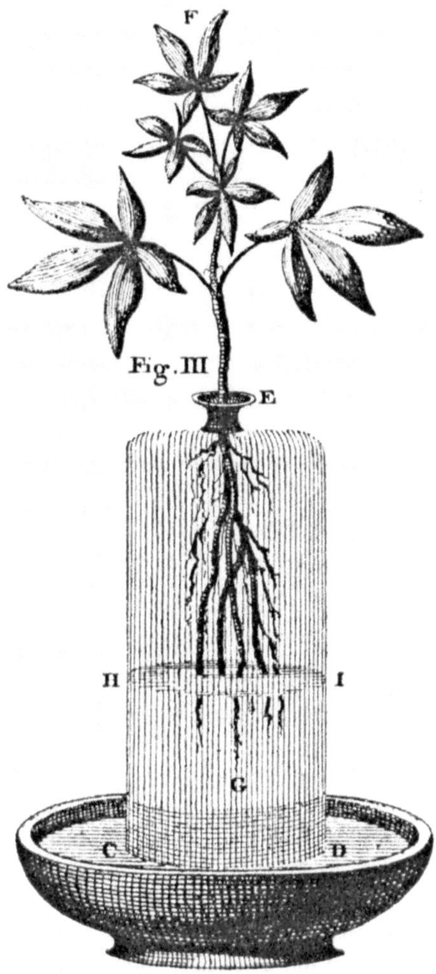

Fig. 4.3 Plant physiological experiments prepared the plant nutrient paradigm and methods of soil-less plant growing (de Saussure, 1804)

bring the yield of his fields, meadows and pastures to a hitherto unknown level!" In the same publication, Sprengel formulated the "law of the minimum" (p. 93): "For it cannot be denied that if a plant needs twelve substances for its development, it will never sprout if only a single one is missing, and it will always grow miserably if only one of them is not present in the quantity required by the nature of the plant" (compare Fig. 4.4).

When Justus von Liebig wrote a textbook in 1840b titled "Organic chemistry in its application to agriculture and physiology," he wrote to his publisher (Liebig, 1840/1986, p. 92 f.): This book will "cause a great sensation. In it, *I developed a theory of plant nutrition (...) that is likely to produce a revolution in the prevailing views*. I have converted the most tenacious humus worshippers at my university to become my most devoted adherents." And in the same year he confessed in a letter

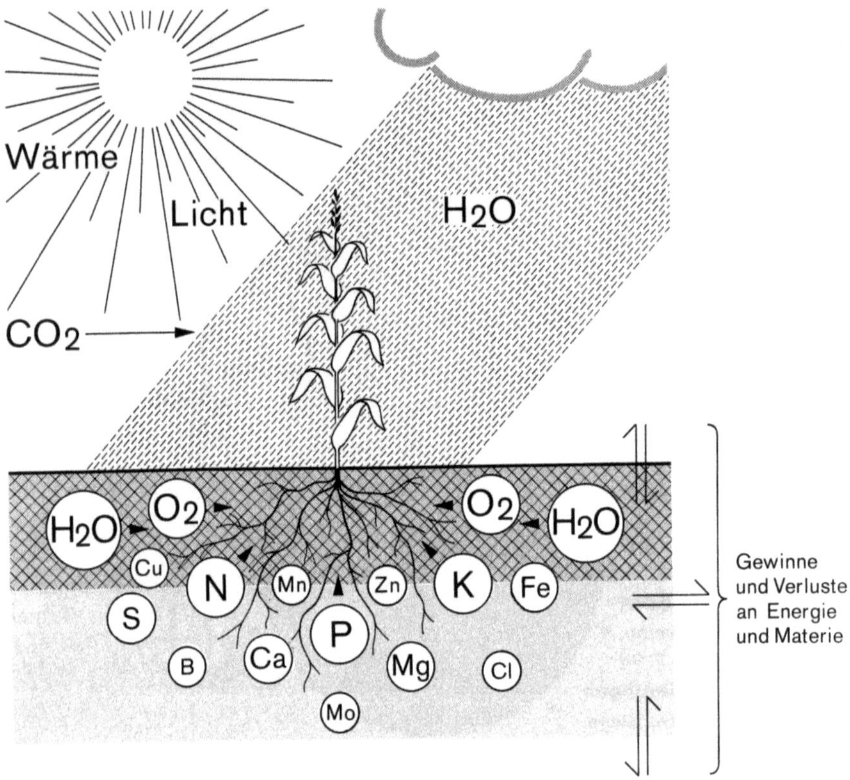

Fig. 4.4 The nutrient elements paradigm in a textbook on soil science. From Schroeder (1969–1992), original title "Scheme of an open plant ecosystem." Side remark "Gains and losses of energy and matter"

to famous Swedish chemist Jöns Berzelius, that to people he wanted "to instill deep conviction that chemistry alone can bring light to the processes of life" (Lewicki, 1982, p. 215). Liebig's follower Adolph Stöckhardt published in 1851 a very widely read textbook entitled the *Chemical Field Sermons for German Farmers*. There he stated that "to a certain extent, chemistry really does have the powers attributed to the Philosopher's Stone" (this was the highest value of alchemy). Liebig supported Stöckhardt in an 1851 letter: "May you never tire of preaching the chemical gospel to the agricultural pagans" (transl. by author, quoted from Schling-Brodersen, 1989, p. 82; see Fig. 4.5).

Scientifically, Liebig assumed that the art of arable farming was above all the restoration of the balance of chemical elements so that the substances removed from the soil with the harvest were returned to it (Liebig, 1845, p. 332). To be able to control the material balance, the chemical analysis of plants should lead to "an exact knowledge of the sum of all the components that we take from the soil in the various harvests" (Liebig, 1845, p. 341, transl. by auth.). After Liebig, in a 1915 textbook on "The Nutrition of Agricultural Crops," Schneidewind discussed the difficulties of

Fig. 4.5 Cover of the German journal *The Chemical Farmer* (1855–1875). It shows a subsurface chemical laboratory with elements of sacred design. Apothecary's cabinet and balance are arranged like an altar and cross under the central light. The vaulted ceiling says "Practice with Science." A gentleman sits with bags of bone meal and guano. Magical elements enter with an owl, a guenon, and a worker in dwarf form. At the top of the picture it says: "All depends on God's blessing." (Credit: Library of the Swiss Federal Institute of Technology, Zurich)

"keeping account of the nutrient balance of the soil" for nitrogen, phosphoric acid, potash and lime (p. 198–202). He summarized experiences of the nineteenth century and pointed out many soil chemical research questions of the twentieth century. This is still done today: measuring plant nutrients to calculate balance, and sometimes also preaching a "chemical gospel".

4.4 The Humus of Life

"Humus" is a Latin word for earth and soil, related to the Greek *chthōn* = earth and the adjective *chthónios* = subterranean. The Greek Goddess Persephone was called the *chthónia* below ground. The Latin word *homo* for man/human is related to it because of the meaning "on/from the earth" (see also the etymology of Adam and Eve in Chap. 3, Sect. 3.2.4).

Used for topsoil in Antiquity and then disappearing from use, "humus" probably re-entered the European scientific vocabulary in the eighteenth century. It appeared in Diderot and d'Alembert's *Encyclopaedia* (1765, vol. 8), with the meaning "mould, garden earth, earth formed by plant decomposition." Wallerius stated in the eighteenth century that humus was "of great importance for the development of plants, as a means of facilitating adsorption and retention of 'soil fertility' in the surrounding medium" (quoted from Krupenikov, 1993, p. 919). De Saussure (1804) ascribed a broad meaning to the word "humus," that is the whole vegetative cover undergoing decomposition, and a narrow meaning that is "mould" which referred to "the black substance plants are imbedded in" (Manlay et al., 2007, p. 220). Grandeau (1878) "suggested that SOM was vital for plant growth, since it increased the solubilization of mineral nutrients and thus their bioavailability to plants—a new concept" (Manlay et al., 2007, p. 220). Berzelius said that the nitrogen source for plants was probably humus, which he defined as "residues of other organisms" mixed with "earth" (1827, p. 211).

The use of the word humus was not standardized until the work of Albrecht Daniel Thaer, who thought that plants mainly fed on humus for their subsistence. Thaer wrote in 1810 that humus "gives food to the organism, thus without it no life on earth can be thought, at least of the higher animals and plants" (vol. 2, § 126, transl. by author).[10]

The humus paradigm of plant nutrition was then overpowered by the nutrient elements paradigm that mainly took over from about the 1840s on. In a combination of description and wishful thinking, Reinsch wrote in 1852 about a "war" between the "humists" and the "ashes men" about who's paradigm of plant nutrition would be the right one, and he told the story (p. 5, transl. by author): *"The humists, after a*

[10] In translation from Shaw & Johnson (1858) it reads in context (p. 167): "*Humus is the product of living matter, and the source of it. It affords food to organization; without it nothing material could have life, at least the most perfect animals and plants could not exist; and, therefore, death and destruction are necessary and accessory to the reproduction of animal and vegetable life.*"

small part of them had gone over to the enemy, withdrew into their indomitable rock castle 'Experience'. (...) and tired of the long war, both fighting parties finally (...) remarked with pleasure that neither humus nor ashes could in themselves sustain the life of the plants, but that this life would unfold to perfection through their mutual union."

A century later, Waksman's (1938) work in the USA on the crucial functions of humus in ecosystems started a new ecological understanding of soil organic matter, which has become increasingly important in science through the present day (Feller et al., 2006, Manlay et al., 2007). A scientific milestone in soil biology was when 1952 Nobel Prize winner Selman Waksman stressed the role of microbes in the development of the complex organic systems making up humus, which serves "as a reserve and a stabilizer for organic life on this planet" (Waksman, 1938, p. XII). Baveye and Wander (2019) asked, "why Waksman's viewpoint might still be considered novel more than 80 years later," and argued that the key reason for that was "the extreme compartmentalization of research and education in soil science, (...) with the result that interdisciplinary efforts that are desperately needed to understand the dynamics of soil humic substances are very hard to launch."

At the same time that there are recurring impulses to strengthen the recognition and valorization of humus, there are recurring waves to question its very existence. Lehmann & Kleber (2015), for example, argue that on closer inspection the humus concept breaks apart into a view of smaller molecules whose progressive mineralization is only slowed down by aggregate formation and mineral adsorption, and conclude (p. 6): "The need for the soil sciences to move away from both the 'humification' model and associated 'humic' language has been much debated. Unfortunately, this objective has not been implemented with rigour ...".

In the history of humus science, it has been important to "recognise this process [how humus is understood and valued] as a complex interaction between the elements of cognitive production, and the societal context, including economic and political facts and ideologies, and the cultural perceptions of environment—whether mainstream or alternative" (Manlay et al., 2007, p. 220); and "there are many cases, often forgotten, of perceptions that prefigure present-day concepts accepted as essential for sound management of natural resources, such as that of sustainability" (Feller et al., 2006, p. 18; Fig. 4.6).

4.5 Economics and Functionalism

In a very early attempt to valorize humus by giving it a virtual price, Sauerlandt wrote in 1929 (transl. by auth., p. 1/434): "The monetary value of the humus present in German arable soils is moderately estimated at 30 billion [$3x10^{10}$] marks." The ideas of government pricing or market pricing to limit or prevent environmental degradation and depletion of natural goods, both which are not actually commodities, developed slowly: an overview is given by Gorman & Solomon (2002). Pigou, in his 1920 "Economics of Welfare," addressed the question of how pollution could

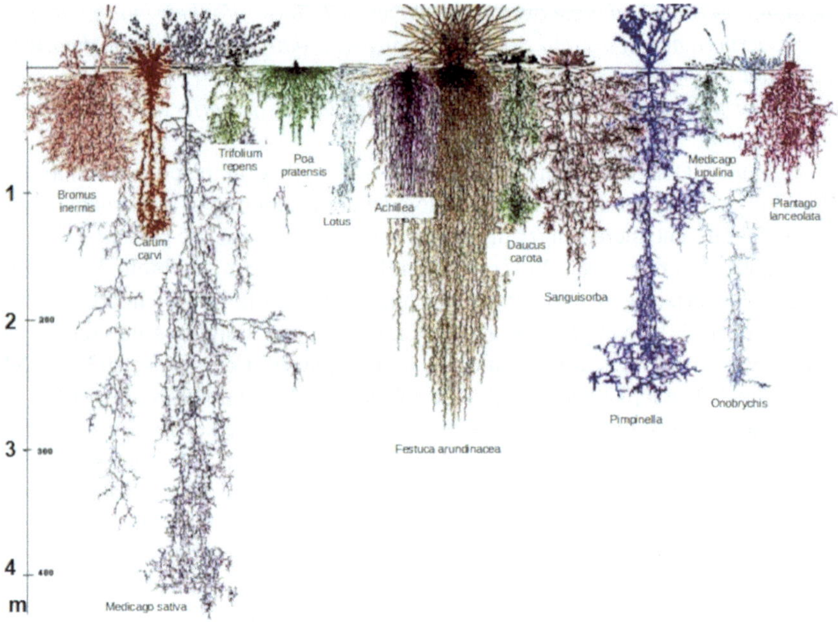

Fig. 4.6 Whether humus exists and how exactly it is built and structured is always disputed in science. Roots play a role. Drawing by the Austrian root researcher Erwin Lichtenegger (with Lore Kutschera) (1928–2004); the author replaced the German plant names with the Latin ones

be dealt with economically. Crocker (1966) worked on air pollution pricing, Dales (1968a, b) on water pollution. The Kyoto Protocol from 1997 (entry into force 2005) brought greenhouse gases into the economic consciousness, and the European Union Emissions Trading System (EU ETS), having started in 2005, has been the first major implementation of this. Currently there is a run on soil, namely humus, in fact any kind of soil organic matter, because for example the airline industry, trading companies and others want to buy themselves out of their emissions by giving money for humus (C_{org}) build-up. In Germany and Switzerland, private-sector virtual prices for humus currently exceed 100 euros per tonne and CO_2 emission prices stated rising in the 2020s EU-wide following the "European Green Deal" and enhanced emission reduction targets, all that boosting offsetting activities based on unsecured re-gaining of humus and its organic pre-stages and maybe surrogates in soils.[11]

The development of agricultural chemistry and the nutrient balance approach coincided with the development of national economy and agricultural business management. In the eighteenth and nineteenth century in Europe, the models of economy changed from a goods-oriented domestic and political economy to a

[11] See also: https://ec.europa.eu/clima/eu-action/forests-and-agriculture/sustainable-carbon-cycles/carbon-farming_en.

theory of the means of production and financial flows. Some ideas of emerging ecology and the theory of ecosystem services were also close to economics. Carl Ferdinand Senft (1888) spoke of the "plant production power of a soil" that is the "regulator, nourisher and nurturer of the plants." He defined the services of soil for the plants to be the rooting place with an appropriate consistency and crumbliness, the storage room of plant food for the present and future from organic and inorganic sources, the requirement that it contains enough air, water and warmth in the right proportions in the rooting zone, and also the regulation of elements considered harmful for plants, such as soluble acids and fungal spores (1888, p. 95–97). We see quite broad approaches; soil science was never completely dominated by one doctrine only. In the twentieth century, the functional approach developed to a more and more functionalistic one with an anthropocentric framing.

From the heyday of communism comes the following quotation (Rosenkranz, 1963, p. 196, transl. by author): "*When detaching forceful from such folksy but scientifically untenable ideas: the soil would be something special, full of life, life-bringing, fertile, maybe even mysterious and unfathomable in its coherence and action, and when replacing these ideas by a dialectic materialistic perspective, then the soil will become a carrier system, one of many means for work, like others, who allow humans, when used by them, to produce plants out of water, sun energy and elements for plant nutrition. Solely man is producing; neither plants nor animals, nor is the soil 'capable' to bring forth a harvest.*" This radical and ruthless devaluation or negation of soil ecosystem properties and natural autonomy apparently was not just an aberration of communist ideology; one may also hear farmers in countries with capitalist structures saying: "*I* produce wheat, milk and meat."

In 1980 (p. 20) Georg Müller et al. from the socialist German Democratic Republic defined the soil functions as "production function," "building ground and building material (site, and raw material function)," "water storage and filter (storage and filter function)," and waste sink ("remediation and disposal function"). Similarly, Sauerbeck wrote in the largely capitalist Federal Republic (West Germany) in 1985: "In the ecological sense, the functions of the soil as a transformer, filter and buffer system run largely parallel with these agricultural fertility criteria." This means that the soil has been understood as a quasi-technical system, its purpose was to provide services to humans – something that to a large extent still applies today.

Under the influence of broadening functionalist concepts, what was previously defined as "soil fertility" changed with scientific fashion and societal values: Being identified with the amount (or permanence or resilience) of yield in the function of internal properties of and external input to soil; or be it more or less an additive sum or resultant list of properties/factors/processes of soil; be it a mere economic production factor, an integrated ecological feature and the self-regulation capacity of soil; or shall it be whatever society wants (Patzel et al., 2000, compare Mizuta et al., 2018)? But as scientists observed the concept of "soil fertility" becoming hardly definable and overstretched, new terms such as "soil quality," "soil health," and "soil security" started competing to provide new functionalist umbrella terms:

Soil quality is a paradigmatic case of a completely relativistic concept in soil science: simply as a scientific sounding portmanteau of socially defined values and desires regarding soil. Soil quality was for example defined by Parr et al. (1992) as "the capability of a soil to produce safe and nutritious crops in a sustained manner over a long period, and to enhance human and animal health, without impairing the natural resource base or harming the environment." Doran et al., 1994 claimed that "Soil quality should not be limited to soil productivity, but should encompass environmental quality, human and animal health, and food safety and quality." Cox (1995) stated that "We need to set national goals for soil quality"; Hortensius and Welling (1996) wrote: "It is recognized that quality is a relative concept; when soil measurements are considered of good quality for one purpose, they may be of very poor quality when considered from another perspective." Harris et al. (1996) stressed that the definition of soil quality and health is just "functional" and "interchangeable." Partly before this initially mainly US-based discussion, there was one in Germany on soil quality in relation to soil fertility, with the outcome that soil quality cannot be defined as a scientific term because of its broadness as well as subjectivity (Linser, 1965; Derbruck, 1981; Bosch, 1991; Schönberger & Wiese, 1991).

Among the new concepts, **soil health** – also reissued first by US authors – stands out because it focuses on an organismic metaphor. "Health" is a typically organismic word that has been applied to humans, animals, then plants: both individuals and populations. Soil, however, when speaking of its "health," is maybe less seen as an ecosystem with its populations of species, but implicitly more perceived like a kind of *meta-organism.* Hence, addressing soil like an organism naturally facilitates the semantic and causal connection to the health and well-being of all organisms that live in it and from it. In 1943 Sekera published about "soil health" (see Sect. 4.8 below); for a recent outlook of the soil health concept see Lehmann et al., 2020, Giri & Varma, 2020, Meena, 2020, Karlen & Rice, 2017; for specific links with human health see for example Brevik and Burgess (2013), Singh et al. (2017), and the organic movement Sect. 4.8 below.

For the smaller literature corpus of "**soil security**," a concept with some focus on human food security in a more holistic context, see for example Koch et al., 2013, McBratney et al., 2014, McLean Bennett et al. (2019), Montanarella and Panagos (2021).

4.6 Soil Systems

There is definitely more than one story to be told in the European context of soil science. Long before the communist victory in Russia, a strongly relational approach was brought to soil science by its most important founder, Vasily Vasilyevich Dokuchaev (1846–1903). Following Oldfield and Shaw (2016, p. 49), Dokuchaev saw that "soils had relevance far beyond soil and soil formation, incorporating as it did deeper notions of the situatedness and interconnectedness of natural phenomena. More specifically, he conceptualized soil as the emergent product of the complex interaction of a range of so-called soil-forming factors." The following statement from his 1883 work, quoted from Dobrovolskii (1996), illustrates Dokuchaev's holistic approach:

One has to see the "*link that exists between forces, bodies, and phenomena; between dead and living nature; between the vegetative, animal and mineral kingdoms on one hand, and between man, his everyday life, and even mental world on the other. It is this relationship, this natural interaction, that makes up the essence of nature cognition (...) —the best and supreme fascination of natural science.*" And further: "*... in the study of these factors and particularly in mastering (if there is such a wish) them, the entire, single, integral, and indivisible nature must be, by all means, reckoned with as far as possible, and not its fragmentary parts; all its principal elements must be revered and studied alike; otherwise, we shall never be able to control them, and we shall never learn what belongs to one factor and what to another.*"

We see that Dokuchaev's highest value was an all-encompassing picture of nature, including its human dimension, and the relationships found in the patterns of all of its elements, with the soil and its forming factors at the center of interest. This approach made Dokuchaev an outstanding ecological pioneer within soil science. We also see here that he has raised the question about the correct human attitude towards soil, triggered by the key words of "mastering" and "control," contrasting "supreme fascination" and "must be revered".

In a quite similar spirit to Dokuchaev, the Swiss-American Hans Jenny (1899–1992) worked out his famous "factors of soil formation" model (Jenny, 1941). In a retrospective interview when he was an old man, Jenny (1984) said: "*I enjoyed seeing field data aligned by equations and derived aesthetic pleasure from the shapes of the curves. ... I could solve the equation. That was the new approach.*" And he also added affection for soil itself: "*... if you are used to thinking of soil as dirt, which is customary in our society, you are not keyed to find beauty in it.*" The soil "*speaks to us through the colors and sculptures of its profile, thereby revealing its personality;* (...) I *am intrigued by the thought that good soils make good people* (...) *Over the years I have acquired a kind of reverence for the soil, for the creature-world inside it, and for its character expressed in the profile features. ... My attitude may be a quirk, or a result of lifelong interest in soil*" (Fig. 4.7a and b).

Jenny admired and loved soils. But nevertheless, the development of opinions on soil in science and technology-based Western agronomic discourse showed the impulse to control soil forming and process factors becoming predominant. That was early and clearly stated by the National Research Council (US), Board on Agriculture (1997, p. 1): "*Precision agriculture is a phrase that captures the imagination of many concerned with the production of food, feed and fiber. ... Precision agriculture conjures up images of farmers overcoming the elements with computerized machinery that is precisely controlled* via *satellites and local sensors and using planning software that accurately predicts crop development. This image has been called the future of agriculture.*" These words express a new program to achieve forceful dominion over the soil and nature's elements, followed by many and filled with many hopes until today. It is often stated as a remarkably strong conviction that

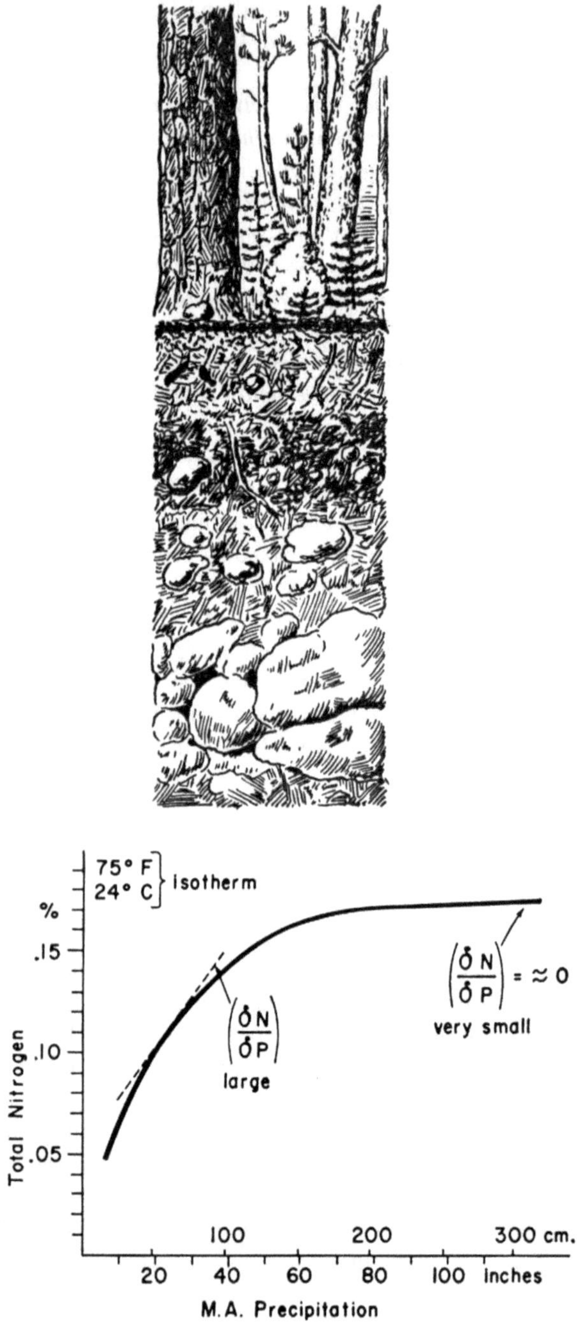

Fig. 4.7 (**a**) Soil profile with forest, an attempt to see the larger systems and soil forming factors. From Jenny (1941). (**b**) Relationship between soil nitrogen and mean annual precipitation in northern India. From Jenny (1961), who "derived aesthetic pleasure from the shapes of the curves"

is at the same time a personal and professional one: "Digitalization and automated systems are the future of agriculture and husbandry." We may see here a devaluation of human labor and human mental capacities regarding soils. There also appears to be a devaluation here of the mostly unknown myriads of soil organisms and their networks of bio-physical interactions that in quantity, quality and possibilities for action play at much higher levels of complexity than supercomputers and "clouds" that are now built with big funding to control soils, using the so-called "artificial intelligence" and robotics (Fig. 4.8). Which way to prefer and to empower is a question of worldview on both the individual and collective levels, which act as drivers in personal behavior, including of scientists. The long-time head of the German government's Advisory Council on the Environment (SRU) and of the government-funded Leibniz Centre for Agricultural Landscape Research (ZALF), Hubert Wiggering, wrote two semi-fictional books that were between research report, novel and dystopia that unfold a deep personal and cultural uncertainty and ambivalence in large parts of the German scientific elite between fascination and hope, fear and fatalism regarding "digitalization": Will man-made technology take "steering" control of humans and nature in agriculture, soil management and "bio-economy," until unforeseen collapses will occur (Wiggering, 2017; Wiggering & Schallwich, 2019)?

Fig. 4.8 "New technologies feed our Earth." Advertisement for an international agricultural machinery fair in Italy

The idea of "having dominion over" soil as a supreme fascination is also used implicitly or explicitly for tractor marketing. For example, the German tractor manufacturer Deutz-Fahr (belonging to an Italian consortium) produces a tractor series with the English name "Warrior" and has marketed it for years with the slogan (in German): *"Feel the power, take over the commando"*[12] (Fig. 4.9 below).

The following parts of this chapter include some of the philosophy and practice of organic farming in Europe. Most of the "inner soil" of organic farming has been split off from popular and scientific mainstream in the 19th and early 20th centuries.[13] The ideological background of organic farming is deeply rooted in European cultural history, together with taking up influences from much older agro-ecological corpora of knowledge from other cultural areas.[14] In Europe and the USA,[15] its development has been driven and strongly fecundated by (political) ecology. Some developers of organic farming were more receptive to the results of scientific ecology and soil science than the broad community of agronomists has been – which favored its convergence with "agro-ecology."

Fig. 4.9 Fantasies of omnipotence seem to strive for the violent subjugation of nature

[12] *"Spüren Sie die Macht. Übernehmen Sie das Kommando."* http://www.deutz-fahr.com/landing/de-de/traktoren-7250-ttv-warrior, July 2020.

[13] See more on the concept of "inner soil" in Chap. 22.

[14] It is sometimes irritating to see European or North American organic practices being promoted or even imposed (by labelling for export) in countries of the Global South. Some of these countries have (had) established similar or locally better practices millennia before of their development in the North.

[15] The foundations and development of organic farming in the United States of America is not subject of this chapter. Just as a hint to its temporal localization: After having met British organic pioneer Sir Albert Howard, Robert Rodale established an experimental organic farm in 1940 in Pennsylvania, and started publishing influential books on organic farming and the journal "The Organic Farmer" with his own publishing house in 1942. Other early important US organic pioneers were Edward Faulkner, Louis Blomfield, and Barry Commoner (Vogt, 2000, p. 195 f.).

4.7 "Vital Forces" in Plants and Soil?

In the Western history of ideas, Greek philosopher Aristotle's concept of *entelechy* founded an important tradition from the fourth century BCE onwards, meaning life having its purpose in itself.[16] The term entelechy means that the goal (*τέλος telos*) of a being is inherent or held (*ἔχειν echein*) in (*ἐν en*) itself, combining to the word *entelecheia* (ἐντελέχεια), that is, striving for self-realization and fulfilment[17]; therefore entelechy implies finality, and that is exactly what Francis Bacon had excluded from natural phenomena (see above), accepting causality only. During the nineteenth century, a new movement reconsidering entelechy and an according vital force was called "vitalism"—first by its adversaries (Engelhardt, 1997, p. 160 f.; Duchesneau, 1997, p. 297), and then by the "vitalists" themselves. As part of the philosophy of nature, the idea of a vital force (Latin *vis vitalis*) in all forms of life split off from European mainstream science[18] during the nineteenth century, and since then became part of organic farming ideas. In Chap. 17 on Bernard Palissy, we saw "salt" as the semi-material "5th element" to be the form-giving factor of plant and other life, and we saw above how the theory of mineral salt was later developed by Wallerius, Sprengel, Liebig and their successors to the mineral nutrition theory that is so powerfully still in use to the present.

In the early nineteenth century, German agronomist Albrecht Thaer called humus *"a creation of the organic force, a compound out of carbon, hydrogen, nitrogen and oxygen, as the inorganic powers of nature cannot bring forth"* (Thaer, 1821, vol. 2, p. 85, § 109; transl. by auth.).[19] A short time after that, the chemist Carl Sprengel, who first formulated the so-called "law of the minimum" quoted above (1828), wrote (transl. by auth.): *"A highly organized body is therefore composed of many life atoms and many chemical atoms. ... The inner nature of the life atoms is as little explainable as that of the chemical atoms."* These life atoms, constituting for Sprengel the *"life principle,"* would stay with the residual products of decaying organisms. The growth of a plant would also depend on the availability of life atoms (Sprengel, 1830, p. 176 f.). Thus, it can be stated that Sprengel's law of the minimum was in reality a pairing: the chemical and life "atoms" would only together form the necessary and sufficient conditions for plant growth.

Justus von Liebig (1803–1873), the well-known developer and promoter of agricultural chemistry, also kept *"vital forces"* somewhere within his image and feeling

[16] For primary sources, see in the Corpus aristotelicum, Aristotle: (*About the soul*) De anima II 1, 412a, (*Metaphysics*) Metaphys. VII.13, 1038b 1–6, IX.8, 1050a 9–16, (*Physics*) Phys. Ill, 1.

[17] In philosophy, the concept of Entelechy was explicitly used for example by Thomas Aquinas (1225–1274) and Gottfried Leibniz (1646–1716) to describe the finality of life.

[18] For this history of ideas see as important examples the works of Jan Baptist van Helmont (1577–1644), Georg Ernst Stahl (1660–1734), Albrecht von Haller (1708–1777), Paul-Joseph Barthez (1734–1806) and Marie Francois Xavier Bichat (1771–1801). For the until present last phase of vitalism discussed in more or less mainstream science, see Bütschli (1901), Wolff (1905), Braeuning (1907), and, at this time quite influential, the bio-philosophical vitalism of Driesch (1922).

[19] In the 1858 translation of Shaw and Johnson it reads (p. 167): Humus "is the produce of organic power — a compound of carbon, hydrogen, nitrogen, and oxygen, such as cannot be chemically composed."

of nature. In his "Familiar Letters on Chemistry," he wrote (von Liebig, 1851, p. 18; see also von Liebig, 1878, p. 14): "*Let us, however, carefully distinguish those effects which belong to the chemical, from those which depend peculiarly upon the vital force, and we shall then be in the right channel for obtaining an insight into the latter. Chemical action will never be able to produce an eye, a hair, or a leaf.*" However according to von Liebig (1878, p. 144, transl. by author), "*The vital forces are not disposable to our will in the same way as are heat, light, gravitation etc.*" Examination of different editions of his works (beginning in the late 1850s) show Liebig was somewhat torn between materialistic and vitalist ideas, and he formulated this hypothetical synthesis (von Liebig, 1878, p. 213, transl. by auth.): "*Exact science has shown, that all the powers of matter really play a part in the organic process. Now, the extreme reactionary forces pretend,* per contra *to the former opinion, that only chemical and physical forces would determine the phenomenon of life, and that no other force would be acting in the body. However, in the same way as former philosophers of nature were not able to prove that their vital force makes everything, yesterday's materialists cannot show a proof that it is done by inorganic forces, that they are sufficient to bring forth organisms, even the mind. ... Truth lies in the middle, which transcends the one-sidedness, acknowledging a formative principle, a ruling idea for organic life, within and together with chemical and physical forces.*"

Contrasting his own statement, Liebig could not keep these eloquently acknowledged two polar parts of his worldview both effective in his theories; in fact there remained the reductive materialist side only. This became an important fork in the history of scientific thinking, leaving soil mainly as "mere matter" or substrate in the framework of the classical mineral nutrition theory, leaving the mainstream of agronomy "purified" from anything spiritual, and for almost a century also devoid of significant valuation for soil organic matter (SOM, especially humus) as being an important constituent of a living soil (however, see Sect. 4.4 above on "the humus of life"). SOM was only brought back into the soil scientific mainstream recently. To wit mainly in relation to the atmospheric carbon content, but also unfolding in the realms of soil biodiversity, water and nutrient availability, harvest resilience and so on.[20]

The spiritual "half," or intrinsic vital quality of Palissy's 5th element prepared and joined the vitalist thread of science, and combined with the (neo-)Aristotelian idea of entelechy, then has become fundamental in some of the founding schools of organic farming in Europe. In a broader view, vitalist concepts may be connected with the cultural motive of the "tree of life" (Lat. *arbor vitae*), that is, with the idea and vegetative symbolism of the enduring strength in the permanence of life.

The actual vitalist ideas found a new home and became significant for some important founders of organic agriculture. For Hans Peter Rusch, the keyword was

[20] Examples from the literature: Batjes (2018), Blanco-Canqui & Lal (2008): Chang et al. (2015), De Stefano & Jacobson (2017), Don et al. (2011), García-Palacios et al. (2017), Gattinger et al. (2012), Houghton (2002), Krauss et al. (2017), Lal (2013), Lorenz et al. (2019), Minasny et al. (2017), Piikki et al. (2019), Sanderman et al. (2017), Spawn et al. (2019).

"living substance" in the soil. First, he identified this "living substance" with free-floating cell organelles and organic macromolecules in the soil solution (Rusch, 1953, p. 15; 1960, p. 53), which he considered to become part of the humus, the latter being considered as the primeval appearance of *"living tissue"* on earth (Rusch, 1955, p. 155). Under the pressure of contradictory scientific evidence, Rusch retired this concrete hypothesis and went back to the core statements of the vitalist tradition, interpreting living matter as the appearance and materialization of a *"really spiritual principle"* (Rusch, 1968, p. 33).

The *living principle* was also important for the co-founder of organic agriculture in Great Britain, Lady Eve Balfour (1943, p. 18): *"In our modern world, which is largely ruled by chemistry, we have tended to overlook this continuity of the living principle in nature,"* being present in the *"organic circle"* (she picked up this Indian term from Howard, see part 10 below). Lady Balfour descended from a family with pronounced spiritual, even spiritistic interest (Inhetveen et al., 2021, p. 39 f.). *"There is extensive evidence that she thought carefully about her religious beliefs, that they were unconventional, and that they had a bearing on how she interpreted events and came to decisions, not least events and decisions relating to the Soil Association"* (Gill, 2010, p. 171). To the present day, "vitality" as well as "health," whatever these terms signify in the concrete case, are important keywords of organic identity concepts as well as of their (often copied) food marketing strategies. In a sense, vitalism is celebrating its first day again in the current concept of *soil health* – if that is not understood as an arbitrary attention-seeking metaphor only.

4.8 Living Soil—Soil Life

The "living soil" is a core meme[21] of organic farming. It is rooted in soil biology and especially in soil microbiology. Like "soil health," "living soil" has an organismic connotation, it is often used together with the term "organic." Its field of associations and network of meanings contains ideas of soil being a "microcosm of life" fertilizing being understood as "feeding the soil" that then subsequently feeds the plant. Early soil microbiologists such as Austrians Franz and Margarete Sekera developed theories of functional soil biodiversity and their importance for soil fertility from the 1930s on (Sekera, 1943/1984). At the beginning of the organic farming movement in the United Kingdom, two foundational books had "health" in their titles: *The living soil. Evidence of the importance to human health of soil vitality, with special reference to post-war planning,* by Lady Eve Balfour (1944); and *The soil and health. A study of organic agriculture*, by Sir Albert Howard (1947).

[21] *Meme* means a content of consciousness (thought, belief, etc.) passed on through communication and internalized through the process of imitation, thus multiplying and being perpetuated socio-culturally in a comparable way as genes are inherited biologically. The word comes from the Greek μίμημα = *mīmēma* = imitated things, to ancient Greek μιμεῖσθαι = *mimeisthai* = to imitate.

By adhering more or less consciously to vitalist concepts, early organic farming theoreticians and practitioners were quite keyed to give special attention to observations and concepts from soil microbiology. In an early stage, the latter also included theories of transmutation of multicellular and unicellular organisms, differences between eukaryotes and prokaryotes not yet known, and before genetics developed as a research field. For almost 100 years, French and German scientists such as Henri Béchamp (1883, esp. p. 375), Gunter Enderlein (1925, esp. pp. 89, 129 ff., 200, 299 ff., 313 f.) and Hugo Schanderl (1947, 1970) promoted ideas of a partly physical continuity of decaying life in soil, that is, of organisms, cells, organelles, tissues, and organic macromolecules in soils, transgressing taxonomic borders and re-combining as suitable, which I witnessed to be believed by some organic farmers still today. A small part of these theories has been proven rather recently by the documentation of endocytosis of large macromolecules including genetic messenger macromolecule ribonucleic acid (RNA) pieces, diverse proteins and other organic macromolecules by plant roots and by intestinal cells of organisms including humans.[22] The organic meme of "healthy soil–healthy plant–healthy animal–healthy human" has received a new boost by these scientific findings.

A second influential development line has been classical soil biology and microbiology (Fig. 4.10). Darwin's earthworm studies in England (first reference 1838) started soil biology/ecology (Feller et al., 2003; comp. Chapter 18). In 1862, Kette (Germany) not only advocated the importance of humus for storing and providing plant nutrients, but wrote about the importance of root exudates for nutrient mobilization and availability (p. 8). In 1888, soil microbial nitrogen fixation was proven by Hellriegel & Wilfarth. In 1938, Waksman (USA) highlighted the function of microbes for the synthesis of humus, which he characterized "as a reserve and a stabilizer for organic life on this planet" (p. XII). In 1943, e.g., Sekera (fascist Austria) published about "healthy and ill soil," and highlighted the role of soil organisms for soil crumb (aggregates) and surface stabilization (called in German *Lebendverbau* = living [soil] support system; 1943/1986).[23] In 1946, Stöckli from Switzerland published about "soil as a living space" (Germ. *Lebensraum*). The Austrian scientists and bio-centrist philosophers Raoul Francé and Annie Francé-Harrar wrote about 100 publications in the years 1910–1967, the most popular books on illustrated soil microbiota and humus, profoundly influencing the emerging organic agriculture and ecological movement in central Europe.[24] These mentioned and other authors (e.g., Johannes Görbing 1877–1946, Ewald Könemann 1899–1976, Maria Müller 1894–1969, Artturi Virtanen 1895–1973) were received by the organic farming theorists and practice pioneers of the time, and memes spread by organic pioneers and formed tenets of a biological doctrine, stating the

[22] See for example Courtoy (1992), Marsh & McMahon (1999), Leborgne-Castel et al. 2010, Goode et al. (2015).

[23] Franz Sekera organized for the National Socialist Party the alignment of the Soil University (BOKU) of Vienna, after having been imposed as their president (Inhetveen et al. 2021, p. 311 f.).

[24] Most relevant references are from Raoul Francé, 1913 and 1922, from Annie Francé-Harrar, 1957 and 1959.

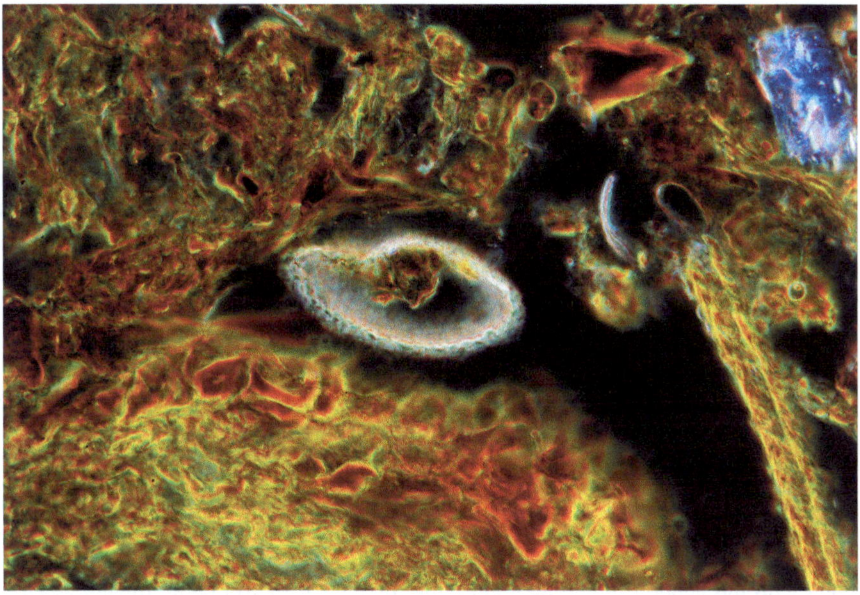

Fig. 4.10 Microscopes opened up new horizons in soil biology for a visual perception of soil life. Picture of a testacee. (Credit: Photo by Otto Ehrmann)

superiority of biological factors over chemical properties of soil until the present, and laying the foundation for the core values of "living soil" and "soil health" in organic farming.

Concerning soil biology and ecology, organic farmers and researchers appear to have a higher-than-average awareness for functional, habitat-driven and taxonomic soil biodiversity. This is also due to their crop's high vulnerability to pests and diseases in case of a lack in regulation in the agroecosystem, chemical hammer-blows not being allowed.

4.9 Mother Earth and Female Spirit

The belief and symbolism of "holy Mother Earth", the "Corn Mother," and generally a female spirit acting in nature is documented from all times and most cultures.[25] It stands more or less consciously behind many statements in favor of the Earth's and soil's integrity and dignity, and there are symbolic bridges to different understandings of tilling. Some early soil scientists as well as pioneers of organic farming stressed this symbolism and framing. This was sometimes explicitly said in contrast to soil being a "mere substrate" for plant growth.

[25] For a global overview see Frazer, 1951, p. 171–199; for source material with European cases see Patzel, 2015, p. 95, 180–183, 193 f., 250–254. Compare also Chap. 2.

Friedrich Albert Fallou, one of the precursors or co-founders of modern soil science and the one who coined the term "pedology" in a book with this title, wrote in his introduction this praise on Mother Earth (1862, p. 3, transl. by auth.): "*Everything ugly and miserable, useless and worthless, man can imagine, he names by the summarising words: dust, dirt, dung and mud. These are the honorary titles he has assigned to his Mother Earth in past and present. No word is breathed that he owes his existence to this cursed dirt and dung. Everybody is pleased to see good old Mother Earth rejuvenating, donning her flowery spring garment. However, nobody remembers the dirt and dung concealed underneath the beautiful clothes, which he avoids touching with hands, and even prefers not to set foot on it. Everybody wishes to walk on flowers, nobody wants to know anything about the soil, without which no flower could emerge.*"

Apparently, Fallou saw the devaluation of and ignorance of soil as a problem which he wanted to counteract by referring to mythical images of Mother Earth. Alternatively, one can interpret that Fallou was a hard-core scientist in the spirit of systematics and classification, but aware that feelings and mythical images had to be evoked to appeal to the readership emotionally. 145 years later, the US geomorphologist David Montgomery was resoundingly successful in the soil science community and somewhat with the general public by addressing the danger of soil being treated as *Dirt: The Erosion of Civilizations* (2007).[26] From other social spheres come groups of perma-culturists, regenerative and community supported farmers (CSA) that use the medium of documentary and testimonial films to promote the value of soil, typically combining statements of farmers, scientists, and engaged people. In my little hometown in southern Germany, more than half a dozen such movies from France, Germany, the UK and the US were publicly staged in the last few years. In Germany, special formations for (mostly organic) farmers that discuss not only how to sustainably care for soil, but also how to communicate about soil and Mother Earth with other stakeholders, are on the rise recently (Patzel et al., 2021). Such issues are also discussed among soil scientists (for example, Brevik et al., 2019).

Going back to cultural history: In the twentieth century, the Austrian philosopher and esoteric Rudolf Steiner, founding "guru" of the biodynamic school of organic farming from 1924 on (Steiner, 1984), said in the tradition of the physician and natural philosopher Paracelsus (1494–1541 CE; Steiner, 1957 ff., vol. 230, p. 122 transl. by auth.): *"The plant's mother is earth, and heaven its father (…), and the fecundation takes place down in earth in wintertime, when the seeds encounter the figures that the gnomes received from the sylphs, (…) this is literally the case."* And further: "*In spring time, the divine creativity is spouting from mother earth*" (Steiner, 1957 ff., vol. 97, p. 262); as well as man was "*born out of Mother Earth herself*" in his spiritual shape (Steiner, 1957 ff., vol. 104, p. 184). Mother Earth was addressed by Steiner as a being with an *"I in her centre"* (Steiner, 1957 ff., vol. 98, p. 153 f.). These ideas are close to animism and polytheism regarding soil and earth. Or said otherwise, Steiner tried

[26] A bit less successful were Montgomery's books with the tiles (with A. Biklé) *The Hidden Half of Nature: The Microbial Roots of Life and Health* (2016) and *Growing A Revolution: Bringing Our Soil Back to Life* (2017), with different co-authors.

consciously to reconnect with an animistic layer of human world perception full of spiritual action, suggesting an alternative to the modern scientific worldview.

Practical consequences of this mystical, or esoteric viewpoint on earth and soil are, beside others, the use of so-called "biodynamic preparations" in their farming practice. These are intended to introduce and modify cosmic and earthly spiritual forces into the soil and to the plants. For this purpose it seems not so important for people to really distinguish between the symbolic, spiritual and material dimensions. Following guidelines of biodynamic farming, cattle dung or crushed berg crystal are filled inside a cow's horn and buried in soil at a special phase of the moon, before recovering and using it as a plant strengthening agent and to regulate fungi. Yarrow blossoms are put into a urinary bladder from red deer, chamomile blossoms into the intestines of cattle, oak bark is stuffed inside the skull of a domesticated animal and buried in the soil, etc. These rituals – they are rather not called rituals by their actors, but "making the biodynamic preparations" to "vivify the soil and stimulate harmonious processes in it" – are grounded in the instructions that Steiner gave in 1924 following his "clairvoyant" perceptions and intuitions on nature and spirit.

Seen from the outside, a distinction from magical practices, which are known from different cultural areas, can hardly be made. Biodynamic farmers mainly reject this framing, and claim that their preparations are a kind of homeopathy. Contested within this community is the relevance and effect of human presence, awareness and action in preparing and applying the biodynamic preparations. Whereas many European biodynamic farmers pass many hours in ruminantly stirring barrels of water with diluted preparations by hand, their Australian colleagues prefer machinery and to spray different preparations all at the same time by airplanes over their very large fields.

The most influential man of the agrarian reform movement "natural farming" that started as a social movement in the twentieth century in Germany earlier than the Steiner school of agriculture, was Ewald Könemann. In 1925, he published about "soil cultivation according to nature," followed by "fundamental insights on fertilizers," etc. In 1939, he published on "Organic soil culture and fertilizer use," about the formation and management of humus in gardening and agriculture. The Nazi government tolerated and even supported and partly integrated biodynamic as well as other organic movements and pioneers into their farming policy (as a kind of "alternative laboratory" for their imagined post-war period) until about 1940, due to the personal protection by members of the inner circle of power, including minister of agriculture Walther Darré and Hitler's vice Rudolf Heß. Coinciding with their removal from power in the years 1940/41, the more or less assimilated or independent organizations of organic farming and land reform in Germany (including Austria at that time) were suspended, but not heavily persecuted by the regime (Vogt, 2000).

Other founders of organic agriculture referred to mother earth in a different cultural context. Swiss organic farming pioneer Mina Hofstetter (1883–1967) wrote (1941, p. 14 ff., transl. by auth.): Men, blinded by economy-driven quantitative thinking, have to become humble again towards Mother Earth, because "*She who is and remains the law, she will be going to teach us again, or to destroy us.*" Humans need to kneel down to find "*the calm, the quietness, and the unity with her. Then suddenly she will start to talk and to be understandable for us (…). Life, life, life is sounding from every little bulk of soil.*" Hofstetter's natural philosophy was that God shaped life from the four elements "*fire (sun/light), air, water, soil,*" and "*their synthesis is the plant that nourishes animal and man.*"

Hofstetter could be labelled a "spiritual feminist". Her background was a "back to nature" movement of the first decades of twentieth century Germany and Switzerland: to eat more raw fruits and vegetables grown with compost and being vegetarian, to profit from the sun by nudism, to reject mechanizing of agriculture, to counteract urbanization because of its effects to "alienate humans from nature," by founding garden colonies and similar. Hofstetter, who referred also to King's "Farmers of forty centuries" (e.g., Hofstetter, 1941), made her farm near Zurich a famous educational site of her time (Schmitt, 2006), giving courses in organic farming since 1929, and their supporting cooperative exists to present.

Lady Balfour (1899–1990), important founder of organic farming in Britain, named the journal of their organic "soil association" "Mother Earth"; and her compatriot Sir Albert Howard (1873–1947), the British pioneer of organic farming who had strong ties to India, introduced his *Agricultural Testament* (1943) with this quotation from Shakespeare's Romeo and Juliet:

> *The Earth, that's Nature's Mother, is her tomb;*
> *What is her burying grave, that is her womb.*

Howard said that agriculture should be done following the principles of nature, invoking the name of "Mother Earth" (1940, p. 4): "*Mother earth never attempts to farm without livestock; she always raises mixed crops; great pains are taken to preserve the soil and prevent erosion; the mixed vegetable and animal wastes are converted into humus; there is no waste; the processes of growth and the processes of decay balance one another; ample provision is made to maintain large reserves of fertility; the greatest care is taken to store the rainfall; both plants and animals are left to protect themselves from disease!*" Here we see an organismic, even personalized concept of nature, and the claim that agriculture shall be oriented on what he called the "rule in Nature," which has to be copied "faithfully" (Fig. 4.11).

The German Hans Peter Rusch, one of the founders of bio-organic farming in continental Europe (based in Switzerland), published 1968 (p. 71, comp. 1955, p. 22, transl. by auth.): "*The farmer of former times greeted 'holy mother earth' as if she were his biological mother, when he came back from abroad. And now she has been explained to him being a mere plant habitat, and the soil is unnecessary for plant growth and only useful as long as there are not enough cultures in gravel and water.*" Rusch was first a gynaecologist and then a microbiologist; his theories about "living substance" were cited above. In a way, he linked back to Fallou and Dokuchaev, claiming reverence for soil, instead of treating it as a mere resource. He also invented a test to determine soil microbial diversity. Only recently, research on soil microbes and their significance for soil formation and fertility have become a bit more popular again, after techniques using automatic genotype diversity analysis

Fig. 4.11 Archetypal feminine notions of abundance and fertility recur time and again (see also Chap. 2). Vessel with modelled breasts from a lakeside Neolithic farm and fisher village of pile dwellings in Zurich (Switzerland). Pottery from about 4000–3800 BCE. (Credit: Kantonsarchäologie Zürich; finding site: Zurich Kleiner Hafner)

(using deoxyribonucleic acid DNA sequencing) allowed new ways to identify taxonomical microbial diversity in soil,[27] thus leading to many new questions concerning their functions and ecology.

4.10 Compost and the Circle of Life and Death

Composting and the use of human feces for fertilization were important elements from the earliest stages of development of organic agriculture. Interest in composting and other organic recycling methods was stimulated by reports from India and China. American agricultural scientist Franklin Hiram King published as his last work in 1911 the book *Farmers of Forty Centuries, or Permanent Agriculture in China, Korea and Japan.* Following Brevik et al. (2018), "King was particularly impressed with the fact that nothing organic went to waste; it was either consumed by humans or animals, worn as clothing, used as fuel, or returned to farm fields as fertilizer." This work quickly became translated to German and gained high attention in reformist garden movements, combined with diverse other Eastern travel reports.

Proponents of "natural farming" drew stimulation and confirmation for their ideas from the Far East, although they were not clearly distinguishing the rather

[27] Geisen et al., 2018, Banerjee et al., 2019, Zumsteg et al., 2012.

anaerobic manure, and applications from the aerobic composting techniques they advocated (Vogt, 2000).

Today, in European countries it is forbidden to apply human manure or sewage sludge on fields due to sanitary and eco-toxicological reasons (in contrast to some practices, for example, in South and East Asia). However, some research and development for regaining at least phosphorus from human feces currently exists, but the problem of organic matter from these resources remains unsolved mainly because of medical residues, heavy metals, etc. Also well-known are problems with parasitic worms in human intestines due to non-sanitized human manure application for legume production, for example, in North Korea.

A more concrete influence of traditional techniques from abroad on organic farming in the global West was effectuated from India, mainly by the works of Sir Howard (1873–1947). Being a farmer's son and an educated botanist, Howard became agricultural advisor in British India, and led a government research farm at Indore. There he learned from the locals, and developed the so-called "Indore composting" in the 1930s. Howard integrated the Hindu religious idea of the "wheel of life" (related also to the *saṃsāra* representation of cyclic existence of living beings in Buddhism) to some extent into his world-view, applied it to plant residue composting and stated (1940, p. 22 f.) that one must look at the *"wheel of life as one great subject ... made up by two processes—growth and decay. The one is the counterpart of the other ... we must adopt a synthetic approach and look at the wheel of life as one great subject and not as if it were a patchwork of unrelated things."* Howard said that humus is the essential link between life, death, and life: *"It is therefore a key material in the life cycle. Without this substance the wheel of life cannot function effectively."*

Lady Balfour (1898–1990), leader of the British Organic Soil Association, adopted Howard's image, when she wrote (1943, p. 18): *"... the organic circle ... This ever recurring cycle of birth, growth, reproduction, death, decay, decay passing once more into birth, is often called the Wheel of Life."* The "circle of life" and the "ever recurring spiral of life and death" is a framing that facilitates the ideas of composting, of nutrient recycling, as well as of endocytosis of macromolecules by plant roots (Fig. 4.12). Besides, already according to English agriculturist Jethro Tull (1674–1741), "the soil particles were actually ingested through openings in the plant roots," called the "mouths of the roots" (Tisdale et al., 1985, p. 10).

The idea of the "circle" has become very important in organic farming. Swiss soil biologist Alois Stöckli wrote in 1946 (p. 1, transl. by auth.): *"Sustainable and growing soil fertility is possible only under condition of the circular flow of the elements,"* facilitated by soil biota. In 1968, Rusch wrote that *"the whole living creation" were one. "This wholeness is inseparable in any respect. ... The model for this interconnectedness is "the circular flow of living substances"* (transl. by auth., p. 90). We see here again the link with vitalism, fitting to the ideals of Antoine

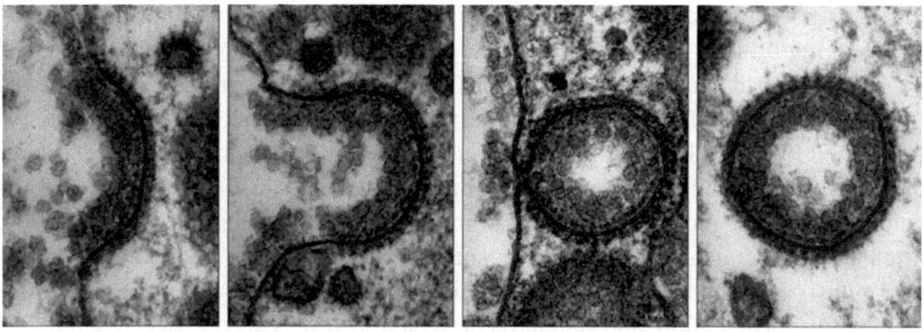

Fig. 4.12 An early image of endocytosis (Pastan & Willingham, 1985). Recently, research has increased on cross-species RNA transfer and the like between different groups of soil organisms and up to humans. This favors notions of "cycles of living substance" and of a "holism of life".

Béchamp who already spoke of a "circle" of life (1883, p. 375) and stated: "*Nothing falls victim to death, everything falls victim to life.*"[28]

At present, the "thought of the circle" (Germ. *Kreislaufgedanke*) is an important meme in the German speaking organic agriculture circles. It is used to underline the necessity to adopt the animal stock of farming to the carrying (feeding) capacity of the farmland, that animals are needed to "feed back" nutrients to soil by manure, that the natural global nitrogen (N) cycle has to be used to nourish plants with N assimilated by microbes (instead of synthetic fertilizer). In the biodynamic concept of the "farm organism," the circulation of elements and organic compounds also has the meaning of best fitting adaption of life forms to the site of the farm. Internationally, the recent concept named "from cradle to cradle" applied circular thinking to general economy and waste management.[29] We see here a model at work which is alternative or complementary to linear development and management models in agronomy and farming, which are held by the majorities of farmers in the global West. As a concept for whole economies it ties back to François Quesnay's (1694–1774) theory of a circular flow of economic goods and values, as described in his *tableau économique* (1758) and other publications.

4.11 Agroecology

Often, the publication of Rachel Carson's *Silent Spring* in 1962 is named a key moment when ecology became part of public awareness, and thus, politically relevant. New and older ecology movements (re-)gained strength. Without the impetus of scientific and political ecology, the environmental and species protection, then water, soil and climate protection would hardly have gained some importance in

[28] *"Rien n'est la proie de la mort, tout est la proie de la vie."* Quoted following Nonclerq, 1977, p. 160.

[29] Braungart & McDonough (2009), Lovins et al. (2014).

modern agriculture; and the varieties of organic agriculture would not have gained a certain importance in some parts of the world.

Today's world's largest "Research Institute of Organic Farming" (FiBL) started in 1972/3 in Switzerland on the initiative of scientists together with a few farmers, influenced by the political ecology movement (being part of the sixty-eight movement in Europe), and the environmental hazards of this time. They wanted to find a valuable scientific recognition, "underpinning" and development for (until that time often not academically accepted) organic farming concepts and practices. This enormously helped to get official recognition of organic farming by state authorities and the European Union, and standards protected by law. A second decisive impulse came from the French engineer Roland Chevriot of the organization "Nature et Progrès," the follow-up organization of "Man and Soil" (L'homme et le sol) founded in 1948. Chevriot started the "International Federation of Organic Agriculture Movements" (IFOAM) in 1972 with the aim of bringing together research and experimental results on organic farming.[30]

A framing similar to organic farming is agro-ecology. In its Latin American meaning, agro-ecology is a holistic concept of sustainable human relationship with nature and for societal change. In its main European meaning, agro-ecology is a scientific approach for designing agriculture by integrating ecological knowledge and to restitute more degrees of nature's freedom and self-regulating action in farming. Like in applied soil science, where concepts such as soil fertility, soil quality, soil health, soil security struggle for status as an umbrella term, it is no different in the field of agricultural reform movements: here we see names such as natural farming or organic farming or agroecology, permaculture (with the less known variant syntropic farming) or regenerative farming, even terms such as carbon farming coming and perhaps going again in the struggle for communicative, agricultural and soil policy dominance. In addition to seeing the people and maybe socio-economic powers supporting them, it can also be important to keep wide-aware eyes on the cultural content and ideological statements of each such concept, most of which have been discussed here somewhat with soil in mind.

4.12 Concluding Remarks

Sciences in the broad sense have existed in all cultural areas of the world, as well as technologies; together they have co-shaped agriculture. It is remarkable that the scientific paradigm and practices of soil management developed in Europe and then also in North America have become globally dominant. This almost global process

[30] Quote from the founding letter of Roland Chevriot from 1972: "*The food quality and ecology crisis is no longer a national problem, but an actual international concern (…) all the scientifical and experimental data we have hardly can cross the borders. Wouldn't it be possible to try to share them? I think that the creation of an international federation of organic farming movements would be of much interest for all of us and for humanity.*"

was driven by strong communicative efforts and the exercise of all kinds of power; largely it has happened without a careful and fruitful interplay of existing and introduced views and techniques. There seems to be a lot to take back and catch up on; it is also a question of intercultural respect and communication skills combined with human attentiveness.

However, a closer look shows that European scientific views on soil and agriculture comprise more than one paradigm – even when giving not much emphasis on the full spectrum of "organic" ideas –, and there is not only one storyline to be told. Similar to the polytheistic and monotheistic contexts of cultural patterns of soil cultivation in Europe, its scientific context is polyvalent. One may say, "by their fruits ye shall know them" (Matthew 7.16); and one may say, all future ways to deal with soils have their origins. All decisions and methods in soil science and farming may follow one or another pattern. These patterns are not just "evidence-based" in a narrow or even naïve sense, but they are part of the cultural history of humankind – including its spiritual, religious, and ideological developments – and influenced even by individual lives of scientists and all involved stakeholders.

In the face of inflated esoterically inspired or ideologically obsessed states of people that become maybe aggressive enemies of science, its methods and results show their strength. To contextualize and relativize science and to recognize some of its shadow and darkness, power drive and fire of the driven, should not give the upper hand to those who despise science, but can actually bring more conscious ethical choice, dignity and credibility to science through enhanced human wholeness of its people, and more self-reflection.

Knowing the cultural rooting spaces and sources of early and modern soil science, and present moves in there and in agriculture can help raise awareness of their socio-cultural and psycho-spiritual context, including ambiguous rewards and threads. This might lead to more contemplation and reflective caution in progress as well as in cultural recollection, reconsideration and renewal; and to more freedom in which ways to try one's hand at soils.

References

Abt, Th. (2014). (*Göbekli Tepe. Cultural memory and the knowledge of nature.*) Göbekli Tepe. Kulturelles Gedächtnis und das Wissen der Natur. *Zeitschrift für orientalische Archäologie, 7*, pp 90–124.

Abt, Th. (2023). *Göbekli Tepe. Cultural memory and the knowledge of nature*. Zurich: Living Human Heritage Publications.

Aristotle (4th century BCE). Corpus aristotelicum: (*About the soul*) De anima II 1, 412a, (*Metaphysics*) Metaphys. VII.13, 1038b 1-6, IX.8, 1050a 9–16, (*Physics*) Phys. Ill, 1.

Arnalds, Ó., & Aradóttir, Á. (2015). (*To read and heal the land.*) Að lesa og lækna landið. Landvernd (Iceland).

Augustine of Hippo (around 400 CE). (*Exposition* of the *Psalms.*) Expositio Psalmorum, p. 565 in the edition printed in Basel in the year 1497 by Johann Amerbach.

Bacon, F. (1623). (*On dignity and increase of science, vol. 3.*) De Dignitate et Augmentis Scientarum, caput III. London.

Balfour, Lady E. (1944). *The living soil. Evidence of the importance of human health of soil vitality, with special reference to post-war planning*. London: Faber.

Banerjee, S., Walder, F., Büchi, L., Meyer, M., Held, A., Gattinger, A., Keller, T., Charles, R., & van der Heijden, M. (2019). Agricultural intensification reduces microbial network complexity and the abundance of keystone taxa in roots. *The ISME Journal*. https://doi.org/10.1038/s41396-019-0383-2

Batjes, N. (2018). Technologically achievable soil organic carbon sequestration in world croplands and grasslands. *Land Degradation and Development, 2018*, 1–8. https://doi.org/10.1002/ldr.3209

Baveye, P., & Wander, M. (2019). The (bio)chemistry of soil humus and humic substances: Why is the "new view" still considered novel after more than 80 years? *Frontiers in Environmental Science*. https://doi.org/10.3389/fenvs.2019.00027

Blanco-Canqui, H., & Lal, R. (2008). No-tillage and soil-profile carbon sequestration: An on-farm assessment. *Soil Science Society of America Journal, 72*(3 May-June), p. 693–701.

Béchamp, A. (1883). (*Microzyma theory*.) Les microzymas dans leurs rapports avec l'hétérogénie, l'histogénie, la physiologie et la pathologie. Paris: Baillière.

Berzelius, J. (1827). (*Textbook on chemistry*.) Lehrbuch der Chemie (transl. from Swedish by F. Wöhler). Vol. 3, part 1. Dresden: Arnold.

Bosch, C. (1991). (*Soil fertility and soil protection in the concept of organic farming*.) Bodenfruchtbarkeit und Bodenschutz im Konzept der ökologischen Landwirtschaft. Berichte über Landwirtschaft, Neue Folge, 203. Sonderheft, pp. 59–63.

Braeuning, K. (1907). (*Mechanism and vitalism in 19th century biology. A historical draft*.) Mechanismus und Vitalismus in der Biologie des neunzehnten Jahrhunderts. Ein geschichtlicher Versuch. Leipzig: Engelmann.

Braungart, M., & McDonough, W. (2009). *Cradle to cradle: Remaking the way we make things. London: Vintage.*

Brevik, E., & Hartemink, A. (2010). Early soil knowledge and the birth and development of soil science. *Catena, 83*(2010), p. 23–33.

Brevik, E., & Burgess, L. (2016 [1st ed. 2013], eds.). Soils and human health. CRC Press.

Brevik, E., Homburg, J., & Sandor J. (2018). Soils, climate, and ancient civilizations. In W. Horwath, & Y. Kuzyakov (Eds.), *Changing soil processes and ecosystem properties in the Anthropocene* (Developments in soil science series, pp. 1–28). Elsevier.

Brevik, E., Steffan, J., Rodrigo-Comino, J., Neubert, D., Burgess, L., & Cerdà, A. (2018). Connecting the public with soil to improve human health. *European Journal of Soil Sciences, 2019*(70), p. 898–910.

Bütschli, O. (1901). (*Mechanism and vitalism*.) Mechanismus und Vitalismus. Leipzig: Engelmann.

Carson, R. (1962). *Silent Sping*. Boston: Houghton Mifflin.

Chang, J., Ciais, P., Viovy, N., Vuichard, N., Sultan, B., & Soussana, J.-F. (2015). The greenhouse gas balance of European grasslands. *Global Change Biology, 21*(10). https://doi.org/10.1111/gcb.12998

Chevriot, R. (1972). *Letter to Lady Eve Balfour and other organic farming pioneers in Europe and the USA*. www.ifoam.bio/sites/default/files/2020-05/founding_letter.pdf

Coulton, J. (1974, November). Lifting in early Greek Architecture. *The Journal of Hellenic Studies, 94*, p. 1–19.

Cox, C. (1995). Soil quality: Goals for national policy. *Journal of Soil and Water Conservation, 50*(3), p. 223.

Crocker, T. (1966). The structuring of atmospheric pollution control systems. In *Harold Wolozin: The economics of air pollution* (pp. 61–86).

Courtoy, P. (1992). *Endocytosis. From cell biology to health, disease and therapy*. Berlin: Springer.

de Saussure, N. T (1804). *Recherches chimiques sur la végétation*. Paris: Nyon.

Dales, J. (1968a). Land, water, and ownership. *Canadian Journal of Economics, 1*, p. 791–804.

Dales, J. (1968b). *Pollution, property and prices*. University of Toronto Press.

Darwin, Ch. (1859). *On the origin of species by means of natural selection, or the preservation of favoured races in the struggle for life*. London: Murray.
Darwin, Ch. (1838). On the formation of mould. *Proceedings of the Geological Society of London, 2*, p. 274–276.
Derbruck, J. (1981). (*The influence of crop rotation on soil fertility*.) Der Einfluß der Fruchtfolge auf die Bodenfruchtbarkeit. *Die Bodenkultur, 32*, p. 215–222.
Descartes, R. (1637). (*Discourse on the method of rightly conducting one's reason and of seeking truth in the sciences*.) Discours de la méthode pour bien conduire sa raison, & chercher la vérité dans les sciences. Leyde.
Descartes, R., Lettre à Morus, 5 février 1649 (*Letter to Morus*), Feb. 5th, 1649. Published on http://www.caute.lautre.net/article.php3?id_article=1431, 12th Sept 2005. Cited July 2022.
De Saussure, N. Th. (1804). *Recherches chimiques sur la végétation*. Paris: Nyon.
De Stefano, A., & Jacobson, M. G. (2017). Soil carbon sequestration in agroforestry systems: A meta-analysis. *Agroforestry Systems*. https://doi.org/10.1007/s10457-017-0147-9
Diderot, D., & Le Rond d'Alembert, J. (1765 [first vol. 1751]). (*Encyclopedia, or reasoned dictionary of Science, Arts and Crafts*.) Encyclopédie ou Dictionnaire Raisonné des Sciences, des Arts et des Métiers: vol. 8.
Dobrovolskii, G. (1996). Dokuchaev and present-day natural science. *Eurasian Soil Science, 29*(2), 105–110. Translation from Pochvovedenie 1996(2), p. 117–123.
Dokuchaev, V. (1883). Russkii chernozem. In Sochineniya (*collected works*), Moscow 1949. Quoted from Dobrovolskii 1996.
Don, A., Schumacher, J., & Freibauer, A. (2011). Impact of tropical land use change on soil organic carbon stocks – A meta-analysis. *Global Change Biology, 17*(4), p. 1658–1670. https://doi.org/10.1111/j.1365-2486.2010.02336.x
Doran, J. W., Coleman, D. C., Bedzidek, D. F., & Stewart, B. A. (Eds.). (1994). *Defining soil quality for a sustainable environment* (SSSA special publication 35). Soil Science Society of America.
Driesch, H. (1922). (*History of vitalism*.) Geschichte des Vitalismus. 2. verb. u. erw. Aufl. [2nd revised and enlarged ed.] (1. ed. 1905). Leipzig: Barth.
Duchesneau, F. (1997). (*Domains and frontiers of vitalism*) Territoires et frontières du vitalisme (1750–1850). In: G. Cimino and F. Duchesneau: *Vitalisms from Haller to the cell theory. Proceedings of the Zaragoza Symposium*, XIXth Int. Congr. of Hist. of Sc., 22–29 Aug. 1993. Firenze: Olschki, p. 297-357.
Enderlein, G. (1925). (*„Cyclogenics" of bacteria*.) Bakterien-Cyclogenie. Berlin: De Gruyter.
Fallou, F. A. (1862). (*Pedology, or general and applied soil science*.) Pedologie oder allgemeine und besondere Bodenkunde. Dresden: Schönfeld.
Feller, C., Raphaël, M., Swift, M. J., & Bernoux, M. (2006). Functions, services and value of soil organic matter for human societies and the environment: A historical perspective. In E. Frossard, W. E. H. Blum, & B. P. Warketin (Eds.), *Function of Soils for Human Societies and the Environment* (Special Publications) (Vol. 266, pp. 9–22). Geological Society.
Feller, C., Brown, G., Blanchart, E., & Chernyanskii, P. D. S. (2003). Charles Darwin, earthworms and the natural sciences: various lessons from past to future. *Agriculture, Ecosystems and Environment, 99*(2003), p. 29–49.
Francé, R. (1913). (*The Edaphone. Studies on the ecology of soil-dwelling microorganisms*.) Das Edaphon. Untersuchungen zur Oekologie der bodenbewohnenden Mikroorganismen. Stuttgart: Franckh.
Francé, R. (1922). (*Life in the arable soil*). Das Leben im Ackerboden. Stuttgart: Franckh.
Francé-Harrar, A. (1957). (*Humus. Life and fertility of soils*.) Humus – Bodenleben und Fruchtbarkeit. München: Bayrischer Landwirtschaftsverlag.
Francé-Harrar, A. (1959). (*The last chance – for a future without misery*.) Die letzte Chance – für eine Zukunft ohne Not. München: Bayrischer Landwirtschaftsverlag.
Frazer, J. (1951). *The golden bough. A study in magic and religion*. Third edition. Part V, Spirits of the Corn and of the Wild, Vol II. New York.

Gattinger, A., Müller, A., Haeni, M., et al. (2012). Enhanced top soil carbon stocks under organic farming. *Proceedings of the National Academy of Sciences (US), 109*(44), p. 18226–18231.

García-Palacios, P., Gattinger, A., Bracht-Jorgensen, H., et al. (2017). Crop traits drive soil carbon sequestration under organic farming. *Journal of Applied Ecology, 2018*, p. 1–10.

Geisen, S., Mitchell, E., Adl, M. B., Dunthorn, M., Ekelund, F., Fernández, L., Jousset, A., Krashevska, V., Singer, D., Spiegel, F., Walochnik, J., & Lara, E. (2018). Soil protists: a fertile frontier in soil biology research. *FEMS Microbiology Reviews, 42*, p. 293–323.

Gill, E. (2010). *Lady eve balfour and the British organic food and farming movement*. PhD thesis Aberystwyth University.

Giri, B., & Varma, A. (2020). *Soil health*. Cham: Springer.

Goode, B., Eskin, J., & Wendland, B. (2015). Actin and endocytosis in budding yeast. *Genetics, 199*(2), p. 315–358.

Gorman, H., & Solomon, B. (2002). The origins and practice of emissions trading. *Journal of Policy History, 14*(3), p. 293–320.

Grandeau, L. (1878). (*Experimental work on the role of soil organic matter in plant nutrition.*) Recherches expérimentales sur le rôle des matières organiques du sol dans la nutrition des plantes. Annales Station Agronomique de l'Est. Nancy, pp. 225–352.

Harris, R. F., Karlen, D. L., & Mulla, D. J. (1996). A conceptual framework for assessment and management of soil quality and health. In J. W. Doran & A. J. Jones (Eds.), *Methods for assessing soil quality* (SSSA special publication no 49) (pp. 61–82).

Hasan, H. (2006): Archimedes, the father of mathematics. New York: Rosen.

Hellriegel, H., & Wilfahrt, H. (1888). (*Studies on the nitrogen nutrition of gramines and legumes.*) Untersuchungen über die Stickstoffnahrung der Gramineen und Leguminosen. – Zeitschrift des Vereins der Rübenzucker-Industrie des Deutschen Reichs, Beilageheft Berlin; Review 1889 in Naturw. Rundschau 4: p. 278–279, 290–292.

Hofstetter, M. (1941). *(Mother, give me bread!)* Mutter, gib mir Brot! In W. Zimmermann, M. Hofstetter, G. Krebs, P. Häusle & E. Bertholet: *(Mother Earth – Wake-up call and practical guide to organic farming.)* Mutter Erde – Weckruf und praktische Anleitung zum biologischen Landbau. Zielbrücke-Thielle.

Hortensius, D., & Welling, R. (1996). International standardization of soil quality measurements. *Communications in Soil Science and Plant Analysis, 27*, p. 387–402.

Houghton, R. (2002). The annual net flux of carbon to the atmosphere from changes in land use 1850–1990. *Tellus B, 51*(2), p. 98–313.

Howard, Sir A. (1943). *An agricultural testament*. Oxford Univ. Press.

Howard, Sir A. (1947). *The soil and health. A study of organic agriculture*. The Devin-Adair Comp.

Inhetveen, H., Schmitt, M., & Spieker, I. (2021). (*Women pioneers of organic farming.*) Passion und Profession. Pionierinnen des ökologischen Landbaus. Munich.

Jenny, H. (1941). *Factors of soil formation: A system of quantitative pedology*. New edition 1994, Dover Publications, Mineola.

Jenny, H. (1961). Derivation of state factor equations of soils and ecosystems. *Soil Science Society of America (SSSA) Proceedings, 25*(5), 385–388.

Jenny, H. (1984). My friend, the soil. (Interview by Hans Kevin Stuart.) *Journal of Soil and Water Conservation 1984 (May–June)*, p. 158–161.

Johansson, P. (2003). *The lure of origins. An inquiry into human-environmental relations, focused on the "Neolithization" of Sweden*. University of Lund.

Jung, C. G., & Pauli, W. (1957, first ed. 1952): (*Nature interpretation and the psyche.*) Naturerklärung und Psyche. Zurich: Rascher.

Jung, C. G. (1995). *Collected works 8, there: Synchronicity: An acausal connecting principle*. Zurich: Walter.

Karlen, D. L., & Rice, C. W. (2017). *Enhancing soil health to mitigate soil degradation*. MDPI.

Kaser, K. (2011). (*The Balkans and the Middle East. Introduction to a common history.*) Balkan und Naher Osten. Einführung in eine gemeinsame Geschichte. Wien: Böhlau.

Kette, W. (1862): (*The fermentation theory versus the humus, mineral and nitrogen theory.*) Die Fermentationstheorie gegenüber der Humus-, Mineral- und Stickstofftheorie. Berlin: Wiegandt.

King, F. H. (1911). *Farmers of Forty centuries, or permanent agriculture in China, Korea and Japan.* Madison.

Koch, A., McBratney, A., Adams, M., Field, D., Hill, R., Crawford, J., Minasny, B., Lal, R., Abbott, L., O'Donnell, A., Angers, D., Baldock, J., Barbier, E., Binkley, D., Parton, W., Wall, D., Bird, M., Bouma, J., Chenu, C., Flora, C. B., Goulding, K., Grunwald, S., Hempel, J., Jastrow, J., Lehmann, J., Lorenz, K., Rice, C. M. C., Whitehead, D., Young, I., & Zimmermann, M. (2013). Soil security: Solving the global soil crisis. *Global Policy, 4*, p. 434–441.

Könemann, E. (1925). (Livestock-less farming – natural soil cultivation) Fiehloser Ackerbau – natürliche Bodenbearbeitung. *TAO Monatsblatt für Verinnerlichung und Selbstgestaltung, 11*, p. 1–20.

Könemann, E. (1939 2nd ed. [1st editions 1931, 1932, 1937]). (*Organic soil culture and fertiliser use*) Biologische Bodenkultur und Düngerwirtschaft (three volumes in one). Tutzing: Siebeneicher.

Krauss, M., Ruser, R., Müller, T., Hansen, S., Mäder, P., & Gattinger, A. (2017). Impact of reduced tillage on greenhouse gas emissions and soil carbon stocks in an organic grass-clover ley – Winter wheat cropping sequence. *Agriculture, Ecosystems and Environment, 239*(2017), p. 324–333.

Krupenikov, I. (1993). *History of soil science. From its inception to the present* (Russian translations series no. 98).

Lal, R. (2013). Soil carbon management and climate change. *Carbon Management, 4*(4), p. 439–462. https://doi.org/10.4155/cmt.13.31

Leborgne-Castel, N., Adam, T., & Bouhidel, K. (2010). Endocytosis in plant–microbe interactions. *Protoplasma (Springer).* https://doi.org/10.1007/s00709-010-0195-8

Lehmann, J., Bossio, D., Kögel-Knabner, I., & Rillig, M. (2020). The concept and future prospects of soil health. *Nature reviews Earth & Environment, 25 Aug 2020, 1*(10), p. 544–553.

Lehmann, J., & Kleber, M. (2015). The contentious nature of soil organic matter. *Nature.* https://doi.org/10.1038/nature16069

Lewicki, W. (1982, ed.). (*Letters between Berzelius and Liebig.*) Berzelius und Liebig. Ihre Briefe 1831–1845 mit gleichzeitigen Briefen von Liebig und Wöhler. Göttingen: Cromm.

Liebig, J. von (1840a). (*Organic chemistry in its application to agriculture and physiology.*) Die organische Chemie in ihrer Anwendung auf Agricultur und Physiologie. Braunschweig: Vieweg.

Liebig, J. von (*Letter of Liebig to Berzelius*) Brief an Berzelius vom 3. September 1840b. In Lewicki (1982), p. 215.

Liebig, J. von (*Letter of Liebig to Vieweg*) Brief von Liebig an Vieweg vom 17. März 1840. In Liebig 1986a, p. 92 f.

Liebig, J. von (1845, and 6th edition 1878). (*Letters on chemistry.*) Chemische Briefe. Leipzig: Winter.

Liebig, J. von (1986b). (*Letters between Vieweg and Liebig.*) Briefe an Vieweg. Herausgegeben und bearbeitet von Margarete und Wolfgang Schneider. Braunschweig: Vieweg.

Liebig, J. von (1851). *Letter to Stöckhardt quoted from Schling-Brodersen 1989.*

Linser, H. (1965). (*Definition and meaning of the term "soil fertility".*) Fassung und Bedeutung des Begriffes "Bodenfruchtbarkeit". Zeitschrift für Pflanzenernährung, Düngung und Bodenkunde 108/2 p. 115-122.

Lorenz, K., Lal, R., & Ehlers, K. (2019). Soil organic carbon stock as an indicator for monitoring land and soil degradation in relation to United Nations' Sustainable Development Goals. *Land Degradation and Development, 2019*(30), p. 824–838.

Lovins, A., Braungart, M., & Stahel, W. (2014). *A new dynamic: Effective business in a circular economy.* MacArthur.

Manen, C., Perrin, T., & Guilaine, J. (2014). (*The Neolithic transition in the Mediterranean.*) La transition néolithique en Méditerranée. Arles: Editions errance, archives d'écologie préhistorique.

Manlay, R., Christian, F., & Swift, M. J. (2007). Historical evolution of soil organic matter concepts and their relationships with the fertility and sustainability of cropping systems. *Agriculture, Ecosystems and Environment, 119*(2007), p. 217–233.

Marsh, M., & McMahon, H. T. (1999). The structural era of endocytosis. *Science, 285*(5425), 215–220.

McLean Bennett, J., McBratney, A., Field, D., Kidd, D., Stockmann, U., Liddicoat, C., & Grover, S. (2019). Soil security for Australia. *Sustainability, 2019*(11), p. 3416. https://doi.org/10.3390/su11123416

McBratney, A., Field, D., & Koch, A. (2014). The dimensions of soil security. *Geoderma, 213*, p. 203–213.

Meena, R. S. (2020). *Soil health restoration and management*. Singapore: Springer.

Minasny, B., Malone, B., McBratney, A., et al. (2017). Soil carbon 4 per mille. *Geoderma, 292*(2017), p. 59–86.

Mizuta, K., Grunwald, S., & Phillips, M. A. (2018). New soil index development and integration with econometric theory. *Soil Science Society of America Journal, 82*, p. 1017–1032.

Montanarella, L., & Panagos, P. (2021). Soil Security for the European Union. *Soil Security, 4*(2021), 100009.

Montgomery, D. (2007). *Dirt: The erosion of civilizations*. University of California Press.

Montgomery, D., Biklé, A. (2016). *The hidden half of nature: The microbial roots of life and health.*

Montgomery, D. (2017). *Growing a revolution: Bringing our soil back to life.*

Müller, G., et al. (1980). (*Plant production. Soil science.*) Pflanzenproduktion. Bodenkunde. Berlin (East).

National Research Council (U.S.), Board on Agriculture (1997). *Precision agriculture in the 21st century*. Geospatial and Information Technologies in Crop Management.

Nonclerq, M. (1977). *(An injustice in the history of science. The case of Antoine Béchamp, the scientist from Lorraine.)* Une injustice dans l'histoire des sciences. Le cas du savant lorrain, Antoine Béchamp. Bulletin Académie et Societé Lorraines des Sciences, 16(4), 137–161.

Oldfield, Jonathan; Denis Shaw (2016): *The development of russian environmental thought. Scientific and geographical perspectives on the natural environment.* London/New York: Routledge.

Parr, J. F., Papendick, R. I., Hornick, S. B., & Meyer, R. E. (1992). Soil quality: Attributes and relationship to alternative and sustainable agriculture. *American Journal of Alternative Agriculture, 7*, p. 5–11.

Pastan, I., Willingham, M. (1985): Endocytosis. New York: Plenum.

Patzel, N., Sticher, H., & Karlen, D. (2000). Soil fertility—Phenomenon and concept. *Journal of Plant Nutrition and Soil Science, 163*, p. 129–142.

Patzel, N. (2010). The soils scientist's hidden beloved: Archetypal images and emotions in the scientist's relationship with soil, chap. 13. In E. R. Landa & Ch. Feller (Eds.), *Soil and culture* (pp. 261–276). New York: Springer.

Patzel, N. (2015). (*Symbols in agriculture. On the spiritual side of nature relation in Swiss agrarian culture.*) Symbole im Landbau. Zur spirituellen Naturbeziehung in der Schweizer Agrarkultur. Munich: Oekom.

Patzel, N., Zellfelder, J., & Griese, S. (2021). (*Soil formation [education]. New alliances for climate and soil protection.*) Boden-Bildung. Neue Allianzen für den Klima- und Bodenschutz. *Der kritische Agrarbericht, 2021*, p. 16–22.

Pigou, A. (1920). *The economics of welfare*. London: Macmillan.

Piikki, K., Söderström, M., Sommer, R., et al. (2019). A boundary plane approach to map hotspots for achievable soil carbon sequestration and soil fertility improvement. *Sustainability, 11*, 4038. https://doi.org/10.3390/su11154038

Plinius, Gaius Secundus Maior [Pliny the Elder] (1995 / 1st cent CE): (*Natural history, vol. 18 on agriculture.*) Naturkunde: lateinisch-deutsch. Buch XVIII Botanik: Ackerbau. Transl. by Roderich König. Munich.

Quesnay F. (1758). (*Economic table, and general maxims of economic governance.*) Tableau économique, et maximes générales du gouvernement économiques. Versailles.

Regenbogen, A., Meyer, U. (2005). (*Dictionary of philosophical terms.*) Wörterbuch der philosophischen Begriffe: *creatio continua.*

Reinsch, H. (1852). (*About the effects of fertilizer, especially of humus.*) Über die Wirkungen des Düngers insbesondere des Humus. Erlangen: Deichert.

Rosenkranz, O. (1963). (*On the economy of soil fertility.*) Zur Ökonomik der Bodenfruchtbarkeit. *Zeitschrift für Agrarökonomik, 6*, p. 195–200.

Rusch, H. P. (1953). (*The method of biological soil investigation.*) Das Verfahren der biologischen Boden-Untersuchung. Kultur und Politik 8/1, pp. 13–18.

Rusch, H. P. (1955). (*Natural science of tomorrow. Lectures on the preservation and cycle of living matter.*) Naturwissenschaft von morgen. Vorlesungen über Erhaltung und Kreislauf lebendiger Substanz. Küsnacht: Hartmann.

Rusch, H. P. (1960). (*On the preservation and circle of living matter.*) Über Erhaltung und Kreislauf lebendiger Substanz. *Zeitschrift für Ganzheitsforschung, 4*, p. 50–63.

Rusch, H. P. (1968). (*Soil fertility. A study of biological thinking.*) Bodenfruchtbarkeit. Eine Studie biologischen Denkens. Heidelberg: Haug.

Sanderman, J., Hengl, T., & Fiske, G. (2017). Soil carbon debt of 12,000 years of human land use. *PNAS, 114*(36), p. 9575–9580. www.pnas.org/cgi/doi/10.1073/pnas.1706103114

Sauerbeck, D. (1985). (*Functions, quality and resilience of the soil from an agro-chemical point of view.*) Funktionen, Güte und Belastbarkeit des Bodens aus agrikulturchemischer Sicht. Materialien zur Umweltforschung. Stuttgart.

Sauerlandt, W. (1929). (*Studies on formation and decomposition of humus in manure and soil.*) Untersuchungen über Bildung und Zersetzung von Humus im Stalldünger und im Boden. PhD th. Univ. Leipzig.

Schanderl, H. (1947). (*A new basis for botanical bacteriology and nitrogen balance of plants.*) Botanische Bakteriologie und Stickstoffhaushalt der Pflanzen auf neuer Grundlage. Stuttgart.

Schanderl, H. (1970). (*A new perspective on soil bacteria.*) Bodenbakterien in neuer Sicht. In: *Boden und Gesundheit, Zeitschrift für angewandte Ökologie, 68*(3).

Schling-Brodersen, U. (1989). (*Development and institutionalisation of agricultural chemistry in the 19th century: Liebig and the agricultural experimental stations.*) Entwicklung und Institutionalisierung der Agrikulturchemie im 19. Jahrhundert: Liebig und die landwirtschaftlichen Versuchsstationen. Serie Braunschweiger Veröffentlichungen zur Geschichte der Pharmazie und der Naturwissenschaften, vol. 31.

Schmidt, K. (2012). *Göbekli Tepe. A Stone Age Sanctuary in South-Eastern Anatolia.* Berlin: Ex Oriente.

Schmitt, M. (2006). Fertile minds and friendly pens: Early women pioneers. In G. Holt & M. Reed (Eds.), *Sociological perspectives on organic agriculture. From pioneer to policy* (pp. 56–69). Wallingford: CABI.

Schneidewind, W. (1915). (*The nutrition of agricultural crops. Textbook based on scientific research and practical experience.*) Die Ernährung der landwirtschaftl. Kulturpflanzen. Lehrbuch auf der Grundlage wissenschaftlicher Forschung und praktischer Erfahrung. Berlin: Parey.

Schönberger, H., & Wiese, J. (1991). (*Need for research in connection with the objectives, measurability and predictability of elements and processes of soil fertility.*) Forschungsbedarf im Zusammenhang mit den Zielvorstellungen, der Meß- und Voraussagbarkeit von Elementen und Prozessen der Bodenfruchtbarkeit. Berichte über Landwirtschaft, Neue Folge, 203. Sonderheft, pp. 144–157.

Schroeder, D. (1969 until 1992). (*Soil science in brief.*) Bodenkunde in Stichworten. Berlin: Hirt.

Sekera, F. (1943; 5th revised ed. by M. Sekera 1984). (*Healthy and sick soil. A practical guide to keeping the field healthy.*) Gesunder und kranker Boden. Ein praktischer Wegweiser zur Gesunderhaltung des Ackers. Wien: Stocker.

Seidel, U. (2016). (*Early farmers on best soils. Neolithic settlements away from lakes and moors.*) Frühe Bauern auf besten Böden. Jungsteinzeitliche Siedlungen abseits von Seen und Mooren.

In *Archäologisches Landesmuseum Baden-Württemberg, Landesamt für Denkmalpflege: 4000 Jahre Pfahlbauten* (pp. 52-58). Ostfildern: Thorbecke.

Senft, K. (1888). (*The soil according to its origin, characteristics and behaviour towards the plant world. A textbook for all friends of the plant kingdom, but especially for foresters and farmers.*) Der Erdboden nach Entstehung, Eigenschaften und Verhalten zur Pflanzenwelt. Ein Lehrbuch für alle Freunde des Pflanzenreiches, namentlich aber für Forst- und Landwirthe. Hannover.

Singh, B. R., McLauglin, M., & Brevik, E. (2017). *The nexus of soils, plants, animals and human health.* Stuttgart: Catena.

Spawn, S., Lark, T., & Gibbs, H. (2019). Carbon emissions from cropland expansion in the United States. Environ. Res. Lett. 14045009 https://doi.org/10.1088/1748-9326/ab0399

Sprengel, C. (1828). (*Of the substances of the topsoil and subsoil, especially how such can be detected and separated from each other by chemical analysis; in what cases they are beneficial or detrimental to plant growth, and what decompositions they undergo in the soil.*) Von den Substanzen der Ackerkrume und des Untergrundes, insbesondere, wie solche durch die chemische Analyse entdeckt und von einander geschieden werden können; in welchen Fällen sie dem Pflanzenwachsthume förderlich oder hinderlich sind und welche Zersetzungen sie im Boden erleiden. – In: Erdmann O. L. (Ed.): J. Technische & Oeconom. Chemie 2: pages 423–472; 3: 42–99, 313–351, 397–421.

Sprengel, C. (1830). XIII. (*On cattle urine.*) Ueber Rindviehharn. Schluss einer mehrteiligen Abhandlung. In: Erdmann O. L. (Ed.): J. Technische & Oeconom. *Chemie, 3/7*, p. 171–195.

Steiner, R. (1957 ff.). (*Collected works in 345 volumes.*) Gesamtausgabe in 345 Bänden. Dornach.

Steiner, R. (1984). (*Humanistic [spiritual] basics for the prosperity of agriculture. Agricultural course. –According to transcripts of 8 lectures held around Whitsun 1924 that were not reviewed by the lecturer.*) Geisteswissenschaftliche Grundlagen zum Gedeihen der Landwirtschaft. Landwirtschaftlicher Kurs. – Nach vom Vortragenden nicht durchgesehenen Nachschr. von 8 Vorträgen auf dem Gut Koberwitz bei Breslau um Pfingsten 1924. Dornach.

Stöckhardt, A. (1851). (*Chemical field sermons for German farmers.*) Chemische Feldpredigten für deutsche Landwirthe. Leipzig: Wigand.

Stöckli, A. (1946). (*Soil as a living space.*) Der Boden als Lebensraum. *Vierteljahresschrift der Naturforschenden Gesellschaft in Zürich, 91*(1), p. 1–17.

Thaer, A. D. (1810). (*Principles of Practical Agriculture.*) Grundsätze der rationellen Landwirthschaft (Vol. 2). Berlin: Realschulbuchhandlung.

Thaer, A. D. (1821). (*Principles of practical agriculture.*) Grundsätze der rationellen Landwirthschaft (2nd ed., vol. 3). Berlin: Reimer.

Thaer, A. D. (1858). *Principles of practical agriculture* (W. Shaw & C. Johnson, Trans.). Moore publisher.

Tisdale, S., Nelson, W., & Beaton, J. (1985). *Soil Fertility and Fertilizers.* New York: Macmillan.

van Bertalanffy, L. (1950). An outline of general system theory. *The British Journal of the Philosophy Science, 1*(2), p. 134–165.

von Engelhardt, D. (1997). Vitalism between science and philosophy in Germany around 1800. In G. Cimino & F. Duchesneau (Eds.), Vitalisms from Haller to the Cell Theory. Proceedings of the Zaragoza Symposium, *XIXth Int. Congr. of Hist. of Sc., 22-29 Aug. 1993* (pp. 157–174). Firenze: Olschki.

Vierzig, S. (2009). (*Myths of the Stone Age. The religious world view of early humans.*) Mythen der Steinzeit. Das religiöse Weltbild der frühen Menschen. University of Oldenburg.

Vogt, G. (2000). (*Origin and development of organic farming in German-speaking countries.*) Entstehung und Entwicklung des ökologischen Landbaus im deutschsprachigen Raum. Ökologische Konzepte 99. Bad Dürkheim: Metropolis.

Waksman, S. (1938). *Humus. Origin, chemical composition, and importance in nature* (2nd ed.). London: Williams & Wilkins.

Wallerius, J. G. (1765). (*Chemical principles of arable farming.*) Chymische Grundsäze des Feldbaues. First published as Wallerius (1761): Agri culturae fundamenta chemica, Uppsala.

Weissert, H., & Stössel, I. (2015, 1st ed. 2009). (*The ocean in the mountains. A geological time journey through Switzerland.*) Der Ozean im Gebirge. Eine geologische Zeitreise durch die Schweiz. ETH Zürich.

Wiggering, H. (2017). (*Land, Landscape, Agriculture 2071. A story between dream and fallacy, which would like to be a fiction and yet is caught up with reality.*) Land, Landschaft, Landwirtschaft 2071. Eine Geschichte zwischen Traum und Trugschluss, die gerne eine Fiktion wäre und doch von der Realität eingeholt wird. Series Agrarkultur im 21. Jahrhundert. Marburg: Metropolis.

Wiggering, H., & Schallwich, D. (2019). (*A war in the cornfield. Cyberwar and digitalisation – field robots and tablets, cyborgs and farmers. A narrative, fictional and realistic, today and anticipating tomorrow.*) Ein Krieg im Kornfeld. Cyberkrieg und Digitalisierung – Feldroboter und Tablets, Cyborgs und Landwirte. Eine Erzählung, fiktiv und realistisch, heute und mit einem Vorgriff auf morgen. Series Agrarkultur im 21. Jahrhundert. Marburg: Metropolis.

Wolff, G. (1905). (*Mechanism and vitalism.*) Mechanismus und Vitalismus. 2. ed. (1. ed. 1902). Leipzig: Thieme.

Zumsteg, A., Luster, J., Göransson, H., Smittenberg, R., Brunner, I., Bernasconi, S., Zeyer, J., & Frey, B. (2012). Bacterial, Archaeal and Fungal Succession in the Forefield of a Receding Glacier. *Microbial Ecology, 63*(3), p. 552–564.

Chapter 5
Soil in the Shamanism Mythology and Rituals in Northern Eurasia

Pavel Krasilnikov, Nelli Agadzhanova, Iuliia Kryukova, and Elizaveta Smirnova

5.1 Introduction

Shamanism is a common trait of Siberian autochthonous religions and is practised in animistic religions worldwide. A common pattern consists of ecstatic practices of the priests (shamans), while the mythological systems may differ from one religion to the other (Eliade, 1964). In this text, we use this word for labelling the entire religious systems. The common view on the origin of Siberian shamanism is that it originated from a Proto-Turkic religion that is called Tengerism, the cult of the sky and earth (Relic, 2015). Some researchers believe that Tengerism originated in Central Asia and Mongolia, while others point out that it could have been imported much earlier from Mesopotamia (Dugarov, 1991). Other researchers believe that shamanism was a Paleolithic religion that once existed worldwide before the introduction of the religions of agricultural societies, pointing out the striking similarity of the pagan cults around the world. Despite these ambiguities in regard to geographical origin, there is much evidence that the mythology of shamanism roots in the pre-agrarian epoch of the history of humankind (Nachtigall, 2011). In this respect, it is interesting to find out what was the place of soil in the world view of the people who did not yet cultivate the land in Siberia.

We already know a lot about the role of soil in the agrarian culture (see Hillel, 1992). Since the Neolithic agrarian revolution, people regarded soil as the source of food, and a large part of the mythological system was constructed around the concept of the motherhood of earth. As Plato, a representative of a mature agricultural civilization in ancient Greece, stated, "the woman in her conception and generation is but the imitation of the earth, and not the earth of the woman" (Schofield, 2009). Many other mythological and religious concepts originated from the tribute to the

---

P. Krasilnikov (✉) · N. Agadzhanova · I. Kryukova · E. Smirnova
Faculty of Soil Science of Lomonosov Moscow State University, Moscow, Russia
e-mail: krasilnikov@soil.msu.ru

N. Patzel et al. (eds.), *Cultural Understanding of Soils*,
https://doi.org/10.1007/978-3-031-13169-1_5

agricultural cycles, for example, the concept of dying and resurrecting gods (Frazer, 1939).

If the hypothesis that shamanism represents a pre-agrarian religion is correct, we have a unique chance to see if soils were valued and respected by people whose lives did not depend directly on land management. The analysis of folk soil classifications showed that the number of soil names is much lower in hunting or nomad cultures than in agrarian civilizations (Holman, 2005). We believe the analysis of the myths and traditions of the people of Northern Eurasia will help us understand the origin of shamanism and maybe recognize the transformation of the concepts we know from other cultures.

5.2 The Distribution of Shamanism in Northern Eurasia

Shamanism is widespread in many ethnic groups in Eurasia, South and North America, Africa and Australia. In Russia, it may be found among many ethnic groups. For example, the peoples of the Volga and Ural regions, such as the Udmurts, Mari, Chuvash, and Mordovians have a hierarchy of Supreme gods, the distinguished patrons of the plant world, atmosphere, and sky. The elements of ancient shamanic cults are still alive among the majority of the ethnic groups even despite the broad distribution of world religions such as Christianity, Islam and Buddhism across Russia (Tishkov et al., 2008). The main distribution of shamanism is across the vast areas of Siberia and the Far East, crossing slightly to the west of the Ural Mountains (Fig. 5.1). Most of the Nenets, Nganasans, Nivkh, Udege, Chukchi, Eskimos (Inuit), some Selkups, Shors, Dolgans, Koryaks, Yakuts, and representatives of other aboriginal peoples living in northern Russia adhere to shamanistic beliefs (Tishkov et al., 2008). As one can see on the map (Fig. 5.1), a plethora of ethnic groups inhabit Siberia and the Far East of Russia, and practically all of them have at least elements of shamanic cults in their mythology and beliefs. The diversity of languages complicates the correspondence of the gods and concepts, but they still show a striking similarity.

5.3 Mythology: The Creation of Soils and Humans

The basis of shamanism is an animistic cult that recognizes the existence of spirits living in landscape or its element, such as rivers, lakes, forests, mountains, etc. The priests – shamans – are the personalities who can best communicate with these beings and interact with this invisible world, influencing their behavior. Frazer (1939) considered that the animistic view is a natural stage in the development of the religious systems. However, apart from the belief in the existence of spirits, shamanism has more complex mythology that establishes the existence of deities of higher level, which are responsible for the creation and functioning of the world and

Fig. 5.1 The distribution of ethnic groups of Northern Eurasia that practice shamanism. (Modified after Dallmann, 2005)

its components. The similarity of these myths across the vast area of Northern Eurasia might allow us to assume the origin of the entire mythological system is from the same source, i.e., ancient Tengerism religion (Relic, 2015). However, a striking similarity in the animistic religions among different ethnic groups across the world may be also explained by the uniformity of human metaphysical understanding of the world, while practices may differ in different cultures (Walsh, 2007).

The creator of the world in most shamanic cults is *Ul'gen* – the head of the gods, who made, among other things, the pastures and rangelands, as believed among many ethnic groups in Siberia[1] (Alekseev, 1984; Potapov, 1991). It is curious to note that the Buryats describe *Ul'gen* as a female deity, while the Yakut and people of the Altai region believe that it is male. This cultural phenomenon may be ascribed to the influence of the transition from matriarchate to the patriarchate in ancient society. Initially, in the matriarchal society, the goddess was female, but when the social organization of the society changed, some ethnic groups changed the gender of the god, because it seemed natural for them that the head of the gods should be male. The other groups, such as the Buryats, did not change the gender of the goddess-creator. The other explanation is that initially, the deity was a hermaphrodite, but later on, different ethnic groups selected one of his/her genders. The same deity is

[1] Including the Altai people, the Yakuts, the Shors, the Khakass people, the Teleuts and the Buryats (see Fig. 5.1).

called *Kurbustan,*[2] *Tengri,*[3] *Yes'*,[4]and *Saveki*, *Enduri* or *Main,*[5] all these deities being male. A female deity *Nylytytya-nguo*[6] is considered to be the mother of nature, the creator of the universe, and the goddess of fire and earth. In some cultures the creator of the universe is zoomorphic: in some myths, it is a hare,[7] in the others, it is a raven-creator of the Earth *Kutkh.*[8] In rare cases, the entire universe was created by multiple deities[9] named *anan*.

Most of the people in Siberia have similar views on the organization of the universe. For example, the Turkic people of the Altai region[10] consider the world to consist of three layers: the upper world with the Sacred Sky (*Tengri*) and good spirits; the middle world, where all sentient beings exist together with the spirits of nature; and the lower world, where the lord of death *Erlik* lives with his demons and where the souls of the dead usually go. The kingdom of *Erlik* is located under the ground at five to seven levels resembling Dante's hell among the viscous swamps and bogs, covered with scant vegetation (Baskakov & Yaimova, 1993). In other ethnic groups,[11] there is also an opposition of the upper world, where *Numi-Torum* is the supreme deity, with the underworld ruled by *Kul-otyr* (Zherebina, 2009). In Turkic mythology, the idea of the sacred tree, which connected the three worlds, is important, and the spirit of the shaman symbolically travelled through this tree (Kharitonova, 2012). Peoples of the other ethnic groups in Northern Eurasia also have similar beliefs.

In many cultures, the three worlds have not only vertical but also a horizontal position. The northern region is associated with the lower world of the afterlife, the south is the upper world – the fantastic country of *Mortim Maa*, and the middle world is the land where the people live.[12] The horizontal sphere is connected with the idea of a river: the lower world in the lower reaches of the Ob River and the upper in its upper reaches (Sokolova, 2009). Other people believed that the entrance to the world of the dead was located in the mouth of the rivers falling into the Polar Ocean, but this world itself was located under the earth (Fedorova, 2007b).

Another aspect of the cosmogonic myths is the creation of the land that may also be interpreted as the origin of the soil. Two brothers – the heads of the lower (older named *Khargi*) and upper (younger named *Seveki*) worlds participated in the

[2] The name is used among some Altai people (Baskakov & Yaimova, 1993).

[3] This name is common among some groups of the Khakass people and Tuvinians (Potapov, 1991).

[4] This name is used a kind god-creator of heaven by the Kets (Zherebina, 2009).

[5] All these names are used among the Evens (Zherebina, 2009).

[6] This goddess is recognized by the Nganasans (Zherebina, 2009).

[7] This myth was recorded among the Yukagirs (Zherebina, 2009).

[8] This myth is common among the Itelmens (Zherebina, 2009).

[9] These deities are recognized by the Koryaks (Zherebina, 2009).

[10] The Teleuts, the Sors, the Tuvinians, etc.

[11] In the beliefs of the Khanty.

[12] This belief is typical for the Khanty people (Sokolova, 2009).

creation of the world.[13] One of the options for creation: initially only water existed, and *Seveki* with the help of waterfowl pulled a piece of earth from the bottom, then created the earth, which began to grow but was still very small. *Khargi* was angry at his younger brother, and while the latter slept, he began to pull the earth out from under *Seveki*. But he could not pull it out, the land only stretched out and became larger (Yermolova, 2007). A similar myth reports that *Ul'gen* created a land the size of a bed, lay down on it to rest and fell asleep.[14] The servants of the devil wanted to steal the bed and pulled at it with iron claws. They wanted to sink the land but stretched it instead, and the earth was formed (Sleptsov, 2009). In some myths, the birth of life comes from the lower world, associated with the image of Mother Earth, the Earth's gut.[15] The cosmogonic myths that describe the emerging of the land from the down to the surface may have a psychoanalytical interpretation: land creation is associated with the emergence and progress of conscious knowledge that originates from unconscious impulses in this case (von Franz, 2017). Although the statement can be hardly proved, a possible interpretation is that in the severe environments of Siberia people receive knowledge mainly from their empirical experience.

The close relation between water and soil is notable in the myths, where the land is depicted as a living creature, a sprawling dragon, which lies on another dragon floating in the water.[16] In other myths, the surface of the Earth is the skin of the divine being *Mou-nyama*.[17] As *Kou-nyama* – "Sun-mother" comes close, the snow leaves, and shedding of fur by *Mou-nyama* occurs. The old fur is the old moss, and the new fur is the new vegetation (Sleptsov, 2009). In some myths, soil cover and hummocks are identified with the body of the goddess of the earth *Etugen* (*Ul'gen*).[18]

Like in many other parts of the world, the people of Siberia and the Far East believe that humans were made of soil material, for example, *Kudai* (aka *Ulgen* or *Seveki*) moulded a man from clay.[19] Some variations of these myths strongly resemble the Christian mythology: it is stated that *Ul'gen* created the world in 6 days, created the sky, sun, moon, stars and the first man of reeds and clay.[20] However, it is not clear, if it is an original myth or recent borrowing from the Bible. Other cosmogonic myths specify the supreme deity created the first people from clay and water, and the heart from iron, breathing life in the form of heat from the fire.[21] Similarly, the creators of people and the whole universe *anan* (or *atyny*) made people of clay and inserted hearts made of hard material, maybe iron or stone. However, a mali-

[13] According to the Evenk myths (Yermolova, 2007).

[14] According to the Yakut myth (Sleptsov, 2009).

[15] In Selkup myths (Stepanova, 2007).

[16] In the myths of the Udege people (Zherebina, 2009).

[17] In the Nganasan mythology (Sleptsov, 2009).

[18] In the traditional worldview of the Buryats (Sleptsov, 2009).

[19] According to the mythology of the Shors (Alekseev, 1984) and the Evenks (Yermolova, 2007; Zherebina, 2009).

[20] In the myths of the peoples of the Altaic group (Toboev, 2018).

[21] In the myths of the Trans-Baikalian Evenks (Sem, 2017).

cious creature – *nen'eeteshn'yn* – interfered in the process of creation, it took their hard hearts out of the original people and replaced them with clay hearts; therefore, people die.[22] The idea that people originate from the earth, or from clay, is reflected in these people's self-naming: Selkups call themselves *Tshumel-gor* or *Tjuje-gom* from *tscu, tju* – clay, earth (Stepanova, 2007). Human souls also emerged from the earth, i.e., the lower world is the source of life (Stepanova, 2007).

Unlike more recent religions, shamanism does not postulate antagonism between heaven and the underworld. As mentioned above, in some myths the masters of the upper world and underworld are brothers who compete but do not fight: the elder brother *Khargi* is doing small mischief rather than really bad things (Yermolova, 2007). In the mythology of the Kets *Hoseda*, the mistress of the underworld is the wife of the good god of heaven *Yes'* (Zherebina, 2009).

5.4 Beliefs Connected with Soil and Related Elements

A common feature of all the peoples in Siberia is respect for Father Heaven and Mother Earth. The latter is often associated with soils in the mythology of shamanism. It may be shown at the example of the myths of the Selkups, an important ethnic group in Eastern Siberia. Earth is the progenitress of all the beings in the world and has two hypostases: spring *Tomem* and winter *Tegem*. Life originates under the earth, which concentrates all the power that gives life to all creatures (Stepanova, 2007). Mother Earth is personified – the sacred mother-ancestor of all living things – the mistress of the upper world and the mistress of the dungeon. The Northern Selkups imagine a maternal ancestor being the old woman *Ilintili-kota*, who controls life and death. The Southern Selkups believe in Old Woman Earth *Payaga*, a cunning old woman, the creator of the inhabited Earth, humans and deer. Zoomorphic deities of the Selkups, the Great (female) Moose and the Great (female) Spider live in a house under the earth. The Great Spider (*Varga pakula kota* – "grandmother spider"), lives in an underground building near the "navel" of the Earth, she is the mistress of the kingdom of the dead (Stepanova, 2007). Khanty and Mansi people believe in a female deity *Kaltas*, or *Ioli torum San*, earthly mother, a cult associated with the lower world (Sokolova, 2009). She is also the mother of the sun that resembles many other myths all around the world.

The beliefs vary even within the same ethnic group; in places, we can observe gender variation of the same deities as we mentioned above for the god(dess) *Ul'gen*. The Evenks living on the Bayaki River imagine the owner of the earth as a female deity: *Grandma Kandika* is the mistress of the universe. Among the Evenks who live on the neighboring Kagan River, the owner of the universe is an old man (Sem, 2017).

[22] In the Koryaks mythology (Zherebina, 2009).

In the beliefs of the Evenks Mother Earth (*Gazar Iej*), like Father Heaven, does not have a human form, but is visible to any person — like the earth that feeds people. It has another name – *Itugen*, and the names of many shamans, especially women, represent different variations of this word, for example, Yadgan, Uthan, Udaga, etc. With this, the shaman wants to show that he or she very much honors Mother Earth. The important feature of shamanism is that it does not postulate an opposition between the male and female nature. In this respect, it differs both from mostly masculine Abrahamic religions and from "yin-yang" oriental philosophy that implies opposition of the gender natures. In shamanism, the gods easily vary in their gender in different ethnic groups. This means that basic natural objects and phenomena may be associated with both genders. Partly it may be a reason for gender equality among shamans: both men and women are allowed to become shamans, and in places, the women are considered even stronger shamans than men. However, there may be an opposite dependence: the absence of gender misbalance among shamans leads to more gender-neutral mythology.

The goddesses of the earth or their close relatives are commonly regarded responsible for fertility as well. For example, *Dia-menu'o* – "The old woman of earth" – is the patroness of birth and women,[23] *Geri-su* is the spirit of earth, of fertility and protector of animals,[24] the spirit-mistress of the earth *Aan Alakhchyn Khotun* in some places is equated to the goddess of fertility and childbirth *Aiyysyt*[25] and the daughter of *Itugen*, *Umai*, is a goddess of fertility who takes care of the souls of the bodies of living beings when they sleep in the Tree of Peace at night. *Umai* also has another name – *Tenger Nyanian*, which is derived from the Evenk word for soil (Breusova, 2009).

Most ethnic groups believe that a hierarchy of gods and spirits exist. For example, the goddess of fertility *Aiyysyt*[25] manages *aiyy* – deities, bearers of various blessings (Zherebina, 2009). Similarly, *Bua*[26] is the major spirit, master of the taiga and the universe, while *odzyan* are the minor master spirits of nature. This reflects one of the most important components of shamanism: its animism, a belief in the existence of numerous spirits, "masters of the places," responsible for particular objects and places. The main gods typically do not care about the human business, but less significant ubiquitous spirits continuously interact with people and may either help or harm depending on their nature and established relationships with humans.

The Buryats believe in the existence of *ezhins* – spirits of the earth (Mikhailov and Asalkhanov, 1987). Since the Buryats evolved from the beginning of the seventeenth century to a developed agricultural society, they included their old beliefs into the new cultural context, and currently have special *ezhins* responsible for tillage (Mukhailov, 1987). Other ethnic groups where land management was

[23] In the beliefs of the Entsi (Sleptsov, 2009).

[24] In the myths of the people of the Altaic group (Sleptsov, 2009).

[25] In the myths of the Yakuts (Sleptsov, 2009).

[26] In the mythology of Udege people (Zherebina, 2009).

introduced in recent centuries also believed in the existence of a spirit, the patron of grain plants, in the middle world.[27] Religions such as Christianity and Buddhism also strongly influenced their beliefs: the cult of the "white old man" *Sagaan ubgen*, the master of the whole earth, stones, mountains, forests, herbs, water, and animals merged with the image of Saint Nicholas (Shaglanova, 2003).

5.5 Practical Rituals Related to Soils

The ancient Turks believed in the spirits of the landscape and its components such as mountains, rivers, forests and even individual trees. To maintain a good relationship with the spirits, people performed various sacrifices and communicated with the spirits through shamans (Kharitonova, 2012). These traditions are mostly alive even today among the people belonging to the ethnic groups habitually confessing shamanism, even if they live in the city and are well educated. It is an admirable fact considering that shamanism was oppressed for a long time. In the Russian Empire, where the Orthodox Church was a part of the establishment, people following traditional religions were treated as pagans, savages, who had to be converted to Christianity. Many people in various ethnic groups of Siberia have Russian names because they were forcibly baptized. From the point of view of the church, the rituals of shamans were dangerous witchcraft. In the Soviet Union, the situation got even worse, because the official ideology of the state was dialectic materialism, and all the religions were declared superstition and obscurantism. The followers of the world religions in the Soviet Union suffered from this policy but survived, because the state could not convince all the believers. But the shamans were few, and many of them were arrested for "religious propaganda" or forced to leave their practices. Only in the 1970s with certain liberalization in the Soviet Union shamanism started to return back as a part of the cultural heritage thanks to the enthusiastic effort of such people as Mongush Kenin-Lopsan, a Tuvinian writer, ethnographer and shaman. This outstanding personality practically reanimated shamanism in Siberia and made it known to the international community (see Kenin-Lopsan, 1997). Currently, shamanism is not oppressed, but until now it has no registration as a religious organization. This reflects the difficulties with self-identification and cooperation between different communities rather than the lack of cooperation from the side of the state.[28] Shamanism is on the rise now: to some extent, it reflects the development of ethnographic tourism business, although there is also an evident honest movement of the people to their cultural roots. There is a trade-union of shamans in Russia.

[27] In the beliefs of the Kachin group of Khakass people (Alekseev, 1984).

[28] The Federal Law "On freedom of conscience and religious associations" declares that a local religious association may be registered if there are more than 10 members in this group, and a centralized religious association may be registered if there are more than three local religious associations in it; the package of registration documents is not very extensive.

The key figure in these beliefs is the priest, the chosen – the shaman. Shamans were known to all the people of Siberia (the Evenks, Yakuts, Buryats, Turks of Sayano-Altai, etc.) (Potapov, 1978). Some people of the northeast, for example, the Chukchi, also had "unprofessional" shamanism, i.e., in the absence of shaman people could perform shamanic rites themselves. However, this is an exception rather than a rule. For most ethnic groups, the shaman had to be chosen by the spirits, commonly after a "shamanic disease," a fever that changes him or her completely. People believed that a special spirit (*kut*) was inside a shaman, who could leave the body of shaman during his sleep, loss of consciousness, illness or special rite (Potapov, 1991). Strictly speaking, a shaman is not a human being anymore: his/her soul is replaced by a spirit from outside. Some communities believed that obsession is the only way to become a shaman; nobody can learn shamanic rituals only from explanations given by another person. However, others allowed for shaman apprenticeship and further initiation. The shaman could send his soul out of the body to travel to the spirit world during a ceremony in a state of ecstasy (Kharitonova, 2012). The rite includes singing, rhythmic drumbeats and dancing. In most places in Siberia shamans also use toxic mushroom *Amanita muscaria* to reach the ecstatic state (Batyanova and Bronshtein, 2016). It is interesting that the use of mushrooms was restricted to shamans in Western Siberia, while in the extreme North-East of Eurasia there was no restriction for their consumption. The researchers describe three stages of intoxication with *Amanita muscaria*: at the first stage people are in the state of noisy gaiety, at the second stage they have visual and auditory hallucinations, and at the third stage they lose consciousness, their souls separate from the body and travel in the world of the dead (Batyanova & Bronshtein, 2016).

For a professional shaman, the soil is an important interface between the middle and lower worlds, an important portal to the dangerous place where important spirits dwell. Traditionally, there were two main groups of shamans in Siberia: the 'white' and 'black' shamans (Zhambalova & Suvorova, 2013). This division is far from the Western dualism of good and bad, white and black magic, although they appeal to the upper and lower worlds, respectively. The concept of the 'white' shaman is closer to a priest who prays to the gods. However, the high gods have little interest in humans and their minor issues. The 'black' shaman tries to solve practical issues of the community through negotiation with more dangerous but more effective spirits of the underworld. In many ethnic groups, shamans may apply both to the upper and lower worlds depending on the aim of the rite. The entrance to the lower world may be located in the far north at the mouth of the great rivers falling into the Polar Ocean, or in the forest, or in a grave. Shamans also can penetrate the soil and enter the underworld spiritually and even physically. Sleptsov (2009) described the rite of cleansing the seat of the shaman in Yakutia. The place where the shaman sat after summoning the spirits began to fall, and he sank into the ground. To remove the spirits that drug him down, the shaman uttered a spell to the sounds of a tambourine, turned around his axis three times in the direction against the sun, then made a horse-like neighing sound and kicked back like a horse. It was believed that in this way the shaman drove the spirits down into the earth. In the northern Yakuts, the shaman turned to the spirit-master of his sitting with a spell, in which he begged the spirit-master to protect him and not to hide evil spirits within him. After the spell, he kicked the mat three times.

There are holidays and ceremonies of the calendar cycle that are known to the people in Northern Eurasia even if they do not practice arable farming; their lives depend on reindeer grazing, mushrooms, berries and other vegetation originating from the soil, thus their life also follows seasonal cycles. For example, people celebrate "Meeting of the Sun" and "Flowering of the Tundra," accompanied by the rites of feeding the "spirit of fire," "feeding mother tundra," and "coaxing the spirits."[29] The Nentsi considered the sacrifice of a reindeer the most effective way to propitiate the spirits and secure their help in the seasonal rituals and rites of the life cycle. Among the sacrifices are the following: the offering of the first prey, the dedication of live reindeer to the spirits of sacred places, and the feeding of household spirits (Kharitonova, 2012).

In places soil, being associated with fertility, was used as a mediator for stimulating pregnancy. Nganasan women made and wore little magic pouches *(simi)* filled with grass and earth mixed with fat – "parts of *Mou-nyama* body with its fur," and coals from the hearth. *Simi* were made every spring by women who wanted to get pregnant. The next spring, these bags were left somewhere on a hummock in the tundra. If a woman gave birth, then after the birth she kept the old *simi* (Sleptsov, 2009).

Although shamanism was not initially connected with tillage, the farmers combined shamanic and agrarian rituals. For example, Mikhailov and Asalkhanov (1987) report the traditions documented in a Buryat village at the end of the nineteenth century. Local people practised *talaghani*, sacrifice to the spirits of land and tillage, several times a year: celebrating the completion of sowing, after ploughing the fallows, after the end of the hay harvest, in honor of the end of grain-harvesting works, and before the beginning of the winter housing season. The sacrifices included food and wine, sometimes purification with smoke using firewood of juniper, thyme creeping and fir bark. Some people celebrated *Ursn Hurty* — the feast of the slaughter of the grain worm, in spring after planting.[30] Others arranged prayers to Tengri seasonally to avoid disasters – fire, drought – and with requests – harvest, good grass growth, an abundance of pine nuts and animals.[31] Usually, it happened in early summer or once every few years – if everything went well (Potapov, 1991). The Yakuts practice the rituals of the cult of fertility, turned to the spirit – mistress of the land *Aan Alakhchin Khotun* and the goddess of fertility and childbirth *Aiyysyt* (Sleptsov, 2009).

Some peoples applied the soil in medical magic. Shamans applied to the Earth-Mother in the rituals of treatment with a request to leave the person alive by offering a sacrifice of an animal or utensils.[32] Some shamans use the phrase "Grandmother

[29] In the Evens culture and some similar events among the Chukchis (Romanova et al., 2012).

[30] The Khakass people (Tishkov et al., 2008).

[31] The Tuvinians (Potapov, 1991).

[32] In the Kets culture (Sleptsov, 2009).

Earth, let go of the patient" in their rituals.[33] Sleptsov (2009) considered that this phrase illustrated the fact that shamanism followed the ancient idea of the essential unity of the earth and the person, as Eliade (1964) believed. The statement is doubtful, because this single phrase cannot be analyzed without its context, and the unity of humans and earth is not a common idea in shamanism.

In many ethnic groups, the soil is associated with burial rituals and thus commonly associated with death. There is an exotic belief that when a child is born, somewhere a handful of soil appears from the depths of the Earth, forming a hill; in this place, a man would die.[34] In a somewhat similar way, death was imagined in the form of a fountain of earth (*cher burt deir*), which occurs in front of a person, marking the end of his or her life.[35]

Although the soil was associated with death, many people of Siberia do not practice burial in the ground. Some peoples believe that it is not proper to dig, prick, or pick soil; otherwise, it will be offended and do poorly.[36] In the societies that do not know arable farming, this taboo is strict, while those applying ploughings make an exception for the tillage operations. A similar restriction is based on the statement that the body of *Etugen* (*Ul'gen*), the goddess of the soil and the mother of all the people, is covered with soil and protected by prohibitions not to dig the soil with cutting tools – actions that would offend and cause pain.[37] However, the Buryats, who traditionally practised incineration, currently mostly accept burial in the ground as is typical for Russians. Many ethnic groups allow soil disturbance but believe that the damage should be paid for. There is a ceremony of "redemption of the land" (*cher choleseni*) – cuts of meat and dairy products were laid and sprayed on the ground at the future site of burial.[38] There is still a similar tradition of throwing "red" money after lowering the coffin into the grave to "buy the grave."[39] This tradition curiously has certain reflections even in the contemporary soil science community. In Siberia excavation of a soil pit often starts with a small sacrifice (food, drops of spirits or smoke) to the "masters of the place" to ask permission to disturb the soil. We know some field pedologists (outside Siberia) who leave a coin in the first soil pit dug in a given season. As Levi-Strauss (1966) noted, the rituals are universal and strikingly similar in the communities we typically consider "primeval" and in our culture. In the contemporary world numerous rituals exist, both related and not related to religion (Warburg, 2016). Some of them are associated with professional communities such as sailors, pilots, actors, etc., why soil scientists should be an exception?

---

[33] In the Nentsi culture (Sleptsov, 2009).

[34] In the Soyots culture (Pavlinskaya, 2007).

[35] In the Tuvinian culture (Kisel, 2007).

[36] In the Selkups culture (Stepanova, 2007).

[37] This restriction exists among the Buryats (Intigrinova, 2001).

[38] In the Tuvinian culture (Kisel, 2007).

[39] Among the Komi people (Liskevich, 2012).

The Nivkhi practice cremation using larch or spruce woods, for children – willow. They put a tombstone house at the site of the burning of the corpse, setting a human figure inside. However, they also practice burials in the ground (Taksami, 2007).

There is no unique burial ritual among the people of Northern Eurasia. In literature, we found the following options:

- Burial in the ground[40].
- Leaving the corpses on the ground, with or without a covering over the body[41].
- Leaving the corpses on the trees, in hollows (mainly dead children) and lifted over the ground[42].
- Giving the corpses to the dogs[43].
- Incineration of the bodies[44].
- Burial in the sea[45].

The variety of the burial rituals among the people of Northern Eurasia reflects mainly the practical difficulties with the disposal of dead bodies: most of the year the soil is frozen and almost impossible to dig, and for extensive areas of tundra even wood is a scarce resource. However, there are also spiritual and cultural reasons that determine the details of the burial rituals.

Most people consider the dead to be leaving forever. Thus, the aim of the rituals was to make the way to the next world comfortable and to prevent the dead from returning back to the world of the living. This idea explains the meaning of some actions done by Siberian people. Some groups of Mansi first buried the dead in the woods, and then dug the bodies back up and transported them to the permanent cemetery. They believe that burying in the grave symbolizes the final death of a person (Fedorova, 2007a; Sokolova, 2009); earth and stone do not allow the deceased to return. Therefore, a stone was placed in the mouth or on the heart of stillborn children (Fedorova, 2007a). These peoples typically place some belongings of the deceased or pieces of food (fish, meat) in the coffin. Sometimes they hang possessions of the dead on trees near a grave. All these objects are expected to be used by the diseased on his way to the country of the dead. For the same reason in places as soon as the funeral is completed, the skin of the deer killed on the eve of the funeral, a hunting rifle, a kettle, a cup, a spoon, a knife, sometimes an axe, and

[40]The Khanty and Mansi (Fedorova, 2007a), the Udege, Orochi, currently the Evens and Selkup people (Zherebina, 2009), the Kereks, in the case if a person died a natural death (Zherebina, 2009), the Evenks (Evenks of Buryatia, 2016), the Khakass and Sagai people (Potapov, 1991) and many others.

[41]The Chukchi and the Udege (Zherebina, 2009), the Khanty and Mansi (Fedorova, 2007a).

[42]The Khanty and Mansi (Fedorova, 2007a), the Evens (until the nineteenth century) the Evenks (only dead infants) (Khasanova, 2007; Yermolova, 2007) and the Udege (Zherebina, 2009).

[43]The Kereks and Itelmens (Zherebina, 2009).

[44]The Buryats (Intigrinova, 2001).

[45]The Kereks (Zherebina, 2009).

a saddle are put on the grave of the deceased.[46] Other burial rituals include walking around the grave following the way of the sun,[47] and then a series of feedings (commemorations) during the year, when food is thrown into a bonfire at the grave: in this way the shaman sends the deceased forever to the "land of the dead."[47] Some people believed that the kingdom of the dead is located in the lower world, but the others – that the body's essence (shadow) goes to the underworld and the *kut* (soul) to the upper world.[48] During the burial ceremony, they could sacrifice a reindeer at the grave; the coffin was fumigated with juniper, and some pieces of food were inserted there.

In this respect, the burial of a shaman is an important procedure, because he or she has supernatural power and thus may easily return back to the world of the living. Shamans are buried in sacred places where cutting down of trees is forbidden (Mikhailov & Asalkhanov, 1987). Also, especially in the case of 'black' shamans, it is important to immobilize him: usually he is buried face down, and the soil is inserted in his hands to anchor him in the grave.

However, some ethnic groups consider reincarnation, which may be borrowed from Buddhism. According to Orochi mythology, the souls of dead people travelled to the habitats of mythical old women – owners of mystic rivers, who lived on the beach, in a place where the mountains towered. The old women fed the souls with ashes from the hearth and then dumped them on the ground, where they were introduced into the bodies of pregnant women (Sleptsov, 2009). The same Orochi people, who used burial in the ground along with other methods, believed that in the case of burial in the ground the soul goes to the underworld on foot (Zherebina, 2009). The idea that the dead bodies can move under the earth did not seem strange to many people; the mole (*yer chychkan*) was considered to be a horse of the underground world in the mythology of the Shors (Zherebina, 2009).

5.6 Ecoshamanism

The term "ecoshamanism" was introduced by Endredy (2005) but did not receive much attention in the serious scientific community evidently because of the popular character of the book. However, the monograph makes the important statement that shamanism includes a significant element of ecological thinking. For the Nentsi, for example, the Earth was thought of as a living being, man as a part of nature (Kharitonova, 2012). There is a whole complex of prohibitions, rituals, the purpose of which was to protect nature and wildlife. Some are rituals aimed at protecting nature and man. Thus, we can consider the existence of sacred ecology (Kharitonova, 2012). For the Chukchi ideas about sacred land (sacred landscape) and protected

[46] In the Evenk culture (Evenks of Buryatia, 2016).

[47] Among the Khakass and Sagai people (Potapov, 1991).

[48] The Dolgans (Dyachenko, 2007).

areas (reserves, ethnic and national parks) coincide in the ecological consciousness (Romanova et al., 2012).

People are losing their connection with their ecosystems and peculiar cultures, with their land. Losing this connection is a not a new process, it started with the Neolithic agrarian revolution, which is blamed by some researchers for "spoiling" the human race provoking overpopulation and degradation of natural resources (Manning, 2005). It is difficult to share the opinion that the whole agrarian civilization was a mistake, but one can evidently see that the expansion of contemporary civilization destroys natural habitats. Recent studies showed that soil and related ecosystem degradation in Northern Eurasia is progressing more than we expected, endangering the environments of the native people of the region (Krasilnikov et al., 2016). We should understand that shamanism is currently alive in the most fragile ecosystems and maintained by even more fragile cultures. It helps people feel in unity with nature and to sustain themselves in the extreme environments of Siberia.

The shaman is the central point in this way of life. Training for a future shaman included the acquisition of knowledge about natural phenomena through nature experience and spiritual experience. A mature shaman and a student went to the tundra, where the teacher gave the student knowledge about the state of the soil, the patterns of migration of deer and birds, etc. (Sleptsov, 2009). The shaman was not just a traveler to the spiritual worlds, but also a daily assistant of the community in their routine life, and advisor in issues such as medicine, agriculture, livestock, security and many others. The loss of the shamanism culture may have a strong negative effect on the life of the peoples of the north.

5.7 Conclusions

Shamanism is one of the oldest human religions, and it evidently traces its roots to the pre-agrarian hunters and nomad cultures. Only a few ethnic groups in Northern Eurasia that maintain elements of shamanism have typical agrarian myths and rituals (the Buryats, Khakass people, and south Yakuts) that reflect their current involvement in agricultural practices. Other people dedicated mainly to hunting and reindeer grazing do not have major elements of agrarian mythology. It shows that the agrarian content came to the culture together with the corresponding practices and lifestyle, as well as with the introduction of world religions.

Shamanism does not have such a strong antagonism between heaven and the lower world, like, e.g., Christianity. On the one hand, in shamanism, the underworld may be a habitat for some evil spirits and the place where the dead go. On the other hand, these spirits may be negotiated with and even forced to serve for the benefit of humans by an experienced shaman. Also, shamanism does not oppose male and female elements, even the same deities may vary in gender in different ethnic groups. The shamanic myths and legends might reflect Paleolithic matriarchate: the universe was created by a goddess, and soil was also associated with a female deity. The more recent patriarchate tradition transformed female demiurges into male ones. Gender neutrality of shamanism might be ascribed also to the fact that both men and women could be shamans.

Although pre-agrarian cultures did not cultivate plants, their life completely depended on products naturally produced by soil, such as mushrooms and berries, trees, herbs and lichens eaten by reindeer, etc. Thus shamanism recognizes soil for its fertility, and even includes it in some rituals related to human reproduction. The gods and spirits were asked to maintain soil productivity by praying and through various forms of sacrifices. The soil was seen as the source of life, the symbol of fertility, as a mother or grandmother.

Soil for pre-agrarian cultures was a sacral object, and thus for them digging and even tillage is associated with hurting Mother Earth. This may be seen even in their burial ceremonies, which in most cultures do not involve disturbing the ground. Generally, burial rituals are complex, and the soil is a major component of the beliefs related to death. In most cultures, it is the place of the "kingdom of the dead," or at least the place where the physical bodies stay.

Shamanism is the cult that is much closer to nature than any other religion. It is based on the concept of the intimate connection between the humans and the natural world, both visible and invisible, is a pre-requisite for the survival of people in the harsh environment of the north. It helps to maintain balance in these fragile ecosystems.

References

Alekseev, N. A. (1984). *Shamanism of Turkic-speaking people of Siberia (an experience of areal comparative study)* (233 pp.). Novosibirsk Branch of Nauka Publ. [In Russian].

Baskakov, N. A., Yaimova, N. A. (1993): *Shamanic mysteries of the Altai mountains* (122 pp.). Altai Book Publ. [In Russian].

Batyanova, E. P., & Bronshtein, M. M. (2016). *Amanita muscaria* in the household, beliefs, rituals and art of the peoples of the North. *Siberian Historical Studies, 1*, 46–58. [In Russian, English summary].

Breusova, A. I. (2009). *Intercontinental theory of shamanism* (93 pp.). Astrakhan University Publ. [In Russian].

Dallmann, W. K. (2005). *Indigenous peoples of the North, Siberia and the Far East of the Russian Federation*. Norwegian Polar Institute.

Dugarov, D. S. (1991). *Historical roots of white shamanism*. Nauka. [In Russian].

Dyachenko, V. I. (2007). Representations of the Dolgans about the soul and death. Why do "real people" die? In L. R. Pavlinskaya (Ed.), *Death mythology: Structure, function and semantics of the burial rite of the peoples of Siberia* (pp. 108–133). Nauka Publ.. [In Russian].

Eliade, M. (1964). *Shamanism: Archaic techniques of ecstasy* (Bollingen Series 76). Pantheon Books.

Endredy, J. (2005). *Ecoshamanism: Sacred practices of Unity, power, and earth healing* (336 p). Llewellyn Publications.

Evenks of Buryatia (2016). Ulan-Ude: NovaPrint Publ., 194 pp. [In Russian].

Fedorova, E. G. (2007a). Representations of death, the world of the dead and the burial rite of the Ob Ugrians. In L. R. Pavlinskaya (Ed.), *Death Mythology: Structure, Function and Semantics of the Burial Rite of the Peoples of Siberia* (pp. 197–221). Nauka Publ.. [In Russian].

Fedorova, E. G. (2007b). River in the burial rite of the peoples of Siberia. In L. R. Pavlinskaya (Ed.), *Rivers and peoples of Siberia* (pp. 216–237). Nauka Publ.. [In Russian].
Frazer, J. G. (1939). *The Golden bough: A study in magic and religion* (823 pp). Macmillan.
Hillel, D. (1992). *Out of the earth: Civilization and the life of the soil*. University of California Press.
Holman, E. W. (2005). Domain-specific and general properties of folk classifications. *Journal of Ethnobiology, 25*, 71–91.
Intigrinova, T. (2001). Les Bouriat et les sol: des rapports en évolution. In R. Lahmar & J. P. Ribaut (Eds.), *Sols et sociétés. Regards pluriculturels* (pp. 57–63). Éditions Charles Léopold Mayer.
Kenin-Lopsan M. B. (1997). *Shamanic songs and myths of Tuva*. Translated and edited by M. Hoppál, co-editor Chr. Buckbee. Budapest & Los Angeles: Akadémiai Kiadó. ISTOR Books 7, 146 pp.
Kharitonova, V. I. (Ed.) (2012). *The religious life of the peoples of Central Eurasia: Digest of articles* (Ethnological studies on shamanism and other traditional beliefs and practices. V. 16). M: Institute of Ethnography and Anthropology of RAS. 336 pp. [In Russian].
Khasanova, M. M. (2007). The way of the soul to the "world of the dead according to the conceptions of the peoples of the Amur". In L. R. Pavlinskaya (Ed.), *Death mythology: Structure, function and semantics of the burial rite of the peoples of Siberia* (pp. 134–154). Nauka Publ.. [In Russian].
Kisel, V. A. (2007). Innovations in the burial rite of the Tuvinians. In L. R. Pavlinskaya (Ed.), *Death mythology: Structure, function and semantics of the burial rite of the peoples of Siberia* (pp. 220–277). Nauka Publ.. [In Russian].
Krasilnikov, P., Makarov, O., Alyabina, I., & Nachtergaele, F. (2016). Assessing soil degradation in Northern Eurasia. *Geoderma Regional, 7*, 1–10.
Levi-Strauss, C. (1966). *The savage mind*. The University of Chicago Press.
Liskevich, N. A. (2012). Features of the burial and memorial rites among the Komi of the south of Western Siberia. *Bulletin of Archeology, Anthropology and Ethnography, 3*, 120–130. [In Russian].
Manning, R. O. (2005). *Against the grain: How agriculture has hijacked civilization*. North Point Press.
Mikhailov, T. M. (1987). Structure of Buryat shamanism. In I. N. Gemuev & A. M. Sagalaev (Eds.), *Traditional beliefs and life of the peoples of Siberia* (pp. 36–51). Nauka Publ.. [In Russian].
Mikhailov, T. M., & Asalkhanov, I. A. (1987). *Buryat shamanism: History, structure, and social functions*. Nauka Publ.. [In Russian].
Nachtigall, H. (2011). The culture-historical origin of shamanism. In A. Bharati (ed.) The Realm of the Extra Human: Agents and Audiences, World Anthropology, 315–323. In L. R. Pavlinskaya (Ed.), *Death mythology: Structure, function and semantics of the burial rite of the peoples of Siberia*. Mouton Publishers.
Pavlinskaya, L. R. (2007). Myths of the Siberian peoples about the "origin" of death. In L. R. Pavlinskaya (Ed.), *Death mythology: Structure, function and semantics of the burial rite of the peoples of Siberia* (pp. 8–29). Nauka Publ.. [In Russian].
Potapov, L. P. (1978). Ancient Turkic features of the worship of the Sky among the Sayano-Altai peoples. In Okladnikov A. P. (Ed.), *Ethnography of the Peoples of Altai and Western Siberia* (pp. 50-64). Nauka Publ.
Potapov, L. P. (1991). *Altai Shamanism*. Nauka Publ.. [In Russian].
Relic, R. (2015). Genesis and origin of the esoteric culture in white shamanism: A historical-cultural analysis. *Journal of Human Values, 21*, 99–105.
Romanova, E. N., Alekseeva, E. K., & Ignatieva, V. B. (2012). Chapter 7. The Republic of Sakha (Yakutia). In N. I. Novikova & D. A. Funk (Eds.), *North and northerners. The Current Situation of the Indigenous Peoples of the North, Siberia and the Far East of Russia* (pp. 71–85). Institute for Ethnography and Anthropology of RAS. [In Russian].
Schofield, M. (Ed.). (2009). *Plato: Gorgias, Menexenus, Protagoras (Cambridge texts in the history of political thought)* (T. Griffith, Trans., 264 pp). Cambridge University Press.

Sem, T. Y. (2017). *Shamanism of the Evenks based on the materials of the Russian ethnographic museum*. Humanities Academy. [In Russian].

Shaglanova, O. A. (2003). *Shamanism among the Buryats of Tunkinskaya valley: the second half of the 19th-20th centuries. Dissertation of historical sciences* (218 pp.). Ulan-Ude. [In Russian].

Sleptsov, E. P. (2009). *Yakut shamanism in the context of archaic symbolic systems: Mythology and ritual: Dissertation of candidate of historical sciences* (208 pp.). Yakutsk . [In Russian].

Sokolova, Z. P. (2009). *Khanty and Mansi: A view from the 21st century*. Nauka Publ.. [In Russian].

Stepanova, O. B. (2007). The world of the dead and the funeral rite of the Selkup. In L. R. Pavlinskaya (Ed.), *Death mythology: Structure, function and semantics of the burial rite of the peoples of Siberia* (pp. 182–197). Nauka Publ.. [In Russian].

Taksami, C. M. (2007). Nivkh views on the universe and the world of the dead. In L. R. Pavlinskaya (Ed.), *Death mythology: Structure, function and semantics of the burial rite of the peoples of Siberia* (pp. 154–182). Nauka Publ.. [In Russian].

Tishkov, B. A., Zhuravsky, A. V., & Kazmina, O. E. (Eds.) (2008). *Peoples of Russia: Atlas of cultures and religions* (320 pp.) Feoria Publ. [In Russian].

Toboev, A. I. (2018) *Shamanism in traditional Altai culture* (112 pp). Altyn-Tuu. [In Russian].

Von Franz, M. L. (2017): *Creation myths* (Rev. ed., 384 pp.). Shambhala Publications.

Walsh, R. N. (2007). *The world of shamanism: New views of an ancient tradition* (336 p). Llewellyn Publications.

Warburg, M. (2016). Secular Rituals. In A. J. Strathern & P. J. Stewart (Eds.), *The Ashgate research companion to anthropology* (pp. 127–143). Ashgate.

Yermolova, N. V. (2007). On the characterization of the early ideas about the origin of death in the mythopoetic tradition of the Evenki. In L. R. Pavlinskaya (Ed.), *Death mythology: Structure, function and semantics of the burial rite of the peoples of Siberia* (pp. 85–108). Nauka Publ. [In Russian].

Zhambalova, S. G., & Suvorova, A. S. (2013). Ritual funerary rite stereotypes of Buryat shamans. *Humanitarian Vector, 34*, 132–143. [In Russian, English summary].

Zherebina, T. (2009). *Siberian shamanism: An ethnocultural atlas*. Amphora. [In Russian].

Chapter 6
The Importance of Soil in Islam: A Study Based on Statements in the *Quran*

Mohsen Makki and Jérôme Juilleret

6.1 Introduction

All great world religions dispose of a corpus of revelation texts, sacred texts, and theological texts. For believers, the holy book of Muslims, the *Quran,* represents the Divine message transmitted by the archangel Gabriel to the prophet Muhammad. Like other religions that are mainly based on holy scriptures, Islam has a long theological tradition made by scholars to interpret the text and show believers implications of the Islamic precepts for their everyday lives that should be governing individual behavior. Hence, the need to find answers to the current climate and environmental crisis inaugurates new perspectives regarding the canonical Islamic text corpus, particularly in terms of its interpretations and practical consequences.

Generally, it is a mission of religion to strengthen the sense of responsibility of humans, not only for mankind itself, but also towards their environment. However, it is difficult for soil scientists to gain awareness from societal organizations (including religious actors) that respect soil education and protection. Hence, it is of uttermost importance for soil scientists to have some statements and proposed actions for religious actors ready, and in this way to create awareness about the importance of soil protection also in the frame of a religious worldview. We see some approaches in the Islamic faith as having potential for the extension of existing methodologies and for the practical application of soil protection, primarily but not only in countries with a large Muslim population. Future generations will and have to intensify their efforts to protect the environment and to use resources more sustainably than in the present day.

M. Makki (✉)
Geography Department, Humboldt-Universität zu Berlin, Berlin, Germany
e-mail: mohsen.makki@hu-berlin.de

J. Juilleret
Luxembourg Institute of Science and Technology, Belvaux, Grand Duchy of Luxembourg

N. Patzel et al. (eds.), *Cultural Understanding of Soils*,
https://doi.org/10.1007/978-3-031-13169-1_6

For Muslims, the *Quran* is known to contain answers to all past and future questions, answers to be found through hermeneutics. However, in the attempt to establish an "Islamic soil ethic," reference to individual verses can be problematic. Indeed, the *Quran* is often used as a quarry where religious arguments are found to justify some actions while well-established ways of interpreting the text are ignored. This involves the risk of an interpretation driven by certain interests; however, the authors of this essay wish only to show the potential that layman scientists, specialists in soil science only, can offer to the dialogue and to the development of an Islamic environmental ethic, which is desirable. To reach our goal, we cited some chapters (*Surahs*) and verses (*Āyāt*) of the *Quran* to highlight the importance of soil consideration in this holy book. In this chapter, we have tried to include the systematic context of the contents of the *Quran* in relation to the topic of "soil" and related terms. We are aware that our efforts relating to individual verses are not sufficient, nevertheless they serve as a call and a starting point for an "Islamic soil ethic."

It appears to us that the best way to introduce and illustrate the status of soil in Islam is expressed by the following verse (Quran translation, 2004):

> *It is He who created you from clay*[1] *and then decreed a term and a specified time [known] to Him; then [still] you are in dispute.* [Al-An'am: 2]

Thereby man is granted a limited time for his stay on earth and at the same time the origin of his creation from clay is indicated.

6.2 The Importance of Soil in the Daily Life of the Muslim

In all religions there is an awareness of soil, but their understanding differs from one to the next (Patzel & Blum, 2017; Szulczewski, 2015; Katsuyuki, 2009). Even if it seems trivial, because of its daily contact, it should be remembered that the soil has a fundamental role in the life of the Muslim. Indeed, natural soil or similar materials are the place of prostration during the five mandatory daily prayers. For example, among the conditions to validate prayer according to the Shi'ah Ja'fari jurisprudence school, prostration should occur on natural soil named *Turbah,* or what grows on the soil. In Islam, there is no specific place to pray like a temple or other building, and the whole earth is considered as a mosque where the Muslim can perform his/her daily prayer on time. Even if most Muslims in western countries pray in a mosque or at home, when the Muslim prays outside, the prayer has another "taste." In parallel to the spiritual experience, the prayer is accompanied by a more material experience because the praying man or woman during their prostration has direct contact with soil. Hence, when prayer is accomplished in nature, the praying person feels the soil's slope, texture, temperature, and humidity with their forehead, hands, knees and toes.

[1] The Arabic word طين (tin) used in the *Quran* is translated in different languages as clay, loam, silt or soil.

Among the conditions for the validity of the prayer, according to Islamic jurisprudence, is that the soil on which Muslims prostate must be free of impurities such as excrements, urine, vomit, blood or sperm. Of equal importance is the cleansing of the physical body from the same impurities, and furthermore the maintenance of this state of purity as recommended by God and stated in the *Surah*. This state of purification is of a general nature, at first physical, but its bearings are of course spiritual. Physical purification is necessary to perform the prayers that, once fulfilled, are sources of purification of sins and a means of preventing them.

> Allah loves those who are constantly repentant and loves those who purify themselves. [Al-Baqarah, 222]

Ritual washing (*Wudu)* is performed with water on the face, hands and feet. Interestingly, in the event that there is no or limited water available for performing *Wudu* or any other reason for which the praying person cannot use water, such as disease or other obstacle where the use of water is harmful, dry ablution made with "pure soil" (*Tayammum*) is permitted. It must be emphasized that "pure soil" represents, for most Islamic scholars, the "natural" soil. If the soil has been changed by industrial processes such as mortar, concrete, or other processes, its purifying ability is no longer valid (Fig. 6.1).

> *But if you are ill or on a journey or one of you comes from the place of relieving himself or you have contacted women and do not find water, then seek clean soil and wipe over your faces and hands with it. Allah does not intend to make difficulty for you, but He intends to purify you and complete His favor upon you that you may be grateful.* [Al-Maidah: 6]

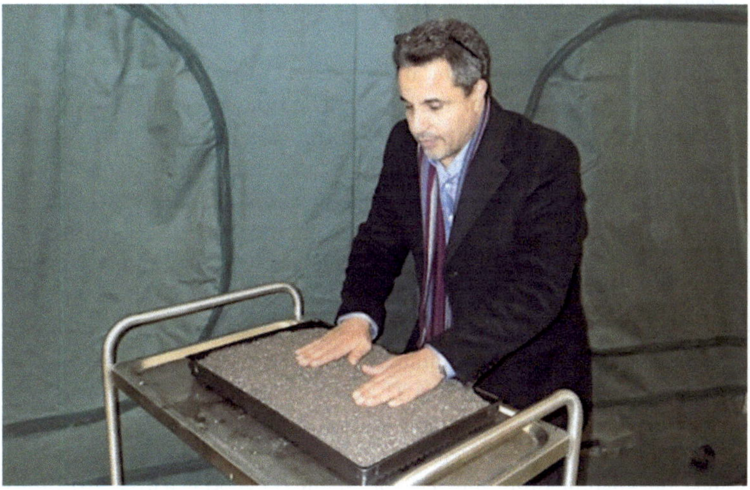

Fig. 6.1 A Muslim performing the dry ablution called Tayammum by tapping natural soil with his hands, then rubbing them together before wiping the palms of both hands over his face and rubbing his hands, photo by M. Makki

In the context of dry ablution (*Tayammum*) we note that soil, as a substitute for water, plays an important purifying role. Interestingly, this spiritual purifying function echoes that of the purifying function that the soil plays, for example, its function to purify water or that of an ointment treating certain diseases. As an example, it is recommended in Muslim tradition to use soil and water to clean vessels licked by a dog (Ahmad & Mahmud, 2019).

6.3 Protection of Soils in Islamic Countries in Terms of Sustainability

For many years, scientists concerned with soil have dealt with a diverse array of phenomena which impact soils. Land grabbing (illegitimate or illegal appropriation of land which is often driven by economical and/or political actors), deforestation and soil degradation cause desertification as a result of land use. Consequently, the actions of humans are the focus of discussions. Out of the examples given, some, such as desertification and contaminated soils as a consequence of war, occur especially in north African and west Asian Muslim countries where they harm humans and their basis of existence. Furthermore, many countries in Asia and Africa are threatened by soil erosion and degradation, due to factors such as salinization and overgrazing (e.g., Borrelli et al., 2017). Most Muslim countries are in mid-latitudes. They have arid subtropical to tropical climates and very limited fertile soil resources, hence sustainable management and protection of soils is necessary to face these problems (Makki & Köhler, 2013).

> *Erosion and salinization in particular are serious problems that cause irreversible damages to soils.* (WBGU, 2014)

Sustainability is a widely used term, in the context of ecological good practise and the preservation of resources for the future. The first comprehensive definition of the term "sustainability" was elaborated by the Brundtland Commission in 1987 (WCED, 1987). Article 1 of the so called Brundlandt Report defines sustainable development as development that "meets the needs of the present without compromising the ability of future generations to meet their own needs." Another key factor is "the demand of a universally designed solidarity as condition for establishing social justice" (SRU, 1994). In 2015, the United Nations General Assembly passed a resolution ("Transforming our world: the 2030 Agenda for Sustainable Development") on 17 global Sustainable Development Goals (SDGs). This was based on the principles of the Agenda 21 action plan of 1992, which was developed in the framework of the United Nations Conference on Environment and Development (UNCED) in Rio de Janeiro and its follow-up, the United Nations Conference on Sustainability and Development (UNCSD; Rio + 20) in 2012. In some Islamic countries sustainable management has been incorporated into the Constitution. In the Islamic Republic of Iran, for example, environmental protection for the recent, as well as for future generations, has been implemented as a public duty in Article 50 (1979) of the Constitution (Papan-Matin, 2013) to provide prospects for a better life. The German Advisory Council on Global Change (WBGU) stated that soils, in terms of environmental protection, do not get the attention they need and that raising public consciousness for soils is a tedious process (WBGU, 1994, 2011). If we compare the content and attitude of the Brundlandt Report with the attitude of religions, we find that most of them have an anthropocentric view of the world.

Despite the existing problems of soil as a natural body and finite resource, the interest in soil protection is, compared with other geoecofactors, in most countries generally low and lacks attention in the public. There are crucial deficits regarding the perception of soils as a source for food and indispensable part in global and local environmental cycles and processes. The path of better perception consists of perspective (focus, intensify, change), processes (recognize, understand, love) and problems (approach, solve, prevent). Very often our actions and intentions are controlled by emotions, among others. Therefore, it is crucial to give meaning to our "emotional horizon" and to expand it. Religions generate both individual and collective emotions in relation to various subjects related to their dogmas. The energy released by these emotions directed at various subjects must be channelled and harnessed for the collective good. In Muslim countries there is a strong potential to set up an environmental ethic based on respect for divine creation manifested by nature, which was granted to humans by God. Now it is up to us to use these potentials, especially on soil protection.

6.3.1 Role of Islam in Terms of Global Soil Protection

6.3.1.1 Islamic Faith Around the World

One significant element of socioeconomic factors for the development and governance of human behavior and for the assessment of interactions, is religion. By 2010, Islam (all orientations) was the second largest religion in the world, having 1.6 billion believers worldwide, and it is expected to increase to 2.8 billion in 2050[2] (Fig. 6.2).

Countries from North Africa and the Middle East to the Indian subcontinent and Indonesia have a long historic link to Islam, and immigration of people from these countries has strongly affected the increase of Muslims in western Europe and the USA. Many Arabic Islamic countries, but also Iran and Indonesia, refer to the *Quran* either in their constitution or use it as a legal source of legislation. Consequently, the question has risen, how can Islamic teachings work to protect soil and whether religious regulations can serve as an effective instrument to sensitise or strengthen the consciousness of Muslims towards the environment and soil?

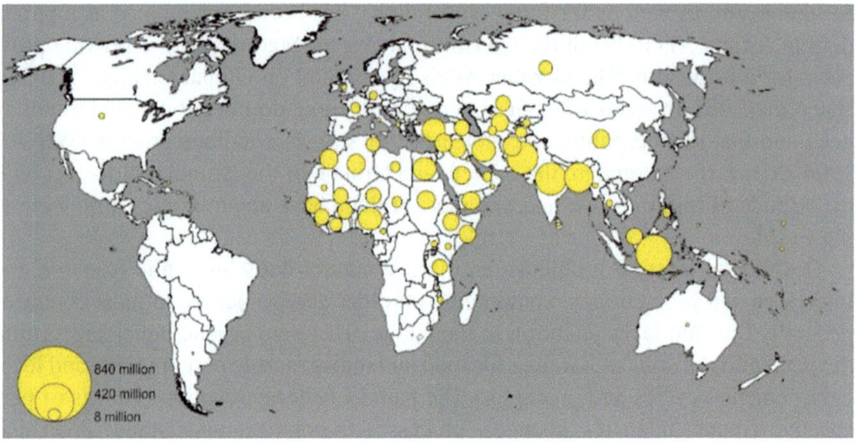

Fig. 6.2 Distribution of Muslim people by country in 2020, Based on data from www.globalreligiousfutures.org/religions/muslims

6.3.1.2 Essential Foundations of the Islamic Faith

The *Quran* is believed to be the written message of God and the most important basis for the Muslims' way of life. The Muslims' Holy book consists of 114 chapters (*Surahs*), each made from a varying number of verses, the *Āyat*. They form the foundation of the Islamic law, *Sharia* (literally, "the *path* leading to the watering place"), in which the Muslim draws the behavioral bases governing their life. Frenchl (2010) notes that: "Islam considers itself as comprehensive order of life, which structures all areas of life and does not provide a separation into sacred and profane."

The second important fundamental source of the *Sharia* is the *Sunna* (tradition) by which the commendable habits and ways of life of the prophet Muhammad and his contemporaries were meant. When the *Quran* gives no direct statements on a specific problem, then further essentials of the Islamic faith are adduced, such as oral or written traditions of deeds and utterances of the prophet Muhammad and his contemporaries. These are named with the term *Hadith* which means report, account or narrative, except Quranic traditions (Ferchl, 2010; Patzel, 2017). For Muslims, the *Quran* is the unmediated word of God and as such it contains some commandments, prohibitions, legal regulations and regulations for penalties of illicit deeds, but it is not solely a legal framework. Since the *Quran*, as well as *Sunna* and *Hadith,* yield no ultimate answer to secular questions and problems, Islamic jurists (*Ulamā*) evolved methodological tools for jurisdiction (Halm, 2014). "*Sharia* is no tidied-up code of law, no work of paragraphs, that is available as a book which could be taken from the shelf. […] This is how Islamic jurisprudence (*Fiqh*) with its principles like consensus (*Ijma*), deductive analogies (*Qiyas*) etc. emerged" (Halm, 2014). In more than 50 countries worldwide the principles of *Sharia* are an essential and inalienable source for legislation. This applies to all aspects of life, regardless of whether it concerns political, educational, economic, social, family or environmental issues.

6.3.1.3 Relation Between God and Man

Understanding the relationship between God and man and between humans and their environment from an Islamic perspective requires an introduction into the basic principles and terms of Islamic faith. According to Wilms (2006) the word *Khalaq*, literally meaning creation or genesis, is used 261 times in the *Quran* to describe the emergence of our living space. In several surahs the creation of humans by God is depicted:

> *And certainly, We created man from extracted clay.* [Al-Muʾminūn: 12]

> *Then We made the sperm-drop into a clinging clot, and We made the clot into a lump [of flesh], and We made [from] the lump, bones, and We covered the bones with flesh; then We developed him into another creation. So blessed is Allah, the best of creators.* [Al-Muʾminūn: 14]

> *And [recall] when Moses prayed for water for his people, so We said, 'Strike with your staff the stone.' And there gushed forth from it twelve springs, and every people knew its watering place. 'Eat and drink from the provision of Allah, and do not commit abuse on the earth, spreading corruption'.* [Al-Baqarah: 60]

From the above verses we conclude that:

- All stages of creation are under the supervision of God.
- The creation of man took place in several stages, starting from clay.
- A just distribution of the water resource was his intention.
- Just use of God's blessings avoids mischief.
- He creates according to mercy, and all His creations are purposeful.

6.3.1.4 Basic Principles of Faith in Islam

The foundation of the Islamic faith must first be established to illustrate the potential for soil conservation based on religious standards. In Islam, there are four columns of faith, which are essential for the everyday life of Muslims: *Tawhid*, *Mizan*, *Fitrah*, and *Caliph*.

These columns will be stated by quotations from the *Quran*:

Tawhid is the fundamental concept of the oneness of God. For Muslims Allah is nonesuch and was begotten without beginning and end. This implies that God is the uncaused cause of the universe. For Muslims he is seen as a superordinate entity, by which the world is ordered and governed. It is the primordial testimony of the whole creation and the intertwined natural order of which humans are a part.

> Say, 'He is Allah, the One. Allah the Eternal Refuge. He neither begets nor is born, Nor is there to Him any equivalent'. [Al-Ikhlās]

> *And of His signs is, that the heaven and earth remain by His command. Then, when He calls you with a [single] call from the earth, immediately you will come forth.* [Ar-Rum: 25]

Mizan is the principle of balance in life. God created humankind and gave it free will and intellect. Although humans are free in the mind, humans, as everything else in the creation, must be subordinate to His commands and distinct purposes. If the sun, the moon, the stars, the trees and the other creations would not defer to this behavioral pattern, life on earth could not exist. Therefore, Muslims have an obligation not to renounce the "gifts of God" and to actively recognize the God-given order for humans as well as for everything else.

Fitrah is the primal state in the history of creation which describes the part that God gave to humans.

> *So direct your face toward the religion, inclining to truth. [Adhere to] the fitrah of Allah upon which He has created [all] people. No change should there be in the creation of Allah. That is the correct religion, but most of the people do not know.* [Ar-Rūm: 30]

Caliph is the duty, which God has given to humans to represent God on earth. Many *Āyats* describe humans' responsibilities. The *Quran* summarizes the human role as follows:

> *And it is Him who has made you successors (Caliphs) upon the earth.* [Fāṭir: 39]

That means humans have been commissioned by God to maintain the earth as his governors, but mankind does not possess it.

6.3.1.5 Fundamental Relationship Between God and Man According to the Islamic View

From the Islamic perspective, God is regarded as the owner of the universe (God as the absolute) with a direct relationship to his creatures. Human beings are designated as a "summit of his creation" to be his representatives on earth and have the duty to nourish themselves by cultivating earth. Islamic faith is based on the belief that the Earth as a planet is just a temporary place for humans, who will enter the eternal world after their death. During their time on earth, humans should take care of the earth as their living space and there should be a fair relationship between man and their environment (Naghizadeh, 2005). Men and women, as God's representatives on earth, are obliged to create an environment in which they can remain pure in their faith. The closer the relationship of humankind to its surrounding living space and environment, the closer the way to God and thus humans' destination to Paradise.

In principle, there should be a harmonious balance between man and nature in God's creation. As God also manifests himself in nature, a re-spiritualization of nature is connected with it and with it an attentiveness towards nature. It is necessary to design these relationships in justice and dignity. When it says in the *Quran* that plants and animals also form communities, that everything is animated and has a prayer for God, this is also a fact that must be observed by everyone who confesses God in his cultivation of relationships and to bring certain oppressive relationships back into balance. This means that the relationship with God is also shown in how just and emancipatory people act in their relationships with other living beings and with nature.

> *And there is no creature on [or within] the earth or bird that flies with its wings except [that they are] communities like you. We have not neglected in the Register a thing. Then unto their Lord they will be gathered.* [Al-Anam: 38]

6.3.2 Islamic Basic Values and the Discourse of Soil Protection

6.3.2.1 The Position of Soil in Islam

In the *Quran*, five terms are used for the description of the earth's surface and soil: *al'Ard* الأَرْض (soil/earth), *Turab* تراب (dry soil, dust or earth), *Saeida* صَعِيدا (pure soil), *Tin* طين (clay), and *Salsala* صلصال (clinking clay). In the sense of soil, these terms appear 46 times in the *Quran*. As mentioned above, the Earth and the soil are defined as the living space for mankind. Many *Āyats* contain the statement that humankind, as the perfect creation and representative of God, is responsible for the maintenance of his living space and especially for the soil as a source of food. The idea of soil protection is mainly anchored in the *Quran* according to the principle "keep intact." Nevertheless, environmental and soil protection are not named as such and must be understood in the figurative sense. In the following sections, some of these aspects will be further explained by citing *surahs* from the *Quran*:

- Evolution of the living space: the Genesis
- Soil/earth as fundamental to life: existence
- Use and conservation of soils: soil protection

6.3.2.2 Genesis of the Earth As a (Physical) Living Space

In the *Quran*, earth or soil play an essential role in life for the existence of all creatures, and the creation of soil itself is considered as God's property:

> *And the earth He laid [out] for the creatures.* [Ar-Rahmān: 10]

Man, as part of God's creation, is not above it, he exists as a part of it ruled by God:

> *To Allah belongs the dominion of the heavens and the earth and whatever is within them. And He is over all things competent.* [Al-Ma'idah: 120]

> *And sufficient is Allah as Disposer of affairs.* [An-Nisā': 171]

> *And it was Him who created and multiplied you throughout on earth, and to Him you will be gathered.* [Al-Muʾminūn: 79]

Humans and soil are intimately connected. Indeed, man was created from soil into which God breathed a soul.

Indeed, I am going to create a human being from clay. So when I have proportioned him and breathed into him of My [created] soul, then fall down to him in prostration. [Sad: 71-72]

Soil is the material origin of humankind but also the place where human bodies will return after their death. Soil represents an entry/exit door of the bodies after death until the judgment day. Hence God illustrates that the cycle of life and death is through soil.

And those who disbelieve say, "When we have become dust as well as our forefathers, will we indeed be brought out [of the graves]?" [Al-Nalm: 67]

And Allah has caused you to grow from the earth a [progressive] growth. Then He will return you (into it) and extract you [another] extraction. [Nūḥ: 17,18]

And of His signs that He created you from dust; then, suddenly you were human beings dispersing [throughout the earth]. [Ar-Rūm: 20]

6.3.2.3 Earth and Soil As Foundation of Life

Islam originated in the desert regions of the Arabian Peninsula. Due to the limited resources (in terms of water and fertile soil), the inhabitants of this region cherish a particularly close relationship to their natural environment. They have an especially intense relationship with water and vegetation as the inherent sources of life. In the *Quran* there is a distinct understanding of the earth as [physical] living space. Furthermore, it is stated that all resources which are important in our lives have been given by God as a gift entrusted to the care of the people.

And it is He who sends down rain from the sky, and We produce thereby the growth of all things. We produce from it greenery from which We produce grains arranged in layers. And from the palm trees – of its emerging fruit are clusters hanging low. And [We produce] gardens of grapevines and olives and pomegranates, similar yet varied. Look at [each of] its fruit when it yields and [at] its ripening. Indeed, in that are signs for a people who believe. [Al-Anám: 99]

The *Quran* describes the natural foundations of human life. It not only determines the general extent of human possibilities for action, but also de facto limits to political and economic power. In this way man is reminded again and again of his moral responsibility towards the Creator and creation. It is expressly emphasized that man must not create any abuse [Al-Baqarah: 60]. Thus, the whole earth can be interpreted as a place of prayer. A holy space in which one can reflect on the divine. Muslims are expected to perform every action in the spirit of prayer. Prayer and nature are thus irrevocably linked. In his religious practice, the Muslim is concerned with the welfare of his fellow human beings and care for his co-creatures and environment. In the *Quran*, humans are reminded that they are not the master of Creation but the servant and representative of God in the world.

The creation of the heavens and earth is greater than the creation of mankind, but most of the people do not know. [Al-Ghafir: 57]

The following verse depicts the rating of soils:

> *And the good land – its vegetation emerges by permission of its Lord; but that which is bad – nothing emerges except sparsely. Thus do We diversify the signs for a people who are grateful.* [Al-A'raf: 58]

6.3.2.4 Usage and Protection of Soils As a Resource

The following verses are a particular admonition regarding the unbeliever's ignorance of the functions of habitat and especially soil fertility.

> *And the earth – We spread it out and cast therein firmly set mountains and made grow therein [something] of every beautiful kind, giving insight and a reminder for every servant who turns [to Allah], and We have sent down blessed rain from the sky and made grow thereby gardens and grain for the harvest and lofty palm trees having fruit arranged in layers, – as provision for the servants, and We have given life thereby to a dead land, thus is the resurrection.* [Qāf: 7 -11]

Furthermore, the *Quran* gives examples of some soil functions, such as the functions of soil as habitat and its utility functions and emphasizes their importance with exhortative messaging.

> *Then let mankind look at his food – how We poured down water in torrents, then We broke open the earth, splitting [it with sprouts], and caused to grow within grain, and grapes and herbage, and olive and palm trees, and gardens of dense shrubbery, and fruit and grass – [as] enjoyment for you and your grazing livestock. But when there comes the deafening blast on the day the man will flee from his brother, and his mother and his father, and his wife and his children, for every man, that day, will be a matter adequate for him. [Some] faces, that day, will be bright. – Laughing, rejoicing at good news. And [other] faces, that day, will have upon them dust. Blackness will cover them, those are the disbelievers, the wicked ones.* [Abasa: 24-42]

The turning away from God and His laws will consequently lead to disorder on earth and the destruction of fertile land. Selfishness and the destruction of the environment are described as something evil, which is not wanted by God:

> And when he goes away, he strives throughout the land to cause corruption therein and destroys crops and animals. And Allah does not like corruption. [Al-Baqara 205]

These examples show that within the *Quran* environmental preservation and soil protection are advised and regarded as a human obligation being part of the connection to God.

Muslims also use numerous *Hadith* (refers to words and actions of the prophet reported by companions) and lore to shape their lives. The analysis of the *Hadith* requires its own execution. At this point two quotations from Muhammad the Prophet and Ali should give a first impression about it. Many Muslims also know the *Hadith*, in which the prophet Muhammad is quoted analogously.

Anas ibn Malik reported that the Prophet said:

> *When the end of the world begins and you have a seedling in your hand, you have to plant it.* Source: Musnad Aḥmad Ibn Hanbal Hadith N°12512

Alī ibn Abī Tālib also places the human species in the focus of life and action on earth: "Fear the God, respect the rights of his creatures and the soil, for you are responsible for land and animals!" (Naghizadeh, 2005).

6.4 Results and Discussion

For a long time, the natural environment was understood as a purely scientific concept, explored in the systematic study of the interdependence of life forms in a given habitat. There has been a search for solutions to solve the world's problems. Under the increased presence of religions (post-secular societies, e.g., in the case of formally more secular countries such as Syria, Libya, etc.) the questions were increasingly asked, as to what extent religions can provide impulses for environmental protection and whether religious communities can be of strategic importance for the dissemination of sustainable environmental action. At the international level religious ethics and organizations have long been integrated as actors in environmental protection programs. In Germany, one of the first conferences on Islam and environmental protection was held in November 2010 by the Protestant Loccum Academy,[2] which led to the formation of HIMA e. V., an association of young Muslim and non-Muslim environmentalists (EAL, 2010; HIMA, 2020).

The results from this study can be summarized as follows:

- In Islam, the position of the human being is defined as God's vicegerent (*Caliph*), and the protection of the environment is prescribed to him with emphasis. Man is only a tiny part of the Divine Order, but within this order they are incredibly powerful beings.
- God created the earth in a perfect equilibrium (*Mīzān*).
- Disturbing the balance in nature is forbidden to humans (*Fitrah*).
- Environmental degradation (of land and water) is classified as disaster (*Fisād*).
- Soil has a central meaning in the creation of life and mankind.
- The purifying function of soil is symbolically anchored in the everyday life of the believers in the case of dry ablution (*Tayammum*).
- The transformation processes and material cycle of substances are clearly illustrated by the example of the emergence and death of mankind.

[2]Conference named "Wie ‚grün' ist der Islam? Umwelt und Klimaschutz mit muslimischer Perspektive" (Environmental and Climate Protection with Muslim Perspective).

6.4.1 Theological-Philosophical Contextualization of Sustainability

In the Islamic history of creation there are many factors connected to sustainable action and thinking. The *Quran* gives numerous references and interpretations to Islamic traditions regarding the universe, the use of natural resources and furthermore to the relation of God and man. In the Islamic perspective, the universe and planet Earth have been created by God and everything has been created in the right measure and relation, such that all should be in equilibrium. The anthropocentric view of Islam is that our universe and its elements care for the commonwealth of humankind and witness the magnificence of God. At the same time, man is described as a creation of God, who is separated from the rest of the universe, having a special status compared to all other parts. Trusting in his wisdom, God has created man as his representative (*Caliph*) on earth. That means that man is both part of the universe and the creation and furthermore an enforcer for God, to whom were given instructions and orders. Mankind is the beneficiary of its status, but not the commander of the universe; heaven and earth belong to God alone. God made humankind his vicegerent on earth to manage it according to His principles, targets and purposes, thus humans can utilize it for their own benefit and those of all other creations in addition to the fulfilment of their own and God's interests. Proof of trust includes the prompt to deal moderately with the earth's resources to improve all areas of life on earth: including health, food, all psychological and spiritual dimensions for the advantage of mankind, the maintenance of his welfare and the improvement of living conditions for future generations. In Islam, the conservation and development of the environment represents a universal good for all creatures, as was shown in the prophetic exemplifications.

Afrasiabi (2003) considers that the development of an Islamic eco-theology and an alternative Islamic theology is indispensable. New theological approaches should be neither oriented on the traditional, medieval interpretation of the *Quran* nor should it be guided by an essentialist and anti-Western ideology. In Afrasiabi's opinion, there is a need among young people for a dynamic Islamic perspective that is contemporary, theoretically appealing, action-oriented and idealistic, but still not utopian or dogmatic. Alternative Islamic theology should be created as a synthesis of systematic theology and modern Islamic mind (Nökel, 2009).

Another perspective on the relationship between man and nature, which can be viewed as Islamic eco-theology, is given by theologist Seyyed Hossein Nasr (Nasr, 1996), who represents the traditional Islam ideology. According to him, "everything modern and secular, that is a threat to Islam, is the wrong way" (Nökel, 2009). Nevertheless, he tries to find concepts for the development of an Islamic ecology. Nasr's theocentric philosophy sees the ecological crisis as a direct consequence of a spiritual crisis. In his point of view, humanity is disconnected from the eternal Islamic principles and verities as well as from the permanently related entity of God, world and man, which also affects Muslim societies. If the earth is no longer

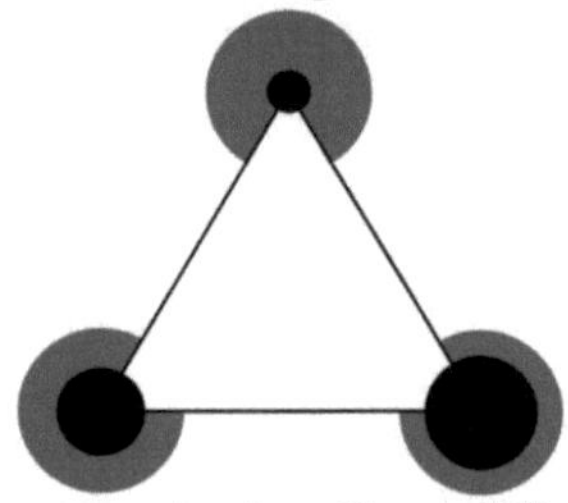

Fig. 6.3 Area and strength of influence (black) of Islam in ecology, socio-cultural affairs, and economics

regarded as an expression of the divine and when order and nature are not seen as something sacred, then limitless and ruthless exploitation of the creation will happen, because man thinks he is the measure of all things (Nökel, 2009).

6.4.2 *Applied and Practice-Oriented Contextualization*

It is a new religious philosophical approach of Islamic societies to put maintenance of the creation into the spotlight, for a practice-oriented commitment towards the environment and the protection of soils. If we look closely at the engagement of his actors and believers, we realize that they, like in other religions, care more about the societal and economical concerns of believers (Fig. 6.3).

Until recent times, the Islamic world showed only slight interest in environmental problems. The supposed reason for this is that many of the Islamic countries face a pervasive social metamorphosis. Due to the rapidly growing population the number of social crises will further increase, which, in the near future, will lead to the perpetuation of the focus on traditional societal and monetary problem fields such as the prevention of poverty. In Islamic and Islam-influenced countries, existing theological approaches to overcoming environmental problems and to protecting soils are, to date, not sustainably implemented. There is no clear awareness for soils among the population, governments or religious actors. There is a prevalent opinion that society does not need to worry about a problem until it arises. Moreover, preventive thinking using principles of sustainability for soil needs more development, based on Islamic values and values already demonstrated in European countries.

At this point, religious actors (*Ulamā*) in general can make a major contribution to soil protection. There are different ways of doing this. However, due to more pressing environmental problems, such as climate change and famines, some believers and religious actors have become aware of these problems, but are not yet convinced enough to open the debate at a theological-philosophical level. A greater focus on soil sustainability is thus needed so that these aspects can be implemented in the everyday lives of believers. Furthermore, the stipulated actions for environmental protection named in the *Quran* are only indirectly associated with soil protection. Consequently, the existing confessions in the *Quran* in the context of sustainability, soil protection and soil awareness are no guarantee that an Islamic society necessarily must meet the requirements of soil protection. Nonetheless, it is important that political actors and governments, as relevant decision makers, should view the Sustainable Development Goals and the implementation of soil conservation strategies with a higher priority, and they can strongly draw on the *Quran* when legitimizing this for Muslims.

6.5 Perspectives

In our own studies (Makki, 2008a, b; Makki & Köhler, 2013; Ramezani & Soufali, 2008; Safaei-Shahverdi, 2011), which were not solely of a natural-scientific character, we deduced that religion has not played a role in terms of soil protection so far. The question arises as to how religious actors (*Ulamā*) can be informed and enthused about the importance of soil protection. We believe that soil and environmental scientists together with religious actors (*Ulamā*) must act on three administrative levels:

6.5.1 Local Level

Islamic authorities are well versed in interacting with people at the local level. They are especially active in the social sector and are well connected with the local population. Their influence as well as their political and social sphere of action could be used to thematize the subject of soil and soil protection. With a more profound technical knowledge about the significance of soil protection, the religious scholars could reach the people and engender their attention for soils. At this level, education and promotion of NGO's, which can also have religious character, are recommended.

6.5.2 Regional Level

The regional level is the level for planning processes, decision-making and execution in terms of environmental protection, urban and landscape planning or remediation. Many religious actors and Islamic foundations with large financial power are operating commercially on this level. Islamic foundations could certainly play a decisive role in the implementation of objectives for soil protection, especially because they own extensive lands and agricultural holdings and co-operations. If these organizations could be convinced to support environmental and soil protection objectives, a major step would have been taken towards achieving the Sustainable Development Goals and protecting soils.

6.5.3 National Level

The laws and legal regulations for the implementation of soil protection objectives and compliance with the Sustainable Development Goals must be enacted at the national level. As an important prerequisite, it is necessary to establish a new Islamic eco-theology based on recent discussions and with special regard to soils as one of the main objectives of sustainable development. At this level, the promotion of a soil protection laws should be put on the agenda, which is underpinned with facts from the Islamic doctrine.

6.5.4 Societal and Cross Level Measures

Apart from the different administrative levels, the involvement and engagement of society is necessary. The decisive factor for successful soil protection is above all the conviction of the believers – this applies to Muslims as well as to believers of other religions. As already mentioned, our actions and intentions are controlled by emotions, among others. Islam can play a major role in giving meaning to and expanding our "emotional horizon," because Islam can reach the feelings and love of its believers. In reality there are serious discrepancies between Islamic principles of action in the field of environmental awareness and the actual actions of Muslims. To this end, environmental discourse in the Muslim community should be made a current topic and the community members should be sensitized to their responsibility in this area. The *Quran* and the Islamic legal system provide a solid basis for enforcing strict laws regarding environmental protection (Alsoufi, 2020). In the process, however, the above suggestions and approaches should be implemented in the frame of an "Islamic environmental ethics," or "Islamic ethics on human-nature relations." Successful environmental communication needs ethics as an orientation for individual and social decision-making conflicts. It requires a precise analysis of interdependencies, risks and opportunities. It requires motivation through positive guiding principles instead of unworldly utopias.

References

Afrasiabi, K. L. (2003). Toward an Islamic ecotheology. In R. C. Foltz, F. M. Denny, & A. Baharuddin (Eds.), *Islam and ecology: A bestowed trust (religions of the world and ecology)* (Vol. 2003, pp. 281–298). Harvard University Press.

Ahmad, A., & Mahmud, A. (2019). The hadith on purifying dog licks: In search of the authenticity from scientific perspective. *Mutawatir: Jurnal Keilmuan Tafsir Hadith, 9*(1), 24–43. https://doi.org/10.15642/mutawatir.2019.9.1.24-43

Alsoufi, R. (2020). An Islamic normative response to the debate on climate change. *Cardo, 18*(2020), 36–40.

Borrelli, P., Robinson, D. A., Fleischer, L. R., et al. (2017). An assessment of the global impact of 21st century land use change on soil erosion. *Nature Communications, 8*, 2013. https://doi.org/10.1038/s41467-017-02142-7

EAL (2010). Evangelische Akademie Loccum; (*How "green" is Islam? Environmental and climate protection with a Muslim perspective*) Wie "grün" ist der Islam? Umwelt- und Klimaschutz mit muslimischer Perspektive, Conference proceedings, Loccum.de. Last access: 2011.

Ferchl, D. (Ed.). (2010). *Al-Buhari, (collection of hadiths) die Sammlung der Hadithe* (512 p). Reclam.

Halm, H. (2014). (*Islam, past and present*) Der Islam, Geschichte und Gegenwart, 9. Auflage, 2014.

HIMA. (2020). (*History of the HIMA association*) Geschichte des Vereins HIMA, https://www.hima-umweltschutz.de/geschichte/. Last access: 01 Jan 2020.

Katsuyuki, M. (2009). Soil and humanity: Culture, civilization, livelihood and health. *Soil Science and Plant Nutrition, 55*, 5. https://doi.org/10.1111/j.1747-0765.2009.00401.x

Makki, M. (2008a). (*Islam and Sustainable Development. In: Socio-ecological problems of fast growing cities in Iran – The example of the city of Arak.*) Islam und Nachhaltige Entwicklung. In: Makki 2008b, pp. 74–79.

Makki, M. (2008b). (*Socio-ecological problems of fast-growing cities in Iran using the example of the city of Arak.*) Sozial-ökologische Probleme schnell wachsender Städte im Iran am Beispiel der Stadt Arak: 39-74. Geographisches Institut, Humboldt Universität zu Berlin.

Makki, M., & Köhler, P. (2013). *Rapidly growing cities in Iran and its geoecological challenges using the city of Arak*, In: Berliner Geographische Arbeiten, 119, Humboldt-Universität zu Berlin, 311 p.

Naghizadeh, M. (2005). *The significance of nature in Iranian culture and cities* (270 p). Iranian National Library.

Nasr, S. H. (1996). *Religion & the order of nature* (310 p). Oxford University Press.

Nökel, S. (2009). (*The environmental crisis is a spiritual crisis too.*) Die Umweltkrise ist auch eine spirituelle Krise. https://de.qantara.de/inhalt/islam-natur-und-nachhaltigkeit-die-umweltkrise-ist-auch-eine-spirituelle-krise

Papan-Matin, F. (2013). The constitution of the Islamic Republic of Iran (1989 edition). *Iranian Studies, 47/1*, 159–200.

Patzel, N. (2017). ("*... which resembles your mother." Soil and Land in Islam.*) «... die eurer Mutter gleicht.» Boden und Land im Islam. Kultur und Politik 3/2017, p. 26.

Patzel, N., & Blum, W. (2017). *Religions and Soil, IUSS Fact Sheet*. https://www.iuss.org/index.php?rex_media_type=download&rex_media_file=fact_sheet_soil_and_religions_final.pdf.

Ramezani, A., & Soufali, A. (2008). Qurans view to nature. In M. Makki, M. Safaei-Shahverdi, & M. Leineweber (Eds.), *Sozial-ökologische Probleme schnell wachsende Städte im Iran am Beispiel der Stadt Arak* (Tagungsbericht, Arbeitsberichte – Heft 141) (127 p). Geographisches Institut, Humboldt-Universität zu Berlin.

Safaei-Shahverdi, M. (2011). (*Possibilities of sustainable urban development in Iran – illustrated by the example of the city of Arak.*) Möglichkeiten der nachhaltigen Stadtentwicklung im Iran – dargestellt am Beispiel der Stadt Arak – Dissertation 2010 – [pdf] [online]: https://eldorado.tu-dortmund.de/handle/2003/27604. Last access: 01 Jan 2020.

SRU (1994). (*German Advisory Council on the Environment: Environmental Report 1994: Towards Sustainable Development.*) Rat von Sachverständigen für Umweltfragen: Umweltgutachten 1994. Für eine dauerhaft-umweltgerechte Entwicklung. Bonn: Drucksache des Dt. Bundestages 12/6995, 1994.

Szulczewski, M. (2015). Soil's social and cultural connections. *Soil Horizons, 56*(6), 1–3. https://doi.org/10.2136/sh2015-56-6-gc

The Quran. Saheeh International Translation (2004). *Abulqasim publishing house 1997.* Al-Muntada al-Islami.

WBGU (Ed.) (1994). *World in transition – The threat to soils, annual report 1994, German advisory council on global change* (252 p). Economica Verlag.

WBGU (Ed.) (2011). *World in transition – A social contract for sustainability, flagship report 2011* (396 p). German Advisory Council on Global Change.
WBGU (Ed.) (2014). *Human progress within planetary guard rails – A contribution to the SDG debate* (Policy paper no. 8) (48 p). German Advisory Council on Global Change.
WCED (1987). World Commission on Environment and Development, *Our Common Future*, UN Document A/42/427, 374 p.
Wilms, S. S. (2006). *(Islam and ecology, how do we understand nature and how do we deal with it?) Islam und Ökologie, wie verstehen wir die Natur und wie gehen wir mit ihr um?* [online]. http://islam.de/5132.php. Last access 01 Jan 2020.

Chapter 7
Soil in Rumi's Thoughts

Seyed Kazem Alavipanah, Jafar Jafarzadeh, and Kolsoum Ghazanfari

7.1 Introduction

Rumi, also called Mowlana ("our master"), is one of the most influential poets and mystics of the Islamic world. He was originally named Jalāl ad-Dīn Muhammad Balkhī: born in Balkh, a city of Greater Khorasan in ancient Persia (today mid-northern Afghanistan). He was a Persian poet and mystic as well as an Islamic *faqih* (jurist) and theologian who lived in the seventh century AH[1] (1207–1273 CE).

Rumi's influence transcends national borders and ethnic divisions: Iranians, Tajiks, Turks, Greeks, Pashtuns, Central and South Asian people have greatly appreciated his spiritual legacy for the past seven centuries. Rumi's work has been translated into many languages and his spiritual poetry, which was very close to earthen nature, has gained respect worldwide. For instance, Mahatma Gandhi used one verse from Rumi (about Moses and the Shepherd's story) several times in *Masnavi* (Yoosefi, 2004, p. 229). The Catholic Pope John XXIII stated in 1958 CE: "In the

[1] AH refers to the Islamic Lunar Hijri calendar. It means "Anno Hegirae," Latin for Arabic "in the year of the Hijra" (هجرة), that is the emigration of Prophet Muhammad from Mecca to Medina and his arrival on 24 September of the year 622 CE. This date marks the beginning of the Islamic lunar calendar as well as of the solar Iranian calendar and the Ottoman-Turkish Rumi calendar. – CE means "common era," a neutral name for the globally spread Gregorian Calendar.

S. K. Alavipanah (✉) · J. Jafarzadeh
Department of Remote Sensing and GIS, Faculty of Geography, University of Tehran, Tehran, Iran
e-mail: salavipa@ut.ac.ir

K. Ghazanfari
Department of History, Faculty of Literature & Humanities, University of Tehran, Tehran, Iran

N. Patzel et al. (eds.), *Cultural Understanding of Soils*,
https://doi.org/10.1007/978-3-031-13169-1_7

name of the Catholic World, I bow with respect before the memory of Rumi" (Halman, 1988, p. 191).

Rumi's thoughts and poetry strongly refer to the divine and to the four elements of nature, including the soil (or earth). It is therefore interesting to analyze the thoughts of this Sufi Islamic mystic concerning nature and the soil. We will first give a short summary of Rumi's life and work before analyzing the place of the soil (earth) in his work, especially in the *Divan-e Kabir* where 484 verses mention the word "soil."

7.2 Rumi's Life and Work

Rumi was born in 604 AH (1207 CE) in Balkh. The title Mowlana was given to Rumi by his disciples and followers after his death. His father was Baha-ud-din Walad, also known as Soltan Al-Olama, that is, Sultan of Scholars. He was a great Sufi and religious preacher who argued against the philosophers of his time (Forouzanfar, 2015, p. 38). Because of either a dispute with the ruler or the threat of the approaching Mongols, Mowlana's family left Balkh and went to Konya in Anatolia (from that came the name "Rumi" because he dwelled in a former area of the East Roman Empire). Mowlana's encounter with the mystic Shams-e Tabrizi in 642 AH (1244 CE) created a revolution in his life which caused him to turn even more towards introversion and spiritual refinement (Molavi, 2009). Following the legend, Rumi died as the sky turned deep red at sunset on December 17, 1273 (Barks, 2005). There were minor tremors, like stomach grumblings. "Patience, old earth!" Rumi called out. "You'll have your sweet morsel soon!"

Whether Rumi used the masculine term "Allah," or referred to God as "The Beloved," it is nearness and closeness to God that Rumi was expressing. The English language has no neutral personal pronoun for God. In Rumi's Persian language, "God" has no gender, and Rumi's symbolic portrayal of God used images of the Lover, and the Ocean, as often as the King (Arberry, 2000). Rumi's *Masnavi* is probably the longest mystical poem ever written by a single author from any religious tradition. It consists of about 26,000 verses and is divided into six books.

Some of Mowlana's poetry is like a deep-sea in terms of mysticism and philosophy of nature, consciousness and intuition, and this spectrum is what Einstein regarded as the basis of the "true knowledge" of man (Planck, 1932). This is why Canadian psychiatrist and philosopher Richard Maurice attributed a kind of "cosmic consciousness" to Rumi (Alavipanah, 2016).

Rumi's impulse goes earthwards, he was no escapist. He was in search of a "grounded spirituality" and an immanent transcendence, and his loving attitude towards nature can be understood to be more *Jamal* (feminine, receptive) than *Jalal* (masculine, commanding). But as Rumi himself says hundreds of times, there is little one can *say* about love. It has to be lived, and it is always in motion (Barks, 2005). Rumi has been perceived like a sparkling spring that is welling over every moment with the help of God and divine revelation. Such a passion in Rumi's works has been inspiring people for centuries.

7.3 The Four Elements

Rumi's interest in nature is particularly pronounced. The four elements of water, fire, wind, and soil (earth) and their variations occur with high frequency in his works (Ali-Aghdam, 2013). His approach was an integrative one, aimed at the convergence (and implying basic unity) of what we today call science and technology, and art and mysticism. He simultaneously addressed the mystical and divine dimension and the concrete material dimension, with no emphasis on distinguishing between them (Fig. 7.1).

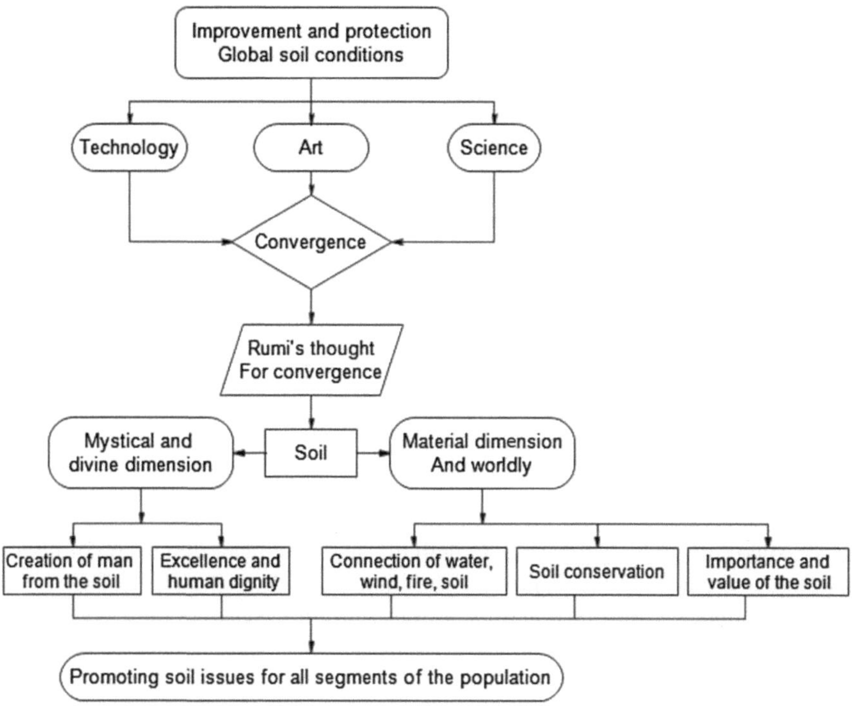

Fig. 7.1 Omnipresence of the immaterial dimension and inherence of art in Rumi's thoughts on soil. (Figure first published in Seyed Kazem Alavipanah, Jafar Jafarzadeh, Kolsoum Ghazanfari (2018): Soil in Rumi's thought, Conference paper of 21st World Congress of Soil Science at Rio de Janeiro, Brazil, August 2018)

The word soil (*khāk*) is used in 484 verses of the *Divan-e Kabir*. It has several meanings like soil, mud, earth, Earth (the whole physical world), and the human body; the latter was based on Rumi's Islamic belief that the body is a form of soil. Rumi referred to the fertility and color of soil in spring and considered its capacity to support growth (Dezfulian & Rashidi, 2013, p. 80). Rumi believed that soil was one of the main elements of nature that should be preserved and *respected* due to its central role in the growth of plants and other living organisms. In Rumi's verses in *Divan*, the combination and mixture of different elements with soil creates food for humans and other beings:

> *And after rain falls on the soil, we see*
> *Fruit, vegetables, and herbs grow healthily;*
> *When men eat vegetables, this doesn't cease,*
> *They gain contentment, joy, and inner peace.*
> *When joy and pleasure fill your soul, then you*
> *Will gain beneficence and kindness too—*
> *Our bodies will gain food that's spiritual,*
> *If we aim higher than what's sensual.*
> (Masnavi, Book 2, p. 65, verses 1097–1100, Rumi, 2007)

For Rumi, the fruits of the earth could be "Manna" in the spiritual sense, in addition to physical nourishment. In these verses there is a fusion (or non-separateness) of symbolic and concrete expression.

Ancient Persians believed that the world consists of four basic elements, one of which is the soil (earth). In Mowlana's works we find soil called the mother:

> *I [Allah] am above all four mothers (elements) and also taller than the seven abba (heavens). I was a precious gem that came to earth.* (Molawi, *Shams Tabriz Ghazaliat*, verse 1390, transl. by authors)

In another instance, soil is highlighted most prominently amongst the four elements, the soil that preserves nature and heavenly glory (Fig. 7.2):

> *Water and wind, and soil and fire are all drunk in divine love*
> *... Soul, intellect and soil and secrets are all drunk of divine love*
> *You should leave the arrogance, and be like the soil,*
> *To see that the particles of the dust of the earth are all drunk in divine love.*
> (Molavi, *Shams Tabriz Ghazaliat,* verse 390, transl. by authors)

Here, soil refers to body and soul, to both the outer and inner dimension of human existence. From Rumi's viewpoint, everything is drunk on the glory of God. The fact we cannot see the spirit and its secrets with the outer eyes does not mean it does not exist and is effective in nature. Rumi indirectly refers to one of the Quranic verses: "They ask you about 'the spirit'. Say: 'The spirit descends by the command of my Lord, but you have been given only a little knowledge'" (Quran [al-Qur'ān/ 17 [القرآن, p. 85).

From Mowlana's point of view, the cosmos is steered and animated by God (Hassanzadeh & Raouf, 2013).

Fig. 7.2 Conceptual graph of the connection of the four elements with the human being. (https://article.tebyan.net/128460)

Earth, water, wind, and fire, his faithful slaves,
Alive with him [God], to us seem dead as graves.
(*Masnavi*, book 1, p. 54, verse 842, Rumi, 2004)

The four elements are all interconnected and create phenomena with the help of each other, and the absence of any of these elements leads to a defect in the functioning of other elements:

All the ivory elements in this world boil like boiled cocoons
No fire, no soil, no air in this world is not sustainable
Sometimes the soil became a plant
And sometimes the water went off and went to the sky.
(*Masnavi Maanavi,* Ghazal 202, transl. by authors[2])

[2] The authors of the article have also used the following website to translate the verses of Masnavi: http://www.masnavi.net/

The combination of various elements with soil creates food for humans and other beings, as well as the creation of happy colors that can overcome human grief.

7.4 Soil and Mankind

Soil is part of the human life cycle. After death, we return to the soil (Alavipanah et al., 2017). From Rumi's point of view, a man in this world must act very consciously, because this world is like a farm where whatever you plant will produce result in the hereafter.

In this context, using a subtle and elegant allegory, Rumi referred to the aspect of trusting in soil and how it multiplies what grows in it:

> *Such ground is pure no matter what you sow,*
> *Without defect plants will quickly grow,*
> *Its soundness comes from the primordial trust*
> *That's why it gets its light from the All-Just,*
> *But spring must give a signal from the Lord*
> *Before the soil reveals its hidden horde.*
> (*Masnavi*, book 1, p. 34, verses 513–515, Rumi, 2004)

Soil, the divine creation, is the sphere where all the beings of nature develop and grow through it and eventually return to it in the process of decomposition and restoration. For Mowlana, the soil was the beginning and the end of creation. Rumi celebrated this existential fact, in contrast to which all human pride is meaningless:

> *My original lineage (is derived) from earth and water and clay: God gave unto water and clay a soul and heart.* (In many verses of the *Holy Quran*, including verse 27–28 of chapter 15, verse 54 of chapter 20, verse 70–71 of chapter 38 and similar verses, the foundation of human development is considered to be soil and mud.)

> *To earth also will return this earthen body of mine; to earth thou likewise wilt return, O terrible one.*
> *Our origin and the origin of all the proud is from a piece of earth, and (there are) a hundred signs thereof;*
> *For thy body receives support (sustenance) from the earth, and from earthly nourishment, thy neck is contorted (with scorn and arrogance).*
> *When the spirit departs, it (the body) will again become earth in the dreaded and horrible grave.*
> *Both thou and we and all who resemble thee will become earth, and thy power will remain no more.*
> (*Masnavi*, Book 4, verses 2313–2317[3])

We may conclude that soil (and water) are so essential that everything is originated from and will return to it.

> *O thou who art a part of this earth, do not lift up thy head (in rebellion); when thou seest the decree of God, do not withdraw (from it disobediently).*

[3] https://books.google.com/books?hl=en&lr=&id=F4c0DwAAQBAJ&oi=fnd&pg=PP1&dq=rumi+masnavi&ots=LWs1Hzv98y&sig=QvFEvPF2ZWWf2qNkBrX_V-zxJm8#v=onepage&q=rumi%20masnavi&f=false

Since thou hast heard "We created thee of dust," (know that) He (God) hath required thee to be (humble and submissive as) dust: do not avert thy face (from Him). (God saith), "Mark how I have sown a seed in the earth: thou art dust of the earth, and I have raised it aloft.
(*Masnavi*, Book 3, verses 453–456)

And:

What is the link between us and that place? (...)
It radiated beams out from the earth.
We were on earth, but of earth unaware,
Heedless of all the treasure buried there.
(*Masnavi*, book 1, p. 164, verses 2661–2665, Rumi, 2004)

Mowlana's attempt to discover the truth of humans and the world emphasized on the one hand their earthly form and their feeding from the soil. On the other hand, he aimed at transcending the material forms and pointed to the spiritual aspect. This implies renouncing material benefits (except enough to live) and avoidance of extravagance and exploitation of the material world.

7.5 Human Soil Eating

The notion of human soil eating (*Khak-Khwari-e Ensan*, eater of earth) that Rumi proposed also has a symbolic meaning of connecting humans with the whole world. Rumi asserted that nutrients are passed from soil to plant to human, or from soil to plant to animal to humans as part of the food web. And it seems that he saw in that a connection between the idea of primordial matter of the created universe, and the life sustaining function of soil.

Rumi also promoted the soil as a cleanser, believing that soil cleans the seeds, and this purity causes germination of the seeds (Fig. 7.3). Indeed, the ideas of humbleness and receptivity are addressed here whilst merging concrete and symbolic meaning.

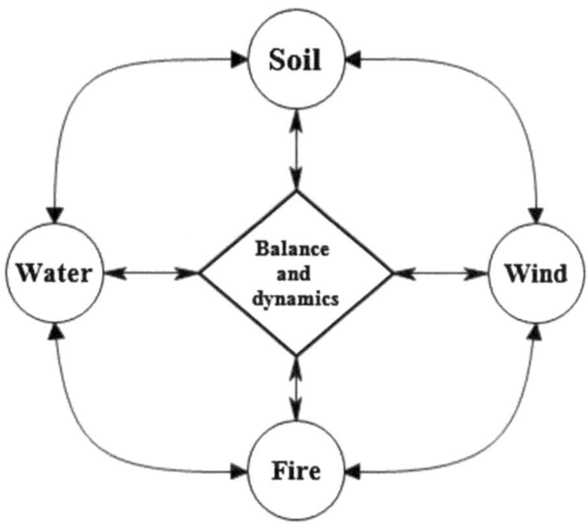

Fig. 7.3 Balance and dynamics of the four elements

Clean yourself and be humble like soil
So that plants grow from your soil.
(*Masnavi Maanavi*, verse 251, transl. by authors)

In Rumi's poetic view, the sweat-soaked earth is flourishing in gold and in the presence of humans, the soil takes in the soul and is blessed with the blessing of the soul. The human seed *daneh-ie adami* (the essence of human existence) is then concealed in the matrix of his/her soil, and rain and water should be brought to it – the water being meant physically to wet the soil and symbolically to soak the germs of human essence with love to let them grow and emerge from the general matrix named their soil:

Have mercy on your being soil because you are
Weaker than these four elements that you
Can create in the soil in a revolutionary way.
(Molavi, *Shams Tabriz Ghazaliat,* verse 308, transl. by authors)

These grains are kept in the basement / Waiting for gentle rains from the Saba wind.
(*Shams Tabriz Ghazaliat*, verse 11, transl. by authors)

Oh God, your kindness gives us life-like water / It's time to make us gracious and kind.
(Molawi, *Shams Tabriz Ghazaliat*, verse 399, transl. by authors)

The consistency and persistence of human life on earth implies the unification of all the elements of life. Not only humans but all other creatures are dependent on the soil because of their vital needs. Therefore, they are all together in the web of life. In this regard, humans are no better than all other creatures, their soil-ness being the basis of their required humbleness. Rumi was against unrealistic ideas of human supremacy. Instead, there was the profound belonging together of all life on earth and the unity of life in the mind of Allah and in their ecological interactions. In Mowlana's implicit theology and philosophy, only God is superior, and all creatures are basically equal and thus have to be humble and collaborate in a "collegial" way, with the ecological and social implications that this may have.

In Rumi's view, soil is the main substance of all creatures of the natural world, the bedrock of their development and the only source of physical nutrition. Humans came from soil, have been sustained by soil through the food it produces, and humans return to become part of the soil after death. The person that eats soil, the *eater of earth* (*Khak-Khwari-e Ensan*), is connected with the universe. There is no distinction and inequality among creatures that implies human beings have a right to dominate other creatures and destroy them.

In his understanding of human's double nature of body and soul, Mowlana was influenced by the *Holy Quran*, in which the soul of God is blown into Adam (chapter 15, 29):

This mouth, indeed, has (always) been an eater of the earth (soil); but the earth (soil) that has been colored. This roast-meat and this wine and this sugar are (merely) colored and painted earth, O son.
When you have eaten or drunk (them) and they have become flesh and skin, He gives them the color of flesh, but they are still the earth of (His) street.

'Tis from a bit of earth that He stitches the (body of) clay, and then makes the whole (fabric) a bit of earth again.
(*Masnavi*, Book 6, verses 4705–4710)

It is permissible for humans to use the material world, according to the verses of the Quran (chapter 2, 60; chapter 7, 31). However, Rumi points out that the foundations should not be violated:

Hence (the command) 'Eat ye' is for the sake of the snare (temptation) of appetite; after that (comes) 'Do not exceed': that is temperance.
Eat less of the grain, do not patch (the body) so much (with food): after having recited eat you, recite (also) do not exceed.
(*Masnavi*, Book 5, verses 582–583)

7.6 Conclusion

According to Rumi as based on the above, the existential nature of humans is defined by the earthly and heavenly dimensions that are present in all phenomena (Chittick, 2004; Hayati, 2009). In Mowlana's thought, the components of nature, including soil and the other three basic elements, are considered manifestations of God's existence.

Given the current state of human vision and its destructive effects on soil and the whole of nature, we need attitudes that can improve the relationship between man, soil, and the universe and help us establish a respectful attitude towards it. In Mowlana's metaphysical system, humans include material and spiritual realms. The dependence of human life and other natural beings on soil indicates that humans are dependent on other beings and in this sense, none is superior to the other.

In conjunction with Mowlana's mystical thoughts, man is responsible for preserving, upgrading, and cultivating all creatures towards their true purpose. Human beings are related to nature and coordinated with it; in that they are to be guided by the eternal love of the creator. This chapter describes a cultural effect of Shiite-Sufi Islam on cultural symbolism and metaphors of soil. Such a pattern can change present human views of nature and change our attitude facing nature.

In the poems of Rumi, the interdependent four elements of soil, water, air, and fire (energy) lived vividly. Most prominent among them is soil. In Rumi's view, soil is the mother, the basic substance of all living beings, the substrate of their development and sole source of physical nutrition. He viewed soil as one of the fundamental elements of nature that must be preserved and respected as it sustains all physical life on earth. For him, soil is earthly nature and heavenly glory, both at once. He called soil a "heavenly element," through that hinting at the spiritual factors acting in it. Rumi, also called Mowlana, fused a deep religious attitude into his poetic works with a loving philosophy of nature. It showed the intimate human connection with the material world and the nourishment derived from it, and at the same time, honored the vital force of divine origin in it.

Rumi considered soil to purify everything from real dirt, being the sphere to which all the creatures of the natural world will return. For Rumi, soil was the beginning and the end of all creation. As we approach 750 solar years after his passing, one may be touched by his mystical poetry, and let one's own attitude towards soil and nature be influenced by it.

References

Alavipanah, S. K., Taghavibayat, A., & Behifar, M. (2017). *Soil in Persian poetry and culture.* EGU Conference.

Alavipanah, S. K. (2016). *Demystifying fire and heat phenomenology, from ancient to nowadays* (1st ed.). Tehran.

Ali-Aghdam, M. (2013). *Elements of the Arbeits in Rumi Poetry*. Master's thesis in Persian Literature. Shahid Madani University of Azerbaijan.

Arberry, A. J. (2000). *Discourses of Rumi* (Or Fihi Ma Fihi). Reprinted with permission of the copyright holder. Little Known Publications.

Barks, C. (2005). *Rumi – The book of love. Poems of ecstasy and longing*. Harper Collins Publishers (Australia).

Chittick, C. W. (2004). *Me and Rumi. The Autobiography of Shams-I Tabrizi*. Fons Vitae.

Dezfulian, K., & Rashidi, M. (2013). *Mowlana's look at nature in the sonnets of Shams Tabrizi. The Old Testament of Persian Literature* (pp. 63–83). Research Institute for Human Sciences and Cultural Studies.

Foruzanfar, Badi[c] al-Zaman (1394/2015). *Treaties in search of Mowlana Jalal al-Din Mohammad Mowlav's Life*. Ed. by Joneidi. Tehran.

Halman, T. S. (1988). Jalāl al-Din Rumi: Passions of the Mystic Mind. In E. Yarshater (Ed.), *Persian literature*.

Hassanzadeh, M., & Rauf, S. H. R. (2013). Mowlana's mystical connection with the environment. *Pajooheshname Erfan, 11*, 66–88.

Hayati, Z. (2009). The semiotic study of mutual elements in the depiction of Rumi's poems. *Literary Criticism, 2*(6), 7–24.

Molavi, Jalal al-Din Balkhi (2009). *Shams Tabriz Ghazaliat*, ed. Mohammad Reza Shafiee Kadkani.

Planck, M. (1932). *Where is science going? Planck philosophical essays in English P2P custom Ed V4 [UL]*. W. W. Norton Publisher. OCLC 594139429; LCCN 33027009; IA whereissciencego00plan_0.

Quran (2004). *The Holy Quran* (M. S. Ali, English transl.). Published by Islamic International Publication Ltd.

Rumi (2004). *The Masnavi: Book* one (J. Mojaddedi, English trans.). Oxford University Press.

Rumi (2007). *The Masnavi: Book two* (J. Mojaddedi, English trans.). Oxford University Press.

Yoosefi, G. (2004/1383). *Bright spring; A meeting with poets*. Elmi Publisher.

Chapter 8
The Dogon: The Creation of Consciousness Through Agriculture

Eva Wertenschlag-Birkhäuser

8.1 Introduction

In West African Mali, the Dogon practice the cultivation of the soil related to mythical representations that have been known and used for centuries. Their central myth, shaping the society and culture of this ethnic group until the present day, is kept alive in their agriculture and their rituals. The Dogon's vision of creation symbolically refers to the Neolithic revolution, uniting therefore the present Dogon people to the transition from the culture of hunting and gathering to the culture of farming. According to Dogon's belief, man was recreated by God in order to become a farmer. This brings forth spiritual awakening, seen as a quantum leap of consciousness.

At the dawn of creation, God was expressed in the Dogon myth as a male-female (or hermaphroditic) oneness (see Fig. 8.1); awakening, the male flung the female earth out of himself and made her into his consort. He became the first to open the earth forcefully and copulate with her. Thus, it was that God himself became a farmer. But the creator is equally seen as the millet-seed, which opens its eyes every time the crops ripen. In mythical time, which is also "timelessly" alive in the present, the divine Word entered the human sphere. It penetrated the soil like rain, fertilized it, and transformed man, who until that time, existed only as a raw, uncultivated being. This moment of awakening is called the "upheaval of the time" in Marcel Griaule's authentic report (Griaule, 1970b, p. 24 ff.). In the farmer's daily work and creation-bringing rituals, this Word, seen as light and fertilizing energy, must always be set in motion through participation.

E. Wertenschlag-Birkhäuser (✉)
Research and Training Centre for Depth Psychology According to C.G. Jung
and Marie-Louise von Franz, Zurich, Switzerland
e-mail: eva.wert@gmx.ch

N. Patzel et al. (eds.), *Cultural Understanding of Soils*,
https://doi.org/10.1007/978-3-031-13169-1_8

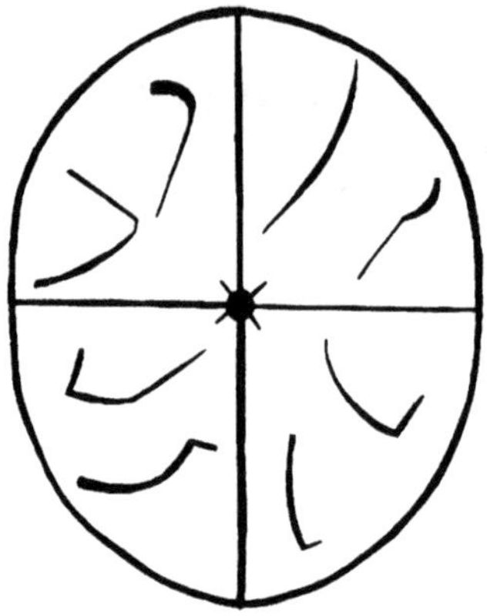

Fig. 8.1 The graffiti shows god's placenta, the state of Amma at the very origin, the male-female oneness, I mentioned at the beginning. Other names are brain, millet-seed, egg, or lap. Each of the four compartments refers to one of the four elements and contains the 64 primordial ideas of Amma. It is said that these ideas manifested when god awakened and became conscious of himself. (Image from Griaule & Dieterlen, 1991; see pp. 62 and 72)

As a consequence, the Dogon agriculture is penetrated and shaped to an infinite degree by mythical images. It is absolutely obligatory to follow these images because they guide the people in keeping the world, in its most objective sense, turning, and moving. For example, sacred graffiti expressing the myth, painted onto the walls of the sanctuary, must be refreshed and repainted regularly so that they remain alive through human participation. It is said that:

> If man neglects to paint it [the image], the rain would cease to fall and culture would end. Men would starve and suffer from thirst being without voice and without Word.[1]

Thus, the millet seeds that the Dogon leave on the roof of this sanctuary become impregnated with the divine Word and are finally mixed with the new seeds for growing.

Are these rituals a bizarre relict? How do they relate to modern man; who must work the ground under the harshest conditions in order to receive nourishment? And why must the farmer always divide his field into 80 × 80 squares of one ell's length, a measurement corresponding to the length from the elbow to the tip of the middle finger of a man? The fields are still structured like that today (Fig. 8.2). Why must the farmers leave a piece of land fallow in the middle of the field for the invisible serpent, who visits the village every night and with its saliva refreshes the Hogon, the head priest?

The information and research underpinning the source of my work and this book chapter originate in 1947, from a conversation between the old Dogon-hunter Ogotemmeli and the French anthropologist Marcel Griaule. This conversation was published under the title *Dieu d'eau* and soon became a world-renowned

[1] Calame-Griaule, G. (1987), p. 189.

Fig. 8.2 A field divided in squares, each virtually one ell. (Travelscape Images/Alamy Stock Photo AC0XED)

book that appeals to readers up to the present day (Griaule, 1970b). During the conversations, the old man recounted the long and dramatic myth and how it formed man from the far-distant mythic past until the present day. The myth shaped the society, including its agriculture and handicrafts. There are several versions of it but I chose this document, because of its firsthand ethnological value.[2] Ogotemmeli discussed the myth and what he was allowed to share with the French anthropologist every evening with the council of the village Sanga's elder (Griaule, 1970b, p. XVI).

Whether the farmers of today remain conscious of the myth and the meaning of the ancestors' knowledge and rituals by which they shape their fields, and whether they believe in the religious ideas with the same intensity as they did 60 years ago, must be left open. It is also difficult to observe if the farmers follow the rituals and the proscriptions so strictly as in the past. More recent field-investigation question it (van Beek, 1991; Gregor, 2011). What I am investigating is the spiritual background, which has shaped, consciously and unconsciously, the Dogon's way of living and thinking over the centuries, their dominant visions revealing why they cultivate the soil in such ways.

[2]The method of Griaule's exploration and receiving data in this conversation, dating 1947, is today criticized in the light of actual scientific standard (van Beek, 1991).

8.2 Who Are the Dogon?

The Dogon in Mali are an ethnic group living south of the Sahel zone at the border with Burkina Faso. Since 1931 they have captured the attention of scientists, many other interested people, and recently even tourists. The Dogon live as if untouched by time in villages of clay houses in a rocky landscape hard to access – above, within, and below the escarpment of Bandiagara.[3] This sandstone cliff drops about 200 meters, stretching over an area of about 200 kilometers. The environment is difficult to cultivate, and the Dogon live out a harsh life. They cultivate a variety of millet, African rice, onions, peanuts, and also cotton. During a short and intense summer rainy season, the rivers and the water holes fill, and from these the farmers fetch water and irrigate their fields by hand (Figs. 8.3 and 8.4). In recent times wells have been drilled by international development aid, but no modern, efficient irrigation methods are found in this part of Mali.

Between 250,000 and 500,000 Dogon were counted in 2010, forming an ethnic group that includes several tribal cultures related to each other. A large part is living in the old traditional spirit handed down orally. The Dogon initiate its members into the knowledge of the ancestors, sharing a common religious belief. The spiritual world system, the metaphysics, and the myth are practiced through a sophisticated, concretely performed symbolic life. Many rituals, feasts, and even gestures in everyday work and life are an expression of their spiritual outlook, and the elders look after these rituals. This traditional mentality holds their society together, keeps the culture alive, and gives the individual dignity. Their symbolic life has meaning and continuity and unites the past with the present, consciousness with the instinct, and satisfies the invisible powers that they serve. This culture has not yet been dissolved by Western, Cartesian, and profit-oriented thinking.

The reader might receive the impression that the Dogon are a completely isolated society. This does not correspond to reality. Since time immemorial the Dogon have lived in a vivid exchange with other ethnic groups like the Fulbe, the Bambara, the Bozo, and others, without losing their characteristic identity, but also sharing common ideas and values. In many Dogon villages, Islam is present. The Muslim Dogon live peaceably with the so-called animists, though the increasing influence of the Wahabis over the last few decades has led to efforts to forbid the mask dances and ridicule the idols. This fundamentalist development undermines Mali's society in general. Besides there are Koran schools in many villages and also schools with a European standard here and there. Of course, many young Dogon are attracted by the modernized Mali and leave their traditional surroundings.

As described previously, the Dogon cultivate the soil by hand along traditional lines. For many centuries it has been a sustainable method to produce nourishment for the population and keep the soil fertile. In the last few decades, the farmers have suffered from climate change and the increasing drought in the Sahel, and famines are forcing families to migrate to Burkina Faso, where they create new villages.

---

[3] Dogon villages are since 1989 under protection of the UNESCO World Heritage. See *Cliff of Bandiagara, Land of the Dogon*. http://whc.unesco.org/en/list/516/video; http://whc.unesco.org/en/list/516/gallery/

Fig. 8.3 Irrigation by hand. (Jamie Carstairs/Alamy Stock Photo BK8CY0)

For the Dogon, the fertility of the earth and the rains they long for are not dependent on outer climatic factors but from a real belief in their own spiritual tasks and attitude. That is, the Dogon must keep the divine Word in motion, and fertility relies on this participation. The Word then impregnates the soil as in mythical time. The rain began to fall in the far past in a specific moment of the myth, when the third Word – being water, light, copper, and the number 8 – reached mankind.

8.3 Research Questions

I worked for over 10 years on a symbolic interpretation of the spiritual background of the Dogon and tried to build bridges between the mythical ideas and the psychology of man in general (Wertenschlag-Birkhäuser, 2022). I am a practicing psychotherapist rooted in the depth psychology of C.G. Jung. Studying the Dogon

Fig. 8.4 Water hole near Tereli. (Jamie Carstairs/ Alamy Stock Photo BHKKET)

documents, we cannot but feel the enormous clash of this worldview with the way of thinking and experiencing of the modern Western mentality, which is penetrating Africa. When I refer to what I call the Western worldview, I mean the current collective zeitgeist and assume that the paradigm of modern science is spread all over the world.[4] These attitudes have a deep impact on thinking and morality in general,

[4] Jung, C.G. (1964) § 112: "And yet, when we approach the world of archaic man, we have the feeling of something prodigiously strange. As far as I have been able to analyse it, this feeling comes predominantly from the fact that the primary assumptions of archaic man are essentially different from ours, so that he lives in a different world. Until we come to know his presuppositions, he is a hard riddle to read; but when we know them, all is relatively simple. We might equally say that primitive man ceases to be a riddle for us as soon as we get to know our own presuppositions. It is a rational presupposition of ours that everything has a natural and perceptible cause. We are convinced of that right from the start. Causality is one of our most sacred dogmas. There is no legitimate place in our world for invisible, arbitrary, and so-called supernatural powers [...]."

even though there are individuals and groups striking other paths. How can an intellectual trained by scientific Western thinking, along with ordinary people living with this same modern mentality, understand the meaning of the Dogon's actions and rituals? Can these ideas and values even be connected with the experiences accessible to the so-called Western man? What are these African ideas and representations saying about the relation of man to the environment and to the invisible world? They touch on problems that many individuals in the Western world are struggling with. Are these visions saying something about factors in the transpersonal psyche, the dynamism of the archetypes, shaping unconsciously the transformation of culture?

One of my hypotheses assumes that these mythical images express the miracle of the origin and evolution of consciousness, which are concretely and sensually performed by the Dogon in their agriculture. The farmers, men and women with different duties, are always involved and connected to the creative activity of nature within themselves and in the environment. They treat the ground and the millet-seed as if partaking in a constantly moving evolution and thus create a flow of consciousness that supports fertility. Marcel Griaule describes the Dogon sense of man:

> Man is a combination of seeds, symbol of life-energy. He is also distributor of forces: in the agricultural act he puts the grains from his collarbones into the earth. [...] Sower of himself, the farmer harvests his own life and yields it symbolically for the future germinations. Man is a living field and an animated granary for the coming and going of the harvests and sowings (Dieterlen, 1999, p. 81).

In addition, one must know that the seeds are divine. Each one is created by God who mixed the Word in it, even God is in it. Therefore, the farmer works on himself and on the divine, that is, something beyond the personal psyche.

Is this a way of seeing a sign of archaic identity of inside and outside that we experience only as children or in pathological states? Or does it express a meaning, which could be accessible to the observer? Can we reach such a union of inner and outer reality? The viewpoint of modern scientific thinking belongs to our modernity and enlightenment and hinders people to go beyond these categories. But if we understand the Dogon vision symbolically, other dimensions of thinking and feeling related to a spiritual reality open. We can compare it for instance with Christian religious ideas.

The seed that falls to the earth, which dies, resurrects, and multiplies, appears as an image in the Gospels. It means that the biological existence of man transforms into a spiritual, timeless essence. Christ compares himself with the dying and resurrecting grain. Those who follow him also become the seeds of resurrection to an eternal life (John, 12, 23–26). In the Gospel of Mark, Jesus tells the parable of the sower and finally concludes:

> What the sower is sowing is the word (4, 14).

I choose two motifs from the conversation between Ogotemmeli and Griaule and comment on them in detail.

The first motif refers to an incisive event in the past when the ancestor blacksmith descended from the beyond, awoke man to agriculture and taught him sowing, saving the seeds, the sacred measurements, the partition of the fields, and much more.

The second motif follows the meaning and activity of the serpent – an embodiment of the spirit of the ancestor – who sneaks through the village every night. It is the guarantor of fertility.

8.4 The Blacksmith Awakens Man to Work, Teaches Him Sowing and Measuring the Fields

The old hunter talks about the celestial granary, which travelled in mythical time from the beyond into the human world. This is a kind of large basket serving as a sort of space shuttle. It brought down the smith, the fire, the iron, and the eight main seeds. It was god Amma, the encompassing creator, who showed the ancestor this specific granary basket.

> 'All this had to be said in words' said Ogotemmeli, 'but everything on the steps [of the granary] is a symbol, symbolic antelopes, symbolic vultures, symbolic hyenas.' He paused for a moment, and added: 'Any number of symbols could find room on a one-cubit step' (Griaule, 1970b, p. 37).

The granary building in the present day and the basket that is used correspond the most to the design of the celestial granary, which is like a three-dimensional mandala (Fig. 8.5). The ancestor smith is not yet completely human in shape, a trait that changes abruptly with the crash of the granary to the ground. Symbolically, this means that the unconscious archetype transforms and crashes into consciousness as an overwhelming inspiration.

> 'The ground was rapidly approaching. The ancestor was still standing, his arms in front of him and the hammer and anvil hanging across his limbs. The shock of his final impact on the earth when he came to the end of the rainbow, scattered in a cloud of dust the animals, vegetables and men disposed on the steps.
>
> 'When calm was restored, the smith was still standing erect facing towards the north, his tools still in the same position. But in the shock of landing the hammer and the anvil had broken his arms and legs at the level of elbows and knees, which he did not have before. He thus acquired the joints proper to the new human form, which was spread over the earth and to devote itself to toil. It was in order to work that his arm was bent,' [...] (Griaule, 1970b, p. 44).

We receive an image of birth! New man is delivered and penetrates the earlier less conscious man. The fresh generation is equipped with a spiritual concept, the third Word, with imaginative representations – symbols – of different phenomena, with the ability to use tools and to work with nature. New man has jointed limbs. He receives the fire and the art of forging. In psychological terms, an awakening of deeper ego-consciousness begins. A new center of awareness is born, breaking through into the human being symbolized by the mandala shaped granary. In yet other variations of this Dogon myth, at the moment when the granary crashes, the

Fig. 8.5 Dogon women working the millet seeds. They are using the traditional basket in the shape of the celestial granary. (dbtravel/Alamy Stock Photo BA222E)

earth starts revolving, the sun rises for the first time, and the rain falls down. We can say: life has a new center of consciousness, which is now freed from an earlier identity with the archetypal background of the psyche.

These are visions of events that might have appeared as historical forces in evolution. Equally, they depict actual developments of consciousness in a child or spiritual transformations in later life or, even collectively, cultural transitions to any new era.

The ability to do work appears as a dominant goal or vision of the creator and seems to belong to awakened man. In the aptly named book of Genesis in the Old Testament, work is presented as a punishment for Adam and Eve. In the concept of the Dogon, work is a progression, the contribution of man to the realization of God. Yet when the granary landed, man received eight grains that were all God-made and which became the basic seeds the farmers cultivate. From this we see that man becomes a protagonist who helps nature to perfection by cultivating what nature gives him. He turns it into nourishment and demonstrates his creativity.

The duty of the patriarch of the clan is to make the ritual of prayer before sowing of the seeds:

> The seeds are like the not yet uttered Word; when the corn sprouts and ripens it turns into the spoken Word; might Amma show us the germination of the millet (Griaule & Dieterlen, 1991, p. 348).

Ogotemmeli continues to tell the myth:

> On making contact with the soil, therefore, the ancestor was ready for his civilizing work. He came down the north stairway, and marked out a square field, ten times eight cubits on each side, oriented the same way as the flat roof on which he had descended, and on which the measurement of landholding was to be based.
>
> The field was divided into eighty times eighty squares of one cubit a side, which was distributed among the eight families descended from the ancestors whose destiny it was to remain on earth. Along the median line of the square from north to south eight dwelling-houses were built, in which the earth was mixed with mud taken from the granary. The smithy was set up to the north of this line.
>
> 'They put celestial mud in the field', said Ogotemmeli, 'and thus purified the soil; and later, as the land was gradually cleared, the impurity of the earth receded' (Griaule, 1970b, p. 44).

Agriculture for the Dogon shows the spread of consciousness through cultivation of the soil. The farmer's practical work is a symbolic ritual act thus asserting the dominance of the light of consciousness over an earlier, darker, primitive state.

From a psychological perspective, the creation of consciousness by man in general invites a dialogue with the unconscious. In turn, this is a cultivation of the natural state of man because he becomes the raw earth to be worked on. As long as unconsciousness prevails, he is, so to speak, "impure." When the increase of consciousness is set in motion, by whatever agency, awareness grows of the central archetype moving life itself, for the Dogon symbolized by the granary and the Word. This is inner spiritual food coming out of the soil of reality of one's own life. I see the motif of treating the earth reflected in contemporary individuals in analysis who often dream of receiving a piece of land as a precious gift and becoming a farmer. The process of individuation is like a form of cultivation, where the dialogue between the conscious and unconscious is paramount. In the psyche, intuitions, and ideas, the seeds must reach the ground in order to touch the soil that makes them fertile. Then the food that one grows in one's own "field" is a sense experience and fully embodied. Often individuals develop in this process a new feeling and relation to concrete nature.

8.5 Measure and Number

According to the Dogon myth, man serves the fertile earth through the creative act of cultivation. His mind and spirit are now structured by nature's laws, and these came down from the invisible world. Brought by the ancestor smith, they are incorporated in the celestial granary. The granary itself is said to be a man of cosmic dimensions, each room of the building representing a part of the body (Griaule, 1970b, p. 38 f.). In many mythologies, this figure is called the Anthropos, the great cosmic man: Purusha in India, Gayomart in old Persia, and Adam in the Kabbalah. With this new spirit, a divine geometry lands, measure and number come into being. In consequence, the measures of the fields and the work in agriculture are not based on the rational principles of man but rely on inspirations sent from the unconscious or the instinct, that is, the invisible world.

The smith reveals the ell, a measurement corresponding to the length from the elbow to the tip of the middle finger of a man. The ell or cubit is one of the oldest measures of cultivated man, known already in Mesopotamia and Ancient Egypt. In Europe the cubit was commonly valid until the nineteenth century. In cities it was an iron bar fixed at a public building.

When the shock of landing gave man elbow joints, it created this natural, yet divine measure. Working on his fields, the Dogon farmer divides the ground by a multiplication of ells, and so the soil receives a human and equally, the divine measure. It relies on the body and on numbers to express an archetypal order. The land must ideally be divided into 80 × 80 making 6400 plots of land each of an ell's length. This order picks up the powerful number 64, representing the primordial ideas in god Amma's brain or placenta (Fig. 8.1).

Sixty-four is a universally meaningful number. It seems to express the primordial order and memory of nature, and it facilitates the transfer of information in the living cell. The growth of certain plants follows the same number pattern as well as the genetic code of man. The DNA proteins are built on a foundation of three and four and sixty-four. The numbers belong, on the one hand, to the memory of the evolution of the physical body. Yet, on the other hand, these numbers and their multiplications express spiritual patterns, which are intuitively found and described by several cultures. Number 64 is, for instance, the key to the Chinese *I Ching*, an oracle book of wisdom and divination with 8 × 8 = 64 hexagrams. It opens the door to transpersonal knowledge, that is, what is constellated in the collective unconscious. In old China sophisticated theories were developed about the numbers of the world structure, and the *I Ching* was part of it (von Franz, 1974, p. 105 f., 22 ff.). Such conceptions are an expression of an intuitively implied order. There, the objective psyche orients and balances itself; its continuous movement is seen in symbols of the *unus mundus*, the idea of an undivided world whose structure orders both psyche and matter (von Franz, 1974, p. 15 f.).

The symbolic thinking of the Dogon continuously refers to the numbers eight and sixty-four and their multiplications. It is significant that 640 bones were found in the tomb of their ancestor Lébé, who underwent death and transformation when the granary crashed to earth. When this happened, he was penetrated by the third Word and then resurrected. But the 640 bones of his renewed skeleton taught people the new order of society. The end of reckoning numbers in the market is also 640, as we hear from Ogotemmeli (Griaule, 1970a, p. 200). There are no activities and handicrafts in the ethnos of the Dogon, which do not refer to sacred numbers. All is measured by archetypal patterns that man feels embedded in. Numbers recur constantly in the findings of depth psychology and symbolize the objective ordering principles of the spirit. These processes of the psyche equate to the numbers at the foundation of matter, as we spoke about in the principles of DNA.

8.6 A Dream

In dreams numbers often indicate how the unconscious restores the balance in an individual. The following dream of a Swiss woman (not the author) expresses such a healing process by numbers. At that moment, in reality, the woman was suffering from a disordered state, and not feeling at one with herself. She received a hint in the dream for her healing. In the dream she was in her house and had to continuously walk in a circle around the central kitchen. It had to be done with regular rhythmic breathing until she entered into a pulse beat of four. When she did that, all of a sudden, she saw two snakes in the center of her house. They were protecting her and she could talk to them. Each had a geometric meander pattern on its skin.

We have several archetypal number-patterns in this dream. By circumambulation in a specific ordered pace of the number four, the dreamer activates her inner center. Circumambulation is an old meditation technique that collects and concentrates psychic energy towards an inner center and activates the healing spot. In an old Chinese Taoist meditation text, called *The Secret of the Golden Flower*, the divine center awakens by consciously circulating around an invisible inner point (Wilhelm & Jung, 1939). The same pattern happens with the dreamer. The psychic center comes alive in the two serpents, who carry the archetypal geometry on their skin. They symbolize the spirit of nature. The dreamer meditates by circumambulation, or circling, and finds her way back to the instinctive order in herself. The motive of the serpent will be taken up again later in this text.

8.7 Ploughing and Weaving as Cultivation of the Word

Traditional Dogon cultivate the soil in accord with measures and numbers given by their myth. Innumerable rules, shaping their everyday life, are respected.

> 'The land,' said Ogotemmeli, 'is cultivated in squares, eight cubits a side, surrounded by embankments of earth.'
>
> The area of each plot, he explained, is that of the flat roof of the celestial granary: and the plot is oriented so that each side faces a cardinal point of the compass (Griaule, 1970b, p. 76)

The first rule is understandable in a practical way, because with this technique the earth does not dry so quickly, and also wind and water erosion are reduced. However, this rational explanation is not satisfying for the Dogon. What matters is that each plot is oriented like the celestial granary landing on earth.

Order spreads through human work. Consciousness expands in accord with the natural laws that were transmitted by the divine granary. This attitude overcomes unconsciousness and the danger of regression and dissolution. Dividing the space into the four cardinal points relates to the foundation of the created world and the orientation of man in this world. This is symbolically meant by making the earth arable.

> 'The old method of cultivation,' he went on, 'is like weaving; one begins on the north side, moving from east to west and then back from west to east. On each line eight feet are planted and the square has eight lines recalling the eight ancestors and the eight seeds.' Furthermore, inside the line the cultivator advances first on one foot and then on the other, changing his hoe from one hand to the other at each step. When the right foot is in front, the right hand on the handle is nearest to the iron, and vice versa when he changes step.

> Cultivation being thus a form of weaving, a field is like a blanket made of eight strips, the black and white squares being represented by the alternation of the mounds made at each step and the gaps between them; a mound and its shadow represent a black square. The whole collection of fields round the village together with the village itself may also be said to recall a large coverlet, the houses with roofs shining in the sunlight being the white squares, and the courtyards lying in shadow, the black. The streets are the seams joining the strips.
>
> 'If a man clears ground and makes a new square plot and builds a dwelling on the plot, his work is like weaving a cloth' (Griaule, 1970b, p. 77).

We hear about similar mythical representations in traditional agricultural techniques and customs in certain regions of South India. The farmer ploughs the field crosswise. Working the soil in this way produces a pattern of a woven fabric. The weaver and the farmer actualize in their activity the primordial order of creation (Högger, 2004). For this culture like for the Dogon, creation is like a cosmic, woven tissue: "In the *Rig Veda* and the *Upanishads*, the universe is envisioned as a fabric woven by the gods. The cosmos, the ordered universe, is one continuous fabric with its warp and woof making a grid pattern" (Kramrisch, 1989, p. 79). In India, ploughing the grid pattern symbolizes a marriage between heaven and earth and a vivifying of the Naga snake. The Indian farmer follows the ordering and orientation giving spirit, the Rita. Rita is the living spiritual order, the balance in the cosmos. It can be compared to the idea of the Word and numbers for the Dogon, present in the woven fabric and in the ploughed field, activating and strengthening the objective, divine order.

Ogotemmeli:

> 'Moreover weaving is a form of speech, which is imparted to the fabric by the to-and-fro movement of the shuttle on the warp; and in the same way the to-and-fro movement of the peasant on his plot imparts the Word of the ancestors, that is to say, moisture, to the ground on which he works, and thus rids the earth of impurity and extends the area of cultivation round inhabited places' (Griaule, 1970b, p. 77).

Therefore, the Word spreads in the moment of cultivating the ground, as a divine, all moving creative dynamism. It is a mystery, and we have no rational way of grasping it.

8.8 Having Measure Versus Excess

The enormous care of the Dogon for the "right" proportions activates the doubts and concern related to the obsession by the modern, Western society for numbers and measures. In scientific thinking everything is measured and quantified. The shadow of this quantification is the excess. How many individuals are workaholics or suffer from burnout nowadays? Man and the environment are in many respects exploited without measure and are suffering. The exploited individual loses his worldview that brings meaning to life and equally his connection to an invisible reality is lost. Concerning nature in general, it is needless to mention the globally growing problems of climate change, polluted air and water, and denuding of the soil's fertility. These phenomena are based on man's lack of proper measure. As a compensation for this situation, increasing regulations are set in place at international negotiations and conferences. Yet as the labyrinth of measures increases, the feeling for binding values gets lost. Ethical values and morality, resting on spiritual structures all but disappear. We have a certain awareness that we profit for instance from products, which are produced by people who get poisoned by the work they do for us. Strangely one knows it, but is paralyzed and inactive.

In the world vision of the Dogon, what is measured has an extra mundane absolute authority and is not build on human rationality. Each individual is bound into this paradigm through an initiation and the actions of his life. Psychologically, their values and measures are based on archetypal patterns, recognized as a reality through the continuity of the myth. The archetypes of the transpersonal psyche are like crystal lattices, which the Dogon observe in their measures. Nature itself equips man with the proper dimensions that his life should take, and they come to him through vision and intuition. In this way the spirit of the instinct sets the boundaries and can balance lack and excess. All these dimensions are respected due to a religious attitude, a respect to values beyond man's consciousness. In contradistinction the dimensions and quantifications used in the current collective Western worldview are based on unregulated instinct that has lost its connection to a nonmaterial, spiritual reality. Materialism, greed, and ruthlessness prevent real feeling and instinct from regulating society through a natural connection to meaning. The consequences are among many other phenomena manifesting in the worldwide problems of soil fertility, the main issue of this volume. Earth is globally no more respected as divine.

Traditional and natural man cannot do meaningless and measureless work and stay healthy in that. The anthropologist Mircea Eliade concludes that man in archaic cultures experiences his work as a reconnection to the moment of creation. His work expresses archetypal powers (Eliade, 1971). The working individual is connected to the mythic past and unites presence and eternity. He is regulated by measures and values from a spiritual reality and receives meaning in life because he is co-responsible for the course of the world.

Two worldviews clash together when the Western collective zeitgeist meets traditional African mentality, a conflict deepening in modern Mali itself. The region of the Dogon seems to be more or less spared. It feels like a terrible blow when we study international reports about the success and development of agriculture in modernized Mali (LIPortal, 2018). Man is qualified according to his per capita income that matches an international standard. The individual turns into a statistical unit and the importance of the state of his soul falls away. The increase of production is especially highlighted in farming which has been recently modernized by western techniques. Efficient irrigation due to new dams on the river Niger raises production enormously, but exploits the soil too much. At the same time, one learns about the increasing shortage of water. The harvest of cotton is boosted thanks to seed improved by genetic engineering. But we hear too, about organic cotton cultivation, which sounds strange, because cotton cultivation in the region of the Dogon was and is always natural and thus organic. The myth of their worldview did not have such greedy standards. Ogotemmeli describes how the cotton seeds are placed in the house above the entrance which symbolizes the feminine sex. The seeds are connected with women's fertility. The thread that a woman spins is said to be the thread along which the celestial granary descended to mankind. This same thread has the quality of copper, and this copper thread is said to be wound around the sun. Therefore, the spinning woman provides the continuous impulse for the movement of the sun. As the sun is a symbol for consciousness, enlightenment of mankind, the woman doing the work of refining the natural product supports the state of the consciousness of mankind. What an enormous sense of feeling she must have!

Intellect based on the rationale of modern science together with an economic attitude alienated from feeling or morality, disconnected from intuition and the invisible world, is a spiritual loss. But in spite of this collective zeitgeist, the individual has always and in all parts of the globe found a way of reaching intuitive knowledge if he or she looks inside. I refer to the dream discussed above. The dreamer discovers an inner order through which she can become centered and balanced. The dream reaches deep into the transpersonal psyche, a reality for all human beings, where she encounters two serpents. From a psychological viewpoint, they are the symbol of an activated independent spirit that has the power to affect consciousness. This could be a spirit of nature that man intuitively knows as a reality.

Now let's look at how the Dogon respect the serpent, guarantor of fertility.

8.9 The Invisible Serpent in the Field

Traditional Dogon know that the success of the cultivation of the soil is not guaranteed by fulfilling the proscriptions and rituals technically. Besides sacrifices and rituals performed around sowing and all the further activities on the field, an active greeting must always be present for the living spirit of the earth and the ancestors, always embodied in a serpent. In mythical times when the celestial granary landed with the third Word, there was a clash between the divine messenger and the village elder, called Lébé, keeper of the second Word. Lébé was swallowed by the messenger, transformed, and spat out again as the new guiding spirit of the community. In his tomb people found his renewed skeleton of 640 bones together with a living serpent. Since then the serpent has always been alive and present in each village. The farmers leave a fallow space in one of the fields as a seat for Lébé.

We hear this in the conversation between Griaule and Ogotemmeli:

> And the Lébé serpent, omnipresent, one yet many like a God, followed each founder [of a village and an altar].
>
> This is why there may be seen, on the rock-table of Upper Ogol to the north-west, a cave called 'Cave of the Steady Women' [the menstruation-hut], out of which he [the serpent] comes in the evening, and to the east another called 'Cave of God's Baobab', to which he returns at dawn. Some, whose eyes are not as those of ordinary men, have sometimes been able to see him in full daylight, coiled, bright beige in colour, in the piece of land called 'Place of Rest', which is left fallow in the field of his priest.

> He has an enormous task to perform; as the mainspring of germination, the protectors of placentas, the purveyor of vital force, he labours day by day with burst of energy at critical times.
>
> 'Every night', said Ogotemmeli, 'he visits his priest, the Hogon, travelling by the western street, which separates Sodamma from Do. In the early morning he returns by the eastern street between Guêndoumman and Do. That is why the Hogon must be at his house at night, and does not go outside the village.'
>
> 'What does he do at the Hogon's?'
>
> Ogotemmeli was silent for a long time. Then in a still lower voice he said, what all the village knew, that Lébé licked the Hogon's body.
>
> 'This', he whispered, 'gives him the strength to live for one day.'
>
> This strength is for the benefit of all. The Hogon is, as it were, the representative of the serpent towards man; but he is also the representative of man towards the serpent. He is entrusted with all the life necessary for men and the land.
>
> For Lébé's saliva has in it the power of moisture, the voiced breath which issues from every mouth. For the whole of one day the old man is endowed with speech (Griaule, 1970b, p. 117f).

The spiritual leader of the community, the Hogon, and the ancestor who was swallowed and newly born by the divine meet every night in a conjunction.[5] The serpent is the bringer of the Word, of divine spirit and truth, fertilizing man and nature. It can be compared with the *soter*-serpent of the Gnostics in antiquity, with whom the holy communion was celebrated (Leisegang, 1940). Or we might think of the serpent symbolizing Christ, which is in the Gospel of John the divine Word.[6] Psychologically, each night there is a renewal of consciousness by the spirit of the transpersonal psyche. We heard that the Lébé transferred the third Word. He facilitated a quantum leap in consciousness and culture. It therefore incorporates a drive within man to awake and develop culture. In modern science we cannot explain how and why man has this amazing drive towards culture, to cultivate nature or build cathedrals. It is a mystery and the serpent is one of the symbols for this energy. It is an impulse in the transpersonal psyche far away from human reason and not controllable. For the Dogon there is no guarantee that the serpent will continue to dwell within the community. Man receiving the truth-bringing spirit and the soil remaining fertile are not given forever. For Ogotemmeli the living reality and presence of the Lébé serpent is a certainty. He has no doubts. But this is not the case for all Dogon. Michel Leiris, one of the French researchers visiting the region of the Dogon in 1931, wrote at that time in his diary:

> At Bamba, one reports that since the French occupation the serpent of the Hogon doesn't come again to lick him, because there is no real Hogon anymore. The myth has survived. What an admirable end for a mythological serpent: it is neither dead, nor dissolved in any scepticism, but only it hides itself, because the time is no longer favourable. We are at the core of this eclipse (Leiris, 1981, p. 147).

The ethnos of the Dogon, like many other traditional societies, exists in a time of cultural upheaval when old beliefs begin to be questioned. Yet the explanation of this Dogon rapporteur assumes that the serpent cannot die. Psychologically it remains an autonomous authority in the psyche, which cannot be destroyed but disappears from consciousness and falls deeper into the collective unconscious. In

[5] The Hogon was until the colonization the political and religious chief in each village.

[6] John 1, 1–18 and 3, 14; Jung, 1978, § 385.

depth psychology such movements are indeed observed and typical. The ego looking inside can encounter it again as a reality of the psyche. We recall the dream of the two serpents discussed earlier. The dream-ego dialogued with the serpent, thus the natural spirit brought inner truth. Taken as an objective document we learn from such a dream that the individual can follow archetypal inner rhythms and in consequence become centered and whole (Jung, 2004, pp. 43–223; Etter, 2003). The Lébé serpent symbolizes an archetypal reality which can move and vivify a whole community. But it has its own rhythm. In the Dogon's view, man's good will and reasonable mind are not enough to shift the zeitgeist. Movements in the transpersonal psyche therefore, impact strongly the collective conditions.

8.10 Final Thoughts

We discussed the meaning of certain rituals with a symbolic value from the central myth of the Dogon and how they shaped agriculture in this remote region of Mali, which is modernizing rapidly towards Western standards. The ethnos remains more or less stable in spirit, and the community works the land in accordance with the myth, with the knowledge of the ancestors and the presence of the beyond that they still feel. It has been possible for them to nourish the population in a certain balance with nature for centuries. Of course, they are also threatened by famines like the population of the whole region belonging to the Sahel zone. But the experience of the living divine Word, transferred by the myth, told by the village elder, continuously activated by the farmer's work gives meaning to their lives. It strengthens and energizes their lives to engage in this hard work. Agriculture is at the core of the relationship and dialogue with the divine. Man cultivating the earth is indispensable for the continuation of the world and for fertility, which is present in the divine Word. The farming has a deep spiritual meaning.

We learned about the measures and numbers brought by the myth and how they are practiced through agricultural customs and rituals to the present day, how a divine dynamism is transferred into the natural order of man's life. In seeing the relationship between the divine and the human, the Dogon farmer sows and harvests within specific boundaries given to him through the limits of measurement, which are based on divine rules. Men and women respect instinct and the ancestors and are not allowed to exploit the ground and the plants. The myth shows agriculture as awakening from an identity with the unconscious and new consciousness begins by working with the soil. If the men and women do not keep the myth in motion, then the light of the spirit would darken. The Dogon does hard work in the knowledge that it has meaning. Human being is embedded in a greater order from which the evolution of consciousness flows.

While investigating the living culture of the Dogon, we can see the deep clash with modern technology. For a scientific mind, it is of course impossible to go back to this kind of thinking which has similarities to old rural customs in Europe. But even though we cannot understand the Dogon completely, we might enter the psychic processes and realities of the human mind through them. From the point of view of depth psychology, respecting these images as a psychic reality and understanding them symbolically makes deep sense for us. We even find a vision of the creativity of the transpersonal psyche moving towards renewal of culture. It is expressed in the image of the serpent which does not die.

References

Calame-Griaule, G. (1987). *(Ethnology and language, The Word among the Dogon.)* Ethnologie et language, La Parole chez les Dogon. Institut d'Ethnologie.

Dieterlen, G. (1999). *(The Dogon. Concept of person and the creation myth.)* Les Dogon. Notion de personne et mythe de la création. L'Harmattan.

Eliade, M. (1971). *The myth of the eternal return: Cosmos and history*. Princeton University Press.

Etter, F. H. (2003). *Mista'peo – der grosse innere Mensch der Naskapi-Indianer* (Jungiana Reihe A, Bd. 12). Verlag Stiftung für Jung'sche Psychologie

Gregor, L. (2011). *(The land of the Dogon. A world in dangers.)* Das Land der Dogon – Eine Welt in Gefahr. https://vimeo.com/60097747

Griaule, M. (1970a). *(God of water. Conversations with Ogotemmeli.)* Dieu d'eau. Entretiens avec Ogotemmeli. Les Editions du Chêne.

Griaule, M. (1970b). *Conversations with Ogotemmeli. An introduction to Dogon religious ideas*. Oxford University Press.

Griaule, M., & Dieterlen, G. (1991). *(The pale fox.)* Le renard pâle. Institut d'Ethnologie.

Högger, R. (2004). Chapter 7: The cross-ploughed field. In R. Baumgartner, & R. Högger (Eds.), *Search of sustainable livelihood systems, managing resources and change* (pp. 166–186). Sage.

Jung, C. G. (1964). *Archaic Man* (CW 10). Pantheon.

Jung, C. G. (1978). *Aion, researches into the phenomenology of the self* (CW 9 II). Princeton University Press.

Jung, C. G. (2004). *Individual dream symbolism in relation to alchemy* (CW 12). Routledge.

Kramrisch, S. (1989). *Kantha Textiles – Handvowen fabrics of India*. Mapin Publishing Pvt. Ltd.

Leiris, M. (1981). *(Africa as a phantom.)* L'Afrique phantôme. Gallimard.

Leisegang, H. (1940). *(The Mystery of the Serpent, A contribution to the study of the Greek mystery cult and its perpetuation in the Christian world.)* Das Mysterium der Schlange, Ein Beitrag zur Erforschung des griechischen Mysterienkultes und seines Fortlebens in der christlichen Welt. Rhein-Verlag.

van Beek, W. E. A. (1991). Dogon Restudied, a field evaluation of the work of Marcel Griaule. *Current Anthropology, 32*(2), 139–167. The University of Chicago Press.

von Franz, M.-L. (1974). *Number and time*. Northwestern University Press.

Wertenschlag-Birkhäuser, E. (2022). *(The third word. The creation myth of the Dogon.)* Das dritte Wort. Der Schöpfungsmythos der Dogon.

Wilhelm, R., & Jung, C. G. (1939). *(The mystery of the Golden Flower.)* Das Geheimnis der goldenen Blüte.

Websites

LIPortal. (2018). *(Mali, Economy &Development, The Country Information Portal.)* Mali, Wirtschaft & Entwicklung, Das Länderinformationsportalportal. https://www.liportal.de/mali/wirtschaft-entwicklung/. Visited November 2018.

UNESCO World Heritage. (1991). *Cliff of Bandiagara, Land of the Dogon*. http://whc.unesco.org/en/list/516/video; http://whc.unesco.org/en/list/516/gallery/

Chapter 9
From Native American Tradition to Modern-Day America: Native Origin Legends That Involve Soil and Earth

Jeffrey Homburg, Sabine Grunwald, and Eric C. Brevik

9.1 Introduction

The creation myth, or cosmogony (a composite of the Greek words *kosmos* and *genesis* – order and birth), is the most important story humans have to tell. That is because it serves as a model for everything we do. Creation myths, like all myths, are universal expressions that aim to explain a culture's place and role in the world, and they still have anthropological relevance to cognitive processes, logic, and thinking in modern culture (Tychkin, 2015). This paper reviews a selection of creation myths documented in North America, especially ones that highlight soil and the earth as central themes to help explain the origin of humans and their world. The aim is also to critically discuss the relevance of Native American earth spirituality in the context of modern-day culture, its role to serve as inspiration in the contemporary environmental movements, and the religious/spiritual domain in North America. A chronological approach is adopted to first present the creation myths followed by a discussion of their relevance in contemporary North American culture. Specifically, we address why Native American perceptions of soil, nature, and

J. Homburg
Catena Affiliates, Tucson, AZ, USA

School of Anthropology, University of Arizona, Tucson, AZ, USA

S. Grunwald
Soil, Water, and Ecosystem Sciences Department, University of Florida, Gainesville, FL, USA
e-mail: sabgru@ufl.edu

E. C. Brevik (✉)
College of Agricultural, Life, and Physical Sciences, Southern Illinois University, Carbondale, IL, USA
e-mail: Eric.Brevik@siu.edu

N. Patzel et al. (eds.), *Cultural Understanding of Soils*,
https://doi.org/10.1007/978-3-031-13169-1_9

spirit have been marginalized and oppressed in current culture and how such oppression has inspired indigenous environmental activism.

There are striking parallels between dreams and myths (Abraham, 1913; Leeming, 2014). Dreams originate from the unconscious. Myths, by contrast, come from conscious thought, but they are often shaped in elaborate and unworldly ways inspired by unconscious thought. C. G. Jung (1875–1961), in his collected works that were released from 1953 to 1979, argued that both dreams and myths originate from the unconscious. In today's post-Freudian era, few would regard dreams as simply untrue stories unworthy of study. Rather, to the extent that the unconscious is believed real, dreams function as metaphorical or symbolic constructs that contain elemental truths about the dreamer and his or her beliefs (Jung, 1953–1979). Similarly, myths supply valuable information about both the inner psyche and worldview of a particular culture (Levi-Strauss, 1963; Malinowski, 1926; Sepie, 2017). Because no one was there to witness and document our origins, humans have a universal need to construct oral histories to explain the purpose of life and our place in this world.

In the classic *Alpha: The Myths of Creation*, Charles Long (1963) identified five basic types of creation myths: (1) creation from nothing (ex nihilo, usually but not exclusively in monotheistic religions like Judaism, Christianity, and Islam), (2) emergence from a hole in the earth, (3) world-parent myths (involve a union, separation, division, sacrifice, or dismemberment), (4) creation from chaos – preexisting, undifferentiated primal elements, and (5) earth-diver myths, which involve diving deep in primordial waters to collect soil for creation. There are, of course, many variations and combinations of these five types (Leeming, 2010). Ex nihilo creation myths are found around the world and are probably the most common of all types. Emergence and earth-diver myths most directly include soil for many Native American creation stories. These stories often involve the use of clay, dust, or mud to make humans. Even the ex nihilo creator uses clay in the later stages of the creation process, and chaos myths are often based on clay as the raw material to be molded (Leeming, 2010). Clay or some earthen material, however, is essential in all earth-diver myths. The Hopi goddess Spider Woman, for example, thought of human beings by singing and then formed them from clay (Titiev, 1948). That soil is ubiquitous in so many early creation myths is not surprising given that it was so familiar in everyday life experiences; for example, its use as a basic raw material for making buildings, paintings, figurines, pottery, etc. and serving as the foundation of the earth's surface, where humans walk, plant crops, and live their lives. Soils are an integral part of human lives in many cultures around the world to this very day, but the connections are often more pronounced in what many would term "less developed" (technologically speaking) cultures (Brevik et al., 2019).

Earth-diver creation myths can be found in many parts of the world, but they are very important in Native American cultures of North America, and they are especially common for Algonquian, Siouan, Iroquoian, and many California groups (Leeming, 2010). A map that shows the locations of North American tribes that have earth-diver myths is given in Fig. 9.1, and a key to the tribal identities is provided in Table 9.1. Earth-diver myths stress the creation of earth rather than the

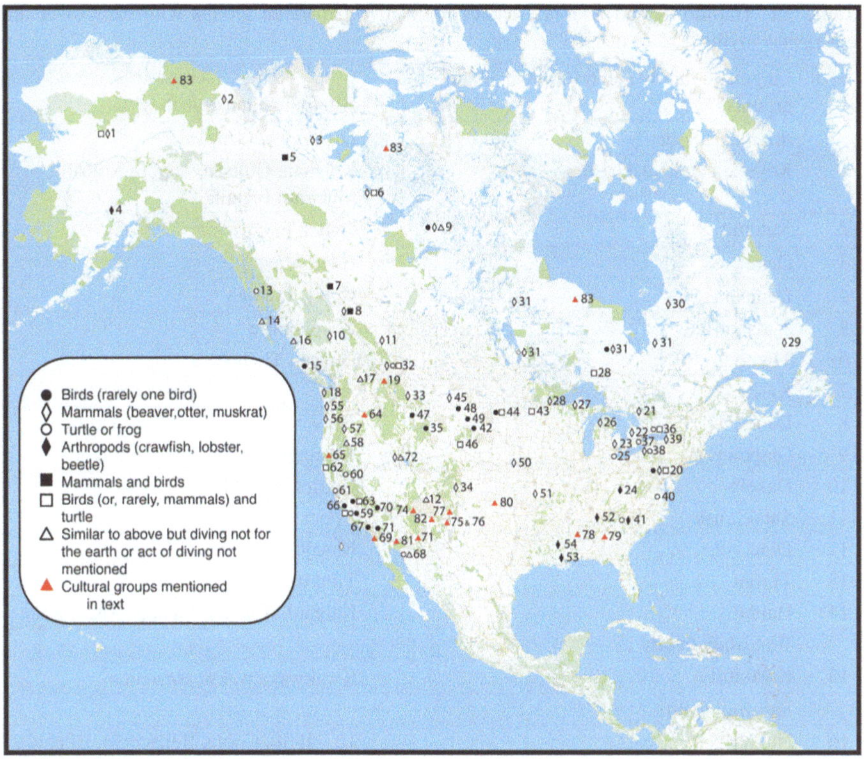

Fig. 9.1 A map showing the locations of various tribal groups in North America that have earth-diver myths. Each symbol with an animal name beside it indicates a tribe with a diver creation myth, with the animal(s) noted doing the diving. The key for which tribe is indicated by each number is given in Table 9.1

larger cosmos, with animals often playing an important role in primeval waters, typically with a dualistic tension of good and evil forces. These myths usually begin with primeval waters and one or more rarely two creators who set out to create earth. Various animals often assist these missions by diving deep to bring up earthen material. North American earth-diver myths emphasize a culture hero or heroine, sometimes a trickster who is mischievous or even amoral. There is sometimes an evil brother who struggles with a good brother for dominance. Algonquian earth-diver myths tend to be post-flood stories, in essence a second creation. The Anishinabe (Ojibwa) have a post-flood story in which the culture hero Nanabozho and a few animals survive to create a new world (White, 1913). Nanabozho dove into the waters to search for mud for a new earth, but the waters were too deep (Leeming, 2010). Several animals tried too but also failed until the lowly muskrat finally took his turn. After a long time, he floated to the surface dead, but carrying bits of earth in his closed paw to make the world. Siouan tribes tell earth-diver myths in which the creative agent is more of a trickster than creator. Coyote is a

Table 9.1 Traditional and alternate names for Native American groups with earth-diver and emergence myths

Map #	Linguistic family and traditional ethnic group(s)	Alternate name(s)/tribes
	Athabascan family	*Dichinanek' Hwt'ana*
1	Koyukon and Kuskoquim	Kolchan, Goltsan, Tundra Kolosh, McGrath Ingalik
2	Kutchin	Gwich'in
3	Dene	Hare, Slavey
4	Dena'ina	Upper Tanaina
5	T'atsaot'ine	Yellowknife
6	Tłı̨chǫ	Dogrib, Tlicho
7	Kaska	
8	Dane-zaa	Beaver, Biever, Dunne-za, Tsattine
9	Chipewyan	Montagnais (not Neenolino Innu)
10	Dakelh	Carrier, Takulli
11	SarsiSalish	Upper Chehalis
12	Dine	Navajo
13	Tlingit	Tlinkit
14	Haida	Kaigani, Xaayda, Xaadas, X̱aad, Xaat
	Wakashan family	
15	Kwakiutl	Kwakwa̱ka̱'wakw, Newettee
	Salishan family	
16	Nuxalk	16 – Bella-Coola, Bellacoola, Bilchula
17	Shuswap and Thompson	Ktunaxa Kinbasket
18	Coast Salish	Upper Chehalis, Tillamook
19	Syilx (or Silx)	Okanogan, Okanagan
	Algonquian-Ritwan family (also includes Cree, Croatan, Gros Ventre, Hathawekela, Kispoko, Lenape, Mahican, Maliseet, Mascouten, Meskwaki, Miami, Mi'kmaq, Missiquoi, Nanticoke, Passamaquoddy, Pennacook, Penobscot, Pequot, Powhatan, Roanoke, Sauk, and Wampanoag)	
20	Delaware	
21	Omàmiwinini	Algonquin, Anicinàpe
22	Ottawa	Odawa, Odaawaa, Adaawe
23	Sauk and Fox	Thakiwaki, Sa ki wa ki, Meskwaki
24	Shawnee	Loyal Shawnee
25	Kickapoo	
26	Neshnabé	Potawatomi, Pottawatomie, Potawatomi
27	Menominee	Menomini, Mamaceqtaw
28	Arishinabe	Ojibwe/Ojibwa, Chippewa, Saulteaux
29	Innu	Montagnais
30	Naskapi	Naskapi Innu (Nascapi, Naskapee, Nascapee)
31	Cree	Swampy, Plain, etc.

(continued)

Table 9.1 (continued)

Map #	Linguistic family and traditional ethnic group(s)	Alternate name(s)/tribes
32	Blackfoot	Siksika (Kainai/Kainah, Peigan/Piikani--, and Amskapi Piikani/Pikuni)
33	Atsina	Gros-Ventres, Aaniiih, A'aninin, Haaninin
34	Hinono'eiteen	Southern Arapaho
35	Cheyenne	Suhtai, Tsitsistas
	Iroquoian family (also includes Cayuga Confederacy, Erie, and Susquehannock)	
36	Huron	Wyandot, Wendat
37	Wyandot	Huron, Petun
38	Seneca	
39	Onondaga, Oneida, Mohawk	
40	Tuscarora	Kautanohakau, Kauwetseka, Skarū'ren'
41	Tsalagi	Cherokee
	Caddoan family	
42	Arikara	Sahnish, Arikaree, Hundi
	Siouan family (also includes Omaha and Yankton)	
43	Santee	Mdewakanton-Dakota
44	Dakota	Minitaree and Wahpeton
45	Nakota	Assiniboin
46	Lakota	Teton Sioux
47	Apsáalooke	Absaroka, Crow
48	Hidatsa	Hiraacá
49	Numakaki	Mandan
50	Iowa	Ioway
51	Quapaw	Arkansas and Ugahxpa
52	Yuchi	Euchee, Uchee
53	Chitimacha	Chetimachan, Sitimacha
54	Alabama-Koasati	Coushatta
	Chinookan family	
55	Chinook	Cathlamet, Clatsop, Lower Chinook, Wahkiakum, and Willapa
56	Cathlamet	Guasámas, Guithlamethl, Kwillu'chini
	Sahaptin family	
57	Molala	Molale, Molalla, Molele
58	Modoc	(Klamoth in OR, Modoc in OK)
	Penutian family	
59	Yokuts	Mariposas (Foothill, Northern Valley, and Southern Valley Yokuts)
60	Miwok	Miwuk, Mi-Wuk, Me-Wuk
61	Patwin	Patween, Southern Wintu
62	Wintu	Northern Wintu

(continued)

Table 9.1 (continued)

Map #	Linguistic family and traditional ethnic group(s)	Alternate name(s)/tribes
63	Maidu	Nisenan, Yamani Maidu, and Konkow (Koyom'kawi/Concow)
64	Mamachatpam	Yakama, Waptailnsim, Pa'kiut'lĕma
65	Specific ethnic group unknown	Northern California coast
	Hokan family	
66	Pomo	
67	Te'po'ta'ahl	Salinan (Miguelino, Antoniano, and Playano)
68	Comcaac	Seri
69	Kumeyaay	Tipai-Ipai, Diegueños, Kamia
	Uto-Aztecan family	
70	Nyyhmy	Western Mono, Mona
71	Tübatulabal and Kawaiisu	
72	Northern Shoshone	
73	Akimel O'odham	Pima
74	Hopi	
	Keres family	
75	Acoma	ákuma, ákomage, Acus, Acux, Aacus, Hacús, Vacus, Vsacus, Yacco, Acco, Acuca, Acogiya, Acuco, Coco, Suco, Akome, Acuo, Ako, and A'ku-me
76	Nawaika	Laguna
77	Zia	Ts'íiy'am'é
	Muskogean family	
78	Choctaw	
79	Muscogee	Mvskoke, Creek, and Muscogee Creek Confederacy
	Kiowa-Tanoan	
80	Ka'igwu	Kiowa, Cáuigù, Gaigwu
	Yuman family	
81	Quechan	Kwtsaan, Yuma
	Zuni	
82	Zuni	A:shiwi
	Eskimo-Aleut family	
83	Inuit	Eskimo

widespread trickster in many Native American myths in the Southwest, Great Basin, and the Plains (Ricketts, 1966). An example is the Navajo, who regard Coyote as an unpredictable and ambivalent being who tests the limits of acceptable behavior (Cooper, 1987). Among the Crow, Coyote sends a duck deep into the primeval sea to find the soil to begin creating the world. Coyote breathed on the mud, and it became earth and then plants, animals, and people (Lowie, 1993). The Kukulik Inuit of St. Lawrence Island in the Bering Sea off Alaska say that Raven acted as an earth-diver to find sand with which to make the earth and separated pebbles from the sand to make humans (Oosten & Laugrand, 2006).

The dive is treacherous, and many divers lose their lives, as plants, animals, and humans are created by Mother Earth. Whether or not the dive is viewed as a metaphor for descending into an unconscious world in search of consciousness, Mother Earth clearly serves as the source of all life (Leeming, 2010). Dundes (1962) views the earth-diver myths as a case of pregnancy envy on the part of males. Versions of the earth-diver myth with the twin motif serve to mediate the duality inherent in life experiences and existence itself. In the Iroquoian myths, for example, the bad twin creates powerful and dangerous animals of the hunt, while the good twin, favored by his mother, creates good domestic animals and edible plants (Fenton, 1962). Conflicts between two opposing forces can thus be expressions of the conflict between choosing a hunter-gatherer versus an agricultural way of life. Myths help humans to mediate between binary oppositions (homogenous ones like summer and winter, land and water, earth and sky, upper and lower, left and right, noble and commoner, strong and weak, elder and younger, etc.; or logically heterogeneous ones, such as stability and change, state and process, being and becoming, synchronic and diachronic, simple and ambiguous, unequivocal and equivocal) (Levi-Strauss, 1955, 1963, 1995).

In contrast, emergence myths involve the gradual creation of particular people in a particular place, a specific emergence site, although these parallel to Mother Earth as in earth-diver myths. For many Native American tribes in the American Southwest, that means the center of the world, the place where their people emerged. In New Mexico, Keresan-speaking people such as the Acoma, Laguna, and Zia say that two sister-spirits lived in a dark underworld where they were taught and fed by a goddess called Thinking Woman (Leeming & Page, 1998). Further west, the Zuni were nurtured by Mother Earth before emerging from the underworld darkness (Cushing, 1896). Hopi, the westernmost Puebloans, believe they were guided by Spider Woman (cf. Arachne in Greek mythology and the Norns in Norse mythology) through three worlds, maturing morally and physically as they traveled, before entering the present world, the Fourth, through a hollow reed and a *sipapu* (a symbolic hole in house floors and kivas, a type of ceremonial structure) through which humans are believed to have emerged from the underworld to the present world (Voth, 1905) (Fig. 9.2). The *Nanih Waiya* Cave, the possible source of the Choctaw emergence myth described by Bushnell (1910; Cherry, 2006), is the end of a long passage where humans and grasshoppers emerged from a large cavern in the earth's interior where Aba, the Great Spirit, created humans by forming them of yellow clay. Creeks of the southeast and other groups are more concerned with their migration account of wanderings after they emerged from the ground (McKee, 1989; Swanton, 1931). Arikara of the northern plains believe they dug their way out, emerging in the east and then traveling west, led and nurtured by Mother Corn to the place where they settled (Parks, 1996). The Kiowa say they were led by ants from the underworld through a hollow log and into the world (Momaday, 1969). An Apache myth describes how buffalo aided underworld beings by using their horns to build a ladder up to a hole so they could enter the world (Leeming, 2010).

Fig. 9.2 A kiva with its various features labeled. Kivas are semisubterranean ceremonial structures used both in prehistory and today by Puebloan groups of the American Southwest. Note the sipapu, the symbolic hole through which humans emerged from the underworld to the present world, in the lower center of the photo

Emergence myths must be ancient, with the mammalian birth process serving as a metaphor. In these myths people gestated in different forms within Mother Earth, often assisted by a midwife-goddess, and struggled through various stages before emerging from the underworld.

9.2 Selected Native American Creation Myths

Selected Native American creation myths that include soil, clay, or earth are summarized in Table 9.2. All tribal names in Table 9.2 are presented as they were recorded in the written records of the myths, but some of these names are no longer used. A key that correlates historical tribal names to modern names is given in Table 9.1.

There are, in a sense, two time periods reflected in Native American creation myths. The "original" myths come from a time when Europeans or other groups from outside the Americas were not known to the myth-makers, and these outside groups do not appear in the myths. Following contact with European explorers and then settlers, some of the myths were altered such that white men and others from outside of North America appear as part of the creation stories (Curtin, 2004). In such cases the white men often end up being regarded as mistakes of creation or socially challenged. See, for example, the Pima myth (Table 9.2), which includes both white (European) and black (African) men in it. Both the white and black men were mistakes made when the creator listened to advice from Coyote, while the

Table 9.2 A summary of selected Native American creation myths that include soil, clay, or earth

Title of myth	Tribe	Type of myth[a]	Summary	References
Tatanka and the Lakota People	Lakota	Emergence	The Great Spirit Skan grew out of stone after the earth was made and created the ancestors of the Lakota. Skan created human bones from stone, human bodies from earth, and human souls from himself, wind, and thunder. The early Lakota lived in the underworld and made their way to the surface through Wind Cave in the Black Hills.	Montileaux (2006)
Creation of the animal people	Okanogan (modern name Syilx)	Chaos	Old One rolled soil, the flesh of Mother Earth, into balls which he shaped into animals. He shaped the final balls into people and blew on the people to bring them to life. Therefore, all living things come from the earth.	Clark (1966)
The well-baked man	Pima	Chaos	The Magician took clay and shaped it into humans who looked like himself. He baked them in an oven, but following the advice of Coyote he did not bake them long enough, so they were underdone. So he placed them on the other side of the ocean; these were white men. He made another batch the same way but over cooked them, again on Coyote's advice. He also placed these on the other side of the ocean; these were the black men. The Magician tried again, and this time he ignored Coyote and cooked them the right amount of time. These became the Pueblo.	Erdoes and Ortiz (1984a)
The good twin and the evil twin	Yuma	Earth-diver	In the beginning, there was only water. Kokomaht, All-Father, the creator, lived deep in the waters. He was two beings, twins. The first twin, the good one, rose from the water, followed by the second evil twin, who Kokomaht named Bakotahl, the blind one, because he was blind. Kokomaht made the four directions – North, west, south, and east. He made the water into a foaming whirlpool and made land. Kokomaht made man and woman by shaping them from mud. These were the Yuma. He then made the Cocopahs, Dieguieños, and Mojaves. Then he rested before also forming the Apaches, Maricopas, Pimas, and Coahuilas. He formed 24 kinds of people in all, leaving the white people for last. The white man was pouting and selfish and had to immediately have anything he saw.	Curtis (1909)

(continued)

Table 9.2 (continued)

Title of myth	Tribe	Type of myth[a]	Summary	References
Old Man Coyote makes the world	Crow	Earth-diver	In the beginning, there was water and Old Man Coyote. Old Man Coyote did not want to be alone, so he walked around and found two ducks with red eyes. Old Man Coyote asked the ducks if there was anything in the world but water, and they replied that they thought there was something deep beneath the water. One of the ducks dived beneath the water and brought back a root. He dived again and brought back a small lump of earth. Old Man Coyote blew on the earth and it grew to become the home for Old Man Coyote and the ducks. He planted the root and it became all forms of plants to provide food. He made hills and mountains, rivers, ponds, and streams. But there were no companions. So Old Man Coyote took up a handful of mud and made people. First, he made men, then he made women.	Erdoes and Ortiz (1984b)
The earth dragon	Northern California coast (exact tribe not known)	Ex nihilo	Two gods, Thunder and Nagaicho, created a world that was suitable for people. Then they made a man from earth and put grass inside him to make a stomach. They formed his heart from more grass, the liver, and kidneys from round pieces of clay, and used a reed for the windpipe. Crushed red stone and water were used for blood. After making man, they made woman by splitting one of man's legs.	Gifford (1930)

Title of myth	Tribe	Type of myth[a]	Summary	References
The white dawn of the Hopi	Hopi	Earth-diver	In the beginning, there was nothing but water on Earth. Two goddesses, Huruing Wuhti, lived on this water world. One lived in the east and the other in the west. Over time the two goddesses made the waters recede eastward and westward, exposing some dry land. In passing over this land the sun noticed there were no living creatures. He mentioned this to the goddesses, and the one in the west invited the one in the east to come talk about it. The two deities created a wren of clay covered with a piece of native cloth. The goddesses sang a song over it and the bird came to life. The deity of the west continued to make many different kinds of birds in the same way. Next the goddesses made different kinds of animals, again in the same way. Then the goddesses decided to create humans. First, the goddess of the east made a white woman and then a white man from clay. The white woman and man were taken over a rainbow to their home in the east. Kóhkang Wuhti, or Spider Woman, who lived at the edge of the water in the southwest, also created a man and woman out of clay and taught them Spanish. Spider Woman continued to create more people and gave a different language to each pair. But she forgot to make a woman for some of the men, which is why there are single men to this day. Spider Woman made one woman that she forgot to make a man for. She told the single woman to go find a man among the single men. The woman did, but the two constantly quarreled. They taught the other couples to quarrel, and that is why there are many arguments between husbands and wives. The eastern and western goddesses tried to create humans who did not quarrel, but every time they came in contact with Spider Woman's people there was trouble. So the goddesses of the east and the west went to live in the middle of the ocean, which is why their kivas are never seen today. But the Hopi still think of the goddesses in their prayers and believe the goddesses still remember them.	Voth (1905)
Creation of the Yakima world	Yakima	Earth-diver	In the beginning, all was water. When Whee-me-me-ow-ah, the Great Chief Above who lived in the sky alone, decided to make the world he went to the shallow places in the water and began to throw up handfuls of mud that became land. The Great Chief made mountains and rocks. He made trees and berries. He made a man out of a ball of mud and told him to hunt and fish. When the man became lonely the Great Chief made a woman and taught her how to dress skins, make baskets from bark and roots, to find edible berries, and to cook the game the man caught.	Clark (1953)

(continued)

Table 9.2 (continued)

Title of myth	Tribe	Type of myth[a]	Summary	References
The Story of Creation	Diegueños (modern name Kumeyaay or Tipai-Ipai)	Earth-diver	Tu-chai-pai, The Maker, and his brother Yo-ko-mat-is, used tobacco to raise the sky because it was too low. Tu-chai-pai then made east (Y-nak) and west (A-uk), and had his brother make north (Ka-tulk) and south (Ya-wak). Tu-chai-pai made hills and valleys with hollows of water so men would have water to drink, and forests so men would not die of cold. Tu-chai-pai dug up some mud and made the Indians first. He did a good job of making the men, but didn't do so well with the women because they were difficult to make. After the Indians, he made the Mexicans, and he finished all his creating.	duBois (1901)
The world's creation	Eskimo	Chaos	Father Raven flew from heaven to Earth, which was sterile. He planted the land with herbs and flowers. Some say that Father Raven created humans from clay. But others say it was by chance, that the first human sprang from a seed pod from a plant that Father Raven had planted, surprising and amusing Father Raven.	Von Franz (1995)

In each case, the soil, clay, or earth connection is emphasized, with most of the myths greatly abbreviated. Additional details can be found by consulting the reference from which each myth was summarized

[a]The myths have been classified as well as possible using the five basic types as identified by Long (1963). However, it must be noted that more information than was available in the myths as recorded and available to be studied for this chapter is necessary to completely utilize Long's (1963) classification

Pima resulted when the creator ignored Coyote and did things properly (Erdoes & Ortiz, 1984a). In the Yuma creation myth (Table 9.2), white men are noted as being pouty, selfish, and greedy (Curtis, 1909).

Some discussions that are common in modern society can also be found in the Native American creation myths. The leading social issue of today that the myths address is probably the age-old battle of the sexes. The Hopi creation myth explains that men and women quarrel because one of the creators made a woman but forgot to make her a man. Because of this, when the woman did find a man they quarreled constantly, and their actions taught other couples to quarrel (Table 9.2) (Voth, 1905). In another example, the stereotypical difficulty men may have understanding women is highlighted in the Diegueños (modern name Kumeyaay or Tipai-Ipai) creation myth, where the male creator did a good job making men, but did not do so well with women because women were difficult to make (Table 9.2) (duBois, 1901). The Yakima creation myth assigns traditional gender roles to men and women, with the men being created to hunt and fish while the women were created to dress skins, make baskets, and cook the game the men caught (Table 9.2) (Clark, 1953). In cases where the creation myths clearly define the order of creation, men are often created first followed by women (Crow, Northern California coast, Yakima) (Erdoes & Ortiz, 1984b; Gifford, 1930; Clark, 1953). There are exceptions, such as the Hopi myth, where women were created first (Voth, 1905) (see Table 9.2), and it is worth noting that some Native American cultures were matriarchal. In the Northern California coast myth, the first woman was created by splitting one of the man's legs (Gifford, 1930) (see Table 9.2), something similar to the Christian view of Eve being created from one of Adam's ribs (woman being created from a part of man).

In addition to the creation myths that have been discussed, there are several quotes attributed to Native Americans and Native American proverbs that recognize the earth as our mother or humans as coming from the earth. Some examples of these quotes and proverbs are given in Table 9.3. While not creation myths, they further reinforce the idea of humans originating from the earth that was held by many Native Americans.

The view of many Native American tribes that they were created from a part of the earth (soil, clay, or mud) or that they emerged from the earth is not surprising, given that Native Americans tend to view themselves as being part of nature. Many in the USA environmental movement have picked up on this and drawn inspiration from Native American views of nature. However, the environmental activists' understanding of Native American perspectives is often quite shallow, missing important aspects of the integrated nature of Native views (Booth & Jacobs, 1990). The traditional Native American view is all -encompassing; they generally see animals, plants, rocks, and other aspects of nature as being alive and equal to humans in the rights and privileges they hold, all aspects of nature are one and interconnected (Corbett, 2006) (also see the Luther Standing Bear quote in Table 9.3). This Native American view of the earth has inspired some in the modern Western environmental movement (Booth & Jacobs, 1990), including the concept of a whole ecosystem approach to sustainable management (Pergola, 1999) and the idea of

Table 9.3 Examples of some quotes and proverbs attributed to Native Americans that recognize the Earth as our mother

Individual	Who they were	Quote/proverb
Chief Joseph	Nez Perce leader who lived from 1840 to 1904, famous for his leadership during the Nez Perce war. His native name was Hin-mah-too-yah-lat-kekt.	"The earth is the mother of all people, and all people should have equal rights upon it."
Mary Brave Bird	A Sicangu Lakota writer and activist who lived from 1954 to 2013.	"The land is sacred. These words are at the core of your being. The land is our mother, the rivers our blood."
Leon Shenandoah	A leader of the Onondaga and Tadadaho (chief of chiefs, the spiritual and political spokesman) of the Iroquois Confederacy beginning in 1969. He lived from 1915 to 1996.	"We are made from Mother Earth and we go back to Mother Earth."
Chief Seattle	A Suquamish and Duwamish chief who pursued a path of accommodation to white settlers. The modern city of Seattle, Washington is named for him. He lived from c. 1786 to 1866, and his native name was Si'ahl.	"All things are connected. Whatever befalls the earth befalls the children of the earth."
Luther Standing Bear	A Lakota chief who fought to preserve the tribe's heritage and sovereignty. He was a noted author, educator, and philosopher. He lived from 1868 to 1939.	"So this land of the Great Plains is claimed by the Lakota as their very own. We are of the soil and the soil is of us. We love the birds and the beasts they grew with us on this soil. They drank the same water as we did and breathed the same air. We are all one in nature. Believing so, there was in our hearts a great peace and a welling kindness for all living, growing things."
Native American proverb		"Mother Earth gave us an abundance of blessings to gather along life's path."
Native American proverb		"The Great Spirit is our father, but the Earth is our mother. She nourishes us."

The examples in this table come from Reicosky et al. (2018), Franzluebbers (2018), and Corbett (2006)

"deep ecology" (Booth & Jacobs, 1990). Deep ecology views humans as just one of many species, with the well-being of all being equally important.

Despite this, to a large degree Native American beliefs of creation and perceptions of soil/earth in contemporary North American culture have been muted and marginalized. The reservation system, established in 1851 by Congress's Indian Appropriation Act, meant that tribes were restricted to the poorest soils in the USA. Re-education was imposed onto Native Americans to teach them the ways of the "white man" and integrate them into white society, which speaks volumes of the catastrophic and devastating effects of acculturation on tribes resulting from alcoholism, poor health, few economic opportunities, and life at the fringes of American society.

Only 2.9 million in the USA identified as Native American (often living in tribal territory) out of a total US population of about 321 million in 2015 (Pew Research Center, 2015); this makes Native Americans less than 1% of the total national population. This small number of Native Americans suggests that socially and culturally they are a marginal subpopulation in the USA, subsumed to a large degree by Euro-American colonists and their descendants who do not share their cultural heritage and spiritual beliefs. The contrasts between perceptions of soils, nature, and the spiritual realm of Native Americans and modern North Americans are stark and will be explored further in the following section.

9.3 Relevance of Native American Earth Spirituality for Modern North American Culture

The European colonization of the Americas occurred in different waves beginning in the fifteenth century and left Native American tribes marginalized in contemporary North American culture (Duran & Duran, 1995). According to Heart and DeBruyn (1998), Native Americans experienced massive losses of lives, culture, land and intergenerational trauma, unresolved grief, and social problems as part of the Native American holocaust (Thornton, 1989; Tafoya & Vecchio, 2005). Concurrently with the destruction of native culture, the beliefs and liveliness of creation myths were largely replaced by, or at least integrated with, Western European worldviews brought across the ocean, although isolated pockets of Native cultures survive, particularly on reservations in the less densely populated Western USA. The thoughts of the Enlightenment and modernity as well as Judeo-Christian religious beliefs were antithetical to the beliefs of Native Americans in North America. In contemporary North American culture, religious fundamentalist movements (e.g., Christian Evangelicals and Mormons) are strongly pronounced (Doktor, 2003; Smith, 2000), while an erosion of institutionalized religion and the rise of "being spiritual, but not religious" have also been observed (Bartunek, 2019; Parsons, 2018). The latter movement of personalized and individualized spirituality is motivated by the search for deeper meaning in a materialistic competitive world that is perceived as masculine with pronounced patriarchal social structures (Bartky, 1990; Fernandes, 2003). Native American casino, bingo, and gaming culture on tribal land under the Indian Gaming Regulatory Act of 1988 stand in sharp contrast to harmonized earth-based Native American culture. Religious leaders in some groups like the Zuni of New Mexico have rejected casinos because they realize that they will lead to erosion of their cultural traditions and beliefs they hold dear. Despite the major overall changes caused by acculturation, Native American views do have a foothold within wider American culture through influences on the environmental movement (Booth & Jacobs, 1990; Corbett, 2006).

Some of the North American creation myths that emphasized male superiority expressed it through the view that women are only a part of men (e.g., the earth dragon myth, Gifford, 1930, Table 9.2) or men were created before women implying dominion of men over women (e.g., Old Man Coyote makes the world, Erdoes & Ortiz, 1984b, Table 9.2). In these cases, a sexualized and gendered hierarchical worldview is expressed. However, it should be noted that there were many Native American cultures, not a single culture, and some were matriarchal such as Puebloan groups of the Southwest USA. Similar to some of the Native American creation myths, Greek and ancient European myths have promoted hierarchies of higher and better (male, masculine) versus lower and inferior (female, feminine), with the latter representing fertility, fruitfulness of the earth, and care (Carlson, 1997). For example, in the Greek Eleusinian myth, the Mother archetype represents the feminine (Neumann, 2015). In this famous myth, the mother-daughter relationship is represented by Demeter-Kore (Persephone) juxtaposed to the patriarchal world represented by Zeus and Hades (see Chap. 2). Kore is raped and subjugated by Hades, God of the underworld (Carlson, 1997). This myth exemplifies Western dualism and its oppression of women and nature by masculine power that harms, subjugates, and destroys nature and the feminine. This dualism between masculine versus feminine raises serious concerns about the earth's renewal and soil protection from an ecofeminine perspective (Dilly, 2009). The masculine power is twofold – "the domination of nature by men" and "the androcentric (i.e., man-centered) domination of women." Such claims to power have inspired movements, such as ecofeminism, to deconstruct power dynamics in the name of gender equality and protection of planet earth (Mickey et al., 2017). In North America, the outrage against women's oppression and feminine values has been articulated in the famous essay "Kinder, Küche, Kirche" (children, kitchen, church) by Naomi Weisstein (Rutherford et al., 2010). Hierarchy and patriarchy have been identified as the root causes for women's subordination, oppression, discrimination, and abuse in contemporary American culture in which women are still perceived as inferior to men, and specifically in Native American communities (Heart & DeBruyn, 1998; Sultana, 2010; Traister, 2018).

Ardelt and Grunwald (2018) pointed out that sociogenic models of human development emphasize the influence of social structure and culture on individual development as well as the individual's liberative transpersonal potentials that extend from ego-, ethno-, humanity- to world-centric (McFarland et al., 2013). Only the human- and world-centric ones extend care toward nature, soil, and the earth (see Chaps. 19, 20 and 21), although they do so due to different motivations or drivers. The social structures and memes (i.e., cultural patterns) from the time creation myths emerged in Native American tribes and at present differ dramatically. A renaissance of creation myths in contemporary America seems unlikely from a social constructivist point of view. According to Sapp (2018), creation myths describe the earliest beginnings of our world as we know it, in North America brought alive in dispersed tribal communities, which differs profoundly from the current information-rich, technology-oriented society concentrated in urbanized centers that distance people from nature, soil, and direct experiences living off the land (Grunwald et al., 2017). Although masses of people are connected via the Internet and social media in the current era, spirit, soul, or deeper spiritual meaning and intimate interpersonal connection may not be found in information technology (Suler, 2016; Zuckerman, 2013). In contrast, evidence shows that a personal and direct connection between people and soil/nature evokes belonging and feelings of fulfillment (Gottlieb, 2019). Approximately 82% of the current US population lived in urban areas in 2018, part of the information-rich culture (University of Michigan, 2018).

Rural populations and farm communities are also a major force in North American culture that have been attributed with closeness to soil and the land (Thompson, 2017). According to Hallam (2019), agricultural production in the USA has changed dramatically in the past few decades with rapid industrialization, intensification, increased farm size, declining numbers of farmers, and reduced significance in terms of the contribution of the agricultural sector to gross national productivity. Contemporary farmers in the USA are embedded in a North American capitalist culture that is profit-oriented and individualistic. In an empirical study, farmers in the breadbasket of the Midwestern USA voiced personal closeness to land and nature and a commitment to ecologically oriented production and stewardship ideals. However, the production-oriented pressures of the agricultural industry and livelihood demands drove farmers to make decisions that maximized crop yields and livestock productivity, often at the expense of environmentally sensitive sustainable management (Yoshida et al., 2018). In a meta-analysis, Carlisle (2016) found that many farmers in the USA fail to adopt soil health practices (e.g., crop rotation, conservation tillage, cover cropping) even though they provide on-site and off-site synergistic economic and environmental benefits. Furthermore, studies have indicated that farmers often underestimate the severity of erosion and do not view erosion as a problem unless it leads to economic loss for the farmer (Osterman & Hicks, 1988). Therefore, despite the idealistic view that farmers are prime land stewards that safeguard the environment, there are concerns that other social and cultural values are viewed as more important than soil management and environmentally friendly (e.g., organic, reduced tillage, crop rotations, use of cover crops, integrated crop-livestock systems, and integrated pest management) farm practices. By contrast, some Native American communities, such as the Zuni, were agricultural and worked with nature to produce successful crops (Muenchrath et al., 2002; Homburg et al., 2005; Norton et al., 2007; Sandor et al., 2007). An example of the Zuni working sustainably with nature, rather than against nature, includes taking advantage of natural erosion processes to direct needed nutrients and water onto their fields. The Hidatsa tribe was also agriculturally focused. They used the natural fertility of floodplain and terrace areas to plant their gardens and used the land's topography to control water flow and minimize soil crusting (Wilson, 1987). There are noteworthy examples of cultures throughout human history that have practiced agricultural management that works with rather than against nature, and several of these cultures developed long-term (lasting thousands of years) sustainable agricultural systems (Brevik et al., 2018).

Sustainability and environmental concerns inspired by indigenous beliefs in the sacredness of the land have spurred environmental protests, such as resistance to the Dakota Access oil pipeline crossing from North Dakota to Illinois, USA. According to Lewis (1995), Native Americans recognize a unity in their physical and spiritual universes, which was infringed by Euro-American settlers who wanted to farm, crop, and use the land for profit. Navajo, Hopi, and other tribes lost land and tribes suffered from oil and gas drilling, contamination of land, radioactive waste disposal, and pumping groundwater for industrial use. Native American beliefs have served as a foundation for environmental consciousness, specifically beliefs in the need for balance and reciprocity and the view of earth as a living being (Booth & Jacobs, 1990; Corbett, 2006). For example, the Chippewa's struggle against mining companies and racist attacks, such as being called "Timber Niggers" when they went spearfishing, motivated resistance against oppression and discrimination (Gedicks, 1998). The indigenous environmental movement was born out of struggles against capitalist Western society, mining and destruction of the land, and corporal profit-seeking, in short, not treating the earth as part of a whole that has equal rights to those held by humans. At stake were sovereignty, treaty rights, and fundamental survival of tribal culture (Clark, 2002). Historically, indigenous-inspired environmental activism was more radical than the more predominant American environmental movements that originated in the Western European upper class with a wilderness-centered ideology, romanticized transcendentalism, stewardship of the earth, and transformation into a diverse amalgam including urban and countercultural environmentalists (Silveira, 2000). Recently, a vision was proposed in the USA to address environmental crises (specifically global climate change) and foster conjoint social and economic reforms; this Green New Deal combines environmental, social, and economic system transformations. However, it lacks linkage to indigenous perceptions of the land and spirituality (Pettifor, 2019). Passmore (1980) eloquently pointed out that the Western Judeo-Christian attitude toward nature is one of domination over nature because everything on earth is considered to be for human use. Nature is not considered sacred like in Native American creation myths, and man is considered as separate from nature (dualistic view of nature versus men). Such worldviews in contemporary American culture stand in contrast to the spiritual connection to nature/earth expressed in Native American creation myths in which the spirit world is considered to be as real as the material world. The shamanic, animistic, polytheistic, pantheistic, or mythical beliefs of Native Americans recognize nature/earth as sacred, which inspires an intimate connection and deep relationship with nature and spirits (Brown, 2001; Komjathy, 2015; Skyfire, 2019).

Spiritual/religious beliefs of people and communities inform how they relate to the environment/nature, which is conditioned by the stage of human development from a psychological perspective. How creation myths and Native American spiritual beliefs are interpreted from a human developmental perspective will be explored in the following paragraphs. Human developmental models to explain differences and hierarchies in cognitive, emotional, values, social, consciousness, self-identity, or other are widespread (Kegan, 1979; Wilber, 2007). In developmental psychology, stages are relatively "stable," with pronounced phases in development from infant to child to adult. Psycho-spiritual developmental models suggest that the developmental stage (level) determines the interpretative lens through which a person views the environment/nature/land/soils, other people, and also the Native American creation myths. Developmental models that map out gradual, but stable changes in relationships between self-other, "I," and self-environment were presented by Piaget, Kohlberg, Loevinger, and Maslow (Kegan, 1979). Kegan's (1982) evolution of self-developmental stages grounded in subject-object theory and cognitive development proposed a hierarchy from impulsive, imperial, interpersonal, institutional, and interindividual (Eriksen, 2006). Such personal developmental models grounded in Western psychology focus on the individual development of self in context with others but discard creation myths as stories (i.e., untrue), spiritual beliefs, or magical nonrational thinking. The model of hierarchical complexity (MHC) ranks reasoning stages of development based on cognitive abilities from lowest "calculatory" to "cross-paradigmatic" (Commons, 2007), and creation myths would be devalued as cognitively irrelevant in the MHC ranking. Wade's (1996) holonomic developmental model uses stages of consciousness from lowest "pre- and perinatal," "reactive," "naïve," "egocentric," "conformist," "achievement," "affiliate," "authentic," "transcendent," "unity" to highest "afterdeath" consciousness. In Wade's model, indigenous beliefs of creation myths would be grouped as "naïve" consciousness characterized by superstition, magical thinking, and spiritual rituals.

Gebser's (1986) stage development of consciousness ranges from lowest to highest as "archaic," "magic," "mythical," "mental," and "integral" with creation myths placed at lower stages (magic or mythical), while Freinacht (2017) suggested effective value memes (eVM) from lowest "archaic," "animistic," "Faustian," "post-Faustian," "modern," "postmodern," and "metamodern" which are more complex than other developmental models. Freinacht's eVM consist of cognitive complexity, symbol stages, available experience of higher states of consciousness, and greater depth (i.e., the spectrum of extreme "good," e.g., meditative transformative nature experience; and "bad," e.g., cancer or a car accident as life experiences). The "animistic" eVM assumes that humans partake in a world of spirits and magic and people are not differentiated from nature (earth, soil, water, etc.); for example, "animistic" eVM are found among indigenous healers or modern urban psychics that romanticize the animistic value memes. The "Faustian" eVM declares independence from nature and its spirits where humans and nonhumans reside in separate worlds. For example, Faustian perceptions are found in ritualistic warlord culture, Wiccan, and pagan spiritualities. In contemporary North America, the "modern" eVM predominates with rational thinking from a scientific perspective interpreting the materialistic world that is viewed as "other." In this meme, the exploitation of soil ecosystems is justified from an anthropocentric perspective. The "postmodern" eVM is likewise dominantly found in contemporary North American culture and is concerned with liberal values of gender equality, multiculturalism, sustainability, pro-environment thinking (e.g., Green New Deal), and opposition to environmental degradation and over-consumption – the latter values hold as important in indigenous cultures. The "metamodern" eVM, in its infancy in North America, combines modern and postmodern values; it values inner and spiritual dimensions (e.g., mindfulness meditation or personal spiritualities) and takes a transpersonal stance in a co-created universe ("I am all that arises; you create me as I create you," Freinacht, 2017, p. 323). In the metamodern view, consciousness co-emerges as an unfolding larger cosmic whole that includes beings, earth, nature, soil, and everything else (Freinacht, 2017). The "metamodern" eVM resembles some of the cultural-spiritual values of Native American tribes, specifically the pro-environmental relationship with nature, soil, and the land.

Wilber (2007) suggested the developmental stages of consciousness from "archaic," "magic," "mythic," "rational," "pluralistic," "integral," and "super-integral," while temporary states of consciousness vary along the spectrum from "gross," "subtle," "causal" to "nondual" typical for nature, deity, formless, and nondual mysticism experiences, respectively. Wilber pointed out that an individual can have profound temporary peak spiritual and religious experiences (e.g., nondual, being one with spirit), but they will interpret the experience from the stage of development they are at. Therefore, a spiritual/religious experience related to a creation myth, for example, Old Man Coyote or earth dragon (Table 9.2), would be interpreted by an individual at the "magic" stage as a personal savior (divine, spirit entity) and at the "mythic" stage as communion with spirit or nature. At the "rational" stage, the spirit/divine Old Man Coyote would be interpreted as a humanized figure and the myth belittled as unreal because it does not fit into the rationalistic-naturalistic worldview. At the "pluralistic" stage, Old Man Coyote would be perceived as one out of many spiritual interpretations and experiences (e.g., Christ consciousness or intimate encounter with Spirit), and at the "integral" stage, Old Man Coyote brings humility to one's being perceived as part of a larger interconnected cosmic whole. Only at higher stage levels (pluralistic and higher) are "green" nature equality and environmentalism valued (Wilber, 2017). The magic and mythic stages emphasize the intuitive empathic lived experiences of relating with nature, while the integral and higher stages combine both – subjective empathic relating with nature and cognitive reasoning about nature to make sense of planet earth (e.g., biogeochemical cycling to support soil quality and health and valuation of shared humanity). Thus, the more embodied forms of environmentalism integrate subjective experiential ways of knowing about nature and spirit as well as cognitive forms of understanding nature, self, and everything (Wilber, 2000).

Importantly, Wilber (2000) pointed out that transpersonal and mystical (temporary) states experienced at higher human developmental stages are not the same as those at lower ones. This *pre/trans* fallacy (i.e., confusing or conflating *trans*personal states of higher development with lower *pre*vious ones) suggests that these states of consciousness are not the same (e.g., a nondual state of consciousness and a magical experience of an indigenous person are not identical) in the hierarchical integral developmental model. The fallacy has another interpretation meaning that transpersonal experiences (e.g., nondual states of consciousness, being one with nature) are interpreted as regression to infantile states (e.g., shamanistic states of consciousness being one with spirit or nature; see Chaps. 5 and 19), while experiences due to the incomplete formation of self-boundaries (e.g., psychotic states or magical thinking) are often labeled as transpersonal. In contemporary North American culture, neo-shamanism, as part of new-age spirituality in urban culture, has seen a revival to connect to altered states with a spirit world (Townsend, 2004; Walsh, 2007). However, a large proportion of Americans remain at the ethnocentric stage with tendencies toward patriarchy, misogyny, fundamentalist religions, and pronounced profit and achievement attitudes that view nature as something to be exploited or controlled to one's own benefit (Wilber, 2017). In this view, active engagement in the material world, scientific understanding of the world, and spiritual dimensions are only integrated at higher stage developmental levels in support of integral ecologies (Allison, 2017).

Berry (2009) proposed an ecologically sensitive spirituality to reunite the material and spiritual realms because Western, and specifically North American, culture is suffering from a disconnection between religion that places the locus of value in a transcendent realm and material needs of the earth and its beings (humans and other organisms). This disconnection between the physical and the spiritual has supposedly caused contemporary ecological crises (e.g., global climate change and soil degradation; Mickey et al., 2017). Creation myths provide access to ecological spiritualities, for example, Mother earth (Clark, 1966, Table 9.2). But they are only acknowledged and/or tolerated by individuals at "pluralistic" or higher stages of development that emanate an open and liberal stance toward different spiritualities (e.g., myths that view the earth as mother and sacred that needs to be cared for and protected). At lower than the "pluralistic" developmental stage (e.g., a fundamentalist religious person), an indigenous creation myth or a practitioner of Native American rituals would be refuted as outrageously "wrong" and inferior. Such view of Native Americans as inferior, who supposedly hold erroneous beliefs of nature and creation, is still common in mainstream contemporary capitalist materialistic North American culture as demonstrated by the separation of tribes into reservations under poor living conditions. Note that the stages and their associated worldviews demonstrate how people would understand and interpret an earth-spirituality (e.g., Native American creation myths); however, these worldviews cannot be substantiated by ontological truth claims (e.g., the proof whether one spirituality is more true than another, such as the claim that Christian religious beliefs are better than or superior to Native Americans spirituality, or vice versa; Ferrer, 2002).

The participatory transpersonal philosophy that is rooted in pluralism, embodiment, and participatory enaction offers an alternative lens to reframe spiritual, social, and cultural discord and hierarchical developmental models (Ferrer & Sherman, 2008; Hartelius & Ferrer, 2013). Ecologies of participation allow one to cocreate life, relationships, and spiritualities focused on deep connection and participation with the environment (Cabot, 2018). The participatory view rejects hierarchies of human development and refutes absolute views (e.g., one spiritual/religious belief as superior to all others; Ferrer, 2017). Participatory knowing occurs through knowing by presence, identity by the virtue of being, and transformation of self, social communities, and the world (Ferrer, 2002). The multiple ways of knowing are supported in participatory theory through the *integration* of all human dimensions (body, vital energy, heart, mind/cognition, and consciousness) to embody wholeness (e.g., embodied spirituality) and fullness of being (e.g., bodyfulness; Ferrer, 2006). Embodiment embraces the equiprimacy principle, which gives equal weight to all human dimensions without allocating supremacy to one or the other. For example, in this view, a Native American shamanic journey to connect with the spirit world and the earth would be equally valued when compared to a spiritual nature experience by a Christian or Buddhist. Any spirituality that evokes care of soils, nature, and the land would be equally valued as long as the spirituality's aim is to bring forth "good" (positive) consequences for the environment and minimize harm for people and society (Ferrer, 2017). A participatory perspective acknowledges equally – immanent spiritualities (e.g., native earth-based spiritualities), subtle energies (e.g., evoked in shamanistic journeying or liminal nature experiences), transcendent spiritualities (e.g., the godhead or spirit), relational spiritualities (e.g., between people and spirit), or any other spirituality (Ferrer, 2011, 2017). In summary, participatory spirituality accepts different spiritual views and nature experiences irrespective of their Native American, Western-European, or other roots.

Adams (2010) provided an ecopsychological account of nature's participatory psyche that goes beyond the dominant traditional Western worldview of nature which views the natural world as less valuable and important than one's own self-centered being. Participatory knowing opens the mind of nature in support of health and well-being in the shared earth community. According to Abram (1996), indigenous people intimately experience the natural world. Being attuned to nature's consciousness is integral to their sense of reality, their place in the community and the cosmos. From a participatory perspective, being one with nature, connecting with spirit or nature, relating to the environment out of deep care for Mother Earth, or viewing clay soils as sacred are multiple possible spiritualities that are not ranked in hierarchical fashion. Hence, the participatory view supports pluralistic perspectives of spiritual ecologies aligned with Native American spirituality and way of life.

9.4 Concluding Statements

Creation myths have existed since the early days of humanity. Many of these myths indicate an intimate connection between humans and nature, including humans and soil or Mother Earth. Such myths are common among the various Native American peoples, with a particular emphasis on earth-diver and emergence myths. The diversity of worldviews, spiritualities, and cultural heritage found in the people native to North America provide abundant opportunities to connect with soil, nature, environment, and the earth. Such spiritually informed connections to nature and the earth are only fully valued by cultures that embrace pluralistic, embodied, and participatory views. These views are becoming more common in the modern USA within the environmental movement, particularly as that movement shifts to holistic ecosystem perspectives of sustainable management. However, views that are less accepting of Native American values (such as fundamentalist religious or the rationalist positivist view that rejects metaphysical/spiritual claims) are still widespread.

References

Abraham, K. (1913). *Dreams and myths: A study in race psychology*. Journal of Nervous and Mental Diseases Publishing Company.

Abram, D. (1996). *The spell of the sensuous*. Vintage Books.

Adams, W. W. (2010). Nature's participatory psyche: A study of consciousness in the shared earth community. *The Humanistic Psychologist, 38*(1), 15–39. https://doi.org/10.1080/08873261003635708

Allison, E. (2017). The relational spiral of integral ecology. In S. Mickey, S. Kelly, & A. Robbert (Eds.), *The variety of integral ecologies* (pp. 161–188). State University of New York Press.

Ardelt, M., & Grunwald, S. (2018). The importance of self-reflection and awareness for human development in hard times. *Research in Human Development, 15*(3–4), 187–199. https://doi.org/10.1080/15427609.2018.1489098

Bartky, S. L. (1990). *Femininity and domination: Studies in the phenomenology of oppression*. Routledge.

Bartunek, J. (2019). *Spiritual but not religious: The search for a meaning in a material world*. Tan Books.

Berry, T. (2009). *The sacred universe: Earth, spirituality, and religion in the twenty-first century*. Columbia University Press.

Booth, A. L., & Jacobs, H. L. (1990). Ties that bind: Native American beliefs as a foundation for environmental consciousness. *Environmental Ethics, 12*(1), 27–43. https://doi.org/10.5840/enviroethics199012114

Brevik, E. C., Homburg, J. A., & Sandor, J. A. (2018). Soils, climate, and ancient civilizations. In W. Horwath & Y. Kuzyakov (Eds.), *Changing soil processes and ecosystem properties in the Anthropocene* (pp. 1–28). Elsevier. https://doi.org/10.1016/B978-0-444-63865-6.00001-6

Brevik, E. C., Pereg, L., Pereira, P., Steffan, J. J., Burgess, L. C., & Gedeon, C. I. (2019). Shelter, clothing, and fuel: Often overlooked links between soils, ecosystem services, and human health. *Science of the Total Environment, 651*, 134–142. https://doi.org/10.1016/j.scitotenv.2018.09.158

Brown, J. (2001). *Teaching spirit: Understanding Native American religious traditions*. Oxford University Press.

Bushnell, D. (1910). Myths of the Louisiana Choctaw. *American Anthropologist, 12*(4), 526–535.

Cabot, Z. (2018). *Ecologies of participation: Agents, shamans, mystics, and diviners*. Lexington Books.

Carlisle, L. (2016). Factors influencing farmer adoption of soil health practices in the United States: A narrative review. *Agroecology and Sustainable Food Systems, 40*(6), 583–613. https://doi.org/10.1080/21683565.2016.1156596

Carlson, K. (1997). *Life's daughter/death's bride: Inner transformations through the goddess Demeter/Persephone*. Shambhala.

Cherry, R. (2006). Insects in the Choctaw emergence mythology. *American Entomologist, 52*(1), 20–22.

Clark, E. (1953). *Indian legends of the Pacific Northwest*. University of California Press.

Clark, E. (1966). *Indian legends from the Northern Rockies*. University of Oklahoma Press.

Clark, B. (2002). The indigenous environmental movement in the United States: Transcending borders in struggles against mining, manufacturing, and the capitalist state. *Organization & Environment, 15*(4), 410–442. https://doi.org/10.1177/1086026602238170

Commons, M. L. (2007). Introduction to the model of hierarchical complexity. *Behavioral Development Bulletin, 13*(1), 1–6. https://doi.org/10.1037/h0100493

Cooper, G. (1987). Coyote in Navajo religion and cosmology. *The Canadian Journal of Native Studies, 7*(2), 181–193.

Corbett, J. B. (2006). *Communicating nature: How we create and understand environmental messages*. Island Press.

Curtin, J. (2004). *Native American creation myths*. Dover Publications, Inc.

Curtis, N. (1909). Creation myths of the Cochans (Yuma Indians). *The Craftsman, 16*, 559–567.

Cushing, F. H. (1896). *Outlines of Zuni creation myths* (Bureau of American Ethnology, Annual Report No. 13). United States Government Printing Office.

Dilly, B. J. (2009). Persephone and Susanna in the garden: Patriarchal seductions of nature and virtue. *Journal of Religion and Society, 5*, 62–75.

Doktor, T. (2003). Religious fundamentalism. In R. L. Piedmont & D. O. Moberg (Eds.), *Research in the social scientific study of religion* (pp. 204–213). Brill Publishers.

duBois, C. G. (1901). The mythology of the Diegueños. *Journal of American Folklore, 14*, 181–185.

Dundes, A. (1962). Earth diver: Creation of the mythopoeic male. *American Anthropologist, 64*, 1032–1051.

Duran, E., & Duran, B. (1995). *Native American: Postcolonial psychology*. State University of New York Press.

Erdoes, R., & Ortiz, A. (1984a). The well-baked man. In R. Erdoes & A. Ortiz (Eds.), *American Indian myths and legends* (pp. 46–47). Pantheon Books.

Erdoes, R., & Ortiz, A. (1984b). Old man coyote makes the world. In R. Erdoes & A. Ortiz (Eds.), *American Indian myths and legends* (pp. 88–92). Pantheon Books.

Eriksen, K. (2006). The constructive developmental theory of Robert Kegan. *The Family Journal, 14*(3), 290–298. https://doi.org/10.1177/1066480706287799

Fenton, W. N. (1962). This island, the world on the turtle's back. *The Journal of American Folklore, 75*(298), 283–300.

Fernandes, L. (2003). *Transforming feminist practice: Non-violence, social justice, and the possibilities of a spiritualized feminism*. Aunt Lute Books.

Ferrer, J. N. (2002). *Revisioning transpersonal theory: A participatory vision of human spirituality*. State University of New York Press.

Ferrer, J. N. (2006). Embodied spirituality, now and then. *Tikkun, 21*(3), 41–64. https://doi.org/10.1215/08879982-2006-3016

Ferrer, J. N. (2011). Participatory spirituality and transpersonal theory: A ten-year retrospective. *The Journal of Transpersonal Psychology, 43*(1), 1–34.

Ferrer, J. N. (2017). *Participation and the mystery: Transpersonal essays in psychology, education, and religion*. State University of New York Press.

Ferrer, J. N., & Sherman, J. H. (2008). Introduction: The participatory turn in spirituality, mysticism, and religious studies. In J. N. Ferrer & J. H. Sherman (Eds.), *The participatory turn: Spirituality, mysticism, religious studies* (pp. 1–78). State University of New York Press.

Franzluebbers, A. J. (2018). Soil proverbs in the United States of America. In J. E. Yang, M. B. Kirkham, R. Lal, & S. Huber (Eds.), *Global soil proverbs: Cultural language of the soil* (pp. 205–207). Schweizerbart Science Publishers.

Freinacht, H. (2017). *The listening society: A metamodern guide to politics*. www.metamoderna.org. Metamoderna ApS. Accessed 28 Nov 2019.

Gebser, J. (1986). *The ever-present origin*. Ohio University Press.

Gedicks, A. (1998). Racism and resource colonization. In D. Faber (Ed.), *The struggle for ecological democracy: Environmental justice movements in the United States* (pp. 272–292). Guilford Press.

Gifford, E. W. (1930). *Californian Indian nights entertainments*. Arthur H. Clark.

Gottlieb, R. S. (2019). *Morality and the environmental crisis*. Cambridge University Press.

Grunwald, S., Clingensmith, C. M., Gavilan, C. P., Mizuta, K., Kastner-Wilcox, R. K., Pinheiro, E. F. M., Ceddia, M. B., & Ross, C. W. (2017). Integrating new perspectives to address global soil security: Ideas from integral ecology. In D. J. Field, C. L. S. Morgan, & A. B. McBratney (Eds.), *Global soil security* (pp. 319–330). Springer.

Hallam, A. (2019). The importance of size and structure in U.S. agriculture. In A. Hallam (Ed.), *Size, structure, and the changing face of American agriculture* (pp. 1–13). Routledge.

Hartelius, G., & Ferrer, J. N. (2013). Transpersonal psychology: The participatory turn. In H. L. Friedman & G. Hartelius (Eds.), *The Wiley-Blackwell handbook of transpersonal psychology* (pp. 187–202). Wiley.

Heart, M. Y. H. B., & DeBruyn, L. M. (1998). The American Indian holocaust: Healing historical unresolved grief. *American Indian and Alaska Native Mental Health Research, 8*(2), 56–78.

Homburg, J. A., Sandor, J. A., & Norton, J. B. (2005). Anthropogenic influences on Zuni agricultural soils. *Geoarchaeology, 20*(7), 661–693.

Jung, C. G. (1953–1979). *The collected works of C. G. Jung* (W. McGuire, Ed., & R. F. C. Hull, Trans., 20 Vols.). Princeton University Press.

Kegan, R. G. (1979). The evolving self: A process conception for ego psychology. *The Counseling Psychologist, 8*(2), 5–34. https://doi.org/10.1177/001100007900800203

Kegan, R. (1982). *The evolving self: Problem and process in human development*. Harvard University Press.

Komjathy, L. (Ed.). (2015). *Contemplative literature: A comparative sourcebook on meditation and contemplative prayer*. State University of New York Press.

Leeming, D. A. (2010). *Creation myths of the world: An Encyclopedia* (2nd ed.). ABC-CLIO.

Leeming, D. A. (Ed.). (2014). *Encyclopedia of religion and psychology*. Springer.

Leeming, D. A., & Page, J. (1998). *The mythology of Native North America*. University of Oklahoma Press.

Levi-Strauss, C. (1955). The structural study of myth. *The Journal of American Folklore, 68*(270), 428–444.

Levi-Strauss, C. (1963). *Structural anthropology*. Anchor Books.

Levi-Strauss, C. (1995). *Myth and meaning: Cracking the code of culture*. Penguin Random House.

Lewis, D. R. (1995). Native Americans and the environment: A survey of twentieth-century issues. *American Indian Quarterly, 19*(3), 423–450. JSTOR. https://doi.org/10.2307/1185599

Long, C. (1963). *Alpha: The myths of creation*. Scholar's Press.

Lowie, R. H. (1993). *Myths and traditions of the crow*. University of Nebraska Press.

Malinowski, B. (1926). *Myth in primitive psychology*. W.W. Norton.

McFarland, S., Brown, D., & Webb, M. (2013). Identification with all humanity as a moral concept and psychological construct. *Current Directions in Psychological Science, 22*(3), 194–198. https://doi.org/10.1177/0963721412471346

McKee, J. (1989). *The Choctaw*. Chelsea House.

Mickey, S., Kelly, S., & Robbert, A. (Eds.). (2017). *The variety of integral ecologies: Nature, culture, and knowledge of the planetary era*. State University of New York Press.

Momaday, N. S. (1969). *The way to Rainy Mountain*. University of New Mexico Press.

Montileau, D. F. (2006). *Tatanka and the Lakota people*. South Dakota State Historical Society Press.

Muenchrath, D. A., Kuratomi, M., Sandor, J. A., & Homburg, J. A. (2002). Observational study of maize production systems of Zuni farmers in semiarid Zuni, New Mexico. *Ethnobiology, 22*(1), 1–33.

Neumann, E. (2015). *The great mother: An analysis of the archetype*. (First published 1955; R. Manheim, Trans.). Princeton University Press.

Norton, J. B., Sandor, J. A., & White, C. S. (2007). Runoff and sediments from hillslope soils within a Native American agroecosystem. *Soil Science Society of America Journal, 71*(2), 476–483.

Oosten, J., & Laugrand, F. (2006). The bringer of light: The raven in Inuit tradition. *Polar Record, 42*(222), 187–204.

Osterman, D. A., & Hicks, T. L. (1988). Highly erodible land: Farmer perceptions versus actual measurements. *Journal of Soil and Water Conservation, 43*(2), 177–182.

Parks, D. R. (Ed., Compiler). (1996). *Myths and traditions of Arikara Indians*. University of Nebraska Press.

Parsons, W. (Ed.). (2018). *Being spiritual but not religious: Past, present, future(s)*. Routledge.

Passmore, J. (1980). *Man's responsibility for nature: Ecological problems and Western tradition*. Duckworth.

Pergola, T. A. (1999). *Managing nature: A look inside the Salmon Arena*. PhD dissertation, University of Washington, Seattle.

Pettifor, A. (2019). *The case for the green new deal*. Verso.

Pew Research Center. (2015). *American Indian and shite, but not "multiracial."* https://www.pewresearch.org/fact-tank/2015/06/11/american-indian-and-white-but-not-multiracial/. Accessed 28 Nov 2019.

Reicosky, D. C., Karlen, D. L., Liebig, M. A., & Levin, M. J. (2018). 21st century perspectives on North American soil proverbs. In J. E. Yang, M. B. Kirkham, R. Lal, & S. Huber (Eds.), *Global soil proverbs: Cultural language of the soil* (pp. 213–230). Schweizerbart Science Publishers.

Ricketts, M. L. (1966). The North American Indian Trickster. *History of Religions, 5*(2), 327–350.

Rutherford, A., Vaughn-Blount, K., & Ball, L. C. (2010). Responsible opposition, disruptive voices: Science, social change, and the history of feminist psychology. *Psychology of Women Quarterly, 34*(4), 460–473. https://doi.org/10.1111/j.1471-6402.2010.01596.x

Sandor, J. A., Norton, J. B., Homburg, J. A., Muenchrath, D. A., White, C. A., Williams, S. E., Havener, C. I., & Stahl, P. D. (2007). Biogeochemical studies of a Native American runoff agroecosystem. *Geoarchaeology, 22*(3), 359–386.

Sapp, R. (2018). *Native Americans state by state*. Chartwell Books.

Sepie, A. J. (2017). More than stories, more than myths: Animal/human/nature (s) in traditional ecological worldviews. *Humanities, 6*, 1–31.

Silveira, S. J. (2000). The American environmental movement: Surviving through diversity comment. *Boston College Environmental Affairs Law Review, 28*(2 & 3), 497–532.

Skyfire, E. (2019). *Journeying between the worlds: Walking with the sacred spirits through Native American shamanic teachings and practices*. Llewellyn.

Smith, C. (2000). *Christian America? What Evangelicals really want*. University of California Press.

Suler, J. R. (2016). *Psychology of the digital age*. Cambridge University Press.

Sultana, A. (2010). Patriarchy and women's subordination: A theoretical analysis. *Arts Faculty Journal, 4*, 1–18. https://doi.org/10.3329/afj.v4i0.12929

Swanton, J. (1931). *Source material for the social and ceremonial life of the Choctaw Indians* (Smithsonian Institute Bureau of American Ethnology Bulletin) (Vol. 103). United States Government Printing Office.

Tafoya, N., & Vecchio, A. D. (2005). Back to the future: An examination of the Native American holocaust experience. In M. McGoldrick, J. Giordano, & N. G. Preto (Eds.), *Ethnicity and family therapy* (pp. 55–63). The Guilford Press.

Thompson, P. B. (2017). *The spirit of the soil: Agriculture and environmental ethics*. Routledge.

Thornton, D. (1989). The Native American Holocaust. *Winds of Change, 4*(4), 23–24, 27-28.

Titiev, M. (1948). Two Hopi myths and rites. *The Journal of American Folklore, 61*(239), 31–43.

Townsend, J. B. (2004). Individualist religious movements: Core and neo-shamanism. *Anthropology of Consciousness, 15*(1), 1–9. https://doi.org/10.1525/ac.2004.15.1.1

Traister, R. (2018). *Good and mad: The revolutionary power of women's anger*. Simon & Schuster.

Tychkin, P. (2015). Myth as an anthropological phenomenon in the context of modern cognitive processes. International conference on research paradigms transformation in social sciences 2014. *Procedia-Social and Behavioral Sciences, 166*, 460–463.

University of Michigan. (2018). *U.S. cities factsheet*. http://css.umich.edu/factsheets/us-cities-factsheet#targetText=Approximately%2082%25%20of%20the%20U.S.,to%20live%20

in%20urban%20areas. & targetText=More%20than%20270%20urban%20areas,million%20inhabitants%2C%20is%20the%20largest. Accessed 28 Nov 2019.
von Franz, M. L. (1995). *Creation myths* (2nd ed.). Shambhala.
Voth, H. (1905). *The traditions of the Hopi. Publications in anthropology* (Vol. 8). Field Colombian Museum.
Wade, J. (1996). *Changes of mind: A holonomic theory of the evolution of consciousness*. State University of New York Press.
Walsh, R. (2007). *The world of shamanism: New views of an ancient tradition*. Llewellyn.
White, J. (Ed.). (1913). *Handbook of Indians of Canada* (pp. 331–335). Published as an Appendix to the Tenth Report of the Geographic Board of Canada.
Wilber, K. (2000). *Integral psychology: Consciousness, spirit, psychology, therapy*. Shambhala.
Wilber, K. (2007). *Integral spirituality: A startling new role for religion in the modern and post-modern world*. Shambhala.
Wilber, K. (2017). *Trump and a post-truth world*. Shambhala.
Wilson, G. L. (1987). *Buffalo bird woman's garden: Agriculture of the Hidatsa Indians*. Minnesota Historical Society Press. (1987 republication of Wilson's 1917 PhD thesis).
Yoshida, Y., Flint, C. G., & Dolan, M. K. (2018). Farming between love and money: US Midwestern farmers' human–nature relationships and impacts on watershed conservation. *Journal of Environmental Planning and Management, 61*(5–6), 1033–1050. https://doi.org/10.1080/09640568.2017.1327423
Zuckerman, E. (2013). *Digital cosmopolitans: Why we think the internet connects us, why it doesn't, and how to rewire it*. W.W. Norton & Company.

Chapter 10
The Yoruba Pedosphere: The Tale of Onilé

Guilherme Augusto Nascimento Sobrinho and Nilton Sousa da Silva

10.1 Introduction

The Yoruba culture has received considerable attention around the Atlantic from both academia and broader society over the last century, mostly due to its prominence in the "Fourth African Diaspora".[1] From the approximately 22 million individuals who were forcefully transported from Africa between 1500 and 1890, over 15.4 million were displaced to the Americas (United Nations Educational, Scientific and Cultural Organization, 2010b). It is estimated that 4.8 million of the 10.7 million slaves that survived the crossing of the Atlantic landed in Brazil, the most relevant arrival location for African slaves (Klein and Luna, 2009). Another estimate indicates that 968,000 Yoruba-speaking Africans were shipped to the Americas from the Bight of Benin between 1651 and 1865, two-thirds of this amount traded between 1776 and 1850 alone (Eltis, 2004).[2] Hence, this relatively recent peak of

[1] For the purposes of this chapter, we are using the term "Fourth African Diaspora" referring to the forced mass transportation of slaves from Africa to the Americas between the sixteenth and nineteenth centuries and its long-lasting global legacies. Other African diasporas have been reported and documented in the literature before and after this period, each with its own genesis and repercussions within and outside Africa (Palmer, 2000).

[2] We recommend the reader visit the website https://www.slavevoyages.org/ where the data used for the quoted estimates can be found, as well as many visual aids to better understand such data.

G. A. N. Sobrinho (✉)
Institute of Radiation Protection and Dosimetry, Rural University of Rio de Janeiro, Rio de Janeiro, Brazil
e-mail: guilhermesobrinho@gmail.com

N. S. da Silva
Laboratory of Psychology and Afrodescendant Information, Psychology Department, Rural Federal University of Rio de Janeiro, Rio de Janeiro, Brazil
e-mail: niltonpsi@ufrrj.br

N. Patzel et al. (eds.), *Cultural Understanding of Soils*,
https://doi.org/10.1007/978-3-031-13169-1_10

Yoruba individuals being transplanted to Brazil was a remarkable event that can shed some light on the relevance of the Yoruba people to the Brazilian cultural "melting pot."

One of the most vibrant and popular Yoruba heritages in Brazil is the *Ifa/Orisha* religious system. Together with the Yoruba language, it constituted an important factor of cohesion and resistance for Afro-Brazilians during slavery and for many years after abolition. Today, it offers a sense of identity and origin for anyone seeking them in the global religious marketplace, despite his or her ancestry (Chirila, 2014). This belief system is based on the concept of *orisha* (Yoruba *Òrìṣà*):[3] apotheosized features of nature or, in some cases, remarkable ancestors (Bascom, 1991; Emanuel, 2000). The *orishas* are quite *archetypal*,[4] representing general aspects, features, or phenomena from nature, as well as personality traits that are common. *Ifa* is the divination system brought to Earth by *Orunmila* (the orisha of wisdom) (Emanuel, 2000). The *Ifa* corpus can be understood as the spiritual philosophy that underlies most Afro-Diaspora faiths (Frisvold, 2016). There are a few Yoruba *orishas* related to chthonic or telluric dimensions that are still being worshiped around the Atlantic. For example, *Nanã* (the queen of the dead), is the lady of the wetlands and stagnant waters, who provided the mud with which *Oxalá* (the creator of the world, commissioned by the even superior God Oludomaré.) formed men. *Omolu* (god of infectious diseases, as well as their cure) is related to the Earth since it is the grave of his victims – the action of this *orisha* is interpreted as a tool of justice from the gods. *Odé* is the god of provision and plenty. Other *orishas* are related to practices performed upon the land. For example, *Ossaim* is the guardian of herbs, forests, and their secrets, needed for most rituals; *Orisa-oko* is an apotheosized ancestor to whom the invention of agriculture is attributed; *Ogum* (a warrior god) is related to iron and everything made or undertaken with it, such as wars, agriculture, and

[3] In Brazil the "èdè Yorùbá" became the "língua Iorubá," with its own spelling and pronunciation characteristics, closer to Brazilian Portuguese orthography and phonetics. For example, given its relevance in Brazil, Rio de Janeiro state and Salvador city granted to this language the legal status of *intangible cultural heritage* (Salvador-BA, 2019; Rio de Janeiro, 2018). We will adopt the Brazilian Iorubá spelling for *orisha's* names, since their narratives and attributes are quite different from those found in Yorubaland, and the English spelling for more general and consensual concepts (such as *Ifa* and *orisha*).

[4] According to C. G. Jung, archetypes are predispositions and tendencies that humans have in their imagination and behavioral patterns. Certain patterns or motifs occur repeatedly and quite universally. Such events or their symbolic representations can affect us powerfully in our body and mind, taking us into an impressing or even "numinous" experience (we may feel and represent them as an undeniable concrete reality). The efficacy of fairy tales, storytelling, mythology, Disney and superhero movies, and virtually anything called classical can be seen as based on this aspect of the human psyche and culture. C. G. Jung (2014) says: "The archetype in itself is empty and purely formal, nothing but a *facultas praeformandi*, a possibility of representation which is given *a priori*. The representations themselves are not inherited, only the forms […]" (CW 9, § 155). And yet: "The archetype is a kind of readiness to produce over and over again the same or similar mythical ideas […]. For when an archetype appears in a dream, in a fantasy, or in life, it always brings with it a certain influence or power by virtue of which it either exercises a numinous or a fascinating effect, or impels to action" (CW 7, § 109).

metalwork; and *Oxaguiã* ("the yam eater") is the *orisha* of livelihood mostly obtained from the land. The *orishas* and their saga, in a broad sense, are used by practitioners and participants of *Ifa/Orisha* cults to explain the order of things and human behavior and, hence, how humans should proceed in life to fulfill their destiny and live longer. These cults pose no soteriological ultimate goals (McKenzie, 1997), as in major western and other religious systems (e.g., Judaism, Islam, Christianity, Hinduism, Buddhism) (Gurtner, 2011; Hassan Khalil, 2011; Maier, 2014; Stetler, 2014).

The *orisha* that we can most closely relate to soils within the Yoruba pantheon is *Onilé*. She is an almost forgotten *orisha* nowadays among the general public.[5] This *orisha* is sometimes regarded as "Mother Earth" (planetary aspect), as the firm ground upon which we stand and to which we land after a flight (geomorphological aspect), as the place where we put our seeds and from which we collect our resources (edaphological aspect), and as the environmental domain of soils ("pedospherical" aspect), just to mention a few.

The present chapter argues in favor of the usefulness of the poetical power of storytelling traditions in order to (1) better imagine our relatedness with soils and (2) communicate our imagination about soils. To do so, the authors will focus on the Yoruba mythic-religious narrative of *Onilé* (a soil deity) as known among Yoruba-descendant religious communities in Brazil.

10.2 The Tale

A story about how the order that we see in nature has been kept since the beginning of time is known among *Candomblé*[6] practitioners in Brazil.[7] A long, long time ago *Olodumaré,* the almighty eternal one and source of all that exists, decided to depart from Earth once and for all. But he could not abandon his creation and leave it on its own. He decided then to leave the Earth under the good care of his children – the *orishas*. The challenge now would be how to properly and fairly distribute the realms of the Earth among them. *Olodumaré* had an idea. He ordered his heralds to announce that the almighty god would have a great feast at his palace, which the *orishas* should attend richly dressed. After all, *Olodumaré* was secretly planning to distribute the treasures of the Earth among them during the party. Afterward there would be lots of food, music, and dance. *Olodumaré*'s heralds spread his announcement everywhere; therefore, everyone prepared themselves diligently.

[5] Despite the prejudice that the *Ifa/Orisha* culture may face by some groups in Brazil, part of the vocabulary related to it – especially the *orishas* names – is quite popular all over the country.

[6] *Candomblé*: An Afro-Brazilian religion derived from the Yoruba *Ifa/Orisha* tradition remarkably characterized by the spiritual possession of its devotees by their respective personal *orisha* (Johnson, 2002).

[7] Other versions can be found to explain such theme, namely, the rule of the *orishas* over specific realms of the Earth, albeit with different "morals" (e.g., Beniste, 2006, p. 65–71).

When the day of the celebration finally came, each *orisha* headed to the palace with great pomposity. Each one was more beautifully dressed than the other, since it was the desire of *Olodumaré*. *Iemanjá* arrived wearing sea foam, her arms adorned with bracelets made from seaweeds, her head girded with a diadem made of pearls and reefs, and her neck bearing a cascading nacre necklace. *Oxóssi* chose a tunic made of soft twigs, embellished with fur and feathers from the most exotic animals on Earth. *Ossaim* dressed himself with a cloth woven with the most fragrant leaves. *Oxum* decided to cover herself with gold, with fresh water on her hair. *Oxumaré's* garment showed all the colors of the rainbow, and he had fresh raindrops on his hands. *Iansã* chose to dress up with a whistling wind, decorating her hair with lightning bolts. *Xangô*, who was very competitive and would never hold back, dressed himself with thunder. Other *orishas* dressed similarly.

Never before had the people of the palace seen such a display of grandeur, luxury, and beauty. Each *orisha* was received with great shouts and amazement. Each *orisha* used the best of their creativity to present themselves in the most beautiful way before the great father. All but *Onilé*, who did not bother to dress herself well. Nothing had interested her, and she did not show off to anybody. *Onilé* was the most discrete of *Olodumaré*'s children. She spent her time alone and was rarely seen by anyone. Every time her brethren gathered at *Olodumaré*'s palace, she dug a hole in the ground and hid herself in it, since it was known that all meetings ended up in feasts, filled with music and dance led by the rhythms of the *atabaques*.[8] *Onilé* was uncomfortable among people. This time was no different.

Once all the *orishas* had arrived, *Olodumaré* ordered that they should be comfortably accommodated on straw mats around his throne. He then told the gathering that they were all most welcome, that all of his children successfully fulfilled his wish so that he could not choose which one was most beautifully dressed. He had all the riches of the world to give to them but did not know how to start distributing them. *Olodumaré* meditated for a moment and announced that each *orisha* had already picked their gifts. By choosing what each considered to be the best in nature to appear before their father, they had themselves already distributed the Earth's realms. Hence, *Iemanjá* would be queen of the oceans; *Oxum*, queen of gold and the rivers; and *Oxóssi*, king of the forests and all its animals, withholding all leaves for *Ossaim*. *Oxumaré* was given the rainbow and the rain, *Iansã* the lightning, *Xangô* the thunder, and so on.

Thus, *Olodumaré* gave each *orisha* a piece of the world, a part of nature, a particular realm. He said then that every time a human had any need, he or she should propitiate to the *orisha* who ruled over the part of nature related to his or her need, with items of their desire. The *orishas*, who so far had listened to everything in silence, started to celebrate, singing and dancing in joy. It was loud in the court, and the feast began. However, *Olodumaré* stood up asking for silence once more, for the division of the world was still unfinished. He said that the most important of the attributions was yet to be bestowed. It was necessary to grant to some of his children

[8] *Atabaque*: traditional tall wooden ritual drum used in *Candomblé*.

rule over the Earth, the world where humans lived and grew their food, drinks, and everything else that was necessary to propitiate the *orishas*. He said then: "The Earth goes to whom is dressed of the Earth." Who would it be? – everyone asked each other. "*Onilé*!" – he answered. "*Onilé*?" – everyone asked, astonished. "How was it possible if she had not even attended?" None of those present had seen her so far. No one had even noticed her absence. "Well, she is among us" – said the father, asking that everyone look to the bottom of the pit. There she was, where she took shelter, dressed in soil, the shy and discrete daughter of *Olodumaré*. There was *Onilé*, with her vest of soil. *Onilé*, who was also named *Ilé*, the country, the planet. *Olodumaré* ordered that all who inhabited the Earth pay tributes to *Onilé*, for she was the mother of all, the shelter, the home. Humankind would not survive without her. Afterall, where did each of the riches bestowed by *Olodumaré* to his children come from? Such was *Olodumaré*'s final command.

Onilé, also called *Aiê* (the Earth), should be always propitiated, so the world of humans would never be destroyed. All present applauded the words of *Olodumaré*. All the *orishas* acclaimed *Onilé*. All humans propitiated Mother Earth. Thereupon, *Olodumaré* departed from the world forever, leaving the governing of all things in the care of his children, the *orishas*.

10.3 Contextualizing the Tale

10.3.1 Arriving at the Myth[9]

The text presented here is our free adaptation into English of the tale as collected by Prandi (2001). A more elaborate English version can be found in Cuoco (2014). This text is a remarkable piece of material since it is a mythological narrative that *explicitly* mentions soils. This is not common in traditional storytelling in Brazil. It is arguably accurate to consider that Brazilian culture is an amalgam of three major cultural groups, namely, Amerindian, Afro-descendant, and West-European descendant. Soils are, at least in part, silent and invisible for most people, especially in today's urban environments, despite the ubiquity of its presence and functions (Food and Agriculture Organization, 2015; Schulte et al., 2014). However, "soil-less cities," as well as urban-rural disconnections, are a relatively recent phenomenon

[9]When we presented this myth at the XXI WCSS (Sobrinho & Silva, 2019), it was argued by a member of the audience whether *more* myths were *needed* or not. We replied that perhaps we do not need *more* myths, but it could be quite useful to remind ourselves of the ones we already have. Most likely this person used the commonsensical meaning of the word myth, which considers myths as anything waiting to be "debunked," i.e., myths are lies. On the contrary, we agree with Joseph Campbell in considering myths to be poetry, instead of lies, designed to bring us into a more deep and meaningful state of consciousness (a.k.a. spiritual), rather than popular superstition (Campbell et al., 1991).

(Elmqvist et al., 2013). It is hard to imagine that the absence of the *pedosphere*[10] in traditional "Brazilian" storytelling and mythology could be an expression of the soil unobtrusiveness. It is well known that most of the territories of Brazil and tropical Africa are covered by vibrant-colored iron-rich soils (EC, 2013, 2014, 2015). Soil perception and conceptualization is as ancient as humanity itself (Brevik & Hartemink, 2010). Although being a formal part of soil-related activities in the modern world, it can be agreed that soils are stable enough to allow exchanges of information (i.e., communication) about their nature between people from the most diverse points of view over long periods of time (Brevik & Arnold, 2015). In other words, soils change slowly enough to give plenty of opportunities for individuals and communities to interact with its properties, certainly influenced by the activities motivating such interactions (e.g., agriculture, governance, building, pottery, painting, mining, rituals) and to communicate with each other about their experiences. Soil systems as well as human-soil interactions can be eventually disrupted due to many natural and anthropogenic reasons. In any community or society, there will always be individuals that have some sort of role-specific interaction with soils (e.g., farmer, builder, potter, priest). It is reasonable to imagine that these individuals are interacting with the same soil systems, although constrained by their specific activities. It is also reasonable to imagine that these individuals interacting with soils and with each other over many generations and within the most diverse emotional/existential frames (daily labor, religious rituals, dangerous situations, burial of the dead, failure of their soil-related activities, etc.) end up with some cultural evidence of such accumulated experiences, for example, in their folklore.

If the lack of explicit soils in such literature is unlikely explainable by a "soil silentness," perhaps there was a bias disfavoring the theme of soils on the social web that produced these tales and storytelling traditions or in the ones that produced the bibliographic content that we had access to – or both. In traditional narratives[11] soils are usually embedded into "framing" or "supporting" telluric/chthonic aspects (e.g., ground, agriculture, landscape, planet) or detached as plastic materials (e.g., raw material for human creation, paints, pottery, construction). But the tale of *Onilé* is something in between: the soil is her vest and hood, dress, and shelter.

[10] *Pedosphere* can be understood as the field of influence of soils – their environmental domain – or simply as the collectivity of the soils of the Earth. Soils can mediate a myriad of fluxes of matter and energy and partake in a plethora of environmental systems but occur only in places where the soil forming factors are active – i.e., wherever the other four major spheres of the Earth meet – that is why some authors use the term pedosphere to highlight the systemic and emergent nature of soils (Lin, 2010; Mattson, 1941; Targulian & Arnold, 2008).

[11] Although the development of both *traditional* and *scientific* narratives shares many similar psychological and sociological mechanisms – science as well is in a sense axiomatic and traditional – we use the term *traditional* to refer to more or less stabilized narratives that function either as foundational or identity symbols, or both, for a given ethos.

As suggested earlier, we consider *Onilé*'s realm as a plausible analogy for the concept of *pedosphere*: the core of the *Earth's Critical Zone*.[12] In fact, both are symbols of our earthen nature. They point to this mesmerizing, unfathomable, and poorly understood thing that is the skin of the Earth, our cradle and tomb (Kutílek & Nielsen, 2015). Although not being completely wrong, the correspondence is quite imprecise since we are trying to compare a scientific concept to a religious narrative. However, two aspects of the authors' reasoning are beautifully synthetized and articulated in the tale: the crucial terrestrial environmental compartment, for humans especially, is also the most discrete one – air and water, for example, are much more "noisy and dynamic."

10.3.2 The Roots of the Myth of Onilé

The tale of *Onilé* as presented by Prandi in his books (2001, 2005) was told to him by *Pai*[13] Agenor Miranda Rocha (Luanda, Angola, September 8th 1907 – Rio de Janeiro, July 17th 2004), considered by many as a legendary *Ifa* priest. The *Ifa* tradition is one of the most organized and professional *orisha* religions both in Yorubaland as well as in Brazil (Bascom, 1991). It is perhaps even more dogmatic in Latin America, since a certain theological and ritualistic "purity" is adopted by *Candomblé* (Brazilian) and *Lucumí Regla de Ocha* (Cuban) priests with no precedent in Yorubaland (Matory, 2005, apud. Apter, 2013). Emanuel (2000) described how the concept of *orisha* was developed and stabilized in Yorubaland over the millennia.[14] According to this author, the Yoruba religious system evolved from (i) the Stone Age (10,000 to 500 B.C.), where their worldview was basically divided into Heaven (*Orun*) and Earth (*Aiê*); through (ii) the Agricultural or Non-Ferriferous Metallic Age (500 B.C to 300 B.C.), where the concept of *orisha* started to develop either by the apotheosis of anthropomorphized forces of nature or remarkable ancestors; (iii)

---

[12] The *Earth's Critical Zone* (*ECZ*) is a region of the planet Earth that concentrates a considerable number of indispensable life-supporting systems and is roughly delimited from the top of the vegetation cover to the bottom of the aquifers (Lin, 2010).

[13] *Babalaô or babalorixá* is the title given to the male chief-priests of a *Candomblé* house (a.k.a. "casa-de-santo"). In Brazil they are also affectionately called *Pai-de-santo* (literally "father of saint"), or simply *Pai* for short. The feminine version being *Ialorixá* or *Mãe-de-santo*. Given the syncretism that took place between Catholicism and the *Ifa/Orisha* religions, "santo" and "orixá" are used interchangeably in popular vernacular. In some regions, due to its better acceptance, the Christian-friendly term is preferably used.

[14] Although this "long durée" narrative is arguably appropriate, we subscribe to the author's synthesis since the formation/evolution of the Yoruba identity is not a trivial matter and such debate would exceed the scope of this chapter. Hence, we use the term "Yoruba culture" referring to either its early elements as well as to its present Pan-Atlantic form. Please see, Apter (2013), Falola and Childs (2004), Falola and Genova (2006), and Peel (2003) for a more detailed discussion.

finally acquiring its current form during the Iron Age (300 BC to 500 A.D.), through the literary refinement of generations of professional *Babalawo*.[15]

Through the millennia the Yoruba became a people of traditional "urban farmers": city dwellers engaged in farming, living in cities surrounded by belts of peripheral farms (Bascom, 1955, 1962; United Nations Educational, Scientific and Cultural Organization, 2010a). This can be seen today in, for example, Oyo, an ancient Yoruba city in Nigeria.[16] This transition from farmers to urban farmers likely took place during the Yoruba Iron Age, in parallel with the development of a professional class dedicated to leisure activities (e.g., poetry, divination, philosophy, sculpture), without ever losing agriculture as the basis of Yoruba economy.

Onilé is not one of the most known *orishas* in Brazil or Yorubaland. In Brazil, *Onilé* occupies a central role in the *Candomblé* houses, despite being an almost forgotten *orisha* to the broader mass of Brazilians. In Nigeria this *orisha*, understood as Mother Earth, is the ultimate divinity worshipped by the *Ogboni,* a *quasi*-secret society ruled by elderly men (Arewa & Stroup, 1977; Roache, 1971). These men have considerable influence due to their role as a bond between Yoruba society and the Earth (Canada, 2012; Idowu, 2005; Lawal, 1995). The *Ogboni* consider that Earth precedes the gods in time, and together with the ancestors, is the very source of moral law (Morton-Williams, 1960).

Onilé barely survived the crossing of the Atlantic, a human trauma that is way too recent to have been overcome either in Brazil or Yorubaland, and that has linked these two territories to the present day. Many *orishas* did not make their way to the new world since the African Diaspora (Chirila, 2014). The ones that survived the trans-Atlantic slave trade or those that possibly were brought to the Americas in later peaceful times did not remain as "geographical" animistic symbols, assuming instead a more psychological and "tribal" or institutionalized meaning (Gordon, 1979; Walker, 1991).

Literature indicates that professional high priests played an important role as reformers and standardizers of the *Ifa/Orisha* theology and cult in urbanizing communities on both sides of the Atlantic (Alonso, 2014; Emanuel, 2000). The impact of such processes on the *orishas'* narratives and terminology, and hence on the myth collected by Professor Prandi, is unclear.

10.3.3 Methodological Considerations

Given all presented so far, if we may, an attempt could be made towards interpreting the scarcity of explicit mentions of soils in the literature about Brazilian folklore and mythological traditions, as well as the insertion of *Onilé*'s myth into such

[15] See footnote n° 13.

[16] The clear distinction between the c.a. 5 km densely populated radius around the city center (7°50'30"N 3°56'09"E), and a wide agricultural area surrounding it, can be observed by satellite imagery (e.g., Google Earth).

context. First, we should acknowledge a few possible methodological limitations that are imposed on our attempt, avoiding their biases.

The material gathered for this chapter includes a considerable amount of literature, as well as a few contacts/interviews with authors, priests, practitioners, and general people, from Brazil and Nigeria. Most of our contacts/interviews were conducted in a semi-structured way via online systems. Therefore, we could not test, for example, the hypothesis of the existence of rural Yoruba or Yoruba-descendant communities (e.g., *Quilombolas*[17]) in which on-farm practices/taboos/rituals related to *Onilé* were explicit. Success using this approach depends on finding someone who (1) has the knowledge; (2) is willing/allowed to share; and (3) is familiar with modern communication tools (social media, email, etc.).

As mentioned earlier, Yoruba communities in Brazil have been mostly urban/urbanoid, and the Yoruba cities in West Africa are quite populous. Therefore, it is very likely that "agricultural revolutions" or disruptions provoked by some historical landmarks such as the Colonization and the Green Revolution, as well as some Muslim-Christian-traditional religious group interactions, "erased" or obscured such on-farm practices and the narratives about them.[18]

Along the eternal saga of the *orishas*, some were replaced by others in related functions or had their "domain" distributed between or absorbed by other *orishas*. In ancient times the "power on Earth" was anthropomorphized as *Ajalaye*. It was probably replaced at some point by *Onilé* (as the "soil" power), while nowadays other aspects have become more relevant to the Yoruba culture. There are many *orishas* historically more recent and more famous related to telluric aspects in Brazil, for example, the ones mentioned in the introduction of this chapter.

The *orisha* religions of the *Ifa* tradition are typically "initiatic" and, hence, knowledge about mythology, theology, and rituals are surrounded by a lot of secrecy. For the noninitiated, such as the authors of this chapter, access to this kind of knowledge could only be made possible through literature or through a relatively long "friendly conviviality." It is worth mentioning that all the literature we were able to access was written either by some initiated or "friend researcher," hence, are generally limited to the religious/theological interests of the former or by the academic questions of the latter. Although in Brazil the "secrecy of religion" can be understood as a resistance strategy against historical moments of religious and political oppression, some priests and practitioners do not agree with the maintenance of such systems nowadays, arguing that they only reinforce prejudice and discrimination. This more "open-minded" position is evidenced, for example, in a myriad of

[17] *Quilombolas* are communities in Brazil that were originated or founded by fugitive or liberated slaves and nowadays have the legal status of traditional communities (Lúcia Andrade, 2020).

[18] Yoruba was an oral language until very recently. During the seventeenth century, the first attempts were made to codify the Yoruba language into a written system using Arabic notation. These attempts did not succeed and were superseded by the use of Romanized symbols in the nineteenth century that have been used and refined for the present day (Falola & Akinyemi, 2016; Ogunbiyi, 2003).

"uncensored"[19] YouTube channels and social media pages dedicated to *Ifa/Orisha* theology, mythology, and rituals, organized by actual priests and practitioners. Unfortunately, it seems that our curiosity (*kosmos*[20] ethnopedology in "pan-Yoruba" culture) spotted a gap in the literature (at least in the languages we can read). In an extreme case, the evidence we are looking for could be protected by vows of secrecy (blood oaths in some cases), and if we were initiates, we would be bound by such vows. There are examples of anthropologists whom, noticing the ancient and deep influence of the *Ogboni* on Yoruba society – the ones that explicitly worship *Onilé* – joined the cult and never published anything "substantial" about it, claiming that they were bound by their oaths – some have published only what they argue to be "common knowledge" (Morton-Williams, 1960; Frisvold, 2016). Although the authors of this chapter are not initiates, this perspective is useful to frame and understand our object of study.

Assuming that none of the hindrances above veils our perception, this "deafening silence" of soils in Brazilian folklore, especially within Yoruba-descendant traditions, is quite intriguing. Although acknowledged as having paramount value for this cultural group, its representations are rather scarce.

10.4 Interpreting the Myth

Myths very often are metaphors for human experiences that are ubiquitous and constantly repeated by all humankind since the "mists of time." The apparent predisposition of human cultures to represent certain "myth-motifs" even being separated in time and/or space led the Swiss psychiatrist Carl Gustav Jung (1875–1961) to theorize about the existence of instinctive ways of imagination, similar to our instinctive ways of behavior. These "basic patterns" (or *archetypes*, as he called them), as with any other human biological (semi)autonomous system, were imprinted in the human body during its biological history on planet Earth.[21] They are not inherited ideas or images but rather simply the possibility or ability of representing them. In general, the most frequent, ordinary, and eternally repeated of human experiences and everyday facts appear to generate the mightiest archetypes of all, the ones more easily noticed and abundant not only in folklore, mythologies, and poetic/mystical experiences but also in our secular rationalistic societies (Campbell, 1971). The encounters between man and woman, experiences with the mother, with the father, etc. are good examples of such incessantly and universally repeated motifs. Also, emotions

[19] Many priests around the world broadcast their rituals and festivals on Facebook or YouTube, even the ones involving animal sacrifice and initiation of youngsters. Others focus on theological and storytelling traditions.

[20] Ethnopedology can be conceptualized as dealing with three main aspects of human-soil relations, known as the "CPK complex": (1) *Corpus*, cognitive systems; (2) *Praxis*, management systems; and (3) *Kosmos*, spiritual and symbolic aspects (Barrera-Bassols & Zinck, 2000, 2003).

[21] See footnote n° 4.

and fantasies inspired by natural phenomena (e.g., day and night, thunder, volcanoes), as well as challenges and hindrances posed by the environment (e.g., crossing seas, famine, passing mountains) can be listed among such experiences represented universally in mythological narratives (Nise da Silveira, 2008). Perhaps the most common experiences with the pedosphere can in fact be characterized by a tacit relationship of dependence, a quiet, but valuable, passivity of the soils. It is reasonable to portray, by way of comparison, the more conspicuous experienced themes as related to "well-developed archetypes," while the more unobtrusive ones as "latent archetypes" – well developed and latent here used with a philosophical sense of presence and absence, respectively, having nothing to do with value judgments. Maybe, this unspoken relevance is the representation of a "soil archetype," a timeless "absent presence" revealed in the mythological context. While The Mother, The Father, The Sage, The Hero, etc. (famous motifs or archetypes) are the protagonists of their play, soils are inevitably the stage on which all of them walk and take place.

Another point worth mentioning is that, in the tale, in a first moment, the soil provides to *Onilé* vest and shelter, while in a second moment, it becomes her identity. Clothes are very commonly used to protect the human body and, at the same time, reveal a place in society. In many cases clothes are also used with expressive intentions. Frequently, individuals identify themselves with the role they perform in society and its related dynamics, history, and paraphernalia. In Brazil many *Ifa/ Orisha* devotees find in the *orishas'* narratives an explanation to their own existence, and, by living a ritual life (which frequently includes a dressing code), they express and reinforce their identities. Just another human universal illustrated in this tale.

10.5 For a Contemplative Approach of Storytelling Traditions

Meaning is the ultimate antidote to chaos (Peterson et al., 2018). Although we as scientists (especially the most academical ones) are typically pursuing apodictic concepts, if we ever achieve them or not is debatable, our first perceptions as general human beings are expressed in a rather poetic form. We apply the best of our efforts in order to build the steps from the perception of a mere possibility towards sound certainty about the nature of things via dialectics and analysis in a quite Aristotelian way (Carvalho, 2005, 2013). It was Aristotle who suggested that the opinion of the wise (*Endoxa*), as well as our unreflected endorsement of such opinions (*magister dixit*), could have some value in helping us to deal with the puzzles posed by the universe to our curiosity (Shields, 2016). In modern academic practice, that would be called literature review, peer review, and scientific method. Many of us, educated scientists and practitioners, would find the opinion of storytellers and myth believers unproductive to say the least. However, the formula of folk and story tales are quite effective if we want to capture the attention of a listener and invite him or her to a fruitful dialog. Indeed, we can enjoy many "aha" moments, so precious to the intellectual life, when we are exposed to such types of structured narratives. After all, we are still a storytelling animal (Gottschall, 2012; Rumschlag, 2016). You may argue that the imprecision and ambiguity of this literary genre may be rather confusing instead of helpful. You may point to the use of this ancient human trait in dull power politics as well as in consumerist advertising, to mention just a few at least dubious examples. To that we must say that weaving the evanescent glimpses of our perceptions with the slippery unscientific language of our daily lives into a structured narrative of universal validity takes way more than a recipe: it is the fruit of genuine and meaningful human experience processed by true literary talent. The efforts of such talented "writers," mostly anonymous in the case of indigenous storytelling traditions, can be of great value not only to open our minds to different perspectives, but also to teach us an effective language.

One must be careful with the propositions we are making for numerous reasons – especially out of respect or tolerance for these traditions. As bridge builders between soil sciences and broader society, we should not use such "techniques" in a pervasive way. After all, this kind of rhetorical tool tries to translate into moral values (right vs. wrong) not only ontological propositions (how things are), but also practical ones (how we should deal with them). This translation is basically the essence of environmental conscientization campaigns and policies. Thus, whenever we use these stories or this style, we do so as an invitation to contemplating the otherness in it. We invite our listeners or readers to notice that we all share the mechanisms that make these narratives so intriguing (Gottschall, 2012), and that perhaps those stories were put up in ancient times as an attempt to face the same problems we do today (e.g., food production, social tensions, soil security, sustainability) without our sophisticated secular language. Then, we leave it to the listener to ponder about our point.

Be that as it may, engaging in soil awareness and soil conservation actions can only be motivated if such initiatives are in alignment with the values of the individuals we wish to involve in them. It is quite unlikely that conceptualizing soils as a god could serve as a pedological model nor as the basis for a soil policy; after all, this is not the purpose of mythology, and gods do not belong to the scientific axiomatic tradition so far. Notwithstanding, we can still invite those willing to listen to consider the possibility of other ways of relating to nature rather than only using it. Storytelling is a lovely way to do so, as the ancient Yoruba people already figured out ages ago.

10.6 Conclusions

As pointed out in the introduction, Brazilian society is a "melting pot," an amalgam of cultures (esp. European, Amerindian, and African). Although Brazilians are typically not bellicose (we are quite the opposite actually), to the point that we have been considered a "racial democracy"[22] by a given sociological tradition,[23] there is nowadays a clear and well-acknowledged tension between many social groups, especially religious ones. This tension, although mild and not formally institutionalized, is ubiquitous and its expressions can be seen even in the political agency of our elected representatives. One such tension can be seen between Christian and Afrodescendant religious groups, between which there is no real dialog. Commonly, the views that one person holds about the members of the other group is the one offered by his or her spiritual leaders. It might be easier for devotees of Afrodescendant religions to dare to listen to what the other is saying, when compared to Christians. Afrodescentant religions have a remarkably adaptative nature in order to accommodate other belief systems within their cosmoview. Christianity is much more rigid in this sense. Many Christians cannot even say out loud the words and names related to the other group's beliefs, so deep is the spiritual terror they feel towards this topic.

There is hope, though. Some members and supporters of both groups are indeed attempting to promote honest dialogs with each other toward religious tolerance. The authors of this chapter represent somehow this kind of effort and have been involved in correlated activities, with special attention to the theme of environmental education and soils. The first author is a white-*ish* soil scientist and Christian, the second, a black analytical psychology professor and Afrodescendant religions sympathizer. We have given talks presenting the symbolic importance of soils for all three cultural groups (i.e., European, Amerindian, and African), and that fearless tolerance is possible. Storytelling has proved useful. The present chapter is in its way an expression of it.

The Brazilian-Yoruba tale of *Onilé* is an interesting piece of literature both in content and form. On the one hand, this narrative seems to resonate with topics dear to soil scientists, such as the importance of the pedosphere for humans and ecosystems, the problem of soil perception and conceptualization, and how we should deal with soils sustainably. On the other hand, such messages are conveyed in a very effective language: a mythical one. Although we are not capable of "living the

[22] Due to the biological feebleness of human racial purity (National Geographic Society, 2018), and the widespread miscegenation that took place in Brazil ever since its colonization (Freyre, 1986), the racial issues in this country are quite peculiar and not as straightforward as in countries that experienced segregation. According to the Brazilian Institute of Geography and Statistics (IBGE) in 2019 the population was 42.7% white, 9.4% black, 1.1% yellow or Amerindian, and 46.8% "*pardo*" (≈mixed races), in terms of self-declared race or skin color (Instituto Brasileiro de Geografia e Estatística, 2020). In the past there were terms in Portuguese for specific combinations between white, black, and Amerindian people – kids had to memorize these categories in school, namely, *mulato* (white and black descendants), *caboclo* or *mameluco* (white and Amerindian descendants), and *cafuzo* (black and Amerindian descendant). Nowadays, given the many generations of widespread miscegenation, these categories hold virtually no practical use and have been replaced simply by the term "*pardo*" – which in most cases means "not enough white and not enough black" – a term with which most Brazilians have identified themselves.

[23] Although Gilberto Freyre (1900–1987) did not coin the term "racial democracy", his book *The Masters and the Slaves* (1986), although polemical, was seminal to a tradition of Brazilian Sociology.

metaphor" contained in the myth without being a proper believer, we can learn from it and the literary style in which it is expressed: storytelling. It is indubitable that there are things all humans have in common and that powerful myths and tales are those that tap into these common dimensions. Therefore, by contemplating the way traditions such as the Yoruba *Ifa/Orisha* tradition represent soils, we can learn new ways in which our insights, imaginations, and knowledge about soils can be informed and expressed more effectively and more authentically with others.

References

Alonso, M. C. (2014). *The development of Yorùbá Candomblé communities in Salvador, Bahia, 1835–1986*. Palgrave Macmillan.

Apter, A. (2013). Yoruba ethnogenesis from within. *Comparative Studies in Society and History, 55*(2), 356–387. https://doi.org/10.1017/S0010417513000066

Arewa, O., & Stroup, K. (1977). The Ogboni Cult Group (Nigeria): Analysis and interpretation of the communicative events which constitute the behavior of its members. *Anthropos, 72*(1/2), 274–287. JSTOR.

Barrera-Bassols, N., & Zinck, J. A. (2000). *Ethnopedology in a worldwide perspective: An annotated bibliography*. International Institute for Aerospace Survey and Earth Sciences (ITC).

Barrera-Bassols, N., & Zinck, J. A. (2003). Ethnopedology: A worldwide view on the soil knowledge of local people. *Geoderma, 111*(3-4), 171–195. https://doi.org/10.1016/S0016-7061(02)00263-X

Bascom, W. (1955). Urbanization among the Yoruba. *American Journal of Sociology, 60*(5), 446–454. https://doi.org/10.1086/221604

Bascom, W. (1962). Some aspects of Yoruba Urbanism. *American Anthropologist, 64*(4), 699–709. JSTOR.

Bascom, W. W. (1991). *Ifa divination: Communication between gods and men in West Africa* (Vol. 638). Indiana University Press.

Beniste, J. (2006). *Mitos yorubás: o outro lado do conhecimento*. Bertrand Brasil.

Brevik, E. C., & Arnold, R. W. (2015). Is the traditional pedologic definition of soil meaningful in the modern context? *Soil Horizons, 56*(3). https://doi.org/10.2136/sh15-01-0002

Brevik, E. C., & Hartemink, A. E. (2010). Early soil knowledge and the birth and development of soil science. *Catena, 83*, 23–33.

Campbell, J. (1971). *The portable Jung* (R. F. C. Hull, Trad.). Penguin Books.

Campbell, J., Moyers, B. D., & Flowers, B. S. (1991): *The power of myth* (1st Anchor Books ed). Anchor Books.

Canada. (2012). *Refworld | Nigeria: Ogboni society, including its history, structure, rituals and ceremonies; information on membership and the consequences of refusing to join*. Refworld. https://www.refworld.org/docid/50c849842.html

Carvalho, O. D. (2005). From poetics to logic: Exploring some neglected aspects of Aristotle's organon. In *Handbook of the first world congress and school on universal logic* (Vol. 1, pp. 57–65).

Carvalho, O. D. (2013). *Aristóteles em nova perspectiva: introdução à teoria dos quatro discursos*.

Chirila, A. (2014). The river that crosses an ocean: Ifá/Orisha in the global spiritual marketplace. *Qualitative Sociology Review, 10*(4).

Cuoco, A. (2014). *African narratives of orishas, spirits and other deities: Stories from West Africa and the African diaspora: a journey into the realm of deities, spirits, mysticism, spiritual roots and ancestral wisdom*.

Elmqvist, T., Redman, C. L., Barthel, S., & Costanza, R. (2013). History of urbanization and the missing ecology. In T. Elmqvist, M. Fragkias, J. Goodness, B. Güneralp, P. J. Marcotullio, R. I. McDonald, S. Parnell, M. Schewenius, M. Sendstad, K. C. Seto, & C. Wilkinson (Orgs.), *Urbanization, biodiversity and ecosystem services: Challenges and opportunities: A global assessment* (pp. 13–30). Springer. https://doi.org/10.1007/978-94-007-7088-1_2

Eltis, D. (2004). The diaspora of Yoruba speakers, 1650–1865: Dimensions and implications. In T. Falola & M. D. Childs (Eds.), *The Yoruba diaspora in the Atlanic world* (pp. 17–39). Indiana University Press.

Emanuel, A. (2000). *Odun Ifa* (Ifa Festival). West African Book Publishers Limited.

European Commission. (2013). *Soil atlas of Africa*. European Commission. https://esdac.jrc.ec.europa.eu/content/soil-map-soil-atlas-africa

European Commission. (2014). *The major soils of Africa*. European Commission; jrc91922_jrc_africa_sheet_v3.pdf. https://ec.europa.eu/jrc/en/publication/major-soil-types-africa

European Commission. (2015). *Soil atlas of Latin America and the Caribbean*. European Commission. https://op.europa.eu/s/n7mB

Falola, T., & Akinyemi, A. (2016). *Encyclopedia of the Yoruba*. Indiana University Press.

Falola, T., & Childs, M. D. (Orgs.). (2004). *The Yoruba diaspora in the Atlantic world*. Indiana University Press.

Falola, T., & Genova, A. (Orgs.). (2006). *Yorubá identity and power politics*. University of Rochester Press.

Food and Agriculture Organization. (2015). *Soil functions – Soils deliver ecosystem services that enable life on Earth* [Infographic]. http://www.fao.org/3/a-ax374e.pdf

Freyre, G. (1986). *The masters and the slaves*. University of California Press.

Frisvold, N. D. M. (2016). *Ifa: A forest of mystery*. Scarlet Imprint.

Gordon, J. U. (1979). Yoruba cosmology and culture in Brazil: A study of African survivals in the new world. *Journal of Black Studies*, *10*(2), 231–244. JSTOR.

Gottschall, J. (2012). *The storytelling animal: How stories make us human*. Houghton Mifflin Harcourt.

Gurtner, D. M. (2011). *This world and the world to come: soteriology in early Judaism*. Bloomsbury Publishing.

Hassan Khalil, M. (2011). Salvation and the 'other' in Islamic thought: The contemporary pluralism debate (in English): Salvation and the 'other' in islamic thought. *Religion Compass*, *5*(9), 511–519. https://doi.org/10.1111/j.1749-8171.2011.00295.x

Idowu, W. (2005). Law, morality and the African cultural heritage: The jurisprudential significance of the Ogboni institution. *Nordic Journal of African Studies*, *14*(2), 175–192.

Instituto Brasileiro de Geografia e Estatística. (2020). *Pesquisa Nacional por Amostra de Domicílios Contínua: Características gerais dos domicílios e dos moradores 2019*. IBGE. https://biblioteca.ibge.gov.br/visualizacao/livros/liv101707_informativo.pdf

Johnson, P. C. (2002). *Secrets, gossip, and gods: The transformation of Brazilian Candomblé*. Oxford University Press.

Jung, C. G. (2014). *The Collected Works of C.G. Jung* (R. F. C. Hull, Trad.; Vol. 1–19). https://press.princeton.edu/books/ebook/9781400851065/the-collected-works-of-cg-jung

Klein, H. S., & Luna, F. V. (2009). *Slavery in Brazil*. Cambridge University Press.

Kutílek, M., & Nielsen, D. R. (2015). *Soil: The skin of the planet Earth*. Springer. https://doi.org/10.1007/978-94-017-9789-4

Lawal, B. (1995): À Yà Gbó, À Yà Tó: New perspectives on Edan Ògbóni. *African Arts*, *28*(1), 37–100. JSTOR. https://doi.org/10.2307/3337249

Lin, H. (2010). Earth's critical zone and hydropedology: Concepts, characteristics, and advances. *Hydrology and Earth System Sciences*, *14*, 25–45.

Lúcia Andrade. (2020, julho 15). Quilombolas Communities in Brazil. *Comissão Pró-Índio de São Paulo*. https://cpisp.org.br/direitosquilombolas/observatorio-terras-quilombolas/quilombolas-communities-in-brazil/

Maier, R. (2014). Salvation in Buddhism. *Journal of Adventist Mission Studies*, *10*(1), 9–42.

Matory, J. L. (2005). *Black Atlantic religion*. Princeton University Press. https://press.princeton.edu/books/paperback/9780691059440/black-atlantic-religion

Mattson, S. (1941). The laws of soil colloidal behavior: Xxiii. the constitution of the pedosphere and soil classification. *Soil Science, 51*(5), 407–426.

McKenzie, P. R. (1997). *Hail Orisha! A phenomenology of a West African religion in the mid-nineteenth century*. Brill.

Morton-Williams, P. (1960). The Yoruba Ogboni cult in Oyo. *Africa, 30*(4), 362–374. https://doi.org/10.2307/1157598

National Geographic Society. (2018, abril). The race issue. *National Geographic, 233*(nº 4).

Nise da Silveira. (2008). *Jung* (16º ed). Paz e Terra.

Ogunbiyi, I. A. (2003). The search for a Yoruba Orthography since the 1840s: Obstacles to the choice of the Arabic script. *Sudanic Africa, 14*, 77–102. JSTOR.

Palmer, C. A. (2000). Defining and studying the modern African diaspora. *The Journal of Negro History, 85*(1/2), 27–32. JSTOR.

Peel, J. D. Y. (2003). *Religious encounter and the making of the Yoruba* (1. paperback ed). Indiana University Press.

Peterson, J. B., Doidge, N., & Van Sciver, E. (2018). *12 rules for life: An antidote to chaos*. Random House Canada.

Prandi, J. R. (2001). *Mitologia dos orixás* (1a. ed., 1a. reimpr). Companhia das Letras.

Prandi, J. R. (2005). *Segredos guardados: orixás na alma brasileira*. Companhia das Letras.

Rio de Janeiro. (2018). *Lei Ordinária nº 8085 – Declara patrimônio imaterial do estado do Rio de Janeiro o idioma em iorubá, praticado nas religiões afro-brasileiras*. https://leisestaduais.com.br/rj/lei-ordinaria-n-8085-2018-rio-de-janeiro-declara-patrimonio-imaterial-do-estado-do-rio-de-janeiro-o-idioma-em-ioruba-praticado-nas-religioes-afro-brasileiras

Roache, L. E. (1971). Psychophysical attributes of the Ogboni Edan. *African Arts, 4*(2), 48. https://doi.org/10.2307/3334522

Rumschlag, S. (2016). The storytelling animal: How stories make us human by Jonathan Gottschall. New York, NY: Mariner Books, 2013. 248 pages. Paperback: $14.95. *Curator: The Museum Journal, 59*(1), 61–63. https://doi.org/10.1111/cura.12146

Salvador-SA. (2019). *Lei Ordinária nº 9503 – Autoriza a declaração do idioma Iorubá, um dos valores da civilização africano-brasileira, como patrimônio imaterial do Município de Salvador*. https://leismunicipais.com.br/a/ba/s/salvador/lei-ordinaria/2019/951/9503/lei-ordinaria-n-9503-2019-autoriza-a-declaracao-do-idioma-ioruba-um-dos-valores-da-civilizacao-africano-brasileira-como-patrimonio-imaterial-do-municipio-de-salvador

Schulte, R. P. O., Creamer, R. E., Donnellan, T., Farrelly, N., Fealy, R., O'Donoghue, C., & O'hUallachain, D. (2014). Functional land management: A framework for managing soil-based ecosystem services for the sustainable intensification of agriculture. *Environmental Science & Policy, 38*, 45–58. https://doi.org/10.1016/j.envsci.2013.10.002

Shields, C. (2016): Aristotle. In *The stanford encyclopedia of philosophy* (Winter 2016 edition). Edward N. Zalta. https://plato.stanford.edu/archives/win2016/entries/aristotle/

Sobrinho, G. A. N., & Silva, N. S. D. (2019). *Soils within the Yoruba mythology: An example of a modern conceptualization from an ancient mythology*. In 21 WCSS: Proceedings of the 21st World Congress of Soil Science; 2018, August 12-17; Rio de Janeiro, Brazil [e-book], I, p. 256. https://www.21wcss.org/docs/Proceedings_of_the_21WCSS_Volume_I.pdf

Stetler, E. (2014): Soteriology. In D. A. Leeming, Encyclopedia of psychology and religion (p. 1688–1689). Springer. https://doi.org/10.1007/978-1-4614-6086-2_654

Targulian, V. O., & Arnold, R. W. (2008). Pedosphere. In *Encyclopedia of ecology* (pp. 2665–2670). Elsevier. https://doi.org/10.1016/B978-008045405-4.00753-9

United Nations Educational, Scientific and Cultural Organization. (2010a). *História geral da África* (Vol. 1–VIII). UNESCO.

United Nations Educational, Scientific and Cultural Organization. (2010b). *História geral da África, V: África do século XVI ao XVIII* (Vol. V). UNESCO.

Walker, S. S. (1991). A Choreography of the Universe: The Afro-Brazilian Candomble as a Microcosm of Yoruba Spiritual Geography. *Anthropology and Humanism Quarterly, 16*(2), 42–50. https://doi.org/10.1525/ahu.1991.16.2.42

Chapter 11
Cultural Understanding of Soil from a Japanese Perspective

Hideaki Hirai and Katsuyuki Minami

11.1 Introduction: Basic Description of Process

Minami (2009, 2014, 2017) and Nara and Minami (2020) have previously described understanding of soil from the perspective of Japanese culture. This includes understanding the meaning of Chinese characters related to soil (e.g., "土," soil), as Japanese writing evolved from written Chinese characters. More complete reviews of ancient Chinese culture and soil include Jikai, Jikun, and Jitsu (Sirakawa, 1995, 1996a, b), "Soil Proverbs in China" (Chen & Zhang, 2018) and a chapter in this book (Chap. 13). These works can provide additional insights regarding the Chinese links in the etymology section of this chapter. The concept of *kami* (Japanese Shinto) from ancient times to the present age was researched using the compiled documents of the shrine's head office (Jingushicho, 2015a, b). Finally, the first author went on a field survey to the oldest shrine and the oldest water shrine in Japan to enable a more complete description of Japanese culture related to the shrine.

It is essential to discuss the influence of ancient Chinese culture on Japan. Ancient Chinese cultural ideas were continuously imported into ancient Japan after the fifth century, resulting in the introduction of kanji (Chinese characters) and Buddhism to Japan. Inukai (2015) forms the basis for this chapter's discussion of the introduction of Chinese characters into Japan. Buddhism is widespread in Japan and has profoundly influenced the country's culture. As an example, Jizo Bodhisattva (地蔵菩薩) images such as those seen in Fig. 11.1 are common at intersections and

H. Hirai (✉)
School of Agriculture, Utsunomiya University, Utsunomiya, Tochigi, Japan
e-mail: hirai@cc.utsunomiya-u.ac.jp

K. Minami
Emeritus of Kitasato University, Tokyo, Japan

N. Patzel et al. (eds.), *Cultural Understanding of Soils*,
https://doi.org/10.1007/978-3-031-13169-1_11

Fig. 11.1 Jizo Bodhisattva along the roadsides and/or intersections in Tochigi, Japan

along roadsides in modern Japan. Jizo (literally means "earth" + "treasury") Bodhisattva means plenty of lovingkindness and compassion to save people living anywhere and support all trees and plants (Shimoizumi, 2019), which are related to mother earth. A Tathagata (e.g., Amitabha Tathagata, one who possesses infinite light or Mahavairocana Tathagata, great one whose light shines everywhere) image is found on the Buddhist altar in many Japanese houses.

Buddhist temples in Japan can be broadly divided into two types (Kido, 2019): those that principally exist to fulfill funerary functions for the deceased and those that principally exist to serve the everyday spiritual needs of living devotees via teaching and ritual. The second type of temple, which focuses on the living, is sometimes called a devotee temple. Priests recite prayers and make requests on behalf of congregants in front of the image of Acala. In addition to funerals and recitation of prayers, the purpose of temples is to play a role in polishing the inherent buddha nature of the people and bringing them to the same state as Tathagata Shakyamuni's buddhahood.

The Lotus Sutra teaches that "Many Buddhas and Shakyamuni Buddha for one great cause alone appear in this world." That is to say they appeared in order that anybody could see these Buddhas and to purify their minds. These buddhas appeared in our world to make us understand what a buddha is and to enable us to become buddhas (TENDAISHU, 2015). In order to explain the appearance of the buddha, Shinjo (1906–1989) cited a passage from the sutra of the Golden Light: "世尊は不思議なり (Seson-wa-fushigi-nari)。実性の体は (Jissho-no-taiwa), 即ち (Sunawachi),

これ真如にして (Kore-shinnyo-ni-shite), 真如の性は (Shinnyo-no-shou-wa), すなわち如来なり (Sunawachi-nyorai-nari), 名づけて涅槃とす (Nazukete-nehanntosu)" (translation: "The Buddha is wondrous and mysterious. The substance of the buddha is suchness (shinnyo). The true nature of shinnyo is Tathagata, and it is named nirvana") (Kido, 2019; Shinjo Kanko Kai, 2016). Kido (2019) explained the meaning of the passage as "The true nature of the world is suchness (shinnyo), and it is nothing less than the appearance of Tathagata, and its inside state is named nirvana."

Buddhism is important to understanding the soil from a Japanese cultural perspective in several ways: (1) "ブディスト・エコロジー(Buddhist ecology)-共生(Kyousei)・環境 (Kankyo)・いのちの思想(Inochi-no-shiso)" (translation: "Philosophy of cooperative existence and/or living together, environment and life") – written by Takemura (2016), a Japanese scholar majoring in Buddhism, (2) "草木成仏の思想 (Soumoku-jyoubutsu-no-shiso)" (translation: "Philosophy that grasses and trees become buddhas") written by Sueki (2017). Furthermore, Japanese cultural ideas, including Buddhism, are a part of daily life. For example, it is customary to fold one's hands before a meal as shown in Fig. 11.2. The book of "摂受心院 (Shojyushinin)-人の心に生きる (Hitono-kokoroni-ikiru)" (translation:

Fig. 11.2 Photo to fold hands before a meal

"Shojyushinin-Live in the human heart") (Shojyushinin Kanko Kai, 2018), in which the words and deeds of a great female Japanese Buddhist monk, Tomoji (1912–1967) were introduced in order to explain the underlying ideas that support the behavior of Japanese people, was also used as a source for this chapter.

Thus, China's ancient culture (Confucian), Shinto, and Buddhism have merged in Japan, forming a unique Japanese culture that continues into the modern society. For this reason, it should be noted that the living environment in which the authors have been nurtured has influenced the discussion in this chapter. To facilitate the cultural understanding of soil from a Japanese perspective, it is necessary to understand the underlying ideas within Japan, but this was a difficult undertaking for the writers who studied natural science and the humanities, including culture, separately. Therefore, we used the evidence proof method of natural science and the demonstration method of social science with field survey and/or documentary investigation. Although there might be points where the descriptions are not always sufficient, we hope that the readers are able to gain meaningful information on the cultural understanding of soil from a Japanese perspective.

11.2 The Etymology of Soil (土: "tu" in Chinese, "tsuchi" in Japanese)

The central concept of the Chinese ideogram 土 ("tu"; Fig. 11.3) is neither that of land with economic value nor the material properties of soil. As we see from the oldest Chinese dictionary, written by Shuowen Jiezi during the Han Dynasty (B.C.E. 206–C.E. 220), "tu" is strictly that which fosters life (Minami, 2014; Wen, this volume). The Han scholar Xu Shen explained in the book of Shuo Wen Jie Zi in the Donghan Dynasty (C.E. 58–C.E. 147) (Chen & Zhang, 2018) that "tu" is that which produces all things from the land (Lin, 1996). "Tu" is expressed as follows; the two horizontal lines represent the topsoil (upper –) and the subsoil (lower _) (Fig. 11.3). In other words, soil is the layer between the upper line representing the meeting of the atmosphere and the lower line indicating the bottom of the root development zone. The vertical line symbolizes a tiny bud together with roots growing from the lower line to the upper line. The character 吐 ("tu"; Fig. 11.4), which indicates emit or emanate, is pronounced "tu" in the Shuowen Jiezi in the same manner as 土 (soil). According to the ancient people's recognition, the ground has a mouth ("口") and is breathing like a human being. That could be because ancient people noticed the ground has a lot of tiny paths and/or holes that lead to the surface, and plant buds and tiny creatures such as insects, earthworms, and beetles crawl out of soil ("土") through these holes (pores). The word "吐" is one of a personified and/or anthropomorphized expression of the tiny paths and/or holes that emanate life from inside the soil (Lin, 1996).

The character 土 ("tu") developed into the characters 生 ("sheng"; Fig. 11.5), 世 ("shi"; Fig. 11.6), and 姓 ("xing"; Fig. 11.7). 土 ("tu") changed to 生("sheng"), indicative of a shape showing a plant with a leaf and/or stem growing, symbolizing

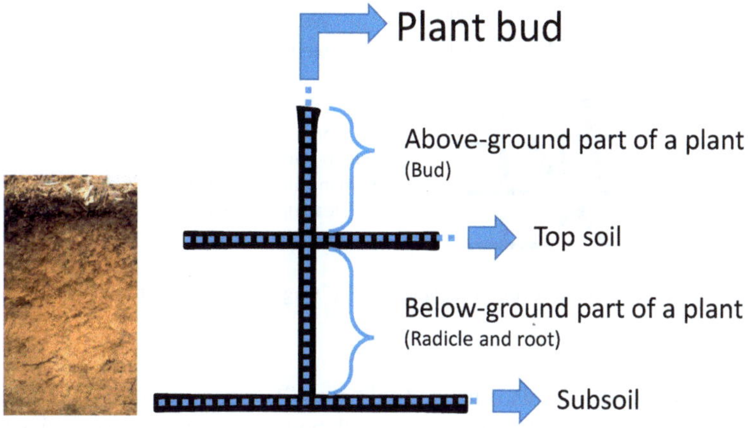

Fig. 11.3 Illustration of the Chinese character 土 ("tu", in Chinese pronunciation, "tuchi" or "do" in Japanese)

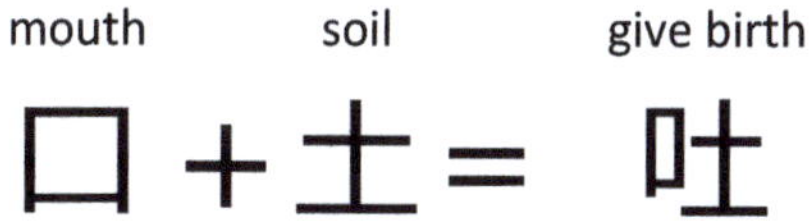

Fig. 11.4 Illustration of the Chinese character 吐 ("tu" in Chinese pronunciation)

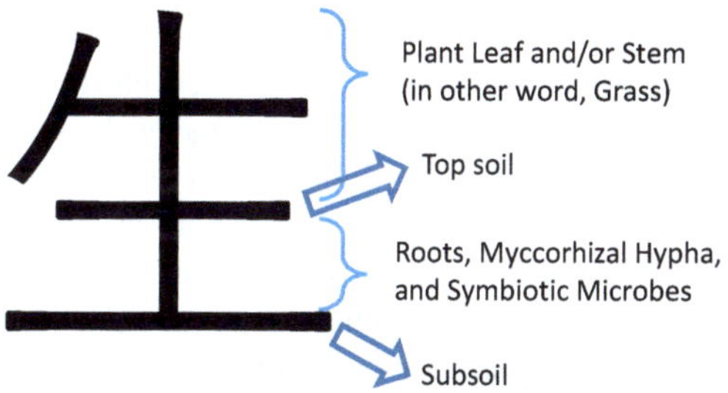

Fig. 11.5 Illustration of the Chinese character 生 ("sheng" in Chinese pronunciation, while "sei", in Japanese)

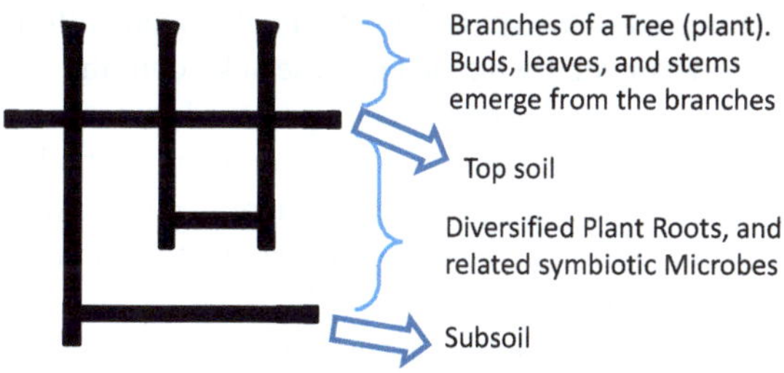

Fig. 11.6 Illustration of the Chinese character 世 ("shi" in Chinese pronunciation, while "yo", in Japanese)

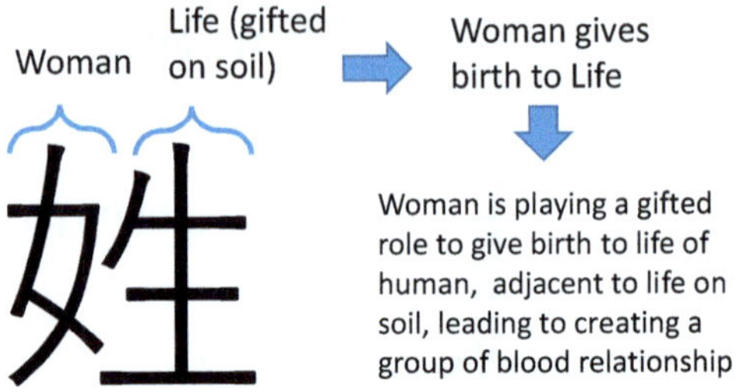

Fig. 11.7 Illustration of the Chinese character 姓 ("xing" in Chinese pronunciation, "sei" in Japanese)

life arising out of soil. 土 also changed to世 ("shi"), suggestive of a shape showing a plant dividing into branches and leaves with buds emerging. This signifies multiplying and diversifying life, leading to a plant community and/or human society. Moreover, 生 ("sheng") advanced to 姓 ("xing"), signifying that women play a gifted role in giving birth to human beings that leads to a family of blood relationships (Sirakawa, 1996a; Minami, 2009).

11.3 Soil and Spirits

According to Sirakawa (1995, 1996b), "tsuchi" in Japanese is derived from the two syllables "tsu" and "chi," which are written 土 ("tu", Fig. 11.3) and 地 ("di" in Chinese, "chi" in Japanese, ground and/or earth, Fig. 11.8), respectively. Hence, "tsuchi" literally means soil and ground and/or earth. The Chinese character 地 is composed of 土 and 也. The ancient Chinese meaning of 也 is a female organ for the path to birth (Sirakawa, 1995). 地 ("di") therefore means "path to birth from soil (incubator of all living creatures)" and is indicative of tiny paths of birth from the soil's surface layer.

In ancient Japan, before the Chinese characters were introduced, the means of communication was oral and syllables in oral pronunciation also have meaning. The Japanese oral pronunciation of 土 is "tsu-chi." The syllable "chi" indicates spirits (Fig. 11.9a), whereas "tsu" is place. Thus, Japanese "tsuchi" did not signify soil in general but was a name that called forth something spiritual concealed in the ground. "Tsuchi" refers to an earthly guardian of a place or a land. In ancient times there were many rites dedicated to the spirits of the land that made offerings to the spirits by pouring Japanese fermented wine or Japanese sake onto the soil before starting to build (Fig. 11.9b). This ritual is practiced in present Japanese society before construction of a house, factory, or building. Moreover, there is another case where the "tsu" in 土 ("tsuchi") indicates the presence of sacred power and "chi" is similar to 霊 ("hi"), which means spirits or souls (Fig. 11.9a, b), so that "tsuchi" refers to the spiritual source of sacred power in life.

A place where people dedicated offerings to the guardian of a land was considered 土 in ancient times and pronounced "sha" not "tsu." Later the meaning of 土

Fig. 11.8 Illustration of the Chinese character 地 ("di" in Chinese pronunciation, while "chi" in Japanese)

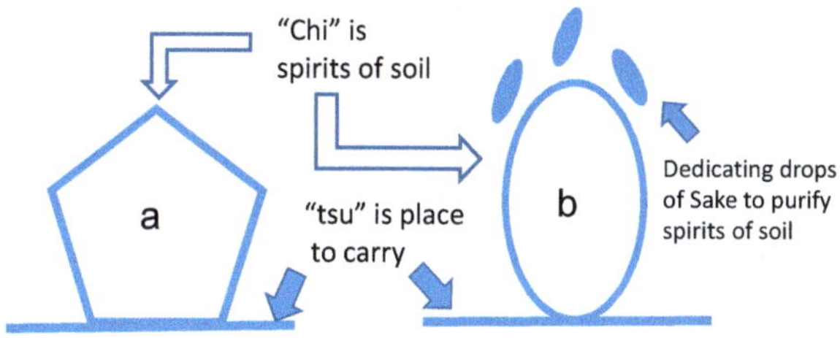

Fig. 11.9 (**a**) Letter for soil in oracle bone script 1, indicative of soil and/or land guardian. (**b**) Letter for soil in oracle bone script 2, indicative of purifying soil and/or land by dedicating sacred Japanese fermented wine (Sake in Japanese) to a guardian

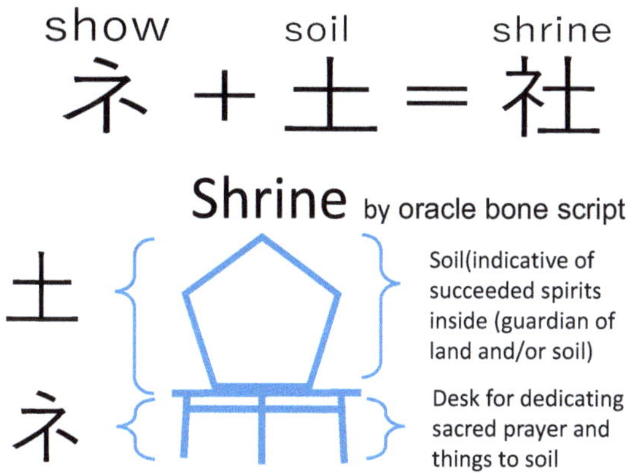

Fig. 11.10 Letter 社(shrine) in oracle bone script, indicative of dedicating sacred prayer together with sacred food and/or something to the guardian of soil

changed to the new concept of soil and/or land. The Chinese character 社 (shrine: "sha" in Japanese, made of the Chinese characters 土 and ネ; Fig. 11.10) came to be used as an expression for a place where the guardian (composed of spirits) of a land resided. At present this concept is called 神社 ("jinja" in Japanese: literally means "jin": guardians + "ja" or "sha": a place where a guardian resides, indicative of a place where guardians reside together).

In summary, there are four meanings of "tsuchi," (1) 土, (2) 土地, (3) 霊, and (4) 社. It is believed that the oldest meaning of "tsuchi" is (3) spirits: the place with sacred power; and the next oldest meaning is (4) shrine: the house and/or desk for the spirits; the third oldest is (1) soil: bringing all living things forth in nature; and the final one is (2) land: to indicate soil and/or ground.

11.4 Natural Spirits to Deities Related to Soil in Japan

The myriad *kami* in Japan – a word that corresponds to the spectrum from "nature spirit" to "deity" and Godhead in English – reside in the natural world. There are the *kami* (divine beings) of the mountains, the *kami* of the sea, etc. *Kami* are all around us in everything and every person (Jingushicho, 2015b), so there are also the *kami* of soils. It is important to introduce what *kami* are. The following is an excerpt from the book "Soul in Japan" (Jingushicho, 2015b):

> Since ancient times, Japanese have expressed the divine energy or life force of the natural world as kami. Kami derived from nature, such as the kami of rain, the kami of wind, the kami of the mountains, the kami of the sea, and the kami of thunder have a deep relationship with our lives and a profound influence over our activities. Individuals who have made a great contribution to the state or society may also be enshrined and revered as kami. Nature's severity does not take human comfort and convenience into consideration. The sun, which gives life to all living things, sometimes parches the earth, causing drought and famine. The oceans, where life first appeared, may suddenly rise, sending violent tidal waves onto the land, causing much destruction and grief. The blossom-scented wind, a harbinger of spring, can become a wild storm. Even the smallest animals can bring harm-the mouse that eats our grain and carries disease, and the locust that devastates our crops. It is to the kami that the Japanese turn to pacify this sometimes calm but at times raging aspect of nature. Through ceremonies, called 'matsuri', they appease the kami and wish for further blessings. Shinto observes no one single, omnipotent Creator. Each kami plays its own role in the ordering of the world, and when faced with a problem, the kami gather to discuss the issue in order to solve it. This is mentioned in records from the eighth century which tell the story of the Divine Age before written history began, and is the basis for Japanese society's emphasis on harmony, and the cooperative utilization of individual strengths.

In ancient times, people regarded the natural world as sacred without the need for special buildings, as the *kami* were believed to exist everywhere. Later, dwellings were built for the *kam*i in the forests, structures to be renewed in perpetuity where rituals could be conducted. This is the origin of the shrines known as "Jinja" (shrine). There are more than 80,000 Jinja in Japan today where various *kami* are enshrined. Based on a field survey of the oldest shrine in Japan, the translated story says that there was no building for the *kami* in ancient times, but the mountain itself was regarded as the *kami*. Later, in the era of Sujin (B.C.E. 97–B.C.E. 30), the *kami* was enshrined in the form of a shrine with deep gratitude for guarding the nation. The Jinjya Honcho was formed with the consensus of 80,000 Shinto shrines throughout the Japanese archipelago in 1946. Since then, the Grand Shrine of Ise has been respected as the most prominent shrine in Japan. In the Grand Shrine of Ise (Ise jingu), there is the Outer Shrine, Geku (also officially known as "Toyouke Daijingū"), dedicated to Toyouke-Ōmikami, the deity of agriculture. There is also a shrine called "Tsuchi"-no-"miya" to enshrine the *kami* who protects the precinct of the Geku (Jingushicho, 2015a). Nara and Minami (2016) suggested that these *kami* are guardians of the land and/or the ground. There is a shrine called "Kaze"-no-"miya" in

Geku to enshrine a couple of the *kami* to rule wind and rain which are necessary for the growth of grains. These *kami* are regarded as guardians of sky and/or heaven. Prayers are delivered to these *kami* with deep gratitude for their blessings such as food and/or water from the land and the sky.

Thus, "地" ("di" in Chinese pronunciation, ground, land and/or earth) is often used in contrast to "天" ("tian" in Chinese pronunciation, sky and/or heaven); "地" manifests its power in the form of bountiful productive capacity and brings forth new life through stimulation by "天." For example, it is believed that rain and wind are proof that 天 (heaven) and 地 (earth) have unified (Minami, 2014) and that thunder and lightning were the effects of this heaven–earth synergy. Likewise, life is gifted by bountiful productive capacity from soil or earth and sky or heaven.

Of the 80,000 shrines in Japan, there are 607 shrines in various parts of Japan that worship the soil. The ancient Japanese had respect for soil as their guardians, that the guardians had diversity, and that the diversity came from the fact that the soil had diversity in different regions (Nara & Minami, 2020). From the contents in "the Ten Thousand Leaves, Man'yoshu," written in the ancient Japanese language and edited around the middle of the eight century, it can be seen that the lives of the ancient Japanese were closely related to the soil peculiar to their local environment which determined the effects of the heaven-earth synergy. The ancient Japanese, therefore, had a longing for the soil and heaven-earth synergy as supporters of their lives (Minami, 2017).

Before the Japanese eat meals, we make it a rule to put our hands together and say "itadakimasu" ("I humbly receive"). It is important for us to say "itadakimasu" in order to express deep gratitude for the meals, even one grain of rice and one drop of water. They were made through the mercy of plentiful life, of living creatures, and we give thanks to the grace given by heavenly and earthly blessings, in other words, the lovingkindness and compassion of buddhas (Shojyushinin Kanko Kai, 2018). If we know all the meals we eat are blessed by "tsuchi" (soil), which is the source of sacred power for spirits that influence our lives as shown in the ancient recognition, then naturally we put our hands together and deliver deep thanks to "tsuchi" and/or the *kami* related to soil in addition to those that are buddhas.

11.5 The Guardians: Paddy Fields, Mountains, Water, and Others

Paddy field's guardians watch over paddy rice cultivation during the seasons of seeding, transplanting, harvesting, and threshing. After that, during autumn, the guardians return to the mountains to reside there as mountain guardians. During the next paddy rice cultivation in spring, the mountain guardians descend to the paddy fields to watch over them again. It is believed that mountainous areas are sacred and purified religious places where the souls of ancestors reside and from which the mountain guardians descend. In reality, the water used to replenish paddy rice fields descends from watersheds in the mountains and supports the lives of villagers. A lack of water is a serious problem for paddy rice cultivation as well as daily life even if irrigation canals have been built in modern society. In ancient times, the ancestors must have struggled to get water for rice cultivation, which led to prayers for water guardians.

Thus, people enshrined paddy field guardians in mountainous areas, and rice farmers at the base of a mountain climbed up the mountain to offer prayers for a good harvest (Miyazaki, 1962). In ancient times, people used trees in the mountains for fuel and to build houses, resulting in clearing of the trees. This caused water shortages in the surrounding rivers. In response, farmers have planted trees in mountainous areas adjacent to paddy rice fields to increase the water holding and supplying capacity of the soils on the mountains (Toyama, 1993). Even though paddy rice fields are not located in mountainous areas, water is so imperative for daily life that people in ancient times looked to mountain watersheds to get the water they needed for life.

These ancient water supply systems still work today in mountainous areas where the *kami* of the water are enshrined and many people, including the large water companies, visit the shrines where the *kami* of water reside. The shrine for guardians that watch over rice cultivation, therefore, stands close to one for guardians of the water supply that comes from the watershed of given sacred sites. This situation can be observed in the vicinity of the Utsunomiya University experimental farm which is located along the Kinu River in Tochigi prefecture, Japan. Here, there is a shrine to enshrine water guardians and rice guardians.

In ancient times the construction of a shrine created implicit rules that made it impossible to develop the land in a watershed area. Due to the success of these implicit rules, modern societies have preserved the forest and the soil integrated with it. Beneath the trees in a forest protected in such a way, soil-forming processes are ongoing to produce good soil structure in the surface layer through the action of soil biota which utilize the dead leaves, branches, soil animals, etc. As a result, many tiny holes and paths (pores) with different diameters are created in the soil surface layer for the emanation of living creatures as shown by "地" in Fig. 11.8 and for infiltrating and holding water from rain drops. This function is considered a source of blessing from mountain guardians and/or water guardians responsible for water and nutrient supply to many kinds of living creatures, including human beings.

At the oldest shrine in Japan, named "Oomiwa," the whole mountain itself is considered guardians (Oomiwa Jinja, 2014). Forest trees have been conserved and maintained for many years dating back to the Sujin era of B.C.E. 97–B.C.E. 30 (Fig. 11.11a, b). There is plenty of water, such as the water that flows beside the road leading to the shrine (Fig. 11.11c) and/or the water for purifying our hands and mouth (Fig. 11.11d). Near the summit of the mountain and behind the shrine (Fig. 11.12a) there is a well in which sacred water springs (Fig. 11.12b), and many people gather to drink the sacred water. Based on the sign (Fig. 11.12c), this mountain and sacred water are also an object of Japanese faith and worship. Trees on the mountain have not been logged since at least the Sujin era. Without permission from the shrine, people cannot enter and climb the mountain. This mountain as a whole has been supplying water for people living in the surrounding area in a sustainable way for more than 2000 years. Water gathers to make a pond (Fig. 11.13a) at the base of the mountain and many kinds of living creatures visit

Fig. 11.11 The oldest shrine in Japan named Oomiwa. (**a**) shows an entrance, while (**b**) shows the road to the shrine with forest trees. The water falls besides the road and the staircase to the shrine (**c**) continuously. (**d**) shows the place to purify the body by washing hands and rinsing the mouth before entering the sacred land of the shrine. The snake image pouring water from the mouth is impressive. It is said that the snake is an incarnation of the guardian (Oomiwa jinjya, 2014). Since the mountain as a whole is a guardian, the snake might be a representative living creature into which water and foods from the mountain are condensed

Fig. 11.12 The oldest shrine in Japan (**a**). Behind it, there is a well in which the sacred water springs. People are gathering in search for it (**b**). Based on the board (**c**), the shrine was enshrined around 2000 years ago. At that time, there was no building like the figure in the left

Fig. 11.13 The guardian for water is enshrined beside the pond in the lower part of the mountain. Many living creatures are brought up in the pond

there. Another function of the pond is to suppress flooding and to serve as a relay point to supply water to people living in the surrounding area. The shrine of guardians for water is located beside the pond (Fig. 11.13b).

Moreover, the oldest water guardians in the shrine of Niukawakami, different from the previously mentioned oldest shrine of Oomiwa, were enshrined in 675. Japanese religious faith is explained in the following text from the shrine's website (http://web1.kcn.jp/niukawakamijinja-kamisha/english.html, accessed 2020-04-09):

> Niukawakami-Jinjya-Kamisha (in Japanese, kami: upper part sha: shrine) Shrine enshrines a dragon. The name of this dragon is Takaokami-no-Okami. In Japan, a dragon is a god (guardian) who controls water or rain. So Japanese worship a dragon because water is important to agriculture. Takaokami-no-Okami is a dragon who lives in the peak of the mountain, while there is another dragon in the bottom of the valley. Its name is Kuraokami-no-Okami, enshrined in Niukawakami-Jinjya-Shimosha (in Japanese shimo: lower part) Shrine. In Japan, a lot of shrines deify Takaokami-no-Okami. The place of Niukawakami has been sacred for a long time. In 675, a shrine was established for controlling rain.

There are two different kinds of dragons that control water or rain: one lives at the summit of the mountain (Fig. 11.14) and the other lives in the bottom of the valley where plenty of water gathers to make a river and a pond (Fig. 11.15). This view is almost identical to the previous example at Oomiwa shrine. From the standpoint of the Sustainable Development Goals (SDGs), the reason why the mountain as a whole deserves worship would be derived from its important function to supply water in a sustainable way. By keeping the mountain forested, resilience is built against drought and/or flooding. If the mountain controls water in a sustainable way, living creatures are brought up without growth inhibitions. This precious sustainable and altruistic function could be viewed as guardians, and people could have deep gratitude to and/or worship the mountain for this service. The series of ecological systems provide functions once or still attributed to guardians that deserve worship by people. The guardians can support people and all other living creatures from the present to the eternal future as long as the ecosystem is managed in ways that do not exploit the mountain and induce degradational processes, such as soil erosion.

Although not practiced based on these cultural and spiritual beliefs, the parts that appear on this surface, i.e., sustainable supply of water from a mountain watershed

Fig. 11.14 The oldest water shrine in Japan dating back to the year 675. The shrine enshrines a dragon (Right). The name of this dragon is Takaokami-no-Okami and he lives in the summit of the mountain

Fig. 11.15 The oldest water shrine in Japan dating back to the year 675. The shrine enshrines a dragon which name is Kuraokami-no-Okami and lives in a valley

Fig. 11.16 Local water supplying system from watershed forest in the northeast of Thailand. (**a**) Water tank adjacent to village. The villagers cooperated with each other to build a water supplying system by providing free work. (**b**) Water pipes from the water tank to households. The number of the pipes is increasing with time, contributing to the villagers' life. (**c**) Water pipes from the water tank to households

where logging is prohibited, may be relatable to other places with different cultural backgrounds. Hirai et al. (2007) investigated the development of villages due to the introduction of water supply systems in northeast Thailand, where shifting cultivation is widespread over sloping lands. Villagers there recognized that continuously cutting trees for agricultural use decreased the water volume supplied from upstream portions of the watershed, inducing a lack of water. After construction of a water supply system to every household in 1992 (Fig. 11.16), the villagers agreed on several rules to prevent excessive cutting of trees from the forest to protect the watershed in the mountains. The water supply systems have worked to support the villagers' needs ever since 1992, even during the dry season, and hence the standard of living has gradually been improving year by year. This indicates that protection of forested watersheds with participatory contribution of local people in a modern society could contribute to sustainable water supply in places with different cultural backgrounds than Japan. In terms of SDGs, this supply of water in a sustainable way would be applicable to the realization of Goal 6 to ensure availability and sustainable management of water and sanitation for all.

11.6 Understanding Soil Based on Buddhist Philosophy from a Japanese Perspective

The function of guardians to supply and control water could be related to soil formation and/or soil genesis. Practices that conserve the forest can also facilitate soil formation (soil genesis) through processes such as decomposing organic matter and the succeeding formation of humus and/or soil organic matter (Fig. 11.17). These processes are driven by many living creatures, such as soil animals and microorganisms living together in the soil or on leaves on the forest floor as a cooperative existence. Through these forming processes, a porous surface horizon is formed, resulting in the development of the function to infiltrate rainwater into the subsoil via the surface horizon. By infiltration of water into the subsoil through the surface soil, primary minerals in the parent rock are transformed into clay minerals and sesquioxides.

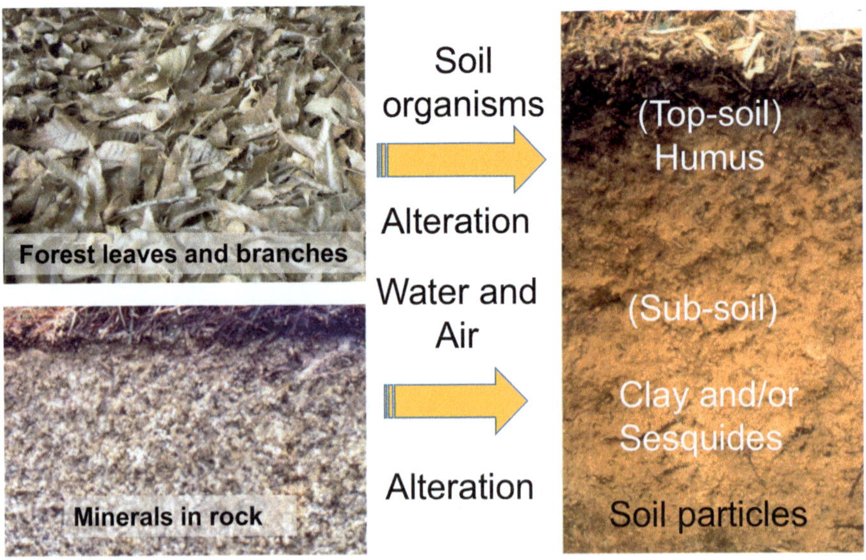

Fig. 11.17 Illustration of soil forming process

It is imperative to discuss views and ideas on mountain trees, leaves, living creatures in soil, rocks, and minerals from a Buddhist perspective in Japan. Sueki (2017) provides a detailed discussion of whether or not plants and trees would become buddhas. In addition, Takemura (2016) has compiled discussions on the idea that plants, trees, and the lands would be buddhas, and then introduced the contents of the book written by Chu-Jin (1065–1138) of TENDAI, a Buddhist monk. Considering the contents, the true nature of roots, stems, branches, and leaves of plants and trees themselves are identical to the substance of the buddha. When buddhas (people who have attained buddhahood) make observations on plants, trees, and lands, what they see on them in prayer is none other than the substance of the buddha. Finally, Takemura (2016) concluded that these arguments would suggest that there is no distinction among trees, plants, lands, and people from the viewpoint of the substance of the buddha. Moreover, Takemura (2016) introduced the statement of Dogen (the founder of the Soto Buddhist school, 1200–1253) that mountain and water are always preaching as buddhas to people and that people possess buddha natures and trees, plants, and lands possess buddha natures.

Sueki (2017) examined in detail the contents of the work 草木成仏論 (Soumoku-jyoubutu-ron) (literal translation is "plants and trees become buddhas") by Annen (841–915). Annen argued that there must be shinnyo in plants and trees that cannot be felt by the human senses or subconsciously by deep consideration in Buddhist prayer with buddhahood. Moreover, when a certain factor or condition works on invisible shinnyo, it turns into visible plants and trees with buddha natures.

From a Buddhist point of view in Japan, based on the literature above, soil generated by the soil-forming factors is what the invisible shinnyo changes due to invisible local environmental factors to form a unique soil with a buddha nature. In other words, when invisible factors (edges) work to change shinnyo into a certain form, various soils with individual buddha natures around the world are created.

Living creatures that support soil-forming processes have buddha natures from creature to creature, based on a concept that all sentient beings possess a buddha nature (Anton, 2009; Sueki, 2014; Kido, 2019). Also, the individual buddha nature within one individual creature makes it a rule to cooperate with each other and live together in harmony to generate collective buddha natures. The following example helps to explain this; "Rafter, pillar, wall, roof, and land are inevitable to build a house." This figurative example means that different construction materials come together to make one house. This example could be utilized in order to explain that plants, trees, and lands are all buddhas (Takemura, 2016). Taking this figurative example into consideration, the tiny holes and paths in the surface soil layer have been generated by cooperative activities of many living creatures with individual buddha nature and/or that are all buddhas. This function can be considered to result from the lovingkindness and compassion of all buddhas or all buddha natures.

11.7 The Importance of Paddy Soil-Based Rice Production from a Japanese Perspective

Cultivated soils are represented by 壌 ("rang" in Chinese) (Fig. 11.18). 壌 is composed of 土 and 襄. The meaning of 襄 is to sandwich waste threads in 衣 of 襄, indicative of alteration by people to cultivate soil and add something to 土 ("tsu", natural soil) suitable for grain production (Fujiwara 1991). To deepen understanding of representative paddy soil characteristics, paddy soil profiles are shown in Fig. 11.19. The water utilization system in paddies results in an alternation between oxidizing and reducing soil-forming processes that work over many years to create iron and/or manganese accumulations in the subsurface layer (Fig. 11.19), particularly around plant roots. In some locations, the Japanese people have been cultivating paddy soils for more than 1000 years in the same field every year. Tiny soil particles such as the humus and/or soil organic matter and clay that formed under the temple forest as shown in Fig. 11.17 are also incorporated into the paddy

Fig. 11.18 Illustration of the Chinese character 壌 ("rang" in Chinese)

(a) (b)

Fig. 11.19 Paddy soil profiles in Kagawa prefecture, south-east Japan, showing A horizon, conspicuous humified surface soil layer, is deeper than 15 cm. (**a**) Lowland paddy soil developed on sandy fulvic material. (**b**) Lowland paddy soil developed on clayey fulvic material top-soil layer to replenish soil fertility. Based on King (1911), rice consumption in Japan was 1370 kg/ha/year/capita in the early 1900s, but today the average rice consumption is less than 600 kg/ha/year/capita. Although rice is the main Japanese crop, students in the Japanese educational system do not have a chance to learn the rice production system and how to grow rice in paddy soils during their elementary, junior high, and high school years.

To change this, soil education in a paddy field was provided and reported in Hirai (2019). The lecturer conveyed that a rice hill (Fig. 11.20) was produced from three rice plant seedlings which were grown from three seeds of unhulled rice that were planted into the paddy soils in May without fertilization. The students recognized that the limited number of seed grains (seeds) planted in the form of seedlings into soil (壌, "rang") led to the seedlings absorbing water, nutrients, and energy from soil, irrigation water, and sunlight and resulted in amazing returns. Just as soil under the forest is important in the supply of water, as discussed earlier, the irrigation water for rice paddy production is also derived from soil (土, "tsu") under forests. The lecturer then took the students to a soil profile to show topsoil and sub-soil so they could touch it. The goal was for the students to feel something about the value and preciousness of soil, so they would respect and give thanks to soil. Using these observations, the lecturer might make note of various kinds of blessings produced from ecological services; in other words, the invisible altruistic power of heavenly and earthly guardians or spirits to support life on earth. Not even one grain or one drop of water can be produced without blessings from top-soil and sub-soil produced by soil-forming processes which have been facilitated for a long time by organisms.

The cultural understanding of soil from a Japanese perspective has been affected by China's ancient culture (Confucian), Buddhism, and Shinto. The soil-forming processes have been occurring for a long time on earth. They are the driving forces

(a) (b) (c) (d)

Fig. 11.20 Paddy rice cultivation from rice seedlings to harvest on paddy soil, showing that planting three rice seedlings in one hill increased up to more than 1000 unhulled rice along the panicle. (**a**) Paddy rice seedlings for planting in May 2018. (**b**) Harvest time in September 2018. No fertilizer was used there. (**c**) Selected rice plant with average panicle number. (**d**) More than 1000 unhulled rice along panicles in one hill produced from three seedlings that allow many functions to flourish and provide soil ecological services and could be recognized as the real source of altruistic soil blessings regenerated by innumerable life with spirit, in other words, the lovingkindness and compassion of *kami* (guardians) and buddhas (buddha natures).

Acknowledgments We would like to deliver sincere thanks to the editors for their untiring suggestions and comments. Without them, this chapter could not have been finalized.

References

Anton, P. (Ed.). (2009). *SHINJO reflections*. Somerset Hall Press.

Chen, N., & Zhang, G. (2018). Soil proverbs in China. In J. E. Yang, M. B. Kirkham, R. Lal, & S. Huber (Eds.), *Global soil proverbs: Cultural language of the soil*. Catena-Schweizerbart.

Fujiwara, A. (1991). 土と日本古代文化, 博友社, 東京 [Soil and Japanese ancient culture]. Hakuyuu-sha (in Japanese).

Hirai, H. (2019). カレーライス一杯の御飯(お米)に必要な表土の面積を計算してみよう, 理科教室 [Let's try to calculate the area of top soil necessary to produce the rice in a bowl of curry]. *The Journal of Science Education, 62*(1), 2–7 (in Japanese).

Hirai, M., Hirai, H., & Chino, J. (2007). Roles and cooperation in a village, NGO and local government on the development of a village community. *Journal of Agricultural Development Studies, 18*(2), 6–18. (in Japanese with English summary).

Inukai, T. (2015). *Spread of Chinese characters: From China to the Japanese Islands through the Korean Peninsula* (Vol. 194, pp. 237–244). National Museum of Japanese History Repository (in Japanese).

Jingushicho (Jingu Administration Office) (2015a). *Precinct map: Illustrated map*. Geku. http://www.isejingu.or.jp/en/map/pdf/map_geku_en.pdf. Accessed 04 Apr 2020.

Jingushicho (Jingu Administration Office) (2015b). *Soul of Japan. An introduction of Shinto and Ise Jingu*. http://www.isejingu.or.jp/en/pdf/soul-of-japan.pdf. Accessed 09 Apr 2020.

Kido, S. (2019). *Sculpting the Buddha within: The life and thought of Shinjo Ito*. Wisdom Publications.

King, F. H. (1911). *Farmers of forty centuries, or permanent agriculture in China, Korea, and Japan*. Organic Gardening Press.

Lin, P. (1996). *Soil classification and land utilization in ancient China*. Science Press (in Chinese). (Cited from Minami (2014)).

Minami, K. (2009). Soil and humanity: Culture, civilization, livelihood and health. *Soil Science and Plant Nutrition, 55*(5), 603–615. https://doi.org/10.1111/j.1747-0765.2009.00401.x

Minami, K. (2014). Soil and cultures: The etymology, Chinese ideogram and proverb of soil. *Bulletin of The Institute for Agriculture, Medicine, and the Environment, 2*, 1–36. (in Japanese).

Minami, K. (2017). The soil appeared in the ten thousand leaves, Man'yoshu: The origin and meaning of beauty in green earth, yellow soil and red soil. *Japanese Journal of Soil Science and Plant Nutrition, 88*(6), 568–573. (in Japanese).

Miyazaki, S. (1962). Paddy field's guardian, mountain's guardian, and water guardian. *Akita, 5*, 51. http://common3.pref.akita.lg.jp/koholib/search/html/005/005_054.html (in Japanese). Accessed 09 Apr 2020.

Nara, Y., & Minami, K. (2016). Ise Jingu and soil. In I. Taniyama (Ed.), The soil and world gods. *Japanese Journal of Soil Science and Plant Nutrition, 87*(2), 147–148. (in Japanese).

Nara, Y., & Minami, K. (2020). The gods of soil found in KOJIKI: Records of ancient matters in Japan. *Japanese Journal of Soil Science and Plant Nutrition, 91*(2), 90–93. (in Japanese).

Oomiwa Jinja. (2014). http://oomiwa.or.jp/jinja/ (in Japanese). Accessed 09 Apr 2020.

Shimoizumi, Z. (2019). 密教の仏がわかる本, 大法輪閣, 東京 [Book to understand the buddhas of esoteric Buddhism]. Dai-Hourinkaku.
Shinjo Kanko Kai (publication society) (2016). 真乗-心に仏を刻む-, 中公文庫, 中央公論新社 [Shinjo-Sculpting the Buddha within, Chukobunko]. Chuokoron-Shinsha Inc (in Japanese).
Shojyushinin Kanko Kai (publication society) (2018). 摂受心院-その人の心に生きる, 中公文庫, 中央公論新社 [Shojyushinin – live within one's mindset, Chukobunko]. Chuokoron-Shinsha Inc, (in Japanese).
Sirakawa, S. (1995). *Jikun*. Heibonsha (in Japanese).
Sirakawa, S. (1996a). *Jitsuu*. Heibonsha (in Japanese).
Sirakawa, S. (1996b). *Jito*. Heibonsha (in Japanese).
Sueki, F. (2014). 日本仏教入門, 角川選書, 東京 [Introduction to Japanese Buddhism]. KADOKAWA (in Japanese).
Sueki, F. (2017). 草木成仏の思想 :安然と日本人の自然観, サンガ文庫, 東京 [Philosophy that grass and trees become Buddha]. Sangha Publication (in Japanese).
Takemura, M. (2016). ブッディスト・エコロジー-共生・環境・いのちの思想, ノンブル社 [Philosophy of cooperative existence and/or living together, environment and life] (pp. 1–309). Nonburusha publishing company (in Japanese).
TENDAISHU (2015). *Dengyo-Daishi's life and teaching*. http://www.tendai.or.jp/english/chapter01.php. Accessed 09 Apr 2020.
Toyama, K. (1993). 日本の米:環境と文化はかく作られた, 中央公論新社, 東京 [Rice of Japan: The ways of creating the environment and the culture]. Chuokoron-Shinsha (in Japanese).
Wen, S. X. (this volume). Cultural understanding of soil in China. In N. Patzel, E. C. Brevik, S. Grunwald, & C. Feller (Eds.), *Cultural understanding of soils*. Springer.

Chapter 12
Ecosattvas and Ecodharma: Modern Buddhist Perspectives of Soil and the Environment

Sabine Grunwald

12.1 Introduction

12.1.1 Are Buddhist Forest Monks Environmentalists?

Theravāda Buddhist forest monks in Thailand, Sri Lanka, and Burma have maintained the ancient tradition to practice meditation in the forests. One may assume because these monks live in the forest that they are closely connected to the trees in the forest and engaged to protect the environment. The destruction of forest ecosystems in these places has been extensive. Surprisingly, many of the forest monks have not actively engaged in conservation and ecological matters and acted rather indifferent toward their environment. Few of the forest monks have developed environmental rituals (e.g., ordaining trees) or engaged in activism to stop the dramatic deforestation (Darlington, 2017). Buddhist monks are assumed to be compassionate and loving of all sentient beings of which many live in the forest and soils beneath them (Fig. 12.1).

S. Grunwald (✉)
Soil, Water, and Ecosystem Sciences Department, University of Florida, Gainesville, FL, USA

University of Florida Mindfulness Program, Gainesville, FL, USA
e-mail: sabgru@ufl.edu

N. Patzel et al. (eds.), *Cultural Understanding of Soils*,
https://doi.org/10.1007/978-3-031-13169-1_12

Fig. 12.1 Monk meditating in the forest (Buddhist Forest tradition). (Creative common license https://commons.wikimedia.org/wiki/File:Buddhist_monk_in_Khao_Luang-Sukhothai.JPG)

Do Forest Monks Lack Compassion Toward the Environment Including Forests and Soils?

According to Kamala (1997), forest monks seek out the forest because historically it was considered wild and unknown and offers seclusion from society and silence and isolation for ascetic meditation practices. In *Theravāda* Buddhism, liberation (*nirvāṇa*)[1] from cyclic existence (*saṃsāra*) is realized through the *arhat* ideal, which represents the purified saintly one who has transcended all desires, attachment, conditioning, and defilements in personal enlightenment (Samuels, 1997). Most of the Thai forest monks are not outwardly concerned with the condition or health of the forest, because their aim is focused on individual meditation practice and enlightenment rather than environmental activism or socio-political engagement (Darlington, 2017). These *Theravāda* monks seek perfect and complete absorption (*samadhi*) aimed at their own individual liberation. They meditate so deeply in a state of intense concentration that they literally have no memory or awareness of their surrounding environment because the earth and everything else *just is*. This attitude has stirred controversy and questions in regard to Buddhists' environmental ethics, or lack thereof. Schmithausen (1997) echoed concerns about early Buddhist traditions, such as *Theravāda*, which lacked environmental awareness.

[1]All original Buddhist terms in this chapter are in Sanskrit.

The early Buddhist *Theravāda* school was attuned to the first turning of the wheel of dharma emphasizing a lack of an inherent existence of the self and promoting belief in the mutual dependent co-arising of all phenomena, while the second turning of the wheel (i.e., emptiness of self and phenomena) was based on the *Mahāyāna bodhisattva* ideal. The third turning of the wheel of dharma points to the realization of Buddha Nature (Buswell & Lopez, 2014; Ray, 2000). These turnings represent major shifts in Buddhist views that have impacted relations with the environment, the collective of all sentient beings, and oneself. The purpose of this paper is to explore select pivotal Buddhist philosophies and their understanding of nature/environment. According to Buddhist nomenclature, "soil" is considered nature/environment, matter, form, phenomena, and the *dharma*[2] which are conceived as an aspect of truth or reality (Bhikkhu, 1993; Ray, 2000; Williams, 2010). Further details in regard to the etiology of soil (±) in the Japanese Buddhist and Chinese Buddhist context can be found in Chaps. 11 and 13 respectively. Buddhism originated in Asia over 2500 years ago and is still practiced and part of cultures in Tibet, China, Japan, and many other countries (Fig. 12.2). In North America Buddhism is relatively young being introduced by Buddhist teachers from the East in the 1970s, commonly referred to as Buddhist Modernism with secular and hybridized Buddhist forms that have since formed (Fig. 12.3).

This chapter was written from the perspective of American/Western Dharma (Gleig, 2019) also referred to as Buddhist Modernism (McMahan, 2008)—hybridized Western Buddhist forms with Asian influence (Oriental Buddhism)[3]. There is not one Buddhism, but many different Buddhist lineages, schools, traditions, and

[2] *Dharma* (Sanskrit) or *dhamma* (Pali) is a polysemous term and has been notoriously difficult to translate into English. According to Buswell and Lopez (2014) in *The Princeton Dictionary of Buddhism* (p. 246): The term dharma derives from the Sanskrit root "dhr," which means "to hold" or "to maintain." In Vedic lit., *dharma* is often used to refer to the sacrifice that maintains the order of the cosmos. …. Many European translators rendered *dharma* into English as "law," similar to Chinese translators. In Buddhism, *dharma* has a number of distinct denotations. One of the most significant and common usages is to refer to "teachings" or "doctrines" (e.g., teachings of the Buddha). A distinction was made between the *dharma* that is studied and heard (called the scriptural *dharma; AGAMA-dharma*); and the *dharma* that is made manifest in the consciousness of the practitioner (called the realized *dharma; ADHIGAMA-dharma*). A second, and very different principal denotation of *dharma* is a physical or mental "factor" or fundamental "constituent element" or simply "phenomenon" (constituent elements of existence). A third denotation of the term *dharma* is that of "quality" or "characteristic." Reference is often made to *dharmas* of the Buddha, referring in this sense not to his teachings but to his various auspicious qualities, whether they be physical, verbal, or mental. Sometimes this is referred to as the meaning of *dharma* in the term *dharmakāya* (truth body; transcendent qualities of the Buddha). The term *dharma* has gained prominence in American Buddhist Modernism to refer to hybridized Oriental-Western conceptions of Buddhist teachings (Gleig, 2019).

[3] Western Buddhism has roots in traditional canonical Buddhism (*Theravāda*) (Bhikkhu, 1993), specifically *Vipassana* as popularized by S.N. Goenka; *Sōtō* Zen (Suzuki, 1970); Vietnamese Buddhism (Thich Nhat Hanh, 2012); and Indo-Tibetan *Mahāyāna* and *Vajrayāna* Buddhism (Chögyam Trungpa Rinpoche, 2013), among others.

Fig. 12.2 Cave of the great Buddha. The Caves of the Thousand Buddhas located at Dunhuang, northwest China. (Free images Buddhism: https://snappygoat.com/s/?q=bestof%3Awisdom+zen+meditation+buddhist+tranquility+face+statue+serenity+buddhism+relaxation)

Fig. 12.3 Contemporary meditator in nature. (Creative Common license https://caregiverdave.com/how-yoga-and-meditation-can-help-caregivers-to-stay-fit/)

practices that have contributed to Buddhist Modernism in the West[4] and are beyond the scope of this chapter. Here, I aimed to focus only on the main dharmic strands that are relevant to express relations between people and nature, soil, the Earth, form, phenomena. I am a long-time meditator and have practiced and studied in various Buddhist traditions, specifically *Mahāyāna* and *Vajrayāna* (*tantric*, esoteric) Buddhism as well as transpersonal psychology with emphasis in Buddhist Modernism.

12.2 Ecodharma

According to Loy (2019), ecodharma combines ecological concerns (eco) with the teachings of Buddhism and related spiritual traditions (*dharma*). Western system theory and Buddhist philosophy share in common the principle of interconnectedness that views planet Earth and sentient life as a magnificent web (The Karmapa Ogyen Trinley Dorje, 2017). Soil is part of this net that is a continuum of change. In Buddhism the doctrine of dependent origination (*pratītyasamutpāda*) refers to the understanding that nothing has self-existence because everything is dependent on other things. This doctrine views reality as an ongoing, impermanent, and interdependent karmic flux (Esposito et al., 2015). *Karma* (action) refers to causal connectedness of life which determines future consequences (Smith & Novak, 2003). "*Karma* of results" refers to assumption that the past and present circumstances of one's life are the results of actions performed at a previous time, while "*karma* of cause" refers to the actions one performs now which will condition one's future (Ray, 2000). Thich Nhat Hanh (1987) expressed the Buddhist understanding of interconnectivity with everything else most poignantly with a new word he coined—*interbeing*. We 'inter-are' with one another and with all life. However, a desacralized rationalistic worldview that favors separation of the natural world and people and freely exploits the natural world for self-centered gain, and thus, has profoundly contributed to today's ecological crisis, is antithetical to the principles and doctrines of Buddhism (Loy, 2019). Loy calls for ecodharma rooted in collective *karma* to respond to the global, collective, and institutionalized suffering *(dukkha)* harming soils, water, and climate.

[4] Usually what is referred to as "the Western World" includes North America, Western Europe, and Australia. In this chapter, "the West" (the Occident, Latin *sunset West*) contrasts the Orient or Eastern world. Buddhism originated in the East more than 2500 years ago and has been introduced in the West only recently (past 150+ years).

Fig. 12.4 Buddha meditating. Source: Alamy

12.3 A Love Letter to the Earth

Love of the Earth—Have you ever been completely in love with someone or something?

Thich Nhat Hanh (2013), one of the best-known Buddhists of this time, poetically shared the profound teachings of *Mahāyāna* Buddhism in his book *A Love Letter to the Earth*. Deeply loving someone or something refers to the experience of a nondual state of being in which there is no separation between oneself and the other person or thing. In Buddhism, this shift toward nonconceptual perception is called emptiness (*sūnyatā*) of self and emptiness of phenomena (i.e., material objects, physical body, mental states, and everything knowable; Khensur Rinpoche Jampa Tegchok, 2012). When we open-heartedly love that way, it evokes compassion (*karuṇā*) and care in us, because anything that would bring harm or suffering to the other person or thing would also mean harm and suffering to one's own self (Fig. 12.4). Buddhism acknowledges that life is unsatisfactory, it is *dukkha*. Suffering is all around us—the global public health crisis of COVID-19, unemployment and economic hardship, loss of a loved one, old age and disease, global climate change, food insecurity, soil degradation, anxiety concerning an uncertain future, or simply "my expectations are not met in this very moment."

Buddhism asserts that personal and collective suffering is unavoidable, entailing the suffering of all sentient beings and separation from what one loves (First Noble Truth), caused by ignorance, desire, and attachment (Second Noble Truth), with the possibility of the cessation of suffering (Third Noble Truth), realized by means of the noble eightfold path (Fourth Noble Truth) (Smith & Novak, 2003). The idealized striving to do no, or realistically the least possible harm extends not only to people but also to the environment and the Earth. Thich Nhat Hanh (1998) pointed out that being in love with the Earth in this nondual way brings great joy and nourishment. According to Buddhist beliefs, the realization of non-separation is liberation (*nirvāṇa*) on the path to Buddhahood. This love of the Earth is all-encompassing:

> "At this very moment, the Earth is above you, below you, all around you, and even inside you. The Earth is everywhere. You may be used to thinking of the Earth as only the ground beneath your feet. But the water, the sea, the sky, and everything around us comes from the Earth. Everything outside us and everything inside us comes from the Earth. We often forget that the planet we are living on has given us all the elements that make up our bodies. The water in our flesh, our bones, and all the microscopic cells inside our bodies all come from the Earth and are part of the Earth. The Earth is not just the environment we live in. We are the Earth, and we are always carrying her within us." (Thich Nhat Hanh, 2013, p. 8)

Buddhist philosophy asserts that each one of us is interconnected with the Earth (The Karmapa Ogyen Trinley Dorje, 2017; McMahan, 2008). In contrast, if we think about the Earth as the environment around us to serve and support us, we experience ourselves and the Earth as separate objects (duality). As Thich Nhat Hanh (2013) elaborates:

> "But don't think that Mother Earth is outside of you. Looking deeply, you can find Mother Earth within you, just as your biological mother who gave birth to you is also within you. She is in each of your cells." (p. 13)

In general, Buddhist traditions assert that ultimately the people and planet Earth are one and the same (The Karmapa Ogyen Trinley Dorje, 2017). The Indo-Tibetan *Mahāyāna* Buddhist view of this micro-macro cosmic mirror implies that the totality of all being (i.e., the universe or "all there is") is mirrored in each phenomenon (e.g., human body, tree, forests, insect, an atom or a specific soil) (Grunwald, 2021; Morley, 2008; Powers, 2007; Ray, 2016). This view moves beyond the dualistic concepts of protecting, managing, cropping, using, exploiting, or conserving the environment or the Earth for our own personal survival or gains.

The earliest Buddhist doctrine, the *Abhidharma of the Theravāda* Buddhist school (e.g., *Abhidharmakośabhāṣya or Lokānuvartanā Sūtra*), asserted that conventional existents (e.g., a table, chair, soil, or a person) are mental constructs constantly changing (impermanence) in a continuous stream as a result of causes and conditions (dependent origination). Here, the nature of reality is considered empty of intrinsic existence (*svabhāva*) meaning that nothing exists independently (Ronkin 2018; Williams 2010). For example, when applied to soils and nature, this means that a soil particle is considered impermanent and constantly changing as hydrologic flow, rainfall, microbial or earthworm activity, or any of the other connected processes in the web of life that touch and transform the soil particle in an ongoing and never-ending stream of change. The use and management of soils may also change soil particles. *Theravāda* Buddhist thought undergirds that the law of *karma (cause and effect) considers* soils part in the universal changes observable in nature, inviting people to respect all life because we are inseparable from the world, entwined, all one evoking compassion and love for planet Earth. *Karma is considered the shaper of both our personal destiny and the destiny of the world (Batchelor & Brown,* 1994).

To what extent Buddhist doctrines, such as *Theravāda*, mis-/inform environmental ethics and led to dis-/engagement have been a matter of scholarly debate. Harris (1995) critiqued the lack of environmental consciousness claiming it is incompatible with early Buddhist *Theravāda* philosophy, while Schmithausen (1997) was more cautionary in his critique of the *Theravāda* in terms of environmental ethics. From the *Theravāda* point of view, deep meditative absorption (*samadhi*) is viewed as the most important goal on the liberative path, while environmental concern and protection are of lesser interest. One may acknowledge that forest monks practicing in the *Theravāda* tradition are personally invested so much in the path toward individual enlightenment that they completely forget about the forest, soil, water, and air—the whole environment.

Later Buddhist traditions, such as the prominent *Mahāyāna*, stressed emptiness and nonattachment, which informed a different relationship in regard to nature/environment/soil. According to Duckworth (2019a, b), there are two major schools in *Mahāyāna* Buddhism—"Mind-Only" school (*Yogācāra*) and "The Middle Way" school (*Mādhyamaka*) with different views of emptiness, metaphysical conceptions, approaches, and practices. The *Yogācāra* Buddhist school claims that reality is inconceivable and there is only the mind as a spiritual ultimate. This school uses meditation and phenomenological interpretation (i.e., direct subjective lived experience of consciousness) to interpret the world which is perceived as irreducible and inexpressible. In essence, from the Mind-Only perspective, conception collapses into perception with nondual self-awareness (called pure or natural awareness). The *Yogācāra* affirms ultimate reality that is considered empty (nonconceptual) (Duckworth, 2019a; Ray, 2000).

In contrast, the *Mādhyamaka* Buddhist school uses deconstructive ontology (logic) and critical inquiry (negations) to arrive at the understanding that denies ultimate existence; in other words, this school denies any concept of an ultimate and posits that intrinsic nature is absent ("emptiness of emptiness" or negation of emptiness). The negation approach rests on tetralemmas to explain the truth of life, nature/Earth, and the universe: (1) existence ("is"); (2) nonexistence ("is not"); (3) both, existence and nonexistence ("is both"); and (4) neither, existence and nonexistence ("is neither"). All four lemmas are negated in the *Mādhyamaka.* The negation approach deconstructs concepts to demonstrate their emptiness (Duckworth, 2019b). Applied to soil/nature/Earth (form, matter) negation looks like "soil is, soil is not, soil is both, and soil is neither"; therefore, all conceptions about soil are negative because soil is emptiness and even emptiness is empty. Applied to soils and nature both—the *Yogācāra* and the *Mādhyamaka* of *Mahāyāna* Buddhism—share in common that ultimately reality is empty irrespective of very different paths. Emptiness of soil or nature touches on ultimate reality that is totally inexpressible in form of language or logic.

This Buddhist view of nature/Earth as empty has evoked passivity and negativity even in the face of blatant ecological crisis (e.g., global climate change) and the destruction of the environment (e.g., soil carbon losses or erosion of soils) (Loy,

2019). Inherently, nature/Earth/soil viewed as empty may invite escapism into an ultimate reality that is nonconceptual yet inconceivable, and thus, does not motivate activism and engagement to take care of the environment. From the viewpoint of Buddhist cosmology, contemporary soil and ecological problems are considered part of human suffering because, after all, reality is considered ultimately empty (Keown, 2007). As Eckel (2010) pointed out, the Buddhist cosmology of emptiness does not motivate environmental movements or protection of the environment, while the Buddhist notions of interdependence (dependent origination) and *karma* serve as incentives to be kind toward the environment because people perceive themselves as part of the constant unfolding of nature and the cosmos.

In summary, emptiness of self and phenomena is reserved for the (few) enlightened ones with nondual awareness who have attained *nirvāṇa.* Swearer (2006) calls this liberative kind of Buddhist eco-philosophy "eco-constructivist," which suggests that one can construct a Buddhist environmental ethic, though not coterminous with Buddhist metaphysics and doctrines, such as *Mahāyāna* conceptions of emptiness which are not as accessible to people. In contrast, the "eco-apologist" is eco-friendly and attuned to the environment due to environmental awareness that arises out of Buddhist philosophy grounded in the Pali canon (e.g., interdependence or dependent origination), understandable for Buddhists and non-Buddhists alike. Swearer's "eco-contextualist" holds that the most effective Buddhist environmental ethics is context and situation based and a "living tradition" that needs to be adapted to new environmental problems such as global climate change. The latter "lived tradition" has been articulated by Gleig's (2019) *American Dharma* emphasizing that canonical and oriental Buddhist traditions have been hybridized into novel contemporary forms of the *dharma* in the West.

Loy (2019) provocatively asked if the ecological crisis is a symptom of a modern Buddhist crisis because many Buddhists have focused on personal meditation practice rather than engagement in social and ecological activism and protection of the environment. According to Loy, the favor of seeking spiritual transcendence and individual liberation as well as "spiritual materialism"[5] have overshadowed social and ecological engagement due to a nihilistic attitude of "nothing is real, therefore nothing is important" (p. 54). Buddhists clinging to emptiness has become a deterrent that devalues the world, nature, and its problems.

Another demotivating factor to engage in worldly problems, such as degradation of soil resources that support food security, is that conventional truth (also called relative truth, *saṁvṛiti*) is understood to be an illusion, whereas ultimate truth provides liberation from all human and worldly suffering (Duckworth, 2019b). This notion breeds an escapist attitude aiming to attain some transcendent state of bliss and escape into rebirth rather than attending to soil and the body, matter (form) that are considered dirty, impure, and ordinary (Gross, 1993). The

[5]The term spiritual materialism was coined by Chögyam Trungpa Rinpoche (2010). It expresses that in materialistic, individualistic, and achievement-oriented Western cultures, even spirituality is hijacked and commodified. In essence, the materialistic orientation turns spirit into a material thing that is consumed rather than practiced, embodied, or a spiritual belief.

liberative escape from earthly life into blissful *sukhāvatī*, which refers to the pure land of the *bodhisattva* Amitābha in *Mahāyāna* Buddhism, is widely practiced in East Asia (Esposito et al., 2015). Phenomena including forms (e.g., living being, ecosystems, or soilscapes) suffer (Loy, 2019). Although from the *Mahāyāna* Buddhist perspective "form is not different from emptiness, emptiness is not other than form" based on the well-known Heart Sutra (*Prajñāpāramitā*) (Loy, 2019, p. 55). The phrase "*Gate gate paragate parasamgate bodhi svaha*" of the Heart Sutra rejoices the form-emptiness paradox, with "*gate*" meaning gone from suffering to liberation, gone from duality to nonduality; "*paragate*" means gone all the way to the other shore (*nirvāṇa*); "*sam*" in "*parasamgate*" means everyone, the entire community of beings; "*bodhi*" means the light inside, enlightenment; and "*svaha*" is a cry of excitement (Thich Nhat Hanh, 2009, p. 43). Applied to soil (form) = emptiness and emptiness = soil (form), which implies that if one completely immerses in the environment, nature, soil, or any other form, one attains full liberation. For example, if one immerses completely in touching the Earth, being one with the warmth of soil, or marinates in the totality of a soilscape—form and formlessness; soil and emptiness—become undiscernible (nondual experiences, such as nature immersion experiences, e.g., in the Grand Canyon or the backyard of a garden). Through mindfulness meditation and other Buddhist practices of nonattachment to objects (e.g., soil particle, soil, rock, or the Earth), one shifts from the observation of soil to "letting go of soil as an object" (duality) toward "letting be soil" (nonduality) where soil = emptiness and complete immersion and transcendence of worldly appearances are attained. Such view is similar to mystical traditions found in Christianity, Judaism, Indigenous Spirituality, and others, in which complete immersion in nature leads to a state of unity with the divine blending immanence and transcendence (Hollenback, 1996; Lanzetta, 2005). According to *Mahāyāna* Buddhism, to attain liberation is to be mindfully present to nature and the environment moment-to-moment, not attached nor detached from soil/nature, with the best intention of a compassionate and caring attitude toward the environment, which brings forth wisdom and liberation.

The no-self (*anātman*) Buddhist doctrine states that the self is illusory and impermanent. Rather the self is conceived as an ever-changing product of five co-conditioning components (i.e., the *skandhas* called heaps or aggregates)[6], which are the physical body (form, matter), feelings that arise from sensory contact, perceptions (good, bad, and neutral), habitual dispositional tendencies that connect *karma*-producing will to mental action, and consciousness (Smith & Novak, 2003). No-self (i.e., emptiness of self) does not mean there is a meaningless void in a nihilistic sense that would evoke hopelessness in regard to the destruction of planet Earth, global climate change, loss in soil quality, or global soil degradation. On the contrary, emptiness of self is associated with impersonal love of ultimate truth *dharmakāya*, called the *dharma* body, cosmic body, or also called the formless truth

[6] The *skandhas* are the material and mental factors that take part in the rise of craving and clinging, and thus, the *skandhas* evoke suffering.

body. Furthermore, deeply personal love is found in the form body (*rūpakāya*), which consists of the *saṃbhogakāya* (the enjoyment body or spiritual body) and the *nirmāṇakāya* (the emanation body, the physical form body) (Ray, 2000; Shāntideva, 2011). The *dharmakāya* is associated with impersonal "spiritual or transcendental" love, while the *rūpakāya* is associated with personal "worldly" love that we experience for a spouse, a child, a beautiful landscape, the land, or nature (e.g., a sunset at a beach with giant Podzols in Australia). In *Mahāyāna* Indo-Tibetan Buddhism, the three *kāyas* are ultimately indivisible, meaning that when one rests all at once in pure Buddha Nature (pure awareness), one experiences the emptiness of the *dharmakāya*, the impermanence of the *saṃbhogakāya*, and the body form of the *nirmaṇakāya* (Powers, 2007; Ray, 2004). In such states of nonduality, there is no distinction between subject and object and between form (body) and formlessness (emptiness) (Loy, 2015). The belief is that as ultimate reality and relative ("worldly") reality are inseparable and so are their companions of love—impersonal love and personal love. "I am the Earth" expresses that I, you, and others are not separate from the Earth. A nondual view undergirds the notion of complete unity which implies that one cannot separate oneself from the Earth, and thus, "not love" the Earth. If one loves, one cares for what is loved—in this case, the Earth, the soil, everything conceivable including oneself. In the *Mahāyāna* nondual view soil is conceived as form body (one of the *skandhas*) and formless (emptiness) that are inseparable. In contrast, Western soil science dual notions of soil reduce it to matter, an object that is described, measured, mapped, managed, and commodified.

The *Mahāyāna's* emphasis on multifaceted love inspires care for oneself, but also for others, the Earth, and the soil (Kaza, 2019). The deep connection to the Earth and nature is expressed poetically and beautifully in the writings of Zen master Eihei Dōgen (1200–1253) (Kaza, 2019). One example of Zen-inspired poetry was provided by Zen master Loori Daida:

> "When you are open and receptive, when you are alive and alert, everything is constantly teaching, constantly nourishing. The mountains and rivers, the great Earth and its boundless oceans are continually manifesting the words of the ancient teachers, continually expressing the truth of the universe." (Loori, 2004, p. 41)

Loori (2004) affirmed in *Hearing with the Eye* that every one of us has this capacity to "hear with the eye and see with the ear" and find primordial clearness in our own insentient nature. Loori's legacy of nondual environmental teachings stress urgency and responsibility that our lives and the life of this planet are rooted in intimacy with nature (Kaza, 2019).

Such love is not limited to human beings but is found in every species on Earth. The intent is to take refuge in the Earth, to connect with the Earth, and restore our energy and balance because *She* unconditionally accepts us (Thich Nhat Hanh, 2013). Both facets of caring—objective caring and subjective caring—are part of Buddhist care for the Earth. Lerner (2019) discerned objective caring as focused on the outer material world (e.g., supplies of food, clean water, healthy soils), while subjective caring focuses on the inner realities of love, empathy, and care for all and everything. According to Thich Nhat Hanh (2013), "if we care about the health and well-being of the planet, we do so for our own sake" (p. 9), and "if we understand the Earth as a living, breathing organism, we can heal ourselves and heal the Earth as well" (p. 10). The Buddhist holistic view of the Earth as a living organism has rarely been shared by Western scientists that predominantly rely on empiricism and determinism. An exception is Lovelock's (1997, 2003) Gaia hypothesis (now referred to as Gaia theory), which views the Earth as a living self-regulating system that has evolved in such a way as to maintain climatic and chemical parameters favorable for life. This hypothesis has been heavily critiqued by contemporary Western scientists refuting the claim that the planet functions as a unified homeostatic entity using deterministic arguments and system theoretical thinking (e.g., Kirchner, 2002; Moody, 2012). Others have defended the hypothesis on spiritual grounds pointing to the emergence of self-consciousness (conceived as more than only an epiphenomena of the brain) expressing the felt interaction of living organisms with their environment (Schwartzman, 2010). Hasenkamp and White (2017) cautioned that the ultimate nature of reality as described in Buddhist philosophy cannot be evidenced through the Western scientific method (e.g., using objective empirical measurements, replication of experiments, verification of hypotheses as valid through falsification). Instead, first-person subjective modes of inquiry (e.g., through perception with the senses, felt experiences, contemplation, and meditation, such as *Yogācāra*) serve as modes of knowing in the Buddhist view that reveal the paradoxical co-existence of relative reality and ultimate reality.

12.4 Bodhisattvas and Ecosattvas

Who Is in Love with the Earth and All Sentient Beings?

In *Mahāyāna* Buddhism, *bodhisattvas* are considered spiritual warriors who long to alleviate suffering, their own, and that of others on the path to liberation, toward Buddhahood (Chödrön, 2018). The *bodhisattva* motivation is rooted in the arising of *bodhicitta* (Sanskrit: *bodhi*, "enlightenment" + citta, "mind," or "heart") (Williams 2010, p. 195). According to Buddhist beliefs, ultimate *bodhicitta* refers to nondual wisdom, emptiness, or the direct cognizance of the true state of phenomena. Importantly, ultimate and relative *bodhicitta* are two interdependent aspects of the same thing. The true realization of ultimate *bodhicitta* is impossible without the practice of compassion, while no compassion in the day-to-day life can ever be perfect without the realization of the wisdom of emptiness (Shāntideva, 2011). Since ultimate *bodhicitta* is considered boundless and all-pervading, it entails compassion for the Earth, nature, soil, and water – everything there is in the world.

Bodhicitta can be actively cultivated through the practice of the six perfections (*pāramitās*), which are (1) generosity (*dāna*), (2) morality or ethical discipline (*śīla*), (3) patience (*kṣānti*), (4) vigor or enthusiasm (*vīrya*), (5) meditation (*dhyāna*), and (6) wisdom (*prajñā*) (Buswell & Lopez, 2014). Through practice, *bodhicitta* becomes a way of life on the path to Buddhahood (Shāntideva, 2011). The

Bodhicharyāvatāra (*A Guide to the Bodhisattva's Way of Life*), written c. 700 AD, is the best-known *Mahāyāna* Buddhist text composed by Shāntideva, a *bodhisattva* and Buddhist monk at Nalanda University in India (see Shāntideva, 1992 and commentaries on the *Bodhicharyāvatāra* by Pelden, 2010).

> 15: "*Bodhicitta, the awakened mind,*
> Is known in brief to have two aspects:
> First, aspiring, *bodhicitta* in intention;
> Then active *bodhicitta*, practical engagement." (Shāntideva, 2011, p. 33)

The aspiration or intention of relative *bodhicitta* is seeking to attain enlightenment and benefit others, while active *bodhicitta* refers to the practical action of getting to work and to help others (Chödrön, 2018). Chödrön pointed out that in Shāntideva's view, the worldly nature, including water, mountains, trees, and soils, cannot be owned but nature can be enjoyed. In the *Mahāyāna* view, nature cannot be possessed because this would imply attachment to nature and a desire to own it. Such objectification of nature inherently implies separation between people and nature, which is considered a delusional view in Buddhism in general. Dualistic notions of "I versus Earth," "I versus soil," or "us versus them" are rejected as illusion because a nondual view undergirds *Mahāyāna* Buddhism. The *bodhisattva*'s attitude is to appreciate the world and nature's abundance, but not greedily exploit nature's gifts to meet selfish desires. The *ecosattva* combines the *bodhisattva* attitude cultivated through inner development (meditation) and outer environmental activism inspired by the Buddha-*dharma. The* ecosattva path is spiritual activism aimed to live harmoniously within the environment, which aims to protect nature, and engages in environmental activism (Loy, 2019).

12.5 Final Remarks

Contemporary Western Buddhist traditions rooted in *Theravāda* and *Mahāyāna* Buddhism, and schools, such as *Sōtō* Zen, *Mādhyamaka*, and *Yogācāra*, have developed different philosophies regarding soil and the environment, metaphysical understanding of reality, and liberative paths. This chapter is not exhaustive to present Buddhism as a whole or all existing Buddhist strands because there are several thousand. Rather the intent was to identify de-/motivating factors in relation with soil, nature, the environment, form, and phenomena in select Buddhist traditions.

In general, *Theravāda* Buddhism ascertains that *samadhi* (complete meditative absorption) and focus on emptiness of self are profoundly important to attain individual liberation. This path emphasizes escape and transcendence from worldly suffering and less so active engagement to sustain and protect the environment. Soil is viewed as a physical body (form, matter), one of the *skandhas*, interconnected with everything else that is undergoing continuous change based on the doctrines of dependent origination and *karma*. In this view, even the degradation of soils or pristine ecosystems may be viewed as another occurrence in the eternal stream of *karma* (i.e., the *karma* of action and the *karma* of results). Interconnectedness and interbeing undergird this kind of Buddhist-inspired ecodharma.

In contrast, *Mahāyāna* Buddhism emphasizes both emptiness of self and phenomena, the *bodhisattva* ideal, *bodhicitta* (awakened mind and heart), the *paramitas* (perfections), and compassion for other beings, soil, and the Earth. Belonging rather than escapism from worldly suffering is stressed in *Mahāyāna*. The emphasis on immanence opens to an expansive, boundless view in the *Yogācāra* from observing of I and soil/nature (separation, duality) → nonattachment to soil/nature ("letting go") → witnessing of I, soil/nature → being soil/nature ("letting be"; nonduality). The *Mādhyamaka* radically deconstructs phenomena, such as soil and nature, through the negation approach to realize emptiness. Earthly (relative form body) and ultimate (cosmic or formless truth body) are considered one and the same. In the *Mahāyāna*, soil (form) = emptiness, and emptiness = soil (form) that stresses non-separation (nonduality) and conjoint personal and impersonal love. Such view implies that if one completely immerses in the environment, nature, soil, or any other form, one attains wisdom and full liberation. The recognition that if we deeply care for this planet, we also care for us and everybody else provides motivation to actively sustain and protect soil, the Earth, and all of humanity. The *bodhisattva*-inspired ecosattva embodies both—meditative inner knowing of the ultimate truth and outer compassion for soil, the Earth, and everything else.

12.6 A Love Letter to the Soil

How Can Buddhist Philosophy Inspire us to Love Soils?

The *ecodharma* is an invitation to move beyond the concept of "the environment" or "the soil." Such constructs lead people to experience themselves and the soil as two separate entities and the Earth only in terms of what it can do for them. The *ecosattva* forgets oneself and is completely empty, one with soil; we inter-are soil, the rain, the clouds. Would you trample on a beautiful flower for no good reason? You are the flower, you are the soil, you are the land. Liminal nonduality. Mindful walking, a gentle and soft touch on the ground. Soil gives us life. Loving soil is loving us and everybody else on this wonderful planet. This love is intimate and indestructible, as is soil. Being soil is bliss, joy, and liberation, as is dust in the wind.

References

Batchelor, M., & Brown, K. (Eds.). (1994). *Buddhism and ecology*. Motilal Banarsidass.

Bhikkhu, B. (Ed.). (1993). *A comprehensive manual of Abhidhamma*. Buddhist Publication Society.

Buswell, R. E., & Lopez, D. S. (2014). *The Princeton dictionary of Buddhism*. Princeton University Press.

Chödrön, P. (2018). *Becoming bodhisattvas: A guidebook for compassionate action* (H. Berliner, Ed.). Shambhala.

Chögyam Trungpa Rinpoche (2010). *Cutting through spiritual materialism*. Shambhala.

Chögyam Trungpa Rinpoche (2013). *The profound treasury of the ocean of dharma: The path to individual liberation* (J. L. Lief, Ed.). Shambhala.

Darlington, S. M. (2017). Contemporary Buddhism and ecology. In M. Jerryson (Ed.), *The Oxford handbook of contemporary Buddhism* (pp. 487–503). Oxford University Press.

Duckworth, D. S. (2019a). *The profound reality of interdependence: An overview of the wisdom chapter of the way of the bodhisattva*. Oxford University Press.

Duckworth, D. S. (2019b). *Tibetan Buddhist philosophy of mind and nature*. Oxford University Press.

Eckel, M. D. (2010). Is "Buddhist environmentalism" a contradiction in terms? In R. K. Payne (Ed.), *How much is enough? Buddhism, consumerism, and the human environment* (pp. 161–170). Wisdom.

Esposito, J. L., Fasching, D. J., & Lewis, T. T. (2015). *World religions today* (5th ed.). Oxford University Press.

Gleig, A. (2019). *American dharma: Buddhism beyond modernity*. Yale University Press.

Gross, R. M. (1993). *Buddhism after patriarchy: A feminist history, analysis, and reconstruction of Buddhism*. State University of New York Press.

Grunwald, S. (2021). Embodied liberation in participatory theory and Buddhist Modernism Vajrayāna. *Journal of Dharma Studies, 4*, 159–177. https://doi.org/10.1007/s42240-021-00092-4

Harris, I. (1995). Getting to grips with Buddhist environmentalism: A provisional typology. *Journal of Buddhist Ethics, 2*, 173–190.

Hasenkamp, W., & White, J. R. (Eds.). (2017). *The monastery and the microscope - Conversations with the Dalai Lama on mind, mindfulness, and the nature of reality*. Yale University Press.

Hollenback, J. B. (1996). *Mysticism: Experience, response, and empowerment*. Pennsylvania State University Press.

Kamala, T. (1997). *Forest recollections: Wandering monks in twentieth-century Thailand*. University of Hawai'i Press.

Kaza, S. (2019). *Green Buddhism: Practice and compassionate action in uncertain times*. Shambhala.

Keown, D. (2007). Buddhism and ecology: A virtue ethics approach. *Contemporary Buddhism, 8*(2), 97–112. https://doi.org/10.1080/14639940701636083

Khensur Rinpoche Jampa Tegchok. (2012). *Insight into emptiness* (Thubten Chodron, Ed.; S. Carlier, Trans.). Wisdom.

Kirchner, J. W. (2002). The Gaia hypothesis: Fact, theory, and wishful thinking. *Climatic Change, 52*(4), 391–408. https://doi.org/10.1023/A:1014237331082

Lanzetta, B. J. (2005). *Radical wisdom: A feminist mystical theology*. Fortress Press.

Lerner, M. (2019). *Revolutionary love: A political manifesto to heal and transform the world*. University of California Press.

Loori, J. D. (2004). *Hearing with the eye: Photographs from point lobos*. Dharma Communications.

Lovelock, J. (1997). The Gaia hypothesis: The Earth as a living organism. In S. B. Scharper (Ed.), *Redeeming the time: A political theology of the environment* (pp. 53–74). Continuum Int. Publ. Group.

Lovelock, J. (2003). Gaia: The living Earth. *Nature, 426*(6968), 769–770. https://doi.org/10.1038/426769a

Loy, D. R. (2015). *A new Buddhist path: Enlightenment, evolution and ethics in the modern world*. Wisdom.

Loy, D. R. (2019). *Ecodharma: Buddhist teachings for the ecological crisis*. Wisdom.

McMahan, D. L. (2008). *The making of Buddhist modernism*. Oxford University Press.

Moody, D. E. (2012). Seven misconceptions regarding the Gaia hypothesis. *Climatic Change, 113*(2), 277–284. https://doi.org/10.1007/s10584-011-0382-4

Morley, J. (2008). Embodied consciousness in tantric yoga and the phenomenology of Merleau-Ponty. *Religion and the Arts, 12*(1), 144–163. https://doi.org/10.1163/156852908X270980
Pelden, K. (2010). *The nectar of Manjushri's speech: A detailed commentary on Shantideva's way of the bodhisattva* (Padmakara Translation Group, Trans.). Shambhala.
Powers, J. (2007). *Introduction to Tibetan Buddhism*. Snow Lion.
Ray, R. A. (2000). *Indestructible truth: The living spirituality of Tibetan Buddhism*. Shambhala.
Ray, R. A. (2004). Three in one: A Buddhist trinity. *Lion's Roar*, 1–8. Retrieved from http://www.lionsroar.com/three-in-one-a-buddhist-trinity/
Ray, R. A. (2016). *The awakening body: Somatic meditation for discovering our deepest life*. Shambhala.
Ronkin, N. (2018). Abhidharma. *The Stanford Encyclopedia of Philosophy.*. https://plato.stanford.edu/archives/sum2018/entries/abhidharma/
Samuels, J. (1997). The bodhisattva ideal in Theravāda Buddhist theory and practice: A reevaluation of the Bodhisattva-Śrāvaka opposition. *Philosophy East and West, 47*(3), 399–415. https://doi.org/10.2307/1399912
Schmithausen, L. (1997). The early Buddhist tradition and ecological ethics. *Journal of Buddhist Ethics, 4*(1), 1–74.
Schwartzman, D. (2010). Is Gaia a theory, hypothesis, or a vision? In E. Crist & H. B. Rinker (Eds.), *Gaia in turmoil: Climate change, biodepletion, and Earth ethics in an age of crisis* (pp. 1–8). MIT Press.
Shāntideva (1992). *A guide to the bodhisattva's way of life* (S. Batchelor, Trans.). Library of Tibetan Works and Archives.
Shāntideva (2011). *The way of the bodhisattva: A translation of the Bodhicharyāvatāra* (Padmkara Translation Group, Trans.). Shambhala.
Smith, H., & Novak, P. (2003). *Buddhism: A concise introduction*. Harper Collins.
Suzuki, S. (1970). *Zen mind, beginner's mind: Informal talks on Zen meditation and practice*. John Weatherhill.
Swearer, D. K. (2006). An assessment of Buddhist eco-philosophy. *Harvard Theological Review, 99*(2), 123–137. https://doi.org/10.1017/S0017816006001179
The Karmapa (Ogyen Trinley Dorje) (2017). *Interconnected: Embracing life in our global society* (Damchö Diana Finnegan, Trans.). Wisdom.
Thich Nhat Hanh (1987). *Interbeing*. Parallax Press.
Thich Nhat Hanh (1998). *The heart of the Buddha's teaching: Transforming suffering into peace, joy and liberation*. Parallax Press.
Thich Nhat Hanh (2009). *The heart of understanding: Commentary on the prajnaparamita heart sutra* (P. Levitt, Ed.). Parallax Press.
Thich Nhat Hanh (2012). *Awakening of the heart: Essential Buddhist sutras and commentaries*. Parallax Press.
Thich Nhat Hanh (2013). *Love letter to the Earth*. Parallax Press.
Williams, P. (2010). *Mahāyāna Buddhism: The doctrinal foundations*. Routledge.

Chapter 13
Cultural Understanding of Soil in China

Sylvia Xiaorui Wen

China has a long history of agriculture, and farming plays an essential role in Chinese culture. Archaeological and molecular evidence has dated the earliest domestication of millet and rice in China back to 10,000 years ago (Liu et al., 2013). Since early on, the Chinese people have shown respect and reverence to land and soil. They have developed intricate belief and knowledge systems around soil and soil cultivation. These systems in turn have shaped Chinese social, cultural, and political lives since ancient times.[1]

13.1 Etymology of the Character for Soil

The original Chinese character for soil, 土 (*tu*), has gone through its own evolution. In the oracle bone and bronze scripts of the Shang (16th–11th century BCE) and Zhou (1046–256 BCE) Dynasties, there were at least 27 distinctive glyphs recorded for the character (Liang, 1980). The early Chinese written system was logographic, in which most characters indicated the shapes or sounds of the objects. The 2nd century dictionary *Shuowen Jiezi* 说文解字 (*Explanations of Simple Graphs and Analyses of Composite Graphs*) explains 土 as "Earth that produces all living things:

[1] As is demonstrated in the *Soil and Governance* section of this chapter, since the time of the *Book of Songs*, soil management has been intertwined with politics.

S. X. Wen (✉)
Fudan University, Shanghai, China

Sofia University, Palo Alto, CA, USA

California Institute of Integral Studies, San Francisco, CA, USA
e-mail: swen@mymail.ciis.edu

N. Patzel et al. (eds.), *Cultural Understanding of Soils*,
https://doi.org/10.1007/978-3-031-13169-1_13

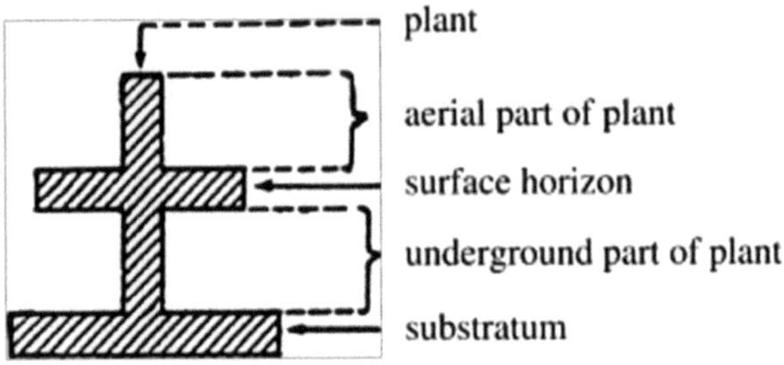

Fig. 13.1 The Chinese character for "soil". (Reproduced from Wang, 1980, p. 18)

the two horizontal lines represent the ground surface (topsoil) and the middle of the ground (subsoil); the upright line represents the things that grow out of the soil" 地之吐生物者也。二象地之下、地之中，丨物出形也 (14.8972). The character illustrates the intimate relationship between soil and plants: Soil nurtures and supports plant growth; it is what makes the growth of vegetation possible (Fig. 13.1). This interpretation of soil, based on close observation of nature, may be considered as pedologic science (Wang, 1980).

The character 土 is also interpreted as the male reproductive organ, with the character 匕 (*bi*) as its female counterpart (Guo, 1931). When the radical part for ox (牛) is added to 匕 and 土, the characters become 牝牡 (*pinmu*), which symbolizes the feminine and masculine, as well as the ancestors. Some other important characters derived from 土 include 生 (*sheng*–life, grow, "represents plants growing out of the soil"), 地 (*di*–the land or earth), 社 (*she*–god of the land), 壤 (*rang*–soft soil), 均 (*jun*–equal), 基 (*ji*–base), 在 (*zai*–exist, alive), 圣 (*sheng*–sage, saint), 里 (*li*–inside, internal), and 金 (*jin*–metal, "represents that metal is formed in the soil") (Shuowen, 7, 14, 15). The 土 and 地 are closely related: while 土 refers both to soil and earth as the matter or element, 地 may mean the land, the soil, or earth as a cosmic half of the universe, as in "heaven and earth."

The modern Chinese[2] word for soil is 土壤 (*turang*), a compound word that consists of the characters 土 (*tu*) and 壤 (*rang*). 壤 contains 土 and 襄, which mean "soil" and "assistance," respectively. The difference between 土 and 壤 is noted in the *Zhouli* 周礼 (Rites of Zhou), a book compiled probably in the 2nd century BCE (Needham & Lu, 1981) to record the institutions of the Zhou Dynasty. The Diguan Situ 地官司徒 (Terrestrial Office with the Overseer of Public Affairs) was required to:

> According to the law appropriate to the land/soil (tu), identify twelve regional products and their names, observe their benefits and harms to local people's residences, in order to let people prosper, birds and beasts proliferate, grasses and trees grow luxuriantly, and to achieve the productivity of the land. Identify the crops of the twelve types of soil (rang) and know their appropriate varieties, in order to educate the people to grow cereals and plant trees. 以土宜之法辨十有二土之名物，以相民宅而知其利害，以阜人民，以蕃鸟兽，以毓草木，以任土事。辨十有二壤之物而知其种，以教稼穑树蓺. (Zhouli, 2.65)

[2] In sinology, "modern Chinese" usually refers to Chinese used after 1911 (end of the last dynasty) when a series of reforming efforts started to standardize the language. The opposite of "modern Chinese" is "ancient Chinese."

The *Shuowen Jiezi* explains 壤 as "soft soil without lumps"[3] 柔土 (14). A 2nd century commentary to *Zhouli* interprets 壤 as a variation of soil: "Rang is also tu, it is a variation of tu. Where myriad things originate from is tu; tu is the same as 吐 (*tu*, to emit). Rang is (soil) ploughed, tilled, and cultivated by people; rang is gentle and soft in appearance" 壤，亦土也，变言耳。以万物自生焉则言土；土，犹吐也。以人所耕稼树艺焉，则言壤；壤，和缓之貌 (Zhouli Zhushu, 10.13). It is therefore reasonable to draw the conclusion that while tu is naturally formed, rang is fertile soil cultivated by humans for the purpose of farming (Wang, 1980).

The etymology and evolution of the Chinese characters for soil reveal a cocreating relationship between humans and the earth, one which involves patient observation and mutual cultivation. This innate affinity has profoundly influenced the Chinese worldview and its spirituality.

13.2 Soil and Creation Myths

Myths describe the "breakthroughs of sacred (or the 'supernatural') into the World." Eliade (1963) pointed out that "as a result of the intervention of Supernatural Beings" human beings become "mortal, sexed, and cultural" beings (p. 6). In many creation myths in China, soil plays an indispensable role. One popular legend is about Pangu 盘古, the creator of the world who was born within the cosmos egg of primeval formlessness. As the first divine human, he separated heaven and earth;[4] thus, the world came into being. In one version of his story, upon his death, the microcosmic body of Pangu transformed into the macrocosm of the physical universe:

> …his breath became the winds and clouds, his voice became the thunder, his left eye became the sun, his right eye became the moon, his four limbs and trunk became the four extremes of the earth and the Five Mountains, his blood became the rivers, his veins became the earth's arteries, his flesh became fields and soil, his hair and beard became the stars, his skin and body hair became plants, his teeth and bones became various metals and rocks, his semen and marrow became pearls and jade, his sweat became the rain and the dew, and the various insects on his body reacted to the wind and turned into human beings. (Yang et al., 2008, p. 177)

The goddess Nüwa 女娲 is another prominent creator figure in Chinese myths. According to legend, in the beginning, when heaven and earth were first created, humans did not exist. There lived a couple with human heads and serpent/dragon bodies, "a brother and his sister" named Fuxi 伏羲 and Nüwa (Yang et al., 2008,

[3] A lump of soil means a mass or aggregation of soil. "Soil without lumps" thus is smooth and fine.

[4] In ancient Chinese understanding, heaven and earth are the two cosmic halves or opposite sides of the universe. The "earth" here is neither the planet Earth, nor the matter earth. It might be interpreted as a principle as yin-yang are the two opposite principles. However, when heaven, earth, and human are used in the same context, each of them is considered as an element. It is a common practice among sinologists to use "earth" with lower case "e" as an element or principle.

p. 121). Out of longing for company, Nüwa created humans by shaping yellow soil with her hands into creatures after herself. In one version of the myth, the meticulous work exhausted her strength, so she eventually gave up the time-consuming molding effort and created the rest of humanity by dipping a rope or cord in the mud and tossing it up to make the soil into human beings.[5] "Thus, rich and noble people were those made by Nüwa's hands, while poor and lowly people were those made by Nüwa dragging a cord through the mud" (Yang et al., 2008, p. 172). Although strictly speaking the Nüwa story is not a creation myth, as it is not about the beginning of the world, nevertheless it is both a myth of the physical creation of humankind as well as a justification for social hierarchy in ancient societies (Bodde, 1961).

There are many other stories related to Nüwa, including myths regarding the Great Flood. The deluge is a common theme shared among many ancient mythologies around the world; myths around the deluge tell the stories of intervention both by the supernatural beings and the first humans (Eliade, 1963). In the Chinese myths, after Nüwa created human beings and brought them to life, she also mended the broken sky with five-colored pebbles and reed ashes to stop the Great Flood (Yang et al., 2008).

The human stories relating to soil start with Huangdi 黄帝 (Yellow Emperor), the legendary ancestral ruler of all ancient tribes inhabiting the land that would become today's central China. The *Shiji* 史记 (Historical Records) of 90 BCE, the first of a series of historical works recording Chinese history since antiquity, explains that "because he (Huangdi) receives the auspicious omen embodied in the virtue of soil, his title was the 'Yellow Emperor'" 有土德之瑞，故号黄帝 (Shiji, 1.3). In the legend told by the *Shanhai Jing* 山海经 (Classic of Mountains and Seas), there existed a certain *xirang* 息壤, mythical living or self-renewing soil, that could help to stop the gushing water from heaven. It was Gun 鲧, grandson of the Yellow Emperor, who stole xirang from the Supreme Deity to control the overflowing floodwater. Gun was killed for his endeavor, and his son Yu 禹 eventually spread the mythical soil and stabilized the world (18.39).

13.3 Soil, Yin-Yang, Wuxing, and the Yijing

The indigenous Chinese concepts of *yin-yang* and *wuxing* (five phases) mutually shape each other, and will be explained in detail in this section. They have permeated every aspect of Chinese culture, including the classical divination text, the *Yijing*. Soil is fundamental to understanding yin-yang, wuxing, and the Yijing.

---

[5] In the Chinese language, there are differences between mud 泥 (*ni*) and soil 土 (*tu*). 泥 is the sludge made of 土 mixed with water. Soil and earth are often used interchangeably in Chinese language, as 土 is both the element and the matter. In the original Chinese story, 俗说开天辟地，未有人民，女娲抟黄土作人，剧务，力不暇供，乃引絙于泥中，举以为人, Nüwa first created humans by shaping yellow soil 黄土 (*huangtu*), then switched to dipping a rope in the mud 泥. It is possible to read mud as diluted and less pure than soil. Dust in Chinese is 塵 (*chen*) (deer鹿 + soil 土) in traditional form, and 尘 (*chen*) (small 小 + soil 土) in simplified form, it literally means small particles of soil (after a herd of deer passed through).

13.3.1 Soil and Yin-Yang

As a foundational doctrine of Chinese cosmology and philosophy, yin-yang is a theory of dialectical dualism in which the world is composed of complementary, opposing forces of *yin* 阴 and *yang* 阳, whose original meanings are darkness and brightness. First formed in antiquity, the concept was already considered common knowledge by the time Qin Shihuangdi unified China in 221 BCE (Peerenboom, 1993). As symbols, yin and yang represent pairs of dichotomous categories observed in nature: female/male, night/day, bad/good, low/high, death/life, earth (soil)/heaven (sky), etc. Yin and yang are interdependent and interconnected; they give rise to each other; together they form a holistic dynamic system.

The understanding of yin-yang by the ancient Chinese people perhaps started with observations of the patterns of nature, including the interplay between soil and the plants that grow out of it. In *Huangdi Neijing: Sunwen* 黄帝内经 · 素问 (Huang Di's Inner Classic: Basic Questions), the Yellow Emperor asked about the correspondence between yin and yang. His advisor Qi Bo 岐伯 answered (6.48.10),

> Heaven covers; earth carries. Thus myriad things are generated. Before they emerge from the soil (which is called the location of yin), it is called yin within yin. After they emerge from the soil, it is called yang within yin. Yang supports straightness [of myriad things]; yin rules [myriad things]. Therefore, generation happens in spring, growth happens in summer, collection happens in autumn, and storage happens in winter. If the regularity is lost, all four [seasonal changes] will be obstructed. 天覆地载，万物方生，未出地者，命曰阴处，名曰阴中之阴；则出地者，命曰阴中之阳。阳予之正，阴为之主。故生因春，长因夏，收因秋，藏因冬，失常则天地四塞。

As discussed previously in the etymology and derivations of soil in Chinese, the character for "generation" in the above passage is 生 (*sheng*), which represents plants growing out of the soil. As a verb, sheng means to grow, to generate, and to give birth. As a noun, it indicates life itself. Just as a mother gives birth to her children and nourishes them for continuous growth, the soil provides the right conditions for seeds to generate and flourish. Both the female/mother in the human realm, and the soil in nature are yin, as opposed to the male/father and the sky, which are both yang. While the yang is "unfolded as an explicit order," the yin is "enfolded in an implicit order," as the background and the hidden (Wang, 2012, pp. 146–147). Chinese cosmology recognizes the movement of yin-yang, together with wuxing, as the dictating force of the oneness in the material and spiritual aspects of reality.

13.3.2 Soil and Wuxing

Wuxing is often translated as the five processes, five phases, or five movements. It applies the five conceptual elements of wood, fire, earth/soil, metal, and water to interpret phenomena in the natural world, both animate and inanimate. The five processes are constantly moving through a two-fold pattern: a productive or generative cycle in which each element benefits the following element and a controlling or destructive cycle where each element overcomes the opposite element. Unschuld (1985) explained the processes in the following way (p. 59):

> Water overcomes fire; fire melts metal; metal—in the form of a knife, for instance—overcomes wood; wood—as a spade—overturns soil; soil—as in a dike—subdues water.
>
> …
>
> Water/watering produces plants and trees, that is, wood; wood brings forth fire; fire produces ashes, that is, soil; soil brings forth metal; when heated, metal produces steam, that is, water.

The myth of Gun and Yu also explains the origin of wuxing. In *Shangshu* 尚书 (The Book of Documents, traditionally attributed to Confucius), perhaps the oldest surviving text mentioning wuxing (Needham & Wang, 1956, p. 242), the chapter *Hongfan* 洪范 (The Great Plan) explains that the reason Gun failed to barricade the flood with the living soil (xirang) is his ignorance of the particular nature and the interactive processes of the five elements. Upon Gun's death, his son Yu was given a great plan comprised of nine enumerated parts necessary for good government, the first two parts of which are *wuxing* 五行 (five processes) and *wushi* 五事 (five things to do) (translated by Graham, 1989, p. 326):

> 1. The five processes. 1, water; 2, fire; 3, wood; 4, metal; 5, soil. Water: wetting, sinking. Fire: flaming, rising. Wood: bending, straight. Metal: yielding to molding. In soil one plants and harvests. 一、五行：一曰水，二曰火，三曰木，四曰金，五曰土。水曰润下，火曰炎上，木曰曲直，金曰从革，土曰稼穑。…
> 2. The five things to do. 1, demeanor; 2, speech; 3, looking; 4, listening; 5, thinking. Demeanor: respectful. Speech: accordant. Looking: seeing clearly. Listening: hearing clearly. Thinking: understanding. 二、五事：一曰貌，二曰言，三曰视，四曰听，五曰思。貌曰恭，言曰从，视曰明，听曰聪，思曰睿。

In both series, each item is envisioned as one process with two aspects, and each process is part of the ceaseless self-generation of natural and human affairs. Furthermore, it suggests that good governance emulates the way of nature: "The five natural processes are presented as basic and the five things for humans to do as secondary" (Parkes, 2003, p. 194).

The perpetual transforming cycle of the five processes is correlated with the four directions of the spatial dimension, the five seasons of the temporal dimension (late or "long" summer is considered as a separate season), and the five colors in nature (Fig. 13.2).

In this system, soil is located in the center, and the south is facing up, a mainstream practice in ancient China. The centrality of soil is expounded in *Wuxing Dayi* 五行大义 (Important Meanings of the Wuxing) of the Sui Dynasty (6th century):

> Wood (plants) does not grow without soil; with soil it flourishes. Fire has no vigor without soil; with wood it comes to being. Metal does not form without soil; with earthen mold it becomes useful. Water does not stop flowing without soil; with dikes it is controlled. Soil supports new things to grow and helps weakened beings; thus achieving the *dao*. Therefore, the cycle of the five processes is dependent on soil. Soil governs the changes of four seasons, it resides at the center. 土也。位在中央。分王四季。(2.15)

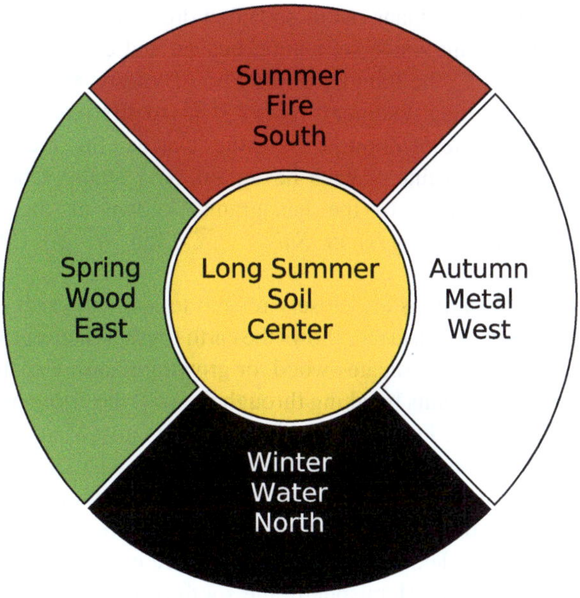

Fig. 13.2 Soil, wuxing, five seasons, four directions, and five colors (Multiple hypotheses have been developed to explain why the element of water is represented by the color black in Chinese culture. A relatively convincing explanation is related to the Four Symbols of the Chinese constellations, namely, the Azure Dragon of the East and spring, the Vermilion Bird of the South and summer, the White Tiger of the West and autumn, and the Black Tortoise of the North and winter. Water is associated with the season winter and the direction north, which is represented by Xuanwu 玄武 [black tortoise], which in fact is a turtle entwined with a snake. Another theory is that in the Chinese medicine system, winter is cold; hence, the blood vessels appear black in winter. The latter theory is endorsed by Paul U. Unschuld, 2003)

13.3.3 Soil and the Yijing

In Chinese cosmology, the myriad things are created through the dynamic polarity of yin and yang. In the *Yijing* 易经 (Book of Changes), both yin and yang are represented by one horizontal line known as *yao* 爻. A solid line (—) is called *yangyao* 阳爻, the symbol for yang; a broken line (--) is *yinyao* 阴爻, the symbol for yin. The combination of three yao is a *gua* 卦 (trigram). The eight (2^3) possible combinations of yangyao and yinyao are known as the *bagua* 八卦 (eight trigrams). The eight trigrams represent eight natural phenomena: *qian* 乾 (☰, heaven), *kan* 坎 (☵, water), *gen* 艮 (☶, mountain), *zhen* 震 (☳, thunder), *xun* 巽 (☴, wind), *li* 离 (☲, fire), *kun* 坤 (☷, earth), and *dui* 兑 (☱, lake or swamp). When two trigrams are combined, they create a hexagram. The Yijing uses the 64 (8^2) hexagrams to first explain natural phenomena and then make predictions of events both in the human realm and myriad things within the universe. Both the human and natural worlds are envisioned as the unity of the yin-yang opposites. According to myth, Fuxi (the previously described legendary sage with human head and dragon body, brother of Nüwa the goddess who created humans out of soil) first drew the eight trigrams based on images and patterns of the universe, nature, heaven and earth, and his own body. Later, King Wen 周文王, the founder of the Zhou Dynasty, arranged the 64 hexagrams and gave interpreting statements (*guaci* 卦辞) to all trigrams and hexagrams. King Wen's son, the Duke of Zhou 周公, further clarified the interpretation (*yaoci* 爻辞) of every individual line of each hexagram. And finally, Confucius himself made ethical commentaries to the hexagrams, known as the *Yizhuan* 易传 (Commentaries on the Changes) or *Shiyi* 十翼 (Ten Wings) (Minford, 2014; Wang, 2012).

Many of the 64 hexagrams are closely related to soil and earth. Hexagram 46, *sheng* 升 (ascend, rise), for instance, shows the earth trigram (kun ☷) above the wind trigram (xun ☴). Xun also indicates wood, or growing plants. Expanding from the natural phenomenon of plants breaking through the soil, the *Yijing* offers the judgement for sheng: "Supreme Fortune. It profits. To see a Great Man. Do not be sad. A Southern Advance is auspicious, *Hoc bonum*"[6] 元亨, 用见大人, 勿恤, 南征吉 (translated by Minford, 2014, p. 360). Minford's interpretation was that the Great Man (大人) is derived from the great earth, a figure that offers protection and nourishment which can be understood as an "inner" figure or an "outer" one. From the perspective of the Northern Hemisphere, the south resembles warmth and brightness, a southern advance, therefore, portrays an image of gradual progress and gentle ascent "into the realm of Truth" through "long and patient Self-Cultivation" (p. 361). Hexagram 30, *li* 离 (radiance of fire, attachment), is formed by doubling the fire trigram (also 离 ☲), hence double brightness. The commentary to the judgement is "Sun and Moon are attached to Heaven [sky]. The Hundred Grains, the grasses and trees, are attached to Earth, to the soil. Double Brightness is attached to Truth, Truth which transforms and perfects All-under-heaven" 日月丽乎天, 百谷草木丽乎土, 重明以丽乎正, 乃化成天下 (Minford, 2014, p. 244),[7] and "A cow is reared. This is auspicious" (Minford, 2014, p. 243). Just as fire and its brightness are attached to, and thus relying on the solid materiality of wood and other fuel sources, and myriad plants are attached to the nourishment of the soil, the outer radiance of enlightenment depends on the inner source of truth. Furthermore, the image of the cow indicates the yin in the center of the trigram, which embodies softness and docility. It can be interpreted as a humble attitude to bow down, and to take and adapt to what comes from the soil, and like the cow, to ruminatingly digest the outcomes of nature, and by that attitude to gain a radiance of truth.

[6] The capitalization of words is slightly altered here from the original translation to avoid unnecessary line breaks.

[7] In the original Chinese 百谷草木丽乎土, there is only one character 土, but Minford translated as both earth and soil. This is because in Chinese, 土 and 地 (soil, earth, and land) are interrelated and interchangeable in many occasions.

13.4 Soil and Chinese Spiritual Traditions

Confucianism and Daoism are two indigenous spiritual and philosophical traditions that originated between the 6th and 5th century BCE. Buddhism was most likely transmitted to China around the beginning of the Common Era and has been Sinicized gradually since its arrival to China (Ge, 2014). From the Eastern Han Dynasty (25–220) onwards, the three traditions have been mutually influencing and shaping each other. There have also been constant efforts by the imperial courts and the literati to incorporate the three teachings (*sanjiao* 三教) of Daoism, Confucianism, and Sinicized Buddhism into a cohesive spiritual framework. The Ming Dynasty (1368–1644) locution *sanjiao heyi* 三教合一 (Unity of The Three Teachings) epitomizes this endeavor.

13.4.1 Daoism and Soil

With its ultimate focus on the notions of *dao* (the Way), *de* 德 (virtuous power), *ziran* 自然 (naturalness), and *wuwei* 无为 (attentive nonaction), the Daoist philosophical system is often viewed as a naturalistic religion. In the *Liezi* 列子, an early Daoist classic attributed to Lie Yukou 列御寇 (ca. 400 BCE), the author asked a rhetorical question in the chapter *Tianrui* 天瑞 (Heaven's Gift): "Crops and seeds, soil and trees, birds and beasts, fish and turtles, they are all begotten by heaven. How can I possess them?" 夫禾稼、土木、禽兽、鱼鳖，皆天之所生，岂吾之所有? (Liezi, 1.15). While the main philosophical discourse of the chapter is about interdependence among myriad things – "Heaven, earth, and the myriad things are not separate from each other" 天地万物不相离也, it is worth noting that in the list of things that belong to heaven (usually understood as nature here, e.g., see, Minford, 2014; Needham & Wang, 1956), soil appears to be equal with other living creatures, both plants and animals. This animistic view of soil is also reflected by a passage in *Zhuangzi* 庄子, another canonical Daoist work:

> In running the government, the ruler should not be rude or reckless; in administering the people, he should not be mean or sloppy. In the past I used to grow crops. When I ploughed the soil rudely, the crops rewarded me recklessly; when I sowed the seeds sloppily, the crops also rewarded me meanly. The following year I changed my methods, ploughing the soil deeply and sowed the seeds with great care; the crops were luxuriant and abundant, and all the year I had more than plenty to eat. 君为政焉勿卤莽，治民焉勿灭裂。昔予为禾，耕而卤莽之，则其实亦卤莽而报予；芸而灭裂之，其实亦灭裂而报予。予来年变齐，深其耕而熟耰之，其禾蘩以滋，予终年厌飧。(25.6)

Although the intention of Zhuangzi here is to caution people not to run away from their innate nature and betray spirits by following the general practices of the crowd 遁其天，离其性，灭其情，亡其神，以众为, the analogy between human and nature is significant. Whether it is the relationship between the ruler and ruled or that between the farmer and crops, they all share the commonality of the dao: "The myriad things might have different principles, yet they are all ruled by the dao" 万

物殊理，道不私 (25.10). In society, the dao is conveyed through government administration; in agricultural practice, the dao is carried out by the mutual productivity between the farmer and the soil.

13.4.2 Sinicized Buddhism and Soil

Buddhism sprouted from Hinduism during the Axial Age in India. It has inherited many ideas from Hinduism, including the concept of *mahābhūta* (great elements) (Buswell & Lopez, 2014). In Sanskrit and Pāli, the mahābhūtas are both the physical manifestation of the four elements of earth/soil, water, fire, and wind, as well as the "major elementary qualities" of the physical world that the elements represent (p. 493). When Buddhism was introduced to China, the mahābhūta was referred to as the *sida* 四大 (four big) or *dazhong* 大种 (great existence). The *Brahmajāla Sūtra* 梵网经 (The Brahma's Net Sutra) asserts that "All soil and water are our former bodies, all fire and wind are our original essence" 一切地水是我先身，一切火风是我本体 (1006b09), from which the Pure Land tradition of Buddhism (*Jingtuzong,* 净土宗) establishes its core teaching of interdependence and mutual-transformation of the inner and the outer. The *Jingtu Shiyao* 净土十要 (Ten Essentials on the Pure Land), an influential text written by the monk Ouyi Zhixu 蕅益智旭 (1599–1655), proclaims that "Silence and illumination are non-dual. Nature and cultivation are non-dual. Body and earth/soil are non-dual" 寂照不二。性修不二。身土不二 (0644b19). Though the original expression "Body and earth/soil are non-dual" 身土不二 (*shentu buer* in Mandarin, *shindo funi* in Japanese, and *sinto buri* in Korean) was to indicate the karmic inseparability between the person and the surroundings,[8] in recent years, it has been adopted and popularized by Japanese and Korean practitioners of local food movements (Kim, 2011; Yamashita, 1998).

Nongchan 农禅 (Agricultural Chan, the integration of farming into Chan [Zen in Japanese]) is another development of Sinicized Buddhism. The Tang Dynasty Buddhist master Baizhang Huaihai 百丈懷海 (720–814) is generally considered to have initiated the monastic practices within the Chan (Zen in Japanese) sect Buddhism. The compiled text *Zutang Ji* 祖堂集 (Patriarch's Hall Anthology) tells the story of Baizhang:

> The master lives his life with such arduous integrity and lofty rectitude that it is difficult to describe. Every day when the practitioners gather for labor, he is always ahead of the rest. The senior monk cannot bear, so he secretly put away his tools and asks him to take a rest. The master says: "I don't have any merit. How can I let others to work on my behalf?" The master searches all over for his tools. When he failed, he refused to eat. Thus, the saying "one day without work, one day without food" starts to circulate widely in the universe. 師平生苦節高行，難以喻言。凡日給執勞，必先於眾。主事不忍，密收作具，而請息焉。師云:「吾無德，爭合勞於人?」師遍求作具，既不獲，而亦忘喰。故有「一日不作，一日不食」之言，流播寰宇矣。(V.14, 0569a09)

[8] See Chap. 12 on Buddhism.

Agricultural Chan played a key role in the history of Chan Buddhism (Ge, 1995). The phrase *yir ibuzuo, yiri bushi* 一日不作，一日不食 (One day without work, one day without food) emphasizes the importance of self-reliance through the integration of farming and Chan practice. It is a clear departure from earlier Buddhist practitioners' dependence on patronage and donations. "Dual Cultivation of Farming and Chan" 农禅双修 is still practiced by many Buddhist temples across China up to the present day.

13.4.3 Confucianism and Soil

For over two millennia from the beginning of the Han Dynasty (206 BCE) to the end of the Qing Dynasty (1911), Confucianism has been the ruling ideology for imperial China. Compared with Daoism and Buddhism, Confucianism puts greater emphasis on social morality and harmony. The Confucian ethical codes are regulated by the *wuchang* 五常 (Five Constants) of *ren* 仁 (benevolence), *yi* 义 (righteousness), *li* 礼 (rites), *zhi* 智 (knowledge), and *xin* 信 (integrity) (*Mengzi or* Mencius, 11.6). The Confucian social order is characterized by the *wulun* 五伦 (Five Relations) of *junchen* 君臣 (sovereign-subject), *fuzi* 父子 (father-son), *fufu* 夫妇 (husband-wife), *xiongdi* 兄弟 (elder brother-younger brother), and *pengyou* 朋友 (friend-friend) (Mencius, 5.4).

Although Confucius himself had few comments about soil, following scholars in the tradition have attributed many Confucian virtues to it. The Song Dynasty Neo-Confucian philosopher Zhu Xi (1130–1200), whose *Zhouyi Benyi* 周易本义 (The Original Meaning of the Yijing) became the standard commentary for the Yijing in later dynasties, explained Hexagram 2, *kun* 坤 (earth, soil) as "The beauty of Yin is contained, in the King's service. Yin never presumes to claim success. This is the Tao of Earth, the Tao of Wife, the Tao of Minister. The Tao of Earth knows of no success; it brings conclusion for others" 阴虽有美"含"之以从王事，弗敢成也。地道也，妻道也，臣道也。地道"无成"而代"有终"也 (translated by Minford, 2014, p. 43). From the Confucian perspective, therefore, heaven and earth/soil are inseparable components of the cosmic macrocosm, just as husband and wife, as well as sovereign and subject, are immanently interconnected parts of the microcosms of human family and government. At the same time, earth/soil is envisioned to conform, submit, and serve the purpose of heaven, embodying the docile and submissive side of the Confucian virtue.

The Han Dynasty Confucian scholar Wang Chong, known for his rationalist approach to Confucius' original teachings, offered extensive discussions on soil in his *Lunheng* 论衡 (Balanced Discourses). In the chapter *Qiguai* 奇怪 (Strange), Wang refuted the mythological non-human origins of two Chinese rulers with the assertion that "Myriad things grow out of the soil, each is like the kind of its own. Those which are not similar to soil are not produced by soil, they are merely nurtured by soil" 万物生于土，各似本种。不类土者，生不出于土，土徒养育之也 (15.7). While acknowledging the generative nature of the soil, Wang also argued for the importance and efficacy of cultivation both for the fertility of the soil and the morality of the human:

> Rich, fertile, stony, and barren, these are the original nature of the soil. Rich and fertile soil by nature is good, trees and crops grow abundant and luxuriant on it. Stony and barren soil by nature is bad, but its energy can be improved through deep ploughing, thorough hoeing, rich application of manure, and diligent employment of human efforts, so that trees and crops can grow similar to those of rich and fertile soil. 夫肥沃硗埆，土地之本性也。肥而沃者性美，树稼丰茂；硗而埆者性恶，深耕细锄，厚加粪壤，勉致人功，以助地力，其树稼与彼肥沃者相似类也. (8.5)

In a later part of the passage, Wang also discussed the parallel between the improvement of the elevation of the land and that of the character of the individual. As Confucianism gained its position as the dominating ideology, it also fundamentally changed the imperial ruler's relationship with soil.

13.5 Soil and Governance

Soil plays an essential role in establishing and securing the legitimacy of ancient Chinese governments, as confirmed by the *Shijing* [The Book of Songs], the oldest poetry collection in China: "Under the whole of heaven, all is the sovereign's land/soil. Within the boundaries of the land, all are the sovereign's servants" 溥天之下、莫非王土。率土之滨、莫非王臣. This passage means that sovereign power was framed as the power over the land (first order principle) and then all its inhabitants (dependent variable).

13.5.1 Dependence on the Spirits of Soil

Throughout the history of imperial China, the state has made offers to an altar dedicated to the spirits of soil and grains – *sheji* 社稷. *The Rites of Zhou* (*Zhouli*) mentions the term sheji at least 12 times. "When a kingdom is being set up, [the ruler] establishes the altars for sheji, determines the borders of the territory, and decides the portion of enfeoffment" 凡建邦国, 立其社稷, 正其畿疆之封 (Zhouli, 2.82). In the occasions of natural and manmade disasters, the sheji shrine was prayed to for blessing, consulted for omens and offered sacrifice with blood, meat, and grains.

The Eastern Han Dynasty historian Ban Gu 班固 (32–92) explained the reasons for the rituals dedicated to the altars for sheji in the *Baihu Tong* 白虎通 (Comprehensive discussions of the White Tiger Hall):

> Why does the king need to establish the sheji altar? To pray for further blessings and to show gratitude for past merits on behalf of all under heaven. Humans cannot stand without the soil; they have nothing to eat without grains. The land is vast and broad, it is impossible to express veneration to all the land; the grains have many kinds, it is impossible to offer sacrifice to all the grains. Therefore a mound of earth is erected as an altar to show reverence for soil. Millet is the first among all grains, therefore an altar of millet is erected for

sacrifices. 王者所以有社稷何？为天下求福报功。人非土不立，非谷不食。土地广博，不可遍敬也；五谷众多，不可一一祭也。故封土立社，示有土尊。稷，五谷之长，故封稷而祭之也。(2.2.1)

Emperors of many dynasties in China have established the land and grains altar for worship and praying. The sacrificial altar built in 1421 by the Yongle Emperor of the Ming Dynasty is still preserved in Beijing at Zhongshan Park. Standing at the center of the altar is a terrace paved with five types of soils in five colors: bluish-green in the east, red in the south, white in the west, black in the north, and yellow at the center (Fig. 13.3). The location of the five-colored soils mirrors both the wuxing (metal, wood, One water, fire, and water) discussed earlier and the geographical distribution of the main soil types in China.[9]

13.5.2 Soil Classification and Management

Ancient Chinese cosmology envisions correlative relations between heaven and earth, and between the natural and the political. According to the Qin Dynasty encyclopedic text *Lüshi Chunqiu* 吕氏春秋 (Master Lü's Spring and Autumn Annals), this sacred relationship manifests itself through a human-soil continuum:

If heaven operates without trustworthiness, it could not complete a year. If the earth operates without trustworthiness, the grasses and trees cannot grow large. The power (*de*) of the spring is wind. If the wind does not blow with trustworthiness, the flowers cannot grow abundantly. If the flowers are not abundant, the fruit cannot be produced. The power of summer is heat. If the heat does not come with trustworthiness, the soil won't be fertile. If the soil is not fertile, the growing of the plants won't be essential. The power of autumn is rain. If the rain does not fall with trustworthiness, then the grain won't grow solid. If the grain

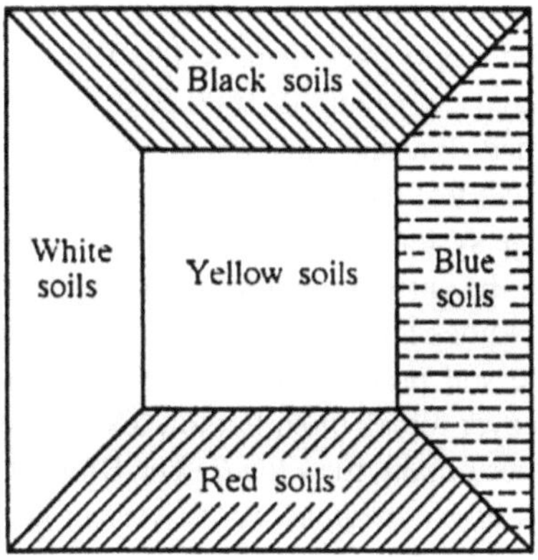

Fig. 13.3 Five-colored sheji altar in Beijing. (Reproduced from Gong et al., 2003) (The map of the altar by Gong et al. follows a modern north-up map direction. In ancient China, maps were often drawn with the south facing up)

[9]The mainstream practice in ancient China for maps and star charts is south facing up. Contemporary popular Chinese wuxing theorists and fengshui practitioners are still using this orientation.

> does not grow solid, then the five seeds won't mature. The power of winter is coldness. If the cold does not come with trustworthiness, then the earth won't be frozen firmly. If the earth is not frozen firmly, then the ice won't melt (in time). Even with the greatness of heaven and earth, and the transformations of the four seasons, still they cannot complete things without trustworthiness. So how much less can human actions do it? 天行不信，不能成岁；地行不信，草木不大。春之德风，风不信，其华不盛，华不盛则果实不生；夏之德暑，暑不信，其土不肥，土不肥则长遂不精；秋之德雨，雨不信，其谷不坚，谷不坚则五种不成；冬之德寒，寒不信，其地不刚，地不刚则冻闭不开。天地之大，四时之化，而犹不能以不信成物，又况乎人事？(19.7, Guixin chapter, translated by Sellman, 2004, p. 123)

The *Jingfa* 经法 (Constancy of Laws) is one of the four appendices to Laozi's *Daodejing* known as the *Huangdi Shu* 皇帝书. It was found in the silk manuscript unearthed at Mawangdui, a tomb sealed in 168 BCE during the Han Dynasty and rediscovered in 1973. In the *Junzheng* 君正 (The Ruler's Government) chapter of Jingfa, one particular passage is dedicated to the connection between good governance of the feudal-bureaucratic rulership and the ruler's impeccable knowledge of the land, including understanding of the soil type and the right use of crops and plants:

> The basis of the humans is the earth, the basis of the earth is suitability, the basis of growing [crops] with suitability is the seasons. The use of seasons depends on the people, the use of the people depends on their capacity, the use of the people's capacity depends on the [ruler's] regulation. If [the ruler] understands the suitability of the earth/soil types, selects the right time to plant, applies appropriate regulation to employ people's capacity, then wealth will grow. … this is the dao of defending borders and defeating enemy in battle. 人之本在地，地之本在宜，宜之生在时，时之用在民，民之用在力，力之用在节。知地宜，须时而树，节民力以使，则财生，… 则守固单（战）朕（胜）之道也。 (Jingfa, 1.3.6, annotated by Chen, 2007, pp. 67-71)

The importance of understanding the diversity of soil types was emphasized by many other classical texts of ancient China. The *Yugong* 禹贡, a geographical survey text perhaps compiled in the 5th century BCE, is considered the oldest document on soil science in the world (Needham & Lu, 1981). The book explains soil classification in the Xia Dynasty (c. 2070–1600 BCE), the first government in the Chinese history of recorded dynastic succession. According to *Yugong*, after the legendary Yu managed to control the Great Flood by combining dikes formed with soil and building a drainage system, he toured all the nine provinces of the country and described the lands of the entire kingdom, including high mountains, great rivers, soil types, the commodities and products of each province, and the taxes and duties associated with the soil productivity of the provinces. Soil characteristics discussed in the *Yugong* cover topics of soil color, texture, fertility, and vegetation. Besides identifying 12 types of soil, the *Zhouli* (The Rites of Zhou) also subdivides the land into mountainous forest areas, areas of rivers or lakes, hilly areas, alluvial plains, and marshy plains (2.61–63). The *Guanzi* 管子, a political text traditionally attributed to the 7th century BCE stateman Guan Zhong 管仲 and edited during the Han Dynasty, classifies soils in ancient China into 18 sublevels and pointed out five manifestations of each of the 18 sublevels. In the section entitled *Diyuan* 地员 (Categories of Land), the text grouped the 90 soil classification units into three classes according to the quality of the soil, recorded soil distribution, and relationship among plants, soil type, moisture conditions, and landscape positions (Xia, 1958).

13.6 Soil in Contemporary China

While inheriting traditional practices of soil, contemporary China also faces its unique challenges and opportunities associated with modernity and industrialization.

13.6.1 Soil and Population Pressure

China has been the most populated country in the world since the beginning of the Common Era (Liu, 1986). To handle the pressure of feeding its large population, the Chinese state and people intensified agriculture, applying various land use and cultivation methods. Labor application, field management, crop rotation, and harvest preservation have all reached such precision that traditional farming in China was often compared with gardening or even embroidery. In the now classic book *Farmers of Forty Centuries; or Permanent Agriculture in China, Korea and Japan*, F. H. King (1911) used the term "permanent agriculture" to describe agricultural practices across the three populous East Asian countries that had been self-sufficient in food production from the same soil for millennia. Although many organic farming and permaculture practitioners today consider the "permanent agriculture" recorded by King as "*ancien régime* organic agriculture" (Paull, 2011, p. 179), others have contested that the Chinese soil has been depleted ever since farmers started plowing up highly erodible loess soils into the Yellow River's headwaters during the Qin and Han Dynasties (Sang, 2005; Shi, 1991). As the early dwellers along the Yellow River's tributaries populated the area, they deforested the great river's headwaters, which in turn escalated the process of erosion. Even the name of the river was changed from The River or Great River to its current name, reflecting the color of soil eroded from the Loess Plateau. The Tang and Song dynasties saw desert expansion through the northern plateau, and large-scale deforestation along the Yangtze River and Pearl River, as well as in the northeastern regions. During the Ming Dynasty, the Columbian Exchange introduced the New World crops of maize, peanuts, potatoes, and sweet potatoes to China, most of which required little cultivation and less fertile soil (Ho, 1959). As these new crops were integrated into primary or staple food sources in various regions of China, the Chinese population doubled from 150 to 300 million during the 18th Century. The ongoing population explosion, on the other hand, forced large scale migration of farmers moving from fertile floodplains to mountainous areas of the lower Yellow River and Yangtze River provinces. By the first half of the 20th century, population density in some coastal districts was so high that it became impossible to retire submarginal land from active service. Subsequently, soil erosion happened at such an alarming rate that crop yields decreased noticeably, and disastrous droughts and floods occurred more frequently (Tieh, 1941). As of 2015, China supports 1/5 of the world's population with just 8.5% of its total arable land (Food and Agriculture Organization, FAO, 2020). The popular Chinese proverb "To the people, food is of primary importance. To the food, soil/land is primary importance" (民以食为天，食以土为本) illustrates the interlocked relationship between population pressure, food security, and soil conservation.

13.6.2 Erosion, Pollution, Depletion

Modern Chinese soil conservation efforts started in the 1920s, drawn primarily from modern scientific principles of soil erosion prevention (Sun & Zhao, 2008). In 1940 the term "soil and water conservation" (水土保持) was coined. After the Chinese Communist Party won the 1949 revolution, the Chinese government constructed a series of large-scale rural infrastructure projects. Inspired by experimental projects in rural areas in the early 1950s, the "Deep Plowing and Soil Improvement Movement" (深翻和改良土壤运动) initially improved both physical and chemical properties of the soil, thus facilitating yield increases. During the "Great Leap Forward" towards the end of the same decade, however, the movement slowly transformed into a national frenzy and in many locations ruined the structure of the topsoil. Hence, the soil became vulnerable to erosion and led to widespread waste of both human and material resources at large (Zhu et al., 2016). The Green Revolution of the mid-20th century, with its focus on maximizing food production, brought high-yielding varieties of grain crops, along with chemical fertilizers and agrochemicals to China. To alleviate ever increasing population pressure, the Chinese government and the Communist Party promoted a diversity of "scientific farming" technologies, along with organic and traditional methods, in the people's communes between 1960 and 1980 (Schmalzer, 2016). After the Household Land Division (分田到户) system was established in the 1980s, field management was redistributed to individual households from the village collective, even though the collectives remained owner of the lands. This new family-based system rekindled rural peasants' enthusiasm in agricultural production, but also lead to increased use of chemical fertilizers and pesticides. China's agricultural support policies and subsidies have also directly and indirectly contributed to the increased use of agrochemical consumption and continuous decrease of organic fertilizer application. Today, China is the world's largest producer and consumer of chemical fertilizers and pesticides. Amid the rapid pace of urbanization in recent years, the Chinese rural collective property rights system has also allowed many arable lands near cities to be converted for urban use and industrial production, both of which are more lucrative for the local governments. The proximity of farmlands to industrial and mining sites across China has led to widespread soil contamination and water pollution. One report showed that during the decade between 2000 and 2010, arable land area in China had decreased by 3.31% due to urbanization, soil erosion, and land degradation (Wang et al., 2019a). Starting from 2010, the total arable land decline has slowed compared to the previous several decades (Liu et al., 2018).

13.6.3 Ecological Civilization and Soil Restoration

Facing the compounded threats of food safety and environmental issues, in 2007 the Chinese Communist Party proposed to build an "ecological civilization." Then in 2012 the Chinese Communist Party wrote the concept into its constitution. In the past several years, "Clear waters and green mountains are as good as mountains of gold and silver" (绿水青山就是金山银山) has turned into a catchphrase to illustrate the importance of green and sustainable development. In his 2016 book entitled *China's Environmental Governing and Ecological Civilization*, the Chinese scholar Pan Jiahua attributed the origin of ecological civilization to "harmony between human and nature" (天人合一), a Chinese "traditional, historical, and cultural concept and practice" of understanding and defining the relationships between human and nature (p. 29).

The ecological civilization envisioned by China requires a holistic approach to the country's intricate situation in regard to soil restoration and environmental protection, both at the policy level and in practice. The "Grain for Green Program" initiative, first introduced in 1999, focuses on returning marginal former farmland to forest, in order to mitigate soil erosion and improve ecological resilience. The 2004 Amendment of the *Land Administration Law of the People's Republic of China* clearly states that "To cherish and give a rational use to the land as well as to give a true protection to the cultivated land are seen as a basic principle of land use in the country" (Article 3). In 2006, the Chinese central government announced the *Outline of the National Overall Planning on Land Use (2006–2020)*, set a "red line" of 120 million hectares of arable land for food production, and a ceiling of 3 million hectares of arable land allowed to be converted to construction land in total by 2020. Since 2011, the Chinese State Council and other administrative bodies have further drawn a set of ecological red lines in forestry, wetland, desert vegetation, and biodiversity for ecological protection and land preservation (Yang et al., 2014).

At the practical level, since the first publication of the Chinese Environmental Quality Standard for Soils in 1995, more than 50 standards and guidelines have been enacted for soil pollution control and soil quality evaluation (Lin & Wang, 2018). A number of soil restoration technologies and related equipment were developed under the "Polluted Soil Restoration Technology and Demonstration" program, a key project for the 12th Five-Year Plan of 2011–2015. In recent years, soil regenerative and natural farming practices such as agroecology, biodynamic, and permaculture have gained popularity among international and domestic practitioners in China.[10] Indigenous soil regeneration initiatives, such as the Loess Plateau Project, have also inspired international ecosystem restoration movements beyond China (Liu, 2016). Driven by both government support and consumer needs, many urban farmers markets and community supported agriculture (CSA) programs feature regenerative products extensively. For the rapidly growing middle-class Chinese consumers and the CSA farmers, organic cultivation is viewed as a production system that simultaneously integrates soil health, human well-being, and agroecosystem sustainability (Shi et al., 2011).

The Chinese efforts at soil restoration and land preservation have generated mixed results. The soil pollution treatment projects, for instance, have been criticized to be piecemeal without national strategy. The "Grain for Green" program, while being successful in increasing soil organic carbon and preventing soil pollution, has not greatly improved biodiversity of the land (Wang et al., 2019b). Even though ecological civilization, material civilization, social culture, spiritual civilization, and political civilization (物质文明、政治文明、精神文明、社会文明、生态文明), are considered as the "five dimensions" of socialist society with "Chinese characteristics" (Pan, 2016), the integration between spiritual traditions of the past and sustainable development in contemporary China has not been as smooth and coordinated as many expected. The task of incorporating spiritual traditions and new approaches towards sustainability, in order to transform Chinese contemporary civilization and soil relations, remains an enormous yet rewarding task for the present and future.

13.7 Conclusion

The Chinese culture is essentially an agrarian culture, emerged from the land, rooted in the earth, and nourished by the soil. Given the length limitation of this chapter, many important aspects of the Chinese cultural understanding of soil, such as soil and folk belief in China, soil and Traditional Chinese Medicine, and soil's role in geomancy, are left untouched. The common Chinese saying "The water and soil in one region nurture a people distinctive in this region" 一方水土养一方人 is often quoted to express the deep embeddedness of a culture in its ecological context. As the increasingly globalized world faces climate change, soil degradation, and other ecological crises, perhaps the ancient Chinese understandings of the human-soil continuum can still shed light for humanity on how to make the shift towards true ecological civilizations.

[10] Permaculture itself is a foreign concept, as are biodynamic farming and agroecology. While they have many overlapping principles and practices, each has its own school of thoughts and promoters. Local practitioners need to be trained by international certified practitioners and teachers, either in China or abroad. There are also foreign students studying sustainable agriculture in Chinese universities and foreign agriculture companies cultivating land in China for organic food items (wine, fruits, etc.).

References

Baihu Tong 白虎通 [Comprehensive discussions of the White Tiger Hall]. Retrieved from https://ctext.org/bai-hu-tong

Bodde, D. (1961). Myths of ancient China. In S. N. Kramer (Ed.), *Mythologies of the ancient world* (pp. 367–408). Anchor Books.

Brahmajāla Sūtra 梵网经 [Brahma's Net Scripture]. CBETA T 24–1484. Retrieved from http://tripitaka.cbeta.org/T24n1484

Buswell, R. E., Jr., & Lopez, D. S., Jr. (2014). *The Princeton dictionary of Buddhism*. Princeton University Press.

Chen, G. (2007). *Huangdi sijing jinzhu jinyi: Mawangdui Hanmu chutu boshu* 黄帝四经今注今译：马王堆汉墓出土帛书 [The four classics of the Yellow Emperor with modern annotations, and translations in modern Chinese]. Commercial Press.

Eliade, M. (1963). *Myth and reality* (W. R. Trask, Trans.). Harper & Row.

Food and Agriculture Organization of the United Nations (FAO) (2020). *China at a glance*. Retrieved from http://www.fao.org/china/fao-in-china/china-at-a-glance/en/

Ge, Z. (1995). Zhongguo chan sixiangshi: Cong 6 shiji dao 9 shiji 中国禅思想史：从6世纪到9世纪 [A history of Chinese Chan Buddhism: From 6th century to 9th century]. Peking University Press.

Ge, Z. (2014). *An intellectual history of China: Volume one: Knowledge, thought and belief before the seventh century CE* (M. S. Duke & J. Chiu-Luke, Trans.). Brill.

Gong, Z., Zhang, X., Chen, J., & Zhang, G. (2003). Origin and development of soil science in ancient China. *Geoderma, 115*, 3–13.

Graham, A. C. (1989). *Disputers of the Tao: Philosophical argument in ancient China*. Open Court.

Guo, M. (1931). *Shi zubi* 释祖妣 [Explanation of zubi]. In *Jiagu wenzi yanjiu* 甲骨文字研究 [Research on the oracle bone scripts] (pp. 1/1a–23b). Dadong Shuju.

Ho, P. (1959). *Studies on the population of China, 1368–1953*. Harvard University Press.

Huangdi Neijing*: Suwen* 黄帝内经・素问 [Huang Di's inner classic: Basic questions] (1983). B. Wang (Ed.). Renmin Weisheng Chubanshe.

Jingtu Shiyao 净土十要 [Ten Essentials on the Pure Land]. CBETA X 61–1164. Retrieved from http://tripitaka.cbeta.org/X61n1164_001

Kim, K. (2011). Rice cuisine and cultural practice in contemporary Korea dietary life. *Korea Journal, 50*(1), 11–35.

King, F. H. (1911). *Farmers of forty centuries, or permanent agriculture in China, Korea and Japan*. Mrs. F. H. King.

Liang, J. (1980). *Zhongguo turang kexue de yunyu ji xingcheng* 中国土壤科学的孕育及形成 [Origin and development of soil science in China]. Mimeographed Copy.

Liezi 列子. Retrieved from http://www.ctext.org/liezi

Lin, Y., & Wang, G. (2018). 20th anniversary for soil environment protection in China. In Y. Luo & T. Chen (Eds.), *20 years of research on soil pollution and remediation in China* (pp. 55–64). Springer Nature.

Liu, T. (1986). Agricultural change and population growth: A brief survey on the case of China in historical perspective. *Academia Economic Papers., 14*(1), 29–68.

Liu, J. D. (2016). Earth restoration peace camps. *Permaculture Magazine, 90*, 42–46.

Liu, L., Bestel, S., Shi, J., Song, Y., & Chen, X. (2013). Paleolithic human exploitation of plant foods during the last glacial maximum in North China. *PNAS, 110*(14), 5380–5385.

Liu, D., Gong, Q., & Yang, W. (2018). The evolution of farmland protection policy and optimization path from 1978 to 2018 改革开放40年来中国耕地保护政策演变及优化路径. *Chinese Rural Economy, 12*(8), 37–51.

Mengzi 孟子 [Mencius]. Retrieved from https://ctext.org/mengzi/

Minford, J. (2014). *I Ching (Yijing): The book of changes, translated with an introduction and commentary by John Minford*. Penguin Books.

Needham, J., & Lu, G.-D. (1981). Chinese geo-botany in statu nascendi. *Journal d'Agriculture Traditionelle et de Botanique Appliqee, 28*(3-4), 199–230.

Needham, J., & Wang, L. (1956). *Science and civilisation in China, Vol. 2: History of scientific thought*. Cambridge University Press.

Pan, J. (2016). *China's environmental governing and ecological civilization*. Springer.

Parkes, G. (2003). Winds, waters, and earth energies: Fengshui and sense of place. In H. Selin & A. Kalland (Eds.), *Nature across cultures: Views of nature and the environment in non-Western cultures* (pp. 185–210). Springer.

Paull, J. (2011). The making of an agricultural classic: Farmers of forty centuries; or, permanent agriculture in China, Korea and Japan, 1911–2011. *Agricultural Sciences, 2*(3), 175–180.

Peerenboom, R. P. (1993). *Law and morality in ancient China: The silk manuscripts of Huang-Lao*. State University of New York Press.

Sang, G. (2005). Vegetation variation of Loess Plateau during human history 黄土高原历史时期植被变化. *Journal of Arid Land Resources and Environment, 19*(4), 54–58.

Schmalzer, S. (2016). *Red revolution, green revolution: Scientific farming in socialist China*. University of Chicago Press.

Sellman, J. D. (2004). *Timing and rulership in Master Lü's Spring and Autumn Annals (Lüshi chunqiu)*. State University of New York Press.

Shangshu [The book of documents]. Retrieved from https://ctext.org/shang-shu

Shanhai Jing 山海经 [The Classic of mountains and seas]. Retrieved from https://ctext.org/shan-hai-jing

Shi, N. (1991). Lun lishi shiqi woguo zhibei de fenbu jiqi bianqian 论历史时期我国植被的分布及其变迁 [Vegetation distribution and changes in China throughout history]. *Journal of Chinese Historical Geography, 3*, 43–73.

Shi, Y., Cheng, C., Lei, P., Wen, T., & Merrifield, C. (2011). Safe food, green food, good food: Chinese Community Supported agriculture and the rising middle class. *International Journal of Agricultural Sustainability, 9*(4), 551–558.

Shiji 史记 [Historical records]. Retrieved from https://ctext.org/shiji

Shijing 诗经 [Book of songs]. Retrieved from https://ctext.org/book-of-poetry/

Shuowen Jiezi 说文解字 [Explanations of simple graphs and analyses of composite graphs]. Retrieved from https://ctext.org/shuo-wen-jie-zi

Sun, H., & Zhao, J. (2008). Idea contrast on soil and water conservation at home and abroad Search Results 中外水土保持理念对比. *Journal of Yangtze River Scientific Research Institute, 3*, 9–13.

Tieh, T. M. (1941). Soil erosion in China. *Geographical Review, 31*, 570–590.

Unschuld, P. U. (1985). *Medicine in China: A history of ideas*. University of California Press.

Unschuld, P. U. (2003). *Huang Di nei jing su wen: Nature, knowledge, imagery in an ancient Chinese medical text*. University of California Press.

Wang, C. *Lunheng* 论衡 [Balanced discourses]. Retrieved from https://ctext.org/lunheng/

Wang, Y. (1980). *Zhongguo gudai turang kexue* 中国古代土壤科学 [Soil science in ancient China]. Kexue Chubanshe.

Wang, R. R. (2012). *Yinyang: The way of heaven and earth in Chinese thought and culture*. Cambridge University Press.

Wang, L., Herzberger, A., Zhang, L., Xiao, Y., Wang, Y., Xiao, Y., Liu, Z., & Ouyang, Z. (2019a). Spatial and temporal changes of arable land driven by urbanization and ecological restoration in China. *Chinese Geographical Science, 29*, 809–819.

Wang, X., Hua, F., Wang, L., Wilcove, D. S., & Yu, D. W. (2019b). The biodiversity benefit of native forests and mixed-species plantations over monoculture plantations. *Diversity and Distributions, 25*(11), 1721–1735.

Wuxing Dayi 五行大义 [Important meanings of the Wuxing]. Retrieved from https://ctext.org/wiki.pl?if=en & res=601464

Xia, W. (1958). *Guanzi Diyuanpian jiaoshi* 管子地员篇校释 [Collation and interpretation of the Guanzi]. Zhonghua Shuju.

Yamashita, S. (1998). *Shindo-fuji no tankyu* 身土不二の探求 [A research into "body and earth are not two"]. Soshinsha.
Yang, L., An, D., & Turner, J. A. (2008). *Handbook of Chinese mythology*. Oxford University Press.
Yang, B., Gao, J., & Zou, C. (2014). The strategic significance of drawing the ecological protection red line 划定生态保护红线的战略意义. *China Development, 14*(1), 1–4.
Yijing [Book of changes]. Retrieved from https://ctext.org/book-of-changes
*Zhouli*周礼 [The Rites of Zhou]. Retrieved from https://ctext.org/rites-of-zhou
Zhouli Zhushu 周礼注疏 [Commentaries and explanations to The Rites of Zhou]. Retrieved from https://ctext.org/wiki.pl?if=en & res=206690
Zhu, X., Ding, Z., & Hu, H. (2016). The deep plowing movement of the "great leap forward". In H. Li & T. Dubois (Eds.), *Agricultural reform and rural transformation in China since 1949* (pp. 74–100). Brill.
Zhuangzi 庄子. Retrieved from https://ctext.org/zhuangzi
Zutang Ji 祖堂集 [Patriarch's Hall Anthology]. CBETA B 25–0144. Retrieved from http://tripitaka.cbeta.org/B25n0144

Chapter 14
Of Planners, Workers, and Activists: A Short History of Urban Soil Cultures (Europe and North America)

Germain Meulemans

14.1 Introduction

In the Global North, urban environments are often seen as antagonistic to soils. In common imagery or discourses about soils, those are assumed to be a characteristic feature of rural or natural, but not urban, landscapes. Far away from the living soil perspective of soil scientists, most urban planners and other city specialists see the urban soil as a two-dimensional abstract ground, the space between buildings. City builders and constructors, for their part, tend to be concerned with the capacity of the land to bear constructions or are only interested in the "topsoil," the upper, fertile layer of soil, which they consider at best as a building material and at worst as a mere waste product. Until the end of the twentieth century, urban soils also remained a marginal topic within the soil sciences themselves, and soil mappers often regarded cities as "non-soil" places (Burghardt et al., 2015; Meulemans, 2020a).

Although city dwellers often think of themselves as being foreign to soils, they do not live without a material connection to the soil. The city cannot exist without a nurturing relationship with the land, and the great development of urban areas over the last few centuries cannot be understood independently of the great transformations in the ways of cultivating the land in the countryside. This is evidenced by the fact that, with the current galloping urbanization, the ecological footprint of cities is continuously increasing on all the planetary ecosystems, including their soils (Wackernagel et al., 2006). In this chapter, therefore, I first address the material, social, and symbolic configurations that conceal these links between the modern city and the soils and allow the city to think of itself as independent from the soil. I then explore two themes in which the separation between soil and cities is

G. Meulemans (✉)
Centre Alexandre-Koyré, UMR 8560 CNRS-EHESS-MNHN, Aubervilliers, France

N. Patzel et al. (eds.), *Cultural Understanding of Soils*,
https://doi.org/10.1007/978-3-031-13169-1_14

challenged and called into question. Furthermore, I examine attempts in the world of urban planning and urban design to reweave symbolic and metabolic links between the city and the soil by attempting to re-loop nutrient cycles and to rethink the principles of urban sanitation. Secondly, I discuss the various movements of gardening and urban agriculture, which in cities of the Global North have sought to respond to the ills generated by urbanization and the alienation of agricultural soils.

14.2 A Threefold "Metabolic Rift"

The industrial revolution and associated urbanization radically transformed production and consumption practices in Europe between the eighteenth and twentieth centuries and, thus, the relationship to soils. In many respects, in the premodern city, the separation between soil and the city was not as clear-cut as it became later. Many premodern cities had been careful to keep vegetable gardens and even fields within their walls to be able to withstand a siege (Mumford, 1956). Furthermore, several historians have shown how, until the early nineteenth century, the soils and fertile materials of the cities were considered a source of wealth and played an important role in the agricultural or artisanal activities of the time (Tarr, 1975). Across the western world, the organic substances produced or transformed by the city – horse and other animal manure, "night soil" (human excreta), food residues – were collected manually for use in fertilizing the fields (Ferguson, 2014). It was common for peri-urban vegetable growers to arrive at downtown markets with a cartload of vegetables and leave with a cartload of town manure. From the end of that era, however, doctors and other scholars began to see urban soils, and the processes of decay and degradation of organic matter, as a source of corruption of the minds and health of city dwellers. At the beginning of the nineteenth century, the concept of cleanliness evolved, and the soil became increasingly seen as something dirty, unclean, which could not fit into the symbolic order of cities. To use Mary Douglas' (1966) definition of impurity, soils become "matter out of place" in cities. Urban soils frequently became presented as a putrid carcass exuding dangerous miasmas (Barles, 1999).

Thus, the modern city developed with a logic of externalizing the vital processes of decomposition and transformation of organic substance. Under the influence of hygienist doctrines, the development of centralized sewerage and waste collection systems profoundly reorganized domestic cleaning and disposal practices, transforming the plurality of materials that are dumped in them into generic "waste" (Barles, 1999). One of the consequences is that waste – especially human excreta – was no longer returned to agricultural land. New technologies implemented on or within urban soil, such as surfacing it with paving stones and asphalt and laying pipes through it, were designed specifically to impair water and organic flows to reach the soil. These sealed, smooth, and dry urban soils became, in the nineteenth century, a prerequisite for any notion of modern civilization and the identities and economies associated with it. As the modern city aspired to become aseptic, it gradually neutralized or concealed the metabolic processes that provided the link with the soil.

Concomitantly, the development of new means of transport has allowed food to come from far away, while the import and then industrial production of mineral fertilizers has allowed agriculture to dispense with the organic matter produced in cities for fertilization. The increase in the price of land for building, compared to the cost of agricultural land, completely pushed agriculture out of the cities. In the twentieth century, the development of the individual car and changes in the culture and lifestyles of the new middle class favored single-family housing and led to the emergence of an economy based on large commercial and logistics centers situated outside of city centers (Melosi, 2005). These elements contributed greatly to making urban sprawl one of the main causes of land degradation in the second half of the twentieth century. Since the beginning of the massive urbanization of the nineteenth century, many urban thinkers have criticized how the growth of cities endangers agricultural soils. One of the most famous is Lewis Mumford (1956), who, in the middle of the twentieth century, pointed out that through urban sprawl, cities eat up their hinterland. Indeed, cities were often established on fertile plains from which they drew their livelihood. It was the production of food surpluses from such fertile soils that encouraged the emergence of the first urban civilizations. Now that cities are spreading out, Mumford insisted, it is often the same very fertile soils that they destroy.

Since 1975, the area of built-up land has doubled in Europe and quadrupled in Africa. At the same time, the urban population has remained relatively constant in Europe, while it has tripled in Africa – leading to very different population density rates (Pesaresi et al., 2016). In 2015, nearly 52% of the world's population lived in urban centers, 33% in small towns and suburbs, and only 15% in rural areas (Pesaresi et al., 2016). This means that the majority of humans now rarely experience the fields and forests where most people locate what they call "soils."

Thus, over the past two centuries, a process that some historians have termed a "metabolic rift" (Foster, 1999) has occurred between the places where food is produced and the places where it is consumed. This break has obscured the links between city and countryside, or between humans and nature, and has led to the emergence of completely new urban soil cultures that are distinct from the approaches prevailing in the countryside. For the geographer Nathan McClintock (2010), the metabolic rift established by the modern city is threefold. First, it produces a biophysical rift by cutting off nitrogen and phosphorus loops, preventing organic matter from returning to fertilize the fields. Secondly, it creates a "social rift" between producers and consumers of food, as agriculture becomes a specialized activity that now occupies only a small percentage of the population. Third, it creates an "individual rift," which McClintock defines as "the perception of self as external to the environment" (2010, p.10).

Urbanization, therefore, has had a significant impact on the way societies understand and connect with soils. In this sense, the sealing of urban soils is intimately linked to the achievement of a kind of cosmic insouciance in which soils and nature are relegated to the background, to the status of virgin space or dumping ground for resources and garbage. For the historian Frank Uekötter (2012), as soils became covered and waste and organic matter were exported out of the cities, issues related to soils and their fertility became largely invisible to a large part of Western society. For Uekötter, the seemingly landless, urbanized landscape with which city dwellers live contributed to the lack of interest in soil and soil degradation in urbanized societies over the last two centuries.

14.3 Attention to Soils in Urbanism

The mainstream principles for the development of the modern city are synthesized in a particularly explicit way in Le Corbusier's Athens Charter (1957), one of the most influential manifestoes of postwar modernist planning. This manifesto recommends that the city be dried up by covering canals, embanking rivers, draining marshes, waterproofing soils, burying the water cycle, and creating well-defined green areas – parks and gardens. In many parts of the world, modernist planning played a tremendous role in shaping the modern "off-ground" city (Njoh & Watson, 2009), even though unplanned districts continued to develop in the poor margins of many cities in the Global South. However, urban planning – from the nineteenth century to the present day – cannot be reduced to this approach. Many town planners – often dissident, but also sometimes very mainstream – have proposed other ways of thinking and organizing the relationship between soils and cities. Indeed, discussions about the return to the land of urban waste and excreta, or the search for a better living environment for the working classes in the garden cities and then for the new middle class, have influenced town planners on both sides of the Atlantic. Many of them have sought to reweave links between the city and the land, as much on a symbolic level – the "connection with nature" – as on alimentary purposes or functional and ecological aspects – sanitation, fertilization, or, more recently, the discourse about the "ecosystem services" rendered by soils (Morel et al., 2015).

Many city planners have approached the problem from an aesthetic perspective, arguing that being close to nature is important for human well-being. Thus, the Garden City movement, which first developed in the United Kingdom through the influence of Ebenezer Howard (1850–1928), claimed to re-establish a link with the soil and nature to improve the quality of urban life, which had been severely degraded by overpopulation and industrialization. Howard promoted the development of more open towns, surrounded by green belts with integrated urban gardens and farms. Some city planners espousing this movement were influenced by discussions over the American Dust Bowl – the massive and disastrous erosion of the great plains of the American Midwest in the 1930s – which had a strong impact on minds at that time and drew attention in many spheres to the question of the relationship to soils in industrial cultures (McNeill, 2000). After the Second World War, the development of a suburban model of urban development, with a private garden for everyone, and based on the generalization of the automobile, was part of the movement to reconnect cities and nature. This also highlights a paradox, since this later movement, in which making a success of one's life equated to owning a private garden, was largely responsible for urban sprawl and the loss of agricultural land in the second half of the twentieth century.

Other city planners were more concerned about the issue of returning urban waste and excreta to the land. Often these approaches build on a critique of sewage systems and the "hydraulic" approach to urban sludge management. The fact that urban sludge was diluted in rivers or in the sea, where the sewage discharges, was soon denounced by prominent figures such as Justus von Liebig (1865) as a shameful waste of nutrients. These criticisms were rooted in a holistic view of nutrient loops and were inspired by the work of chemists and soil scientists in vogue at the time. They were also informed by observers returning from the colonies, where "nightsoil" was still being collected for fertilization. Thus, in an influential book, the American agricultural scientist Franklin Hiram King (1911) called for a complete reconsideration of the Western urban way of life to incorporate traditional Chinese nightsoil collection techniques that he observed during a stay in Asia (Ferguson, 2014). Under these various influences, as early as the nineteenth century, some cities began to organize the return of excreta to the land and bought up infertile plots of land on their outskirts. They then built canal systems for the spreading of urban sludge. Known examples are a system of irrigation canals developed around Berlin in the nineteenth century or the spreading of sludge from Paris on the Pierrelaye area, to fertilize these sandy plains and allow agriculture to develop there. These systems would often appear retrograde in the twentieth century and were dismantled for the most part.

Some urban planners and landscape designers went even further, proposing to reintegrate the practices of composting excrement and soil fertilization into the very heart of urban life. One example is the work of the German landscape architect Leberecht Migge (1881–1935), who sought to rebuild organic links between soils and the city and saw urban soil management and the reclaiming of wasteland as a priority for urban governance. In his designs for neighborhoods and cities, he sought to enable the most disadvantaged to have their own home, a vegetable garden, to lead a healthy existence, and to achieve greater economic independence (Fig. 14.1). Migge was a strong advocate of reusing waste in these vegetable gardens and envisioned collective composters around which social life and the city layout would be organized. As historian David Haney (2010) argues, for Migge and other planners of his time, returning excreta to the ground was not just a technical issue, but a philosophical one: that of recreating the link between waste and consumption, in a holistic understanding of the world where humans and soil were bound by elemental flows that traverse and keep them both alive.

Urban planning discussions on the organic links between cities and soils regularly surfaced until today. From the 1970s onwards, the place of soils was reconsidered in the development of the city, this time from the angle of industrial pollution. This occurred in particular following health scandals such as that of Love Canal in the USA, where numerous cancers and miscarriages occurred in a community living in a district where the soil was heavily polluted by former industrial activities. This scandal left a lasting impression on people's perceptions of soils and led to the

Fig. 14.1 Productive gardens for a self-sufficient housing scheme, drawing by Leberecht Migge (1918, p. 36)

creation of policies to inventory and manage polluted soils in the USA (Colten & Skinner, 1996). In the general public and the world of planners, these scandals contributed to extending, in a new form, the representations of urban soils as dirty, dangerous, and "out of place" environments, which had been popularized by hygienism.

These considerations came at a time when the rich countries of the North were undergoing a wave of deindustrialization. Brownfields, with their polluted soils and ruderal plants, were becoming a common landscape in many cities. They soon became the subject of policies aimed at "reclaiming" them, and a discourse of technical mastery similar to that previously reserved for desert areas or marshlands was applied to them. At the same time, and depending on the region, industrial landscapes were the subject of different policies depending on whether the authorities wished to preserve them as heritage sites, recognizing the historical importance of industrialization (as with the addition of the French Nord-Pas-de-Calais Mining Basin, in France, or the Zollverein Coal Mine Industrial Complex, in Germany, to the UNESCO World Heritage Sites list) or to requalify them as swiftly as possible in order to better forget the industrial past and look forward to a future where coal and steel would be a distant memory. Polluted urban soils became associated, in representations, with this long industrial history which has profoundly transformed the societies of the North and left its mark on their soils. While rusty factories have been leveled, the soil remains in place and prolongs, unavoidably, the memory of these industries.

Since the beginning of the twenty-first century, two major new topics have emerged in the culture of urban and regional planning. The first relates to land take and relates to the long-standing theme of the destruction of fertile land by urban sprawl. The fact that soil is increasingly approached, in collective representations, as a "nonrenewable resource" has emerged largely in the context of discourses on over-urbanization. One must underline the insistence of many soil scientists and the success of popularization operations led by the Food and Agriculture Organization of the United Nations (FAO) and the International Union of Soil Sciences (IUSS) – notably the International Year of Soils in 2015, then International Soil Decade – in raising public awareness about the significance of soils for society and the importance of soil conservation. Thus, many players are now calling for "saving land," or "patrimonializing," and even, in France, "sanctuarizing" agricultural soils in the face of the threat posed by the spread of cities. Although this discourse is mainly aimed at rural soils, it is largely emanating from cities, where the increasing remoteness of production areas is leading to fears of disasters if the economy can no longer rely on fossil resources to bring food from far away (Mason & Knowd, 2010).

At the same time, over the last 20 years, many actors in urban and regional planning have been looking at soils in terms of the "ecosystem services" they can provide, such as runoff and flood control, pollution attenuation, noise reduction, climate and air quality regulation, or the maintenance of biodiversity (Morel et al., 2015). These new representations of soil are part of the development of "green urbanism," which is concerned with nature in the city and the idea that it can take over functions traditionally performed by "gray," cast-concrete infrastructures (De Block, 2016). The new functions that urban soils are intended to perform include infiltration and retention of rainwater – reducing risks of flooding and of overloading sewage treatment systems – the mitigation of the urban "heat island" effect, the storage of atmospheric carbon, and the support of "urban biodiversity," a notion that appeared in the vocabulary of urban planners at the turn of the 1990s (Hinchliffe & Whatmore, 2006). This new functionalist discourse, inspired by economics, and the afferent practices for evaluating soil "efficiency" according to these management objectives are currently developing fast across cities in the Global North.

It is in this context that soil science has come to play a new role within the city. In the past, pedology was largely thought of as a natural science, developing knowledge about natural or agricultural soils, leaving urban areas aside. Today, urban soil science is flourishing and is described as a new front of science (Richter et al., 2014). New soil categories have been introduced in soil classifications to describe soils that have been strongly transformed by human activities, such as "Technosol" and its many subtypes, which was introduced in 2006 in the World reference base for soil resources (IUSS Working Group WRB, 2006) – one of the most widely used soil classification system, edited by the FAO and IUSS. Soil scientists are thus placing themselves more and more at the service of urban planning, aiming to think of urban soils, even highly transformed ones, as an infrastructural capital for the regulation of urban climate, hydrology, or contaminants (Levin et al., 2017). A new ambition of some soil scientists is in particular to develop "constructed Technosols" from urban waste. These are designed both to reuse waste produced by cities – particularly mineral waste – and to provide materials for greening-up cities by building green corridors, planting trees, or creating green roofs on lightweight technical substrates (Séré et al., 2010). The symbolic charge of these new technological fronts should not be overlooked. Soil construction projects, which carry the promise of building soils on demand, according to specifications that allow providing preselected "services," once again disrupt ideas of soil as a natural heritage to be protected. Thus, these emerging socio-technical configurations could participate very concretely in framing future relations with soils as an entity that can be created from scratch, rather than as a fragile environment to be protected (Meulemans, 2020a).

14.4 Urban Allotments for the Working Class

Initiatives to forge new relationships between soil and the city have not been confined to urban planners or developers. Through the theme of urban gardening, soils, and their cultivation have been at the heart of grassroots social movements, associated with the emancipation and food autonomy of the working or disadvantaged classes.

Urban agriculture is as old as the earliest cities. In Mesopotamia, around 3500 BC, agricultural spaces were reserved within the city walls, while in ancient Egypt, fruit trees brought food and shade to the interior of cities. The Aztecs also maintained complex irrigation systems to grow food within the cities. Even today, in many countries, a significant part of agricultural production still takes place within cities, relying largely on nutrient sources produced by the city itself. In 1996, a report published by the United Nations Development Programme (Smit et al., 1996) estimated that more than 800 million people worldwide grow food or raise animals in cities and that this urban agriculture can be up to 15 times more productive per hectare than conventional agriculture. In fact, the terms "urban agriculture" or "gardening" cover a wide variety of situations and approaches to soils, even within one culture. Differences in the forms, functions, motivations, and understandings of urban agriculture – from a means of subsistence in times of crisis, to a simple hobby, or a tool for social reform – depend largely on the social, environmental, and economic contexts in which it takes place.

In the countries of Europe and North America, new forms of urban agriculture developed with the massive urbanization of the industrial era in the nineteenth century, which brought many poor and malnourished workers to the cities. Everywhere, community gardens or allotments became a familiar marker of working-class landscapes (Crouch & Ward, 1988). Their names vary from country to country:

allotment gardens in England, community gardens in the United States, *jardins ouvriers* or *jardins familiaux* in France, and *Schrebergärten* or *Kleingärten* in Germany. The cultural motivations that preside over the establishment of these gardens vary. They were often openly encouraged by wealthy and paternalistic philanthropists as a social hygiene measure to maintain social order and keep the worker away from the pub. At the same time, in a context of class struggle for land, bread, and work, socialist movements regarded these as a tool for the emancipation of the working class, enabling them to feed themselves regardless of their conditions of employment. Early on in the century, local benefactors (often confessional charities) lent plots of land and tools to needy families who arrived en masse in the cities in search of work. It was only at the end of the century that nationwide movements emerged, giving these allotments their enduring form.

In Germany, the first large-scale group gardening movement was the "Schreber movement," launched in 1864 (Wahmann, 1991). This movement was inspired by the ideas of the physician Moritz Schreber (1808–1861), who wrote over thirty books on the consequences of urbanization on children's health. Schreber professed the education and discipline of children through physical exercise. After his death, his followers created numerous collective gardens in Germany, but also in Austria, Holland, and Switzerland. First planned to be exercise grounds, they soon became cultivated. They underwent important development until the middle of the twentieth century – shortly after WWII, Berlin had 200,000 allotment gardens. Today there are still about 80,000. In these countries, allotments are still sometimes known as *Schrebergärten*.

In France, *jardins ouvriers* (workers' gardens) – later called *jardins familiaux* (family gardens) – appeared at the end of the nineteenth century in the industrial cities of northern France. These gardens were mainly supported by social Catholics, philanthropic factory owners, and advocates of the hygienist movement (Cabedoce, 1996). As in Germany, they aimed to improve the living conditions of the working class while also exercising a form of social control over it. Following an approach inspired by Germany, they consisted of plots of land made available to working families by industrial firms or municipalities. One of their main instigators was the abbot Jules Lemire (1853–1928), a Catholic priest of peasant origin and deputy from the North department, who founded the French League of the Coin de Terre et du Foyer in 1896, which later became the Fédération Nationale des Jardins Familiaux et Collectifs (FNJFC – National Federation of Allotment and Community Gardens). Lemire was a follower of "Terrianism," a movement that emphasized the moral importance of the bond to the soil (Cabedoce, 1996). He praised the virtues of agriculture and gardening, against the corrupting "industrialism" of the city and the factory. His *jardins ouvriers* were intended to bring the family together in working the soil. Today there are still about 120,000 allotment gardens in France, with an average surface of 200 m^2. They are generally managed by associations grouped in the FNJFC and are still the object of great enthusiasm of the more modest social classes (Weber, 1998).

The motivations surrounding the cultivation of the urban soil by factory workers also participated in the great social struggles of the industrial era. During the

Fig. 14.2 Landscape (Allotments), oil on board by Frederick Cuming (b.1930), c. 1950. © Royal College of Art

nineteenth century in the United Kingdom, social reformers pressed for government laws that forced local authorities to make land available to workers. These would be the "allotments," "land allocated to workers" under the Commons Act of 1876 and several Allotment Acts (between 1832 and 1950). In the nineteenth century, the allotment movement was primarily rural. It only became predominantly urban in the first quarter of the twentieth century. In the 1930s, the great depression brought about a sharp increase in allotment creation (Acton, 2011). Crouch and Ward (1988), historians of the allotment movement, explain that the cultivation of common land was an important part of the social cohesion of miners' communities in England – even long after the mineshafts had been closed and the miners made redundant. These authors stress the symbolic importance for these working-class communities of being able to create a landscape in their own image, reflecting their social and cultural condition, in contrast to the urban parks and avenues created mainly by the state or the private sector (Fig. 14.2). The aesthetics of the allotments' cultural landscape is not only visual but also embedded in the practice of sharing tools and materials and working together. In the investigations conducted by Crouch and Ward, plot holders emphasize the meaning and pleasure they derive from these relationships of exchange and mutual assistance. This cooperation also extends into negotiations with landlords or in movements to defend sites when they are threatened.

In popular culture, allotments and "community gardens," their equivalent in the USA, gained considerable notoriety as "Victory Gardens" during the two world wars. Civilians were encouraged to grow their own food so that industrial agriculture could serve the war effort. Many city parks, as well as road and railroad sides, were temporarily plowed and turned into allotments. In the USA, it is estimated that Victory Gardens were able to supply nearly 40% of the national vegetable consumption in 1943. In addition to providing food during this critical period, Victory Gardens were also promoted as a way to boost the morale of civilians, who were to become "soldiers of the soil," and to strengthen patriotism (Helphand, 2006). Similar initiatives were conducted by the governments of Nazi Germany and occupied countries such as France, where the number of *jardins ouvriers* tripled between 1939 and 1943 (Cabedoce, 2018).

It seems that major economic and political crises are often matched by a revival of urban agriculture. Cuba is a well-known example of this. In 1989, with the collapse of the Socialist Eastern Bloc (which accounted for 85% of the country's trade outlets) Cuba plunged into an economic crisis known as the "special period." Deprived of imports of fuel, agricultural machinery, pesticides, and cheap fertilizers, the country's Soviet-inspired agriculture collapsed in a short time. Cities across the country turned to urban agriculture, encouraging recycling and local use of resources. Interestingly, this urban agriculture did not disappear with the end of the food crisis and has become part of the Cuban urban identity (Altieri et al., 1999).

In the Global South, which is experiencing high rates of urban growth, urban agriculture often forms an intrinsic part of the urban fabric (Mougeot, 2005). Urban agriculture often takes place without the support of the authorities and is part of informal land and economic regimes based on exchange and/or self-sufficiency. Chemical fertilization practices are less present than in northern countries, and the soil is often amended with food waste. In addition, this agriculture frequently combines vegetable cultivation and livestock farming. In sub-Saharan Africa, although the practice of urban agriculture transcends social or ethnic categories, many studies highlight the important role of women who often carry a significant part of the workload (Gabel, 2005).

14.5 Reclaiming the City and Its Soils

In Europe and the USA, during the 30-year postwar boom, the rising middle class favored private gardening, in the gardens of the many single-family houses that were developing at the time. Furthermore, the democratization of television, restaurants, and shopping malls changed how people spent their leisure time and made allotments appear anachronistic (Acton, 2011). At a time when shortages appeared to be a thing of the past, neglected allotment sites became easy targets for developers, while the turf lawns that are now so typical of the suburban residential areas across the Global North became more and more common. Landscape historians trace the aesthetics associated with these green carpets of grass back to the neo-pastoral aesthetics of eighteenth century wealthy French and English country houses, and they became common in urban parks with the public park movement of

Fig. 14.3 Garden activist Liz Christy in a Lower East Side community garden, New York City, 1975. © Donald Loggins

the mid-19th century. But it was in the postwar period, with the rise of property-owning middle classes that regarded residential lawns as an indicator of social and professional success, that they ceased to be the prerogative of the better-off and became generalized (Jenkins, 2005). In the Global North, turf lawns now cover up to 50–70% of urban green areas and have become an important feature of how urbanites perceive "urban nature" (Ignatieva et al., 2020). At the same time, because they are always made up of the same few plant species, turf lawns represent an important factor in the homogenization of urban environments and are high consumers of pesticides, water, and fertilizers (Wheeler et al., 2017).

Movements advocating collective gardening emerged again in the 1970s. This time they articulated urban gardening with explicitly political and emancipatory objectives. These movements are sometimes grouped under the name of "critical urban gardening" (Certomà, 2015). Their origin is to be found in the reclaiming of urban wasteland by the inhabitants of New York's poor neighborhoods during the "Decade of Fire." Because of the economic crisis that hit these neighborhoods hard, apartment rentals were no longer profitable for landlords, and many of them evicted their tenants before setting fire (arson) to their property to obtain insurance money. These fires resulted in a lot of vacant lots that local communities reclaimed through community gardening. Among the first initiatives to reclaim vacant spaces through gardening were the "green guerillas",[1] among whom were charismatic activists such as Liz Christy in New York (Fig. 14.3). These initiatives built on the legacy of

[1] See http://www.greenguerillas.org/history

earlier allotment and liberty garden movements, but emphasized community activism and empowerment, in connection with the protest movements of the time (Lawson, 2005).

Critical urban gardening movements think of themselves as counter-hegemonic movements that challenge dominant public and private patterns of space use, consumption, and food production. Seeing wastelands and their soils as laboratories for political experiments, these movements first aim to reclaim public space – making privatized land available for common use again. Secondly, they aim to reclaim the right to produce local food – promoting broad and inclusive participation, in response to the social disconnect between food producers and consumers initiated in the nineteenth century, and aggravated by the globalization of the second half of the twentieth century (Tracey, 2007). These movements also focus on agricultural or culinary skills that have been diminished by urbanization and stress the importance of apprenticeship, developing permaculture manuals or workshops to learn how to pay attention to soils.

The gardens developed by these movements have their own aesthetics, both distinct from – and sometimes in opposition to – those of well-tended public gardens and those of the allotments of previous eras. The look of these gardens is primarily the result of the gardening techniques used (Meulemans, 2020b). Soil reconstruction techniques, often inspired by permaculture, result in mounds created by "lasagna beds," created by alternating dry, carbon materials – dry leaves, cardboard, wood – and green nitrogen layers, kitchen waste and fresh leaves. Such techniques make it possible to recycle part of the food waste and to do without commercial fertilizers. Another element is sometimes the presence of vertical structures made of tires or pallets, to respond to the lack of space. In addition, gardens are often maintained in connection with other movements that aim to reclaim public space, such as graffiti artists and squatter communities, which also contribute to creating the aesthetics of these places. In a global movement that challenges moralizing discourses advocating "order" and rejecting the "wild," these new gardens join a new aesthetics of the "Wasteland" that emerged at the end of the twentieth century, celebrating ruderal plants, weeds, ecologies, and spontaneous pathways, instead of horticultural varieties of flowers and well-tended alleys (Myers, 2019).

14.6 Soils in the City Today

Since the beginning of the twenty-first century, urban agriculture and gardening have been undergoing a process of institutionalization. Urban gardening is increasingly recognized as an asset by cities and local authorities in the Global North. It now tends to be encouraged by municipalities, which make plots, tools, and water available to gardeners. Some large cities also provide "clean soil" to circumvent potential soil pollution problems, while urban soil cultivation practices are integrated into the environmental policies of cities as a complement to "green corridors" and other ecological infrastructures. At the same time, the relaxing and sociable character of gardening is being promoted, with it being seen as a hobby that allows contact with nature, and makes children and people more sensitive to environmental, nutritional and food production issues. Therefore, even though initiatives closer to critical urban gardening persist, many of the urban gardening projects that have developed over the last few decades under the patronage of public authorities are based on less overtly social and more recreational, ecological, or commercial motivations (McClintock, 2014).

The fact that urban collective gardening is being appropriated by a part of the urban middle classes, becoming more "mainstream," testifies to a transformation in the perception of gardens – long seen as needing to be ordered – and of the wasteland where they are often created, long seen as a repulsive territory. It also calls into question the conception of cities as a space that is expected to exclude soils and agriculture. At a more individual level, the craze for domestic composting testifies to a new look at the processes of decomposition and transformation of urban materials (Abrahamson & Bertoni, 2014) and reshuffles the boundaries between clean and dirty (Douglas, 1966), whereas the processes of decomposition and putrefaction rarely constituted valued terms or phenomena considered aesthetic in the modern period. Since the end of the 1980s, composting has even become a practice implemented centrally by cities, which see it as a method of reducing the amount of waste produced (Vergara, 2012).

Recently, many studies have addressed the links between urban gardening and health. These studies tend to show that the practice of gardening contributes to physical activity and improves nutrition (through access to vegetables), but it also has an impact on mental health, well-being, and social relationships (Armar-Klemesu & Maxwell, 2000). At the same time, it has been found that urban gardening may involve risks of exposure to toxic substances (Brown & Jameton, 2000). Former industrial activities may have left heavy metals or organic pollutants in soils. Proximity to roads often results in diffuse heavy metal contamination from exhaust fumes. However, it seems that the most common source of pollution comes from the use of pesticides by the gardeners themselves. The risks of pollution – real or not – sometimes lead gardeners to adopt "raised bed" crops or try to emancipate themselves from the existing soil by bringing in supposedly healthier topsoil. However, in general, studies in the USA and Europe show that gardeners are often unaware of these risks – they see their gardens as clean compared to paved streets – or do not know how to manage them (Kim et al., 2014).

Since the beginning of the twenty-first century, a new entrepreneurship of sustainable cities is also emerging. Some operators are developing high-tech tools such as vertical farms or hydroponic greenhouses, made possible by technologies such as LED lamps, which allow lettuce or berries to grow indoors in complete emancipation from the local soil conditions. They justify their social utility by insisting on the relocation of food and the fact that they create employment while stressing the health risks of urban soils on account of their pollution. At the same time, these new actors are moving far away from the countercultural or social emancipation considerations promoted by other urban gardeners. In resonance with the passion of modern societies for technology, they nourish a collective imagination of the city as a place that is totally independent of soils and natural cycles and based solely on high-tech systems. In her analysis of the gigantic greenhouse of the "Garden by the Bay," opened in 2012 in Singapore, Natasha Myers (2019) describes the simulated ecology of this garden that shows an exuberant tropical forest. This garden is made possible only by a set of high energy-consuming technologies and by the continuous maintenance work of many precarious immigrant workers, who must constantly fertilize, clean and replace substrates, and repair the pipes and sensors of the hydroponic systems that allow the plants to live. Myers emphasizes the irony of this garden in a glass case, with its fresh, clean, conditioned air, while outside one only smells the smog of Singapore, largely caused by the real forests that are burning on the neighboring island of Sumatra. Myers concludes, "This garden infrastructure thus performs a magic trick that distracts with spectacle from the slow violence of worldly undoing ongoing on the other side of its gleaming glass exterior" (2019, p.133). Thus, these new technologies contribute to perpetuating the dreams of the city off ground and seem quite foreign to the ambition to combat the extractivist relationship between the city and the soil that has been established with the three-fold urban metabolic rift.

Nevertheless, although many cities continue to expand by consuming agricultural land, others are declining and "shrinking" because of deindustrialization. These shrinking cities, like Detroit, see their abandoned plots being reappropriated by urban populations, who seek to recreate meaning and social ties by working their soil. In other places, although their number has been reduced, allotments continue to be a success with the working class, who find in them a source of pleasure and cheap vegetables, and sometimes a way to escape the urban consumerist culture. In an increasingly interconnected world, associations rely on urban soils and gardening to promote the integration of precarious groups, to help migrants overcome culture shock by putting their agricultural know-how to work or as a meeting place for various age or ethnic groups (Corlett et al., 2003). More than ever, urban gardens and allotments respond to the anxieties of our time around the destruction of soil and nature (Myers, 2019). At a time when humans have become one of the most powerful soil-forming agents in the world (Wilkinson, 2005), and with the risks of a possible "peak soil"[2] (Ahmed, 2013) on the horizon, the question of the place of agricultural activity in the city is a topical one, with all the practical, social, cultural, recreational, heritage, and commercial issues that come with it. More than ever, the relationship with urban soils is part of the broader modern tensions between ecological holism and technical exceptionalism, belief in technical progress and criticism of the destruction of nature, belief in the externalization of ecosystems by man and the search for meaningful links with nature.

---

[2] The expression "peak soil" has recently been popularized by food policy analysts and journalists to indicate how the productivity of global soils is soon likely to peak, then decline, with the maintenance of their fertility growing increasingly costly. To Ahmed (2013), peak soil results from the "perfect storm" created by the combination of current trends in soil erosion, land degradation, energy shortages (due to higher oil prices), water shortages (due to climate change), and a growing world's population.

References

Abrahamson, S., & Bertoni, F. (2014). Compost politics: Experimenting with togetherness in vermicomposting. *Environmental Humanities, 4*, 125–148.

Acton, L. (2011). Allotment gardens: A reflection of history, heritage, community and self. *Papers from the Institute of Archaeology, 21*, 46–58.

Ahmed, N. (2013). Peak soil: Industrial civilisation is on the verge of eating itself. *The Guardian*. https://www.theguardian.com/environment/earth-insight/2013/jun/07/peak-soil-industrial-civilisation-eating-itself. Accessed 4 Feb 2017.

Altieri, M. A., Companioni, N., Cañizares, K., Murphy, C., Rosset, P., Bourque, M., & Nicholls, C. I. (1999). The greening of the "barrios": Urban agriculture for food security in Cuba. *Agriculture and Human Values, 16*, 131–140.

Armar-Klemesu, M., & Maxwell, D. (2000). Urban agriculture as an asset strategy, supplementing income and diets. In N. Bakker, M. Dubbeling, S. Gundel, U. Sabel-Koschella, & H. de Zeeuw (Eds.), *Growing cities, growing food: Urban agriculture on the policy agenda. A reader on urban agriculture* (pp. 183–208). German Foundation for International Development.

Barles, S. (1999). *La ville délétère: Médecins et ingénieurs dans l'espace urbain XVIIIe-XIXe siècles*. Champ Vallon.

Brown, K., & Jameton, A. (2000). Public health implications of urban agriculture. *Journal of Public Health Policy, 21*, 20–39.

Burghardt, W., Morel, J. L., & Zhang, G. (2015). Development of the soil research about urban, industrial, traffic, mining and military areas (SUITMA). *Soil Science and Plant Nutrition, 61*(supl), 3–21.

Cabedoce, B. (1996). Une aventure humaine : les jardins ouvriers. In B. Cabedoce & P. Pierson (Eds.), *Cent ans d'histoire des jardins ouvriers, 1896–1996, La ligue française du Coin de terre et du foyer* (pp. 31–74). Créaphis.

Cabedoce, B. (2018). 1940–1952: une période charnière pour les jardins ouvriers. *In Situ, 37*. https://doi.org/10.4000/insitu.18752

Certomà, C. (2015). Critical urban gardening. In M. Pimbert, R. Shindelar, & H. Schösler (Eds.), *Think global, eat local: Exploring foodways* (pp. 13–17). Rachel Carson Center for Environment and Society.

Colten, C. E., & Skinner, P. N. (1996). *The road to love canal: Managing industrial waste before EPA*. University of Texas Press.

Corlett, J. L., Dean, E. A., & Grivetti, L. E. (2003). Hmong gardens: Botanical diversity in an urban setting. *Economic Botany, 57*(3), 365–379.

Crouch, D., & Ward, C. (1988). *The allotment: A culture and landscape*. Faber and Faber.

De Block, G. (2016). Ecological infrastructure in a critical-historical perspective: From engineering 'social' territory to encoding 'natural' topography. *Environment and Planning A: Economy and Space, 48*(2), 367–390.

Douglas, M. (1966). *Purity and danger: An analysis of concepts of pollution and taboo*. Routledge and Keegan Paul.

Ferguson, D. T. (2014). Nightsoil and the "great divergence": Human waste, the urban economy, and economic productivity, 1500–1900. *Journal of Global History, 9*, 379–402.

Foster, J. B. (1999). Marx's theory of metabolic rift: Classical foundations for environmental sociology. *The American Journal of Sociology, 105*(2), 366–405.

Gabel, S. (2005). Exploring the gender dimensions of urban open-space cultivation in Harare, Zimbabwe. In L. J. A. Mougeot (Ed.), *Agropolis: The social, political and environmental dimensions of urban agriculture* (pp. 107–136). IDRC.

Haney, D. (2010). *When modern was green: Life and work of landscape Architect Leberecht Migge*. Routledge.

Helphand, K. (2006). *Defiant gardens: Making gardens in wartime*. Trinity University Press.

Hinchliffe, S., & Whatmore, S. (2006). Living cities: Towards a politics of conviviality. *Science as Culture, 15*(3), 123–138.

Ignatieva, M., Haase, D., Dushkova, D., & Haase, A. (2020). Lawns in cities: From a globalised urban green space phenomenon to sustainable nature-based solutions. *Land, 9*(73). https://doi.org/10.3390/land9030073

IUSS Working Group WRB (2006). *World reference base for soil resources* (World soil resources reports n. 103) (2nd ed.). FAO.

Jenkins, V. (2005). *The lawn: A history of an American obsession*. Smithsonian Institute.

Kim, B. F., Poulsen, M. N., Margulies, J. D., Dix, K. L., Palmer, A. M., & Nachman, K. E. (2014). Urban community gardeners' knowledge and perceptions of soil contaminant risks. *PLoS One, 9*(2), e87913. https://doi.org/10.1371/journal.pone.0087913

King, F. H. (1911). *Farmers of forty centuries, or permanent agriculture in China, Korea and Japan*. Mrs. F.H. King.

Lawson, L. (2005). *City bountiful a century of community gardening in America*. University of California Press.

Le Corbusier (1957). *La charte d'Athènes*. Minuit.

Levin, M. J., Kim, K. H., Morel, J. L., Burghardt, W., Charzynski, P., & Shaw, R. K. (2017). *Soils within cities. Global approaches to their sustainable management: Composition, properties, and functions of soils of the urban environment*. Schweizerbart.

Liebig, J. V. (1865). *Letters on the subject of the utilization of metropolitan sewage*. W. H. Collingridge.

Mason, D., & Knowd, I. (2010). The emergence of urban agriculture: Sydney, Australia. *International Journal of Agricultural Sustainability, 8*(1–2), 62–71.

McClintock, N. (2010). Why farm the city? Theorizing urban agriculture through a lens of metabolic rift. *Cambridge Journal of Regions, Economy and Society, 3*(2), 191–207.

McClintock, N. (2014). Radical, reformist, and garden-variety neoliberal: Coming to terms with urban agriculture's contradictions. *Local Environment, 19*(2), 147–171.

McNeill, J. (2000). *Something new under the sun. An environmental history of the twentieth-century world*. W. W. Norton & Company.

Melosi, M. (2005). The automobile shapes the city. *Automobile in American Life and Culture*. www.autolife.umd.umich.edu. Accessed 14 Jan 2020.

Meulemans, G. (2020a). Urban pedogeneses: The making of city soils from hard surfacing to the urban soil sciences. *Environmental Humanities, 12*(1).

Meulemans, G. (2020b). Reclaiming freak soils: From conquering to journeying with urban soils. In J. F. Salazar, C. Granjou, A. Krzywoszynska, M. Kearnes, & M. Tironi (Eds.), *Thinking with soils: Material politics and social theory* (pp. 157–174). Bloomsbury.

Migge, L. (1918). *Jedermann Selbst-Versorger*. Diederichs.

Morel, J. L., Chenu, C., & Lorenz, K. (2015). Ecosystem services provided by soils of urban, industrial, traffic, mining, and military areas (SUITMAs). *Journal of Soils and Sediments, 15*(8), 1659–1666.

Mougeot, L. J. A. (Ed.). (2005). *Agropolis: The social, political and environmental dimensions of urban agriculture*. IDRC.

Mumford, L. (1956). The natural history of urbanization. In W. L. Thomas (Ed.), *Changing the face of the earth* (pp. 382–398). University of Chicago Press.

Myers, N. (2019). From edenic apocalypse to gardens against eden: Plants and people in and after the anthropocene. In K. Hetherington (Ed.), *Infrastructure, environment, and life in the anthropocene* (pp. 115–148). Duke University Press.

Njoh, A. J., & Watson, V. (2009). The emergence and spread of contemporary urban planning. In Un-Habitat (Ed.), *Planning sustainable cities: Global report on human settlements 2009* (pp. 47–71). Earthscan.

Pesaresi, M., Melchiorri, M., Siragusa, A., & Kemper, T. (2016). *Atlas of the human planet 2016: Mapping human presence on earth with the global human settlement layer*. Publications Office of the European Union.

Richter, D. B., Bacon, A. R., Brecheisen, Z., & Mobley, M. L. (2014). Soil in the anthropocene. *IOP Conference Series: Earth and Environmental Science, 25*, 24–27.

Séré, G., Schwartz, C., Ouvrard, S., Renat, J.-C., Watteau, F., Villemin, G., & Morel, J.-L. (2010). Early pedogenic evolution of constructed Technosols. *Journal of Soils and Sediments, 10*, 1246–1254.

Smit, J., Ratta, A., & Nasr, J. (1996). *Urban agriculture: Food, jobs and sustainable cities*. United Nations Development Programme (UNDP).

Tarr, J. A. (1975). From city to farm: Urban wastes and the American farmer. *Agricultural History, 49*(4), 598–612.

Tracey, D. (2007). *Guerilla gardening: A Manualfesto*. New Society Publishers.

Uekötter, F. (2012). *Land. EGO European History Online*. Leibniz Institute of European History (IEG). http://www.ieg-ego.eu/uekoetterf-2012-en. Accessed 3 Dec 2019.

Vergara, S. E. (2012). Composting. In C. A. Zimring & W. L. Rathje (Eds.), *Encyclopedia of consumption and waste: The social science of garbage* (pp. 137–140). Sage.

Wackernagel, M., Kitzes, J., Moran, D., Goldfinger, S., & Thomas, M. (2006). The ecological footprint of cities and regions: Comparing resource availability with resource demand. *Environment and Urbanization, 18*(1), 103–112.

Wahmann, B. (1991). Allotments and Schrebergarten in Germany. In M. Mosser & G. Teyssot (Eds.), *The history of garden design: The western tradition from the renaissance to the present day* (pp. 451–454). Thames and Hudson.

Weber, F. (1998). *L'Honneur des jardiniers : Les potagers dans la France du xxe siècle*. Belin.

Wheeler, M. M., Neill, C., Groffman, P. M., Avolio, M., Bettez, N., Cavender-Bares, J., Roy Chowdhury, R., Darling, L., Grove, J. M., Hall, S. J., Heffernan, J. B., Hobbie, S. E., Larson, K. L., Morse, J. L., Nelson, K. C., Ogden, L. A., O'Neil-Dunne, J., Pataki, D. E., Polsky, C., Steele, M., & Trammell, T. L. (2017). Continental-scale homogenization of residential lawn plant communities. *Landscape and Urban Planning, 165*, 54–63.

Wilkinson, B. H. (2005). Humans as geologic agents: A deep-time perspective. *Geology, 33*(3), 161–164.

Part III
Politics on Soil

Chapter 15
Sacred Soil, Divine Labor: Economic Theology and the Agrarian Political in Eastern India

Milinda Banerjee

15.1 Introduction[1]

Over the past few years, soil (in Bengali, *mati*; from Sanskrit *mrittika*) has rapidly become a dominant political concept in the eastern Indian state of West Bengal. The ubiquitous political slogan in West Bengal today is *Ma Mati Manush* (literally, Mother, Soil, Human Being), coined by the current Chief Minister of the state, Mamata Banerjee[2] (1955–present). The slogan was part of a massive political movement which ended the rule of the Communist Party of India (Marxist)-led Left Front government in the state after more than three decades and catapulted Banerjee and her party, the All India Trinamool (literally, Grassroot) Congress, to power in 2011. The regime change came in the aftermath of a wave of popular protests, especially against the Left Front-enabled dispossession of peasants by land-grabbing industrial capitalism. Since 2013, the Trinamool government has held an annual *Mati Utsav* (literally, Soil Festival). Revealingly, the site of this festival in rural Burdwan today is called *Mati Tirtha* (Soil Pilgrimage Place). Precolonial Indian-origin ideas of the sacredness and divinity of soil are rapidly being revived in political lexicons, including in Banerjee's own extensive writings.

---

[1] Unless otherwise stated, non-English words, given in italics, are Bengali words. Most, but not all, of these words have a Sanskritic origin. Hence, the same, or proximate, word-forms can also be found in Sanskrit and in many other Indian languages. I have transliterated Bengali words in ways which would make them recognizable to most South Asians/South Asianists.

[2] Banerjee (Anglicized form of Bandyopadhyay) is a Bengali lineage name; the author is, however, not a relative of Mamata Banerjee.

M. Banerjee (✉)
University of St Andrews, St Andrews, Scotland

N. Patzel et al. (eds.), *Cultural Understanding of Soils*,
https://doi.org/10.1007/978-3-031-13169-1_15

How do we explain this dramatic rise of "soil" as a political and theological concept in an age of predatory capitalism? How do we contextualize discourses about soil in relation to narratives about peasant labor? In seeking an answer, this essay offers a wider theoretical argument about the relation between the formation of political ideas and the mode of production present in a society. While this relation has been long debated by political thinkers, it has recently precipitated a particularly sharp dissensus in the new academic discipline of global intellectual history, as discussed below. We have, on the one hand, a perspective which explains the globalization of ideas through a focus on political conflicts, while insulating the political from the economic and, on the other, a viewpoint which sees the genesis of modern political concepts as overdetermined by the capitalist mode of production and exchange.

This essay goes beyond this polarized impasse. By drawing on case studies from early eighteenth to early twenty-first-century eastern India, I argue that human labor on soil is a prime site for the birth of political concepts. In Bengal, this encompasses the vision of peasant labor on soil as a divinized activity, which is simultaneously the foundation of material wealth, democratic collectivity, and political sovereignty in society. Drawing on the nascent scholarly field of economic theology, I argue that human labor on soil is central to the generation of worldviews which are inextricably political, socio-economic, and theological, and cannot be reduced to any one term within the trinity to the exclusion of others. (In fact, this differentiation of the three categories itself needs to be rigorously historicized, provincialized, and deconstructed, rather than accepted as the a priori of theorization: but that is not the task of this essay.) I also sketch a political theory of agrarian democracy that is rooted in forms of collectivity mediated by human labor on soil.

15.2 The Agrarian Political and Soil

In this essay, I shall use methodologies of intellectual history to analyze agrarian political thought around soil. In recent years, global intellectual history has emerged as a pioneering field which seeks to overcome the Eurocentrism which has often structured intellectual history discussions in Euro-American academia (Moyn & Sartori, 2013). Historians of modern India have played a leading role in shaping the field.

A pressing debate, which is germane for our essay, has revolved around the relation between the political and the economic. Shruti Kapila has argued that concepts of the political should be studied in their own right, as autonomous fields of agonistic conflict, rather than be visualized as fundamentally grounded in socio-economic transformations. "The primacy of the political [...] was [...] an outcome of ideological maneuvers that cannot be explained away by exigent and contingent historical and economic factors alone" (Kapila, 2014, p. 255). This perspective replays the distinction (though, not an absolute one) between the political and the economic– and the anchoring of the political in the friend/enemy distinction – made by the controversial German jurist (and Nazi-apologist) Carl Schmitt (1888–1985) (Schmitt, 2007), whom Kapila indeed refers to (Kapila, 2014, pp. 266, 268). Kapila echoes Schmitt when she argues: "Questions of enmity and violence […] have been central to modern Indian understandings of the political […]" (Kapila, 2014, p. 264). From this perspective of the primacy and autonomy of the political, Kapila critiques the historian Andrew Sartori for supposedly explaining away the global circulation of political ideas in terms of the global circulation of commodities and capital (Kapila, 2014, pp. 259–260).

The roots of this conflict go deeper. Sartori is a historian of colonial Bengal, and his scholarship draws upon long-standing discussions in the intellectual history of the region. Bengali historians have, over the last few decades, centered the region's political and social thought in global discussions of postcolonial theory. This has been driven especially by scholars of the Subaltern Studies collective, such as Ranajit Guha, Gayatri Chakravorty Spivak, Partha Chatterjee, and Dipesh Chakrabarty. Chakrabarty was a strong influence as a teacher on Sartori (Sartori, 2008, p. vii). These Bengali scholars, as well as Sartori, have been inspired by different varieties of Marxian critical theory. They would all agree that the political cannot be disjuncted from the socio-economic (Banerjee, 2019). I situate my work about the agrarian political in relation to this tradition of scholarship. While I appreciate Kapila's emphasis on Indian political agency, I underline that the political is not something that can be disjuncted from the economic.

There is a major difference, however, between the works of the Subaltern Studies collective, especially of Ranajit Guha and Sartori. Guha's scholarship has to be read in connection with the ultra-left Naxalite "Maoist" insurgency in India (and especially in West Bengal) in the late 1960s/early 1970s. In this context, Naxalites premised a concept of the political, rooted in peasant rebellion, which was distinguished from mere "economism" (Banerjee, 2018a). Sartori criticizes the Naxalite-inflected perspective of Guha (Sartori, 2014, pp. 32, 211, 246). For him, the political thought of Indian peasants (at least, of the ones he studies) does not represent a revolutionary other of capitalism (as the Naxalites and Guha thought) but in fact stems from the very logic of the capitalist mode of production. Sartori argues that the capitalist mode of production and commodity exchange produces a "Lockean" viewpoint that sees human labor as constituting property and wealth (John Locke) or, more broadly, value (Karl Marx). "In agrarian Bengal, commodity exchange had become a plausible foundation for normative claims. […] In fact, when agrarian actors organized politically to defend a right to property grounded in labor, they articulated a critique of capital that was imaginable only in relation to commodity exchange" (Sartori, 2014, p. 208).

But even while Sartori's insights are inspiring, I find it problematic to reduce Indian peasant political thought to a superstructure of modern capitalism's logic. If Kapila's focus on the primary and autonomy of the political represents one extreme, Sartori's economic determinism represents another. I underline how the political, the economic, and the theological cannot be separated from each other in such a way. We should not assert the primacy of any one of these categories over the other; they are inextricably interlinked. I take a cue from the nascent field of economic theology, which underlines how economic concepts and arguments are often saturated with theological notions (Dean, 2019). The Italian philosopher Giorgio Agamben has especially been a pioneer in highlighting how modern economic ideas need to be related to far older genealogies of theological discourse (Agamben, 2011). I extend the field of economic theology beyond its European origins, to think about agrarian thought in the non-European world. I ultimately emphasize the creative politics of non-European actors which need to be studied in their own right, instead of being reduced to narrow by-products of modern capitalism. I emphasize the agrarian political as a site for building social solidarity rather than a mere locus for violence and enmity or a simple by-product of capitalist commodity exchange.

In centring soil within this discussion, the German philosopher G. W. F. Hegel (1770–1831) offers provocations to think with and against. For Hegel, tilling the soil was a main source of conceptualizing possession. He asserted in *Outlines of the Philosophy of Right* (1820): "To impose a form on a thing is the mode of taking possession most in conformity with the Idea [*Idee*] to this extent, that it implies a union of subject and object [...]. Examples are the tilling of the soil [...]" (Hegel, 2008, p. 69). He was not the first thinker in this regard. For example, the English philosopher John Locke (1632–1704) observed in *Two Treatises of Government*: "He that in Obedience to this Command of God, subdued, tilled and sowed any part of it [the Earth], thereby annexed to it something that was his *Property* [...]" (Locke, 2003, p. 291). The logic in Locke and in Hegel are comparable. Human beings, by putting their labor (Locke) or will (Hegel) into an object (land/soil), make it theirs. Hegel noted: "A person has as his substantial end the right of putting his will into any and every thing and thereby making it his, because it has no such end in itself and derives its determination and soul from his will. This is the absolute *right of appropriation* which human beings have over all 'things'" (Hegel, 2008, p. 60).

The human-soil relation also had political implications. In *The Philosophy of History* (1822–31), Hegel noted: "Individuals brought into closer relation by the soil which they cultivated, formed among themselves a kind of union, confederation, or *conjuratio.* They agreed to be and to perform on their own behalf that which they had previously been and performed in the service of their feudal lord alone" (Hegel, 1899, p. 384). While Hegel offers here a particular reading of the emergence of autonomous political communities in late medieval Europe (whose historical accuracy is not being discussed here), I would generalize his thesis to interrogate Bengali discourses about the relation between soil, popular solidarity, and the consolidation of political autonomy against lordly elites.

Ultimately, for Hegel, control of soil was central to the construction of political and social life. He asserted in *Outlines of the Philosophy of Right*: "The real beginning and original foundation of states has been rightly ascribed to the introduction of *agriculture* along with *marriage*, because the principle of agriculture brings with it the formation of the land and in consequence exclusively private property [...]. Security, consolidation, lasting satisfaction of needs, and so forth – [...] are the most obvious recommendations of marriage and agriculture [...]" (Hegel, 2008, pp. 193–194). Hegel emphasized the political role of landowners, "whose basis is family life and, so far as its livelihood is concerned, the possession of land. [...] this estate is summoned and entitled to its *political* vocation by *birth* without the hazards of election" (Hegel, 2008, pp. 292–293). Soil was also the foundation of the (patriarchal) family: "The condition of the principle of family life is the earth, the firm and solid ground" (Hegel, 2008, p. 222). Hegel deployed a polysemic phrase, "paternal soil" [German: *väterlicher Boden*] to express this social framework of "stability" (Hegel, 1911, p. 186, 2008, p. 218).

These aspects of Hegel's political theory of soil are irredeemably inegalitarian, and most elements of his historicist argumentation can be challenged. The empirical specificities are not important for my argument in this essay. What I will nevertheless emphasize is that one can dialectically work with and against Hegel to create a theory of agrarian democratic sovereignty: one where engagement with soil helps create sovereignty and political power, but as exercised through the collective political solidarity of the peasant multitudes. I will work out these arguments by drawing on Bengali discourses about soil from the last three centuries.

15.3 Agrarian Labor as Foundation of Wealth, Soil-Oriented Cosmology, and the Critique of Capital

Precolonial Indian ideas about soil are continually invoked in contemporary *Mati Utsav* discussions. This includes the classical Sanskritic idea of *kshiti* – "the earth, soil of the earth" (Monier-Williams, 1986, p. 327) as one of the five elements (*panchabhuta*) which constitute physical creation. In *Mati Utsav* literature, *kshiti* is equated with *mati*. The discourse emphasizes the need for proper care of the soil (and of water and air) to reduce the damages caused by rampant commercial exploitation in the name of "development" (*unnayan*). Ancient Sanskritic ideas about the importance of soil in the cosmos are thus redeployed to critique capitalist development from the standpoint of the environment. The unregulated pursuit of exchange (literally, commercial) value (*vanijyik mulya*) is held responsible for global warming, plant and animal epidemics, and other ecological disasters. New environmentally sensitive techniques of soil care and use for agriculture are introduced and explained within this broader ideology, to enable better cooperation between peasants, who are familiar with the Sanskritic ideas, and scientific experts (Saha-Krishi Adhikarta, 2015).

The eighteenth-century poet-saint Ramprasad Sen's famous lines are also invoked: "Oh mind, you do not know agricultural work. Such a human field has remained fallow; had it been cultivated, gold would have grown" (*Man re krishi-kaj jano na. Eman manav-jamin railo patit, abad karle phalto sona*; Sen, n.d., p. 92). I would historicize this song in the context of the massive expansion of agriculture, including commercialized agriculture, which John F. Richards identifies as one of the hallmarks which makes sixteenth-to-eighteenth-century Indian history a part of global early modernity. Richards highlights "the intensified use of land to expand production," especially in frontier regions of South Asia. "This continuing process of settlement had a windfall effect whereby abundant new resources – soil, timber, wildlife, and minerals – were put into concentrated modes of production for an expanding world economy"

(Richards, 1997, pp. 202–203). As Bengal was integrated into transoceanic networks of early modern global capitalism – owing to South Asian, Indian Ocean, and European mercantile networks – new ideologies of productivity and agrarian growth emerged in the region. These linked the economic and intellectual spheres. The self was visualized as soil to be cultivated, with harvests of gold, in an epoch of early modern commercial expansion. Labor/work (*kaj*) was conceptualized as the foundation of wealth, economically and spiritually. Today, the lines are re-invoked in *Mati Utsav* literature to convey to rural Bengal the need for improving agricultural productivity by caring for the soil (Krishi Adhikar, 2018, p. 1).

Another key eighteenth-century text about soil use known to Bengalis today – though I have not seen it explicitly cited in *Mati Utsav* literature – is Rameshvar Bhattacharya's *Shivayan* or *Shivasankirtan* (written between ca. 1735 and 1750), the most famous in a genre of devotional compositions to the god Shiva which appeared in early modern Bengal. The text was composed under the patronage of the ruler of Karnagarh in southwestern Bengal, where the agrarian frontier was expanding at the time (though large forests exist nearby even today). According to Rameshvar, Shiva became a peasant to appease his wife, the goddess variously identified as Parvati, Durga, and Kali. In Bengal, Shiva was traditionally associated with agriculture. A household saying observes: "A song to Shiva while husking rice" (*dhan bhante Shiver git*). *Shivayan* discusses agrarian work in detail, including techniques of soil care and use (Bhattacharya, 1957, pp. 218–273).

Particularly interesting is a dialogue where the goddess Parvati persuaded the mendicant Shiva to settle down as a peasant and act like a proper householder (*grihi*). Shiva complained that agrarian work was full of hardship; even if the poor peasant was successful in growing crops, the ruler (*raja*) would take away the produce. The god thus considered other occupations. But the goddess dissuaded him. She dismissed commerce, suggesting that "capital and deceit are the root of trade" (*punji ar pravanchana vanijyer mul*). She also did not like the idea of Shiva serving the ruler/state (*rajseva*). It would make the god, who was *sevya* (one worshipped/served by others), approach one who served him (*sevak*). Only as a peasant would Shiva avoid these dangers. The goddess equated cultivation (*chash*) with wealth (*dhan*). She observed: "Commerce with Lanka sits at the corner of the house" (*Lankar vanijya baise bakurir kone*). The line implies that peasant production (in Bengal) fueled long-distance Indian Ocean commerce, signified here by the distant island of (Sri) Lanka. Ultimately, the cultivator (*chashi*) nurtured (*poshe/poshi*) his kin, the ruler, and multitudes. Hence, agrarian work fitted Shiva's cosmological world-protecting role: "Shiva will do cultivation for the sake of beings" (*Jiver nimitta Shiva kariben chash*) (Bhattacharya, 1957, p. 216–217).

While written by a man, the narrative about Shiva's peasantization centers a woman's perspective on agrarian work. The goddess, as in her form Annapurna (literally, the plenitude of food/rice: see, for example, Bhattacharya, 1957, pp. 232, 236), is responsible for agrarian success. But beyond theology, this is also a narrative of a wife persuading her husband to settle down, linking domesticity (and domestication of the "wild man"), sedentarization, and peasantization. Such processes are millennia-old in world history. But they intensified in early modern Bengal, as the agrarian frontier expanded into regions hitherto marked by hunting-gathering, fishing, and pastoral-nomadic formations. In Bengali imagination, divine as well as human women are often presented, in positive and negative ways, as the prime motors behind channeling their husbands into a life of hard work and the pursuit of wealth (Chatterjee, 1993, pp. 62–69), hence, Rameshvar's gendered narrative. Ultimately, agricultural labor is divinized, as a god cultivates the soil. Shiva "wallowed in [literally, devoured] mud and water to prepare the soil [*kshiti*] in the field (*kshete*)" (Bhattacharya, 1957, p. 257). The term *kshiti* recalls the cosmogony aspects of this labor on soil/earth. It hints at the creation of the earth, including the role of the *panchabhuta*, described earlier in *Shivayan* (Bhattacharya, 1957, pp. 21–23).

The human-land/soil interface is central to economic thought. In India, over more than two millennia, sophisticated forms of political, economic, and theological discourse linked notions of self (in Sanskritic discourses, *sva*) to ideas of property (*svatva*) and ownership (*svamitva*) (Lubin, 2018). In seventeenth-century Bengal, as Samuel Wright has recently shown, the nexus became particularly important and was extensively reflected upon, in philosophical discourse as much as in quotidian economic documents. Wright suggests that this resulted from growing commercialization and property transactions in Bengal (Wright, 2017). The early eighteenth century *Shivayan* offers a special twist here. The peasant self, exemplified by Shiva, is celebrated for creating wealth (*dhan*) and for nurturing society, state, and cosmos.

In contrast to this laboring foundation of wealth, capital and commerce – though not trade coming out of the peasant household – are linked with deceit. I interpret the goddess' ideology as a critique of capital and commerce from the standpoint of agrarian labor, much before anti-colonial and socialist-Marxist thought would normalize such perspectives in Bengal. Derived from Sanskrit (and Bengali) *punja* (literally, accumulation), *punji* seems to have acquired the meaning of capital in Bengali during the early modern period: probably in tandem with the growing accumulation of capital in the region. (In Sanskrit, terms like *muladhana* and *muladravya* were more commonly used to denote capital: Monier-Williams, 1986, p. 826). The sentence quoted above from *Shivayan* (*punji ar pravanchana vanijyer mul*) thus acquired canonicity; it is often the only source cited by modern Bengali dictionaries while defining *punji* as capital (Das, 1916, p. 959; Bandyopadhyay, 1988, p. 1338). Moreover, as I will discuss later, the earth/soil is often gendered feminine in India and identified with various goddesses. In *Shivayan*, the goddess is very close to the agrarian soil; in a key episode, she goes to the field to seduce Shiva and ends up with mud all over her body (Bhattacharya, 1957, pp. 255–73). The goddess' critique of commerce and capital arises, thus, in a sense, not only from the standpoint of labor, but from the standpoint of the earth/soil itself.

15.4 Male Cultivation and Female Soil: Agrarian Labor as Foundation of Property and Democratic Sovereignty

The buoyant manufacturing, commercial, and agrarian life of early modern Bengal – and more broadly, India – attracted European traders, especially the Portuguese, English, Dutch, French, and Danish. In the decades following British victory at the Battle of Plassey (1757), Bengal became the launching pad for British imperial conquest of India. The region remained the epicenter of British India at least until 1911, when the capital was shifted from Calcutta to Delhi. Establishing control over the agrarian sphere became central to the British project of ruling and exploiting India. After years of debate, the Permanent Settlement of 1793 vested a class of *zamindars* with full proprietary rights (and revenue-paying obligations) in Bengal, transforming them into a type of absolute landlords which they had never been in precolonial times. The settlement severely challenged the customary rights of peasants over the soil they cultivated, rendering them vulnerable to *zamindari* extortions. Elsewhere, in western and southern India, the British introduced the Ryotwari Settlement, inspired by the notion that the state itself was the supreme landlord over peasants (Banerjee, 2018, pp. 57–58).

In reaction to such colonial interventions, sections of anti-colonial Indian nationalist elites as well as peasants put forward alternative perspectives, some of which rooted property rights in peasant labor on soil. Analyzing Muslim peasant discourses in colonial Bengal, Sartori suggests that the "Lockean" idea of the foundation of property in labor, especially agrarian labor, was a logic generated by the capitalist mode of production itself. He argues: "Marx recognized that commodity exchange ("the productive, real basis of all *equality* and *freedom*") in the sphere of circulation (that "very Eden of the innate rights of man" that forms the logical starting point from which the analysis of *Capital* sets out) generates, as part and parcel of the liberal conception of universal rights, a natural and originary state in which labor is constitutive of, and the true measure of, property" (Sartori, 2014, p. 207). Sartori ultimately emphasizes "the role of Lockean conceptions of property in underwriting the Pakistan demand" (Sartori, 2014, p. 138), that is, in generating the political movement for partition and a separate homeland for Indian Muslims: in Bengal, most Muslims being agrarian actors.

In contrast, I posit that arguments about the foundational role of labor in the creation of property were not mere by-products of modern-colonial capitalism but have a far longer history and more complex logic. In India, a key text is Manu's *Manavadharmashastra* (or *Manusmriti*) (late first millennium BCE/early first millennium CE), which observes (Manu 9.44): "Even as those who know the past regard this earth as the wife of Prthu, so they say that a field belongs to the man who cleared the stumps and the deer to the man who owns the arrow" (Manu, 2006, p. 192). The woman, the field, and the earth are juxtaposed as belonging to laboring and possessing men. Thus (Manu 9:33): "Tradition holds that the woman represents the field (Sanskrit, *kshetra*) and the man the seed (Sanskrit, *bija*)" (Manu, 2006, p. 191). The archetype is the first good king of Sanskrit legend, Prithu, from whom the earth (Sanskrit, *prithivi*) supposedly gets her name. One may compare this founding man to how the Biblical Adam serves as the prototype for all laboring and possessing men in Locke: in both cases, labor is framed in a masculinist way. Patrick Olivelle thus notes "the interesting parallel between arrow, the implied plow, and the penis" (Manu, 2006, p. 325). Male labor/seed on feminized earth/field/soil connects power and property and is indeed the foundation of both. The work of the mythical king Prithu, husband of the earth-goddess Prithivi, serves for Manu as the prototype for the propertied power of non-regal men.

Timothy Lubin interprets the Manu passage as "a version of the labor theory of value" (Lubin, 2018, p. 811), the latter today classically associated with Karl Marx. There are other Sanskrit texts too which compare regal and non-regal work. For example, Jaimini's *Mimamsa Sutra* (6.7.3) (fourth/third c. BCE) suggested that the king could not give away land/earth (Sanskrit, *bhumi*) which was common to all (Jaimini, 1923, p. 390). In the interpretation of Shabara (fourth/fifth c. CE), as Lubin notes, "the inhabitants of the earth exercise dominion over their particular fields and possessions, just as the king does over the whole realm." For Shabara, the king was not alone in his work of protection; human beings as well as the earth performed various important functions, including supporting living creatures (Lubin, 2018, p. 812). Hence, from the perspective of this Sanskrit intellectual tradition, there is a broad comparability between kingly and non-kingly labor. Manu's text also invites comparisons beyond the Sanskritic world. For example, Roman legal thought (Gaius, as cited in *Digest* 41.1.5.5) allowed wild animals like deer to "become the property of the first to seize them" (Starr, 1992, p. 438; Vuković, 2019). Locke later invoked this trope to discuss the foundation of property in labour: "Thus this Law of reason makes the Deer, that *Indian's* who hath killed it; 'tis allowed to be his goods who hath bestowed his labour upon it, though before, it was the common right of every one" (Locke, 2003, pp. 289–90).

I would situate these parallels as part of broader worlds of Eurasian – or at least, Indo-European – legal thought. In Indo-European languages, ideas of property are often indelibly linked to concepts of the self, structuring a deep prehistory of what C. B. Macpherson famously termed (in the more restricted context of early modern England) "possessive individualism" (Macpherson, 1962). (Arguing this does not imply that early modern English capitalist ideas should be universalized into other times and places; it is to point to conceptual genealogies of self and property, including those linked to broader collectives, which however transform with socio-economic changes.) The association is transparent in Latin *suum* and *proprietas*, German *Eigentum*, English "own" and "property," and Sanskrit *svatva* (cognate to *suum*). As Emile Benveniste notes about a key Proto-Indo-European root: "This word **swe* has given rise to an adjective indicating 'personal belonging' Skt. *sva-*, Lat. *suus*, Gr. **swós* […]. All that is ascribable to **swe* becomes **swos,* Lat. *suus* 'his' […] and ownership proper is defined only within the group included within the limits of **swe*" (Benveniste, 1973, pp. 269–271). Further research is needed about the connections, in terms of global intellectual history, between the general idea of the self as foundational to property and the more specific idea of the self's labor (or the laboring self) as the real root of property: but some of the aspects should be clear by now. The gendered framework of the male labor/seed-female soil connection is designed to make the soil/land the man's own, and thus his property.

Manu's dictum was widely known among Bengalis in the colonial era. Bhudev Mukhopadhyay (1827–1894) argued in a book published in 1892 that in Saxon England, the subjects enjoyed property rights over the land/soil (*bhumite prajar svatva*). But with the Norman Conquest (1066), property rights over soil were transferred to the king and the landlords. British colonialism transplanted this structure to India, establishing the idea of the state as the holder of ultimate property rights over soil and undercutting traditional ideas as embodied in Manu's passage; hence, Indians regarded the British as "exploiters" (*shoshak*) (Mukhopadhyay, 1892, pp. 162–64). Manu's line enjoyed such a canonical status in India that even the statist intellectual Benoy Kumar Sarkar (1887–1949) had to admit in 1920: "This is individualistic tenure and jurisdiction in their primitive form." But he hastened to add that "the sacredness of property can be established only by the state through its *danda*" (Sarkar 1920, p. 313. *Danda* is a complex Sanskrit term which Sarkar interpreted as "punishment or sanction," Sarkar 1920, p. 311).

Meanwhile, Jaimini, as interpreted by Shabara and other Sanskrit commentators, was cited by the British Orientalist H. T. Colebrooke (1765–1837). Colebrooke translated the Sanskritic discussion into a terminology of "labor": "The earth is not the king's, but is common to all beings enjoying the fruit of their own labour. It belongs, says Jaimini, to all alike" (Colebrooke, 1827, p. 457). These discussions gained urgency in the twentieth century in the context of demands for land reforms to benefit peasants. The Muslim politician and peasant rights advocate Shah Syed Emdadul Haq cited Manu and Jaimini in the Bengal Legislative Council in 1923 to justify that "those who actually tilled the soil were virtually the owners of the land" (Government of Bengal, 1923, p. 72). The early postcolonial Congress Agrarian Reforms Committee, with J. C. Kumarappa (1892–1960) as Chairman, similarly invoked Jaimini, via Colebrooke, in its Report (1949): "the earth [...] is common to all beings enjoying the fruits of their own labour on it. It belongs to all alike." The conclusion was clear: "In our scheme the right of cultivation will belong to the cultivator, but he must be subject to the control of the community" (Congress Agrarian Reforms Committee, 1951, p. 36).

Modern India's most famous poet (and Nobel Laureate) Rabindranath Tagore (1861–1941) was a key contributor to discussions on the human-soil interface from the 1910s to the 1930s. In Bengal, the (generally high-caste Hindu) *bhadralok* elites, often seen as alienated from soil and the peasantry, sought to reconnect with the agrarian-rural world across the early-mid twentieth century (Iqbal, 2009). Rabindranath undertook extensive projects for agrarian development, including improving soil care techniques, especially at his Sriniketan center. He was concerned by the depletion and exhaustion of soil resources caused by unbridled commercialized agriculture. More broadly, he sought to defend rural-agrarian civilization against industrial capitalism, holding the latter responsible for materially and spiritually devastating India and the wider world. Hence Rabindranath (for example in a 1918 essay) conceptualized soil (*mati*) as mother (*mata*), as *Bhumilakshmi*, the Lakshmi (goddess of economic welfare) of the land. He felt that Sarasvati, the goddess of learning, in the form of modern scientific knowledge, had to be brought into conversation with peasants for environmentally-sensitive soil use and agrarian productivity (Tagore, 1986, pp. 36–40; see also pp. 117–122).

Rabindranath, in a 1939 lecture, reinterpreted the ancient Indian epic tradition of *Ramayana* (narrated in Sanskrit and many vernacular languages over the centuries) as a story of Indo-Aryan agrarian expansion. He read the popular legend of *Ramayana's* hero, the god-king Rama, changing the cursed Ahalya from a stone back into a living woman, as a story of agriculture making the unplowed soil fertile. *Hala* is the Sanskrit word for plow; *a* is the negating prefix; *ahala* is a Sanskrit word meaning "unplowed." Further, Rabindranath saw Sita [literally, "a furrow, the track or line of a ploughshare" (Monier-Williams, 1986, p. 1218)] – Rama's wife, a daughter of the earth, and adopted daughter of the king Janaka, who found her while plowing the soil – as an embodiment of soil and agrarian fertility. Such perspectives encoded Sanskritic patriarchal discourses about the penetrative relation between the male/king/plow/seed and the female/earth/soil/womb. The relation between Prithu and his wife *prithivi* (earth) was reproduced in the relation between Rama and Sita, between any king and the earth, and between a laboring male cultivator and his field. On a more egalitarian note, Rabindranath cast the human-soil interaction as the primordial site for the genesis of human collectivity (*janasamavay*), religion, and philosophy: underlying these was the coming together of people into a unity (*aikyabodh*) through agriculture (Tagore, 1986, pp. 106–107).

In his celebrated play *Raktakarabi* (1924, 1926), Rabindranath cast a woman, Nandini, as agrarian nature's revolt against a masculinist, statist, and exploitative industrial modernity. The poet identified the latter as the demon-king *Ravana*, the main villain in the *Ramayana*, while Nandini was a new Sita who enabled a democratic revolution against the exploitative state (Tagore, 1964, pp. 545–559). A song in the play, widely performed in Bengal, juxtaposed the "*anchal*" (loose fringe of an Indian woman's *sari*) of soil (*matir anchal*) with the beauty of ripe paddy: equating the feminized soil with the fertility of nature (Tagore, 1964, p. 351).

In 1928, the poet helped to initiate the *Halakarshan Utsav* (plow-cultivation festival; originally named *Sitayajna*, Sita's Worship), which involved an annual ritual plowing of the soil. Rabindranath did the plowing himself in 1928 and 1929. The elaborate liturgy comprised chanting of Sanskrit verses – including to earth/land/soil (*prithivi* and *bhumi*) – as well as singing of Rabindranath's Bengali songs. Santal "tribal" women performed agrarian rituals too (Mandal, 1994, pp. 50–56), reflecting Bengali elite male constructions of the "indigenous," the "primitive," and feminized nature/soil. The festival has since become a canonical event in Bengal. In 1980, the (peasant-origin) President of India Neelam Sanjiva Reddy (1913–1996) performed this plowing ritual (Mandal, 1994, p. 54), replicating the relation between the sovereign and the soil at a time when peasant politics was again on the ascendancy across India.

In peasant politics, the links between labor on soil, the critique of capitalist exploitation, and ideas of democratic sovereignty were immediate. The Rajavamshis, one of the largest peasant communities of eastern India, constitute a case in point. The community had emerged out of early modern processes of peasantization, commercialization, and state formation in northern Bengal and Assam. Due to their role in regional state-building, local populations adopted the Rajavamshi (literally, 'of the royal lineage') identity, and claimed descent from the god Shiva. Thus, the *Shivayan* noted how Shiva had gone to the city-state (*nagar*) of the Koches to "do cultivation" (*chash chashite*), but had turned Koch women into "flowering gardens" (*kusum udyan*). Rajavamshi power declined in the nineteenth century due to British colonial political-economic policies and the rise of immigrant Western-educated high-caste Hindu Bengali elites. Confronted with loss of land, state power, and social status, Rajavamshis consolidated into an organized movement, spearheaded by the Kshatriya Samiti: founded in 1910, and led until 1935 by Panchanan Barma (1866–1935) (Banerjee, 2018, pp. 313–315).

For Rajavamshis, plowing the soil was a divine and kingly act. They noted that the god Shiva, the king Janaka, and the demigod Balarama had all plowed the soil. In arguing so, they invoked Indian concepts of sovereign plowing which have influenced South and Southeast Asia from ancient times to today. Somewhat like *Shivayan*, the Rajavamshi politician Nabinchandra Barma suggested in 1919 that peasant labor nourished society and was its true support (*bharsa*). In contrast, doing a job under a boss was comparable to slavery (*golami*). Rajavamshis saw themselves as kingly Kshatriyas. As peasants and soldiers (many fought for the British in the First World War in the Middle-East and Europe), they were guardians of the people (*lokapal*) and guardians of the earth (*bhupal*). In Rajavamshi thought, their work (*karma*) was divine labor because it protected and nourished society (Banerjee, 2018, pp. 315–326).

Rajavamshis criticized the modes of production and exchange which enabled the rich to extract labor (*shram*), land, and raw materials from the poor and to sell finished products to them at high prices. They conceptualized this as "theft" (*apaharan*) and "exploitation" (*nishpeshan*). In resistance against high-caste elites, they emphasized collective cooperation to improve agrarian and business techniques to generate wealth (*dhan*), as well as organization of community education and healthcare. New scientific techniques of soil use were also discussed (as a typical example of this last point, see Samiti, 1919, pp. 12–13).

Rajavamshis demanded – and over the 1920s and 1930s, achieved – political-legislative representation. These demands were premised on ideas of the common divinity of all beings, as well as on theories of dialectical movement from lordship (*prabhushasan*), monarchy (*rajar shasan*), and heteronomy (*parashasan*) to political autonomy (*atmashasan*) and agrarian-economic self-reliance (*atmanirbhar*). Mediating this transition were "feelings of unity" (*milanbhav*), "feelings of sociability" (*samajik bhav*), and collective self-governance.

Indigenous traditions of community self-governance, such as through Panchayats, were also brought to play from the 1910s onwards, to demand formal local self-government and political representation from the British. Agrarian labor, which in the first place necessitated cooperation between peasants, as well as self-conscious collectivist politics together generated a profound sensibility of democratic self-governance. Cooperation in cultivation and animal husbandry, in various related agribusinesses, and in other economic institutions, reinforced cooperation and solidarity in politics, creating one of the most powerful collectivist movements of late colonial India. By the 1940s, Rajavamshis were one of India's most well-organized and militant peasant communities. They took part in the Communist-influenced Tebhaga struggle for improving the rights of poor cultivators, especially sharecroppers, often challenging their own community elites in the process. From the 1940s until today, they have also led repeated campaigns for patriotic-territorial autonomy, often in alliance with other lower-caste Hindu and Muslim communities of the region (Banerjee, 2018, pp. 318–335). In the end, peasant labor on soil has grounded the desire for collective political autonomy. Rajavamshis have successfully cast labor as the foundation of wealth, rights, and democratic sovereignty.

With respect to Bengali Muslim agrarian politics, Sartori has argued, on a theoretical plane, as we noted already, that arguments about the property-constituting nature of labor stemmed from the logic of the modern capitalist mode of production and exchange. Yet his empirical discoveries somewhat undercut his argument. Sartori shows, in brilliant detail, how Bengali Muslim politicians cited several traditions of argumentation (many considerably older than colonial capitalism) about the relation between labor, wealth, and property, including Manu, Jaimini, fourteenth-century English peasant revolt ideas (slogan at that time: "When Adam delved and Eve span, who was then the gentleman?"), and Adam Smith (Sartori, 2014, pp. 146, 148, 155). In a more recent article, Sartori seems to partly revise his earlier economic-deterministic and modernist position. He admits that within classical Islamic law, there was a strong tradition of arguing that human labor constituted property. For example: "Abu Yusuf (eighth century CE), one of Abu Hanifah's most authoritative students, had argued that property derived initially from the primordial right to possession acquired through the original reclamation and cultivation of *res nullius*" (Sartori 2018, p. 6). Though the rise of powerful early modern Muslim states like the Mughals in India and the Mamluks and Ottomans in West Asia and North Africa partly undercut this focus – by rooting the origin of property in state authority – the legacy of the classical tradition remained alive. In the nineteenth century, the Faraizi movement in Bengal thus raised the dictum "*langal jar, jami tar*" (whose plow, his land) (Sartori, 2018, p. 13).

In a fascinating anthropological study on Bangladesh in the 1970s, John Thorp found that Bengali Muslim male peasant proprietors identified themselves with Adam (the primal father in the Islamic tradition too). They saw themselves as *maliks* with *adhikar* (authority/right) over the soil: the term *malik*, an inheritance into Bengali from Arabic, conflating idioms of kingship, mastery, and propertied ownership. The *maliks* were masters of their family/household as well as of the land. Simultaneously, the Bengali peasants saw themselves as originating from the soil (*mati*), given the Islamic narrative about God's creation of Adam. Cultivating the soil, and engaging in labor (*kaj*) in general, was seen as a divine injunction for Adam as well as for his descendants. The earth/soil was seen as endowed with *shakti* (power), which had to be enhanced through right knowledge and scientific techniques (Thorp, 1978).

In Sanskritic discourses, *shakti* is often gendered feminine and associated with goddesses. The Muslims studied by Thorp also shared Sanskritic ideas of the comparability of the woman with the field (*kshetra*), where the man plants his seed just as in the soil (Thorp, 1978, p. 69). But one can also compare these ideas with the Hegelian notion of paternal soil. In the Bengali case, the Muslim male peasant proprietor is both from the soil and the master of the soil: Soil is the foundation of agrarian male authority. The end result of such ways of thinking was to instigate Bengali Muslim peasants to rise up in successive revolutions: first, against Hindu landlord rule, culminating in massive Muslim peasant support in eastern Bengal for an independent Pakistan (achieved in 1947) and the abolition of the Permanent Settlement and second, against the rule of the West Pakistan elites, leading to the independence of eastern Bengal as the new nation-state of Bangladesh in 1971. The Bengali Muslim achievement of sovereignty through Bangladesh was, in a sense, the climax of a specifically agrarian tradition of grounding claims to both property and democratic sovereignty in peasant labor on soil.

15.5 Sowing and Soil Goddesses: Female Economic Theology Against Industrial Capitalism

Following the end of British rule in 1947, the British Indian empire was partitioned into the independent nations of India and Pakistan. The Hindu-majority parts of Bengal became the Indian state of West Bengal. Given the heritage of peasant struggle and radical agrarian thought in the region, the postcolonial Congress government enacted some land reforms, including the abolition of the Permanent Settlement. Following further agrarian militancy across the 1960s and 1970s, and the establishment of a Left Front government in West Bengal in 1977, more substantive land redistribution took place, benefiting subaltern peasants. But from the 2000s, the Left Front changed its position on the land question. To reduce dependence on the agricultural sector and create new jobs, it desired large-scale industrialization and entry of transnational capital. Left-modernist ideologies about industrial progress also had a constitutive role. The government sought to coercively acquire fertile land from peasants to set up industries. Singur and Nandigram became two flashpoints, as peasants opposed such land acquisition. Following state atrocities, including the police killing several people at Nandigram in 2007, and a sustained wave of protests involving peasants and middle classes, the Left Front was voted out of power in 2011. Since then, the coercive land acquisition drive has been largely checked, with acquired land often returned to peasants, as in Singur.

The most important figure in the protests, and the new Chief Minister, was Mamata Banerjee. In fact, Sartori's book begins with an analysis of the "most peculiar alignment (though in no sense alliance) between disaffected leftist intellectuals and the aggressively populist and anticommunist leader of the Trinamul Congress, Mamata Banerjee." For Sartori, the Lockean logic of labor's property-constituting power lay at the heart of this alignment. Hence, "many leftist intellectuals, deeply critical of the neoliberal developmental policies in which the CPM had been dabbling, lined up alongside the populist Trinamul Congress to denounce capital in the name of property rights" (Sartori, 2014, pp. 2–3). Sartori attributes this logic to the modern capitalist mode of production, and the entire book functions, in a sense, as a long historical investigation of the emergence of this logic in colonial Bengal. While Sartori is absolutely right about Banerjee's defense of peasant property, I would argue that the ideological roots of this defense can be found not just in the modern capitalist mode of production and exchange, but also and simultaneously, in far older theologies of labor and property.

Though Mamata Banerjee comes from a Brahmin (and not a peasant) family, her populist politics draw upon agrarian practices and beliefs. Thus, in 2006, she ritually sowed potato seeds at Singur. "Mamata reached Singur […], sowing potato seeds in land that is being acquired by the government for the Tata small-car project. Village women lined up to blow conch shells and ululate as Mamata performed the symbolic ceremony of protest" (The Telegraph, 2006). After the Supreme Court of India allowed her project of returning land to peasants at Singur in 2016, Banerjee again performed the ritual. "Singur teaches us that we should never run from a fight and that rights are not served to you on a platter. You have to snatch them," Mamata said, before scattering the mustard seeds on their plots. Mustard, according to agriculture minister Purnendu Bose, was chosen because it is a dry season now" (Bhattacharya, 2016). These rituals viscerally linked the female sovereign with agriculture and the soil. Given that sowing is often done by women in Singur (Das, 2016, p. 10), the ceremony probably also embodied female labor: a feminized counterpart of male-kingship-plowing rituals (see Fig. 15.1).

Fig. 15.1 Painting of Mamata Banerjee sowing seeds at Singur, by Kaushik Adhikari (2018)

In her poetry, Banerjee draws upon Indian ideas of divine female soil, while erasing the patriarchal elements. Thus, while invoking soil and its female human defenders as Sita and Ahalya, she edits out Rama (Banerjee, 2010, pp. 43, 54; Banerjee, 2014, pp. 78, 88, 174, 176). A poem from 2009 transforms the reanimation of Ahalya into a parable of revolution: "Return o dear mother Ahalya with mother Tebhaga […]. Awake Bengal, awake mother-soil (*mati-ma*), bring the dawn of humanity" (Banerjee, 2014, p. 78). This is a source for her slogan *Jago Bangla* (Awake Bengal). In Indian thought, the goddess Lakshmi connects the welfare of the household, agrarian and economic plenitude, and social and political prosperity. Thus Vastulakshmi (Lakshmi of the household, *vastu* signifying the foundation or ground of the house, or the house itself) in Hindu Bengali homes takes the physical shape of soil and grains. The granary (*marai*) is venerated as Marailakshmi. The courtyard (*uthon*) of the agrarian house is venerated as Uthonlakshmi (Thakur, 2018). Lakshmi thus represents an immanent theology of agrarian household production, property, and wealth. In line with this, Banerjee identifies soil with Lakshmi as an agrarian goddess (Banerjee, 2014, pp. 177–178).

Banerjee understands Vastulakshmi as the embodiment of the peasant household and small peasant property, and links her with Rajlakshmi, the Lakshmi of the ruler/state. As she laments in her poem about peasant dispossession, "Vastulakshmi" (2007): "Vastulakshmi is now without home/foundation (*vastuhara*) […]. Rajlakshmi cries, the ruler laughs, Lakshmi is now bereft of auspiciousness (*lakshmichhara*, literally, bereft of Lakshmi). In the deep waters of globalization, those who have lost all (*sarvahara*: in Bengali Marxist discourse, often used to translate 'proletariat') are now without home/foundation (*vastuhara*)" (Banerjee, 2010, p. 43).

Mamata Banerjee eulogizes peasant women who suffered – like the teenager Tapasi Malik, who was raped and killed – because they stood up against the Left Front government's land acquisition drive. These women become synonymous with the soil, home, and crops they sought to defend (Banerjee, 2010, pp. 27–28, 30, 33, 36, 49–50, 54–56, 68, 71, 111–112, 132, 139; Banerjee, 2014, pp. 60, 62). She sings paeans to soil: "This soil is sacred (*pavitra*), this soil is soil of gold" (Banerjee, 2014, p. 148). It cannot be expropriated from the cultivator: "Even if soil is forcibly occupied, can soil ever be alienated (*kakhana par hay*)?" (Banerjee, 2010, p. 127). Soil brings people together in the struggle for a dignified life: "With soil are the Quran and the Purana, the message of togetherness and union. […] On soil goes ahead the fight for the dignity (*samman*) of living" (Banerjee, 2014, p. 176). Hence, Banerjee condemns the land-grabbing Left Front government as "emperors of capitalism" (*punjibader samrat*: Banerjee, 2010, p. 56). She ultimately draws upon centuries-old agrarian theologies: which have for ages, much before modern capitalism was born, invested peasant labor and peasant property with affective and theological value, and driven peasants to defend themselves against rapacious elites.

It remains unclear whether this defense of peasant property occludes a focus on labor. The sacralization of Vastulakshmi risks fetishizing property: obfuscating the problems faced by property-less peasants, like sharecroppers and wage laborers. In any case, Banerjee's government has created severe dissensus in Bengal in recent years. The leftist intelligentsia as well as sections of the general population have become disenchanted with the Trinamool Congress's Caesarist style of maintaining party dominance. Hence, agrarian theologies have assumed even greater importance in manufacturing hegemony over rural Bengal. Since 2013, the *Mati Utsav* provides an annual focus for harnessing peasant aspirations into a statist framework. The festival sees itself as prefiguring United Nations declarations about World Soil Day (5 December) and International Year of Soils (2015) (Krishi Adhikar, 2018, p. 1). During the festival, agrarian actors showcase and sell their produce and interact with scientific experts. New environmentally sensitive technologies of soil care and use are disseminated too. The festival emphasizes the divinity of soil and the sacredness of peasant labor and property, thereby also legitimating Banerjee and her party. Banerjee's poetry decorates the festival grounds. Soil is ultimately visualized as generative of human collectivity. As a *Mati Utsav* pamphlet from 2018 observes:

> Soil provides us with food, clothing, and shelter. On soil are constructed buildings, and there emerges civilization (*sabhyata*). On soil takes place creation (*srishti*) – there advances agriculture (*krishi*) and culture (*krishti*). We do agricultural labour on soil. Simultaneously, this is a festival of trying to "grow gold on human field". In this festival of union, peasants, fisherfolk, minstrels, holy men [*fakir*, *darbesh*], and people of different occupations of Bengal come together. […] People from outside Bengal celebrate along with all Bengal. Bengal becomes Global Bengal (*vishvabanga*). (Krishi Adhikar, 2018, p. 1)

15.6 Conclusions

Let me now offer general conclusions. First, I have critiqued two reigning models in global intellectual history. The first model emphasizes human political agency and justifiably underscores the intellectual creativity of non-European actors in molding modern worlds of thought. But it creates problems in seeking to divorce the political from the economic sphere. The second model conjuncts the political and the economic and (unlike the first) emphasizes the role of non-elite actors like peasants. But it makes the political and the intellectual an epiphenomenon of the economic: a superstructure of the capitalist mode of commodity production and exchange. In contrast to both paradigms, this essay conceptualizes the agrarian political as simultaneously grounded in economic, political, and theological spheres. It places emphasis on canonical intellectuals and on subaltern actors, on male and female agency. It challenges Eurocentrism by acknowledging the legacies of precolonial/non-European ways of thinking – here, about human-soil interaction – on modern actors.

Second, when I argue that human beings have long regarded labor as the foundation of wealth and property, the stakes are political. Sartori wants to explain the labor theory of property/value as an effect of the capitalist mode of production and exchange. For him, modern Indian (male) peasants are thus not the Others of capitalism – as Subaltern Studies theory argued – but presences within capitalism: their thought-worlds made possible by capitalism. In contrast, I do not think that modern Indian peasant thought can be adequately understood exclusively within a capitalist framework, even if peasants have taken part in, and been exploited by, capitalist structures of production and exchange. Peasant resistance in modern India, and surely elsewhere, has operated through *and* against capitalist structures. This Otherness of peasant resistance – or, more correctly, its dialectical nature: working through and challenging elements of capitalism, be it in the form of colonial landlordism in the past or industrial capitalism today – is a reality that Sartori does not adequately acknowledge. Peasant labor, even when operating through capitalist structures, draws on concepts and practices – models of divinity and cosmology, gender, mastery, rulership, protection, welfare, and wealth – whose roots go beyond the capitalist framework.

Further, the idea that labor is the root of wealth and property is not one provincial to capitalism. It is transhistorical. It can be found, for example, in classical Indian, Roman, and Islamic traditions that have influenced modern actors. Hence a history of capitalism does not provide an adequate standpoint to grasp the wealth/property-constituting power of labor. It follows that to destroy capitalist frameworks of production and exploitation, we should not rely only on intellectual logics internal to capitalist history, such as Lockean perspectives. We can summon centuries-old arguments about the foundation of social wealth and well-being in labor and be inspired by struggles of laboring actors from societies past and present. Labor, not capitalist logic, provides a normative standpoint for critiquing capitalist exploitation. Having said that, a note of caution is necessary. When intellectuals have emphasized agrarian labor – from Manu to Locke, Hegel, and Marx – male labor has been valued far more than female labor. Conventional intellectual histories of labor have been more preoccupied with the standpoint of primal fathers, Prithu and Adam, than of primal mothers like Parvati and Tapasi Malik. A transtemporal perspective invites us to rethink these millennial prejudices that haunt political theory.

Third, this study retheorizes democratic practice. The democratic political is not just about the friend/enemy distinction and violence, as Kapila underlines, drawing on Carl Schmitt: it is also about solidarities. Human labor on soil has not only produced notions of rights, but also of social interdependence. Agrarian cooperation has helped generate political cooperation. In India, and surely elsewhere too, concepts of divinity have expressed these forms of sociality and interdependence engendered by labor. Admittedly, these forms are often expropriated by elites and constricted into definitions of property and sovereignty which privilege the propertied over the property-less and men over women.

Divinization of agrarian property may fetishize possessive individualism. Nevertheless, forms of divinity which reflect the coming together of human beings, men and women, in practices of labor, contain within them a revolutionary excess. To think of peasant labor on soil as sacred – to see peasants as gods and gods as peasants (for Hindus), or peasant labor on soil as mandated by God (for Muslims) – to regard the social interdependence of peasants with each other and with their material environments as something intrinsically charged with divinity: these economic theologies have had revolutionary consequences in India, in successful struggles against colonial landlordism and industrial capitalism. We need to recuperate these theologies, while exorcising their hierarchical aspects, to make cooperation and solidarity between laboring men and women the very beating heart of future models of democracy.

Acknowledgments I would like to thank Guilherme Sobrinho and Nikola Patzel for involving me in the Soil and Culture network, and sparking my engagement with the topic; Simon Cubelic and Nirmalya Guha, for advice on interpreting Manu, Jaimini, and Shabara; Arjundeb Sensarma, for discussions on Manu and *Shivayan*; Kresimir Vuković, for drawing my attention to the Gaius text; Philipp Sperner, for his comments on the essay; and Ilya Afanasyev, for our conversations about political economy.

Bibliography

Agamben, G. (2011). The Kingdom and the Glory: For a theological Genealogy of Economy and Government (L. Chiesa, Trans.). Stanford University Press.

Bandyopadhyay, H. (1988). *Bangiya Shabdakosh* (Vol. 2). Sahitya Akademi.

Banerjee, M. (2010). *Ma-Mati-Manush*. Dey's Publishing.

Banerjee, M. (2014). *Pachhander Kavita*. Dey's Publishing.

Banerjee, M. (2018). *The Mortal God: Imagining the sovereign in Colonial India*. Cambridge University Press.

Banerjee, M. (2018a). "We shall create a new world, a new man, a new society": Globalized horizons among Bengali Naxalites. In T. Chaplin & J. E. Pieper Mooney (Eds.), *The global 1960s: Convention, contest, and counterculture* (pp. 52–71). Routledge.

Banerjee, M. (2019). Transversal histories and transcultural afterlives: Indianized renditions of Jean Bodin in global intellectual history. In L. Abu-Er-Rub, C. Brosius, S. Meurer, D. Panagiotopoulos, & S. Richter (Eds.), *Engaging transculturality: Concepts, key terms, case studies* (pp. 155–169). Routledge.

Benveniste, E. (1973). *Indo-European language and society* (E. Palmer, Trans.). Faber and Faber.

Bhattacharya, R. (1957). *Rameshvarer Shiva-Sankirtan ba Shivayan* (J. Haldar, Ed.). Calcutta University Press.

Bhattacharya, M. (2016, October 21). Evolution of Singur. *The Telegraph*. https://www.telegraphindia.com/states/west-bengal/evolution-of-singur/cid/1320182. Accessed 23 Nov 2019.

Chatterjee, P. (1993). *The nation and its fragments: colonial and postcolonial histories*. Princeton University Press.

Colebrooke, H. T. (1827). On the philosophy of the Hindus. Part III. *Transactions of the Royal Asiatic Society of Great Britain and Ireland, 1*, 439–466.

Congress Agrarian Reforms Committee (1951). *Report*. All-India Congress Committee.

Das, J. (1916). *Bangala Bhashar Abhidhan*. Indian Press.

Das, R. (2016). The politics of land, consent, and negotiation: Revisiting the development-displacement narratives from Singur in West Bengal. *South Asia Multidisciplinary Academic Journal, 13*, 1–20.

Dean, M. (2019). What is economic theology? A new governmental-political paradigm? *Theory, Culture and Society, 36*(3), 3–26.

Government of Bengal (1923). *The Bengal Legislative Council Proceedings, vol. 12*. Bengal Secretariat Book Depot.

Hegel, G. W. F. (1899). *The philosophy of history* (J. Sibree, Trans.). The Colonial Press.

Hegel, G. W. F. (1911). *Naturrecht und Staatswissenschaft im Grundrisse: Grundlinien der Philosophie des Rechts*. Verlag von Felix Meiner.

Hegel, G. W. F. (2008). *Outlines of the philosophy of right* (T. M. Knox, Trans.). Oxford University Press.

Iqbal, I. (2009). Return of the Bhadralok: Ecology and Agrarian relations in Eastern Bengal, c. 1905–1947. *Modern Asian Studies, 43*(6), 1325–1353.
Jaimini. (1923). *The Mimamsa Sutras of Jaimini* (M. L. Sandal, Trans.). The Panini Office.
Kapila, S. (2014). Global intellectual history and the Indian political. In D. M. McMahon & S. Moyn (Eds.), *Rethinking modern European intellectual history* (pp. 253–274). Oxford University Press.
Krishi Adhikar, P. S. (2018). *Mati Utsav 2018 Shuru Aj*. Krishi Adhikar, Pashchimbanga Sarkar.
Locke, J. (2003). *Two treatises of Government* (P. Laslett, Ed.). Cambridge University Press.
Lubin, T. (2018). The theory and practice of property in Premodern South Asia: Disparities and convergences. *Journal of the Economic and Social History of the Orient, 61*(5–6), 803–850.
Macpherson, C. B. (1962). *The political theory of possessive individualism: Hobbes to Locke*. Clarendon Press.
Mandal, S. (1994). *Vishvabharatir Utsav*. Suvarnarekha.
Manu. (2006): *Manu's code of law: A critical edition and translation of the Manava-Dharmasastra* (Edited and translated by P. Olivelle, with S. Olivelle). Oxford University Press.
Monier-Williams, M. (1986). *A Sanskrit-English Dictionary*. Motilal Banarsidass.
Moyn, S., & Sartori, A. (Eds.). (2013). *Global intellectual history*. Columbia University Press.
Mukhopadhyay, B. (1892). *Samajik Pravandha*. K. Bhattacharya.
Richards, J. F. (1997). Early modern India and World History. *Journal of World History, 8*(2), 197–209.
Saha-Krishi Adhikarta (Prashasan), Vistarita Bhumi Samiksha. (2015). *Samrakshita Krishii Amader Bhavishyat*. Saha-Krishi Adhikarta (Prashasan).
Samiti, K. (1919). *Kshatriya Samiti, Ekadash Sammilani, Karya-Vivarani*. Manohar Barma.
Sarkar, B. K. (1920). The theory of property, law, and social order in hindu political philosophy. *International Journal of Ethics, 30*(3), 311–325.
Sartori, A. (2008). *Bengal in global concept history: Culturalism in the age of capital*. University of Chicago Press.
Sartori, A. (2014). *Liberalism in empire: An Alternative History*. University of California Press.
Sartori, A. (2018). *Property and political Norms: Hanafi Juristic Discourse in Agrarian Bengal*. Modern Intellectual History. https://doi.org/10.1017/S1479244318000215
Schmitt, C. (2007). *The concept of the political* (G. Schwab, Trans.). University of Chicago Press.
Sen, R. (n.d.). *Kaviranjan Ramprasad Sener Granthavali* (U. Mukhopadhyay, Ed.). Basumati.
Starr, R. J. (1992). Silvia's Deer (Vergil, Aeneid 7.479-502): Game Parks and Roman Law. *The American Journal of Philology, 113*(3), 435–439.
Tagore, R. (1964). *Ravindra-rachanavali* (Vol. 15). Vishvabharati.
Tagore, R. (1986). *Palliprakriti*. Vishvabharati Granthanvibhag.
Thakur, S. (2018, February 5). Maraike Kendra Kare Ek Samriddha Lokasanskritir Vistar. *Ei Samay*.
The Telegraph. (2006, November 26). Mamata Sows Seeds of Potato, Patch-up. *The Telegraph*, https://www.telegraphindia.com/india/mamata-sows-seeds-of-potato-patch-up/cid/754711. Accessed 23 Nov 2019.
Thorp, J. P. (1978). *Masters of earth: Conceptions of "Power" among Muslims of Rural Bangladesh*. PhD dissertation, University of Chicago, Chicago, USA.
Vuković, K. (2019). Silvia's Stag on the Tiber: The setting of the Aeneid's Casus Belli. *Mnemosyne: A Journal of Classical Studies, 73*, 1–19.
Wright, S. (2017). The practice and theory of property in seventeenth-century Bengal. *The Indian Economic and Social History Review, 54*(2), 147–182.

Chapter 16
A Deity We Belong to or a Property to Own? Relations Between Land and People in the Sayan Mountains in Southern Siberia

Tatiana Intigrinova

16.1 Introduction

This chapter considers the religious beliefs of the Buryats and Soyots that inhabit the Sayan Mountains in two districts of the Republic of Buryatia, Russia—Tunka and Oka. It places the relations between people and land in the context of developing a notion of land being subject to ownership rights introduced by the Russian federal government in late 1991 as part of the agrarian reform ordering privatization of all Soviet collective farm assets and the division of agricultural and pastoral lands into individual parcels. Later on, the state introduced private property for lands used for both production purposes and housing. The scale of de facto land partitioning and its registration into private ownership in different localities of the Sayan Mountains varies; however, private property ownership imposed by government policies is gradually becoming part of local life, challenging local spiritual relations with land.

At the time of the Soviet Union's collapse, Tunka and Oka districts had almost identical land tenure patterns. All land belonged to the state being used by collective and state farms. In Tunka, however, local households could secure de facto use of individual plots for housing. In the 1990s, the two districts followed different concepts of collective and state farm privatization (Intigrinova, 2015), but land was allocated to individual households for haymaking. Grazing lands also were partitioned in Oka with particularly unequal distribution of land in one of the two studied communities. As animal husbandry is the occupation that provides some means of living for the vast majority of the population in the research area, land allocation and individualization of its use affect almost everyone in the communities under discussion.

T. Intigrinova (✉)
Independent scholar, Frankfurt, Germany
e-mail: tintigrinova@gmail.com

N. Patzel et al. (eds.), *Cultural Understanding of Soils*,
https://doi.org/10.1007/978-3-031-13169-1_16

As will be discussed below, the processes observed in the research area, such as legal and actual segmentation of the land into individually used parcels, does not correspond well with religious understanding of the landscape as a totality that controls peoples' fate. The local concept of landscape includes not only soil, geological features, rivers, and vegetation but also people, wildlife, domestic herds, and inhabitants of the spiritual world. All these constituents are interrelated and compose the land's totality.

Highlighting the religious concepts behind human and landscape relations may seem inappropriate at first glance in the context of a society that experienced Soviet state collectivization and campaigns against religion and in which a majority of the people went through a secular education based on atheistic ideology. However, both this research and ethnographies of other Siberian indigenous groups (Kasten, 2002; Jordan, 2004; Ziker, 2003) confirm that native communities in the region still maintain their spiritual relationships with the land in the post-socialist period.

In southern Siberia, the regulating institutions that are based on spiritual relations have been challenged by outside interventions for at least a century. The property consciousness of post-socialist ideology brought new challenges to the spiritual connection of people with land. Land turning into a property or a commodity may result in changing the notion of belonging between people and land. It may be altered from the notion of someone belonging to a place to a notion of land as a property belonging to someone (Peters, 1998, p. 360).

The local case study presented in this chapter allows the raising of a more widely applicable question concerning the influence and resistance of local cultures to changes imposed from outside. It also demonstrates the influence of policies aimed at purely economic transformation on local values and people's attitude towards nature. It is noteworthy, however, that the case study captures the local situation intrinsic to the two districts in the Republic of Buryatia and may not accurately reflect the situation for Buryats residing in other parts of Inner Asia. Buryats, in contrast to the regional minority of the Soyots,[1] are a widely dispersed ethnic group, residing in Russia, Mongolia and China. The diversity of both historic and modern social, religious and economic settings as well as varying environmental conditions resulted in a variety of the locally defined cultural characteristics within this ethnic group. The integration of the Soyot and Buryat traditions also has to be pointed out as a specific characteristic of the local culture.

[1] Soyots—the ethnic group (slightly over 3500 people) living in the territory of two administrative districts (Oka and Tunka) in the Rebublic of Buryatia, Russia.

16.2 Research Methods

The dynamics of local relations with land presented in this chapter are based on the results of ethnographic studies implemented in two districts, Tunka and Oka, in four communities—two in each district, of the Republic of Buryatia (Russia). Data presented in this chapter were collected within a broader project about land and people relations (Intigrinova, 2010, 2015). It considered the complexity of these relations, including state land policies and their local implementation in the context of post-socialist land reform, with particular reference to the process of land allocation, land titling, and the impact of these factors on pastoral resource use and livelihoods as well as religious beliefs and people's perception of their spiritual relations with land. Data were collected starting in 2003. Two shorter preliminary field trips were undertaken in 2003 and 2005; the longer 12-month research trip was conducted in 2006 and 2007.

Residing with local households in all the communities provided the possibility for ongoing implementation of participant observation as one of the main research methods implemented in this study. Various forms of interviewing were also utilized in the framework of the research. These included in-depth interviews with key informants—more knowledgeable members of the communities and religious specialists, life histories, occasional unstructured interviews, and semi-structured interviews conducted on a household basis.

Three hundred twenty-five semi-structured interviews were conducted between July 2006 and July 2007.[2] Initially, the interviews were focused on pastoral land use, allocation, and titling. At a later stage of the research, to better capture local people's attitude to land, questions concerned with religious beliefs were added to the interview. One of the questions added in the process of interviewing concerned the local belief that arguing over land inevitably results in death or bad luck for those involved in the dispute or their relatives. The question on the belief was added to the interview after it was spontaneously mentioned by a number of interviewees. As a result, the belief was discussed with 270 out of 325 interviewees participating in the semi-structured interviews.[3] One may argue that numbers are not important in the evaluation of cultural values. However, the previous history of anti-religious policies in the country encouraged me not only to discuss the belief at length and observe its application in reality but also to investigate how many people were aware of the prohibition. The fact that the question was included in the household interviews provides grounds on the one hand to argue that the belief is well rooted

[2] A sample covering all households in one community in each district was the ultimate goal of the semi-structured interview. In two local communities—one in each district—only few households were not interviewed due to their unwillingness to talk. To provide a basis for the generalization of research conclusions for Tunka and Oka districts, households representing all wealth groups (Grandin, 1986) were interviewed in one more community in each district. Each wealth group was represented in the sample according to their proportion in the general population.

[3] All sampling deficiencies due to interview questions added at a later stage of the research are stated in the text.

in all the research communities and on the other to demonstrate the appearance of its varying interpretations.

The semi-structured nature of the interviews was chosen to allow flexibility during the interview to discuss issues of acute importance for particular informants at length, resulting in a better rapport with local people and a better understanding of their situation. Quotations derived from these interviews were brought into the discussion below to better present local views.

16.3 Altan Gazar Is the Golden Soil, the Land and the Infinite Earth

In Buryat, there is a word *gazar*. *Gazar* can be translated into English as soil, land, ground, territory, and the Earth. The first meaning, however, is rarely highlighted by local people. Instead, the concept of wholeness dominates informants' views. *Altan Gazar*, the Golden Earth, refers to Mother Earth as Nature, encompassing all features of the landscape—mountains, forest, rivers, springs, and all living beings and plants. People also refer to the Earth as *Delkhei Daidyn*, the infinite earth (in Buryat *Delkhei* means the Earth), and *Khangai*, the taiga. The English words soil, land, and earth are used interchangeably in the text below. The choice of English equivalent can be biased as the notion of Buryat gazar is inseparable, representing all the terms.

16.4 The Ancient Cult of Sacred Land

Mongolian people's worship of land has been well recorded since the thirteenth century. Russian geographic expeditions that visited the area under discussion in the late nineteenth century reported that the two ethnic groups that populated it performed identical offerings to the local spirits, masters of mountains, mountain passes, and locations. Libations and offerings were made before meals, drinking, leaving for a trip, etc. The spirit masters were addressed with words of gratitude and asked for help (Ritter, 1894).

In her monograph on the Buryat religious beliefs, Galina Galdanova quoted a ceremonial book written in ancient Mongolian in which the goddess of land (*Etugen*) is listed, among other deities, for whom libations had been performed on the New Year since the time of Genghis Khan. Seventy-seven *etugen ekhe* (spirits and spiritual levels of the land) and *gazar-usan* (Buryat soil-water) had to be addressed among the other deities (Galdanova, 1987, pp. 12–13).

Mikhailov (1980), a local ethnographer, designated *Ul'gen Ekhe* (Buryat mother *Ul'gen*) as the deity of the land, once revered by the majority of Buryats. Menes argued that the "ideological content of the mythological representation connected with the term *Ul'gen* is common for Mongolian and Turkic people" in the region

(Menes, 1986, p. 99) and thus could be inherent in the religious beliefs of both the Buryats and Soyots.[4]

According to Galdanova, Buryats still performed a ritual for *Etugen* (*Ituge*) to improve fertility of all living beings and plants in areas located toward the northwest of Lake Baikal up to the twentieth century. In Transbaikalia or territories southeast of the lake, however, it was already "forgotten" by that time. In the vocabulary of the Buryats living in Transbaikalia, the word *etugen* was used as an exclamation of fright, astonishment, or obscenity with the meaning of female genitals (Galdanova, 1987, pp. 26–27). In Oka and Tunka, the word etugen is also known today as an obscenity that is avoided in speech. The goddess *Etugen* or *Ul'gen Ekhe* is not remembered. But the ancient cult of the goddess is echoed in beliefs and rituals reemphasizing the motif of the earth being a spiritual source of human life and its parental relations to the people (see Fig. 16.1).

An elderly Soyot woman pointed out during an interview that the land feels it, even at a distance, when a person born and raised on it—a person who drank from its rivers—dies. "Elderly people knew in WWII when somebody [from the community] died [in a battle], they could recognize the signs of the earth," she noted. Pavlinskaya (2002, p. 235) recorded from one of her informants a belief that at the moment of childbirth, a handful of soil shoots out of the ground, and the person later ends his/her earth journey at that exact place.

During this research in Oka, a tendency was observed to revive the tradition discontinued in the Soviet period of burying the afterbirth of a newborn child. The

Fig. 16.1 Making offerings to the deities, 2003. (All photos are taken by the author)

[4] Soyots culture has been significantly influenced by both Mongolian and Turkic cultures.

doctors practicing at the district hospital, where local women give birth to their children, offered fathers the choice of taking the afterbirth and burying it. They admitted that they were doing it due to the demand on the side of the parents, which became very high in the years after the collapse of the Soviet Union. According to local custom, the afterbirth should be enveloped in a clean soft cloth, accompanied by grain and coins, and buried near the house on a day defined by a *lama* (a Buddhist clergyperson) as a favorable one. Informants from the elder generation in Tunka told me that at the time when Buryats lived in yurts (Buryat *ger*), the afterbirth had to be buried within the yurt behind the fire that protected it from bad spirits.

The afterbirth in Buryat can be named directly as *khukhenei khoito* (Buryat, child's place) or metaphorically *khuukhei nukher*. *Khuukhei* means child. *Nukher* is polysemic. It is normally used with the meaning of "a friend." Pavlinskaya (2002, p. 219) interprets it as the twin soul of the child, which should be returned into the soil in order to be reborn in future generations. The second meaning of the word is "a spouse" (Pavlinskaya, 2002, p. 219; Batoeva et al., 2002, p. 58). According to Pavlinskaya, this semantic usage recognizes the afterbirth as a symbolic spouse of the newborn child, thus a guarantee of descendants.

Although the author distinguishes the two different meanings, for our analysis, it is important that both possible interpretations link the ritual of inhuming *khuukhei nukher* with continuing people's regeneration. This association suggests that although the goddess Etugen is forgotten today, the connection between the land and people's fertility is still preserved in local rituals. One of my informants, expressing her opinion on the general well-being of the regional population, said: "How could people multiply well if afterbirths were thrown into the garbage for decades?" However, it should be made clear that the parental relation of the land to people is not common knowledge. Few informants could provide information on it.

Galdanova (1987) also considers the worship of caves as a vestige of the lost cult of the goddess *Etugen*. Caves were seen as vulvae of the land. They also were attributed with the ability to bestow fertility on living beings. Today, one of the most powerful sacred objects in Tunka and Oka associated with the ability to bestow fertility is the upright pillar of puddingstone (a conglomerate rock with rounded pebbles) located far in the mountains. The worship of these pillars is linked to the cult of the mountains, which in turn pertains to the ancient cult of the Mother Earth.

The mountain cult of the Turkic people and of the Mongols was also recorded by early Mongolian and Chinese chronicles (Potapov, 1946) and revealed by archeological data. It is continued in contemporary practice that relates directly to the ancient cult of the land (Menes, 1986). The mountain landscape overall is considered sacred, while certain mountains in particular are perceived as especially powerful. It can hardly be distinguished from the notion of *Oron Delkhei* or *Khangai*. "*Oron Delkhei* is everything, everything in the world"—local people repeatedly explained to me. *Oron Delkhei Ezhin* is understood to be the patron of the world and was referred to by several informants in Oka as the one they appeal to with milk libations in their daily morning prayer. *Oron Delkhei Ezhin* is also said to be one of the deities appealed to by *khadashi*, elderly men mediating relations between people and mountain deities through religious rituals. *Oron* in standard Buryat means bed. Informants in Oka interpreted *oron* in a religious context to mean the universal cradle. One of them said:

> *Oron* is a bed and a cradle. When I was a girl, one elderly man told me that the land is our cradle. *Oron Delkhei* nourishes all of us, gives us strength and we can sleep on the ground. That is why it is a cradle for the universe. It is a synonym for *Khangai* (Buryat—taiga and the master of it). It is nature—our mother.

One younger informant directly associated *oron* as a place of birth, the native land: "*Oron* is where you were born—the native land. *Oron Delkhei Ezhin* is the patron of the native land."

"We say *khada* (in Buryat, a mountain) and *Khangai* (in Buryat, taiga/the master of taiga)—it is all one," my informants noted disapprovingly when I attempted to find out the difference between the two words. The wholeness of the landscape as perceived by the local population is represented in the name *Khada Khangai Ezhin* (in Buryat, the deity of the mountains and the taiga).

According to local belief, soil taken from a holy mountain possesses the power to protect people and their homes. This can be interpreted as a sign of the link between the mountain cult and the belief in the holiness of land. The Sayan Mountain range is considered to be sacred and not only for the local population. It is associated with the residence of the highest deities of the all-Buryat pantheon (Gerasimova, 1981, p. 17). However, the belief in mountain masters among the lay population of the research sites coexists with the understanding of the sacred power of the physical mountain. For local people in the research area, mountains and mountain masters are the deities themselves. They control peoples' fates, protect, and vindicate. The mountains' power spreads out to all spheres of people's lives. Some are appealed to only for specific concerns, while others are all-powerful.

The concept of land continuity as the unity of the soil, forest with the entire fauna and flora, mountains, rivers, springs, lakes, etc. also vividly declares itself through the cult of the spiritual masters of locales, patrons to everything that the locality includes. The cult of the spiritual masters of locales is undoubtedly connected with the sacred attitude toward land. The entire territory "known" by local people is viewed as being inhabited by the spiritual masters of canyons, valleys, and smaller locales. These masters require respect, offerings, and the observance of behavioral rules as all other spirits do.

16.5 Attitudes Toward Land and the Influence of Buddhism

Mahayāna ("Great vehicle") Buddhism penetrated the southern Siberian margins of the Russian Empire in approximately the seventeenth century (Abaeva, 2004, pp. 400–401; Galdanova et al., 1983, pp. 12–26). It spread to the study area only in the nineteenth century when the first *datsan* (Buddhist monastery) was built, incorporating local religious practices and beliefs. Its impact on the local population's behavior and worldview is still significant. Buddhist philosophy and practices do not contradict the pre-Buddhist understanding of the world in the research region—that is, the concept of the world as wholeness, as a tangle of subtle interrelations. Furthermore, there is no contradiction between the Buddhist idea concerning land and the pre-Buddhist attitude toward land held by local populations. This allowed a syncretism between the two religious concepts for laymen.

The deities of the land—the goddess of land (*Sa'i lha mo*), the spirit guardians of the land (*sabdaks*), and the four guardians of the world's cardinal points—are described in *Mahayāna* Buddhist tractates and addressed by the rituals propitiating the deities of the land (Gerasimova, 1989). These rituals precede all actions requiring a disruption of the land surface, such as the construction of houses, fencing, grave digging, and tillage. The purpose of the rituals is to ask the deities' permission for the necessary intrusion. Any unnecessary intrusion is considered to be *serte* (in Buryat, a notion that may be translated most closely as sin or taboo).

The explanation of illnesses and failures, well-being and ill-being, poverty, and prosperity through personal interrelations with the macrocosm is incorporated into both religious concepts and is embraced by the population. For many, it is the primary factor defining their relations with the environment.

Buddhist philosophy and its complex understanding of the world, however, are not fully comprehended by laymen. Continuous communication with *lamas* (the Buddhist clergy), providing advice on appropriate personal behavior for any given time or environment, shapes laymen's behavior. Such guidance is based on the philosophy of reciprocal influence of all world elements and processes.

As explained by local Buddhist clergy, there are five *mahābhūtas* (Sanskrit)—the subtle notions, known as earth, space (or "ether"), water, fire, and wind (or air). Various combinations of these *mahābhūtas*, fluctuating in quality and quantity, form the body tissue of living beings, human speech, thought, and the environment in its entirety. The interaction of the outer and inner substances of a person may be in harmony only when they conform to each other—in other words, if the fundamental characteristics of the inner mahābhūtas correspond with the fundamental characteristics of the outer *mahābhūtas*. The interdependence of inner and outer substances is regulated by energies, promoting health or causing illness.

Time is one more component that is integral to the understanding of the interdependence of world elements. The sequence of *mahābhūtas* corresponding to the life cycle of individuals and the cycle of bio-energy of consciousness resonate, intensify, or suppress the effects of each other in a certain point of time (Abaev, 1992, pp. 80–81, 94).

Using their own intuition and taking into consideration all the subtle relations of time and the inner and outer substances of a person, lamas define propitious and unpropitious periods of time for a particular individual's actions. For example, if soil is the determining element for a person for the new lunar year, a lama may advise a believer not to dig soil for the duration of the year, not to uproot any plants, not to move stones, etc. Following these simple rules is believed to help the person to maintain the harmony of the micro- and macrocosm.

Representatives of many local households visit *lamas* at the beginning of a lunar year in order to learn how each household member should behave to maintain his or her personal balance of time and astrological qualities. Besides personal behavior patterns, rituals performed by lamas also serve to harmonize world elements and minimize the possible negative impact of such on an individual. Believers are normally advised to place a *khii-morin* (Buryat *khii*, wind; *morin*, horse) at special sacred places; these are the colored flags symbolizing the elements of the universe and containing the ability to mitigate relations between an individual, the macrocosm, and time (see Fig. 16.2). The flag, each believer is instructed, should be of a particular colored textile and a particular colored print. It should be placed facing a particular cardinal point.

An extract from an interview with one of my informants may well illustrate the perceived logic of the universe by local people:

> According to the Buddhist Buryat lunar calendar, there is [a] 12 year [cycle]. Each year has a meaning for everybody. Everybody [knows] his year [according to birth]. One year one should not dig the ground. I believe in this. One may [dig] for eleven years but should not for [the twelfth]. I may not dig even to put a pole for a fence—it would be bad for me. You'd better hire somebody [for digging], but you should not do it yourself. Some should not cut trees. Even to cut trees to make firewood is forbidden... First, you go to a *datsan*. They [*lamas*] will look and read. If they say not to dig, I should not even plant potato. I am following this. You may harm your life. Losses, wastes, misfortunes, getting sick or cutting your leg with an axe—various things can happen. They [the rules] should be followed....

Fig. 16.2 A *khii-morin* above a sheep shelter to protect the animals, 2007

Since the 1990s, various newspapers and calendars publish advice regarding more general everyday actions, indicating favorable or unfavorable days for certain deeds. Although these are taken less seriously by some informants, others said that they follow the advice. One of the widely known regional newspapers—"InformPolis"—publishes Buddhist calendars weekly with daily advice for the believers to follow. The following text, published online at the website of the newspaper (www.infpol.ru) on May 28, 2008, is a typical example of such recommendation:

> Tuesday, 3 June (29th day according to lunar calendar), the day of a blue dog, three blue *menge*, element—wood. It is good to believe and pray, to learn the alphabet and astrology, to do good to people, to suppress enemies, to plant trees and to make a foundation for a house. It is bad (Russian *nel'zya*) to dig land, pay back a debt and cut wood. Haircutting will result in a separation of the soul from the body and its getting lost.

Today, pre-Buddhist and Buddhist religious traditions form an inseparable whole within the local belief system. The powerful patrons of the local people and of their herds include the mountains and their spirits, the spirit of the household fire, masters and mistresses of locations, rivers, and caves. Repeated libations and offerings are performed daily. These start early in the morning when fresh milk, tea, or butter are offered to spirits. Hunting, herding, and other production activities as well as traveling involve numerous offerings and rituals directed toward the spiritual world. For more significant occasions, rituals are performed by *lamas* (Buddhist monks), shamans, or *hkadashi*,[5] including special libations for the chthonic deities.

16.6 The Soviet State and Religion

The Marxist-Leninist ideology that guided the Soviet period policies considered religion to be an "opiate of the people" and a potentially dangerous irrational superstition (Hann, 2006, p. 2). This ideology was enforced by repressing religious leaders and eliminating religious institutions. In the research area, the antireligious campaign of the 1930s is still remembered as a period when local Buddhist monasteries *datsans* were turned into public buildings and Buddhist *lamas* and shamans were arrested—many of them never to return. As a result of this campaign, religion was removed from the public sphere, concealed and domesticated within families (Hann, 2006, p. 3).

Religious policy and governmental control over the religious sphere relaxed with time. After WWII, two *datsans* were reopened in the country: one near the capital of the Autonomous Republic of Buryatia, Ulan-Ude, and another in the Aga Buryat Autonomous District. Few religious specialists provided their services to rural communities. In the late Soviet period, in districts such as Tunka and Oka that are remote from the urban centers, anti-religious policies overlapped in a peculiar way with

[5] *Khadashi* can be translated from local Buryat as one who accesses mountain deities, acting as a mediator between people and mountains.

local religious understanding. Although collective and state farms were the institutions that had to locally impose Soviet ideology and ensure people's behavioral compliance, their employees performed rituals aimed at improving collective production with the tacit agreement of the farms' management. Domestic rituals, however, remained concealed.

Freedom of religion became an integral part of democratic reforms in the 1990s. Piety no longer had to be concealed from public view and the authorities. Datsans were reconstructed or rebuilt in many locations, including the study area. In 1993, 15 *datsans* already operated in the Republic of Buryatia. All of them were built or reconstructed with funds raised by the faithful. Buddhist educational institutions were reopened in Ulan-Ude and St. Petersburg. Gradually, *lamas* appeared in many large rural settlements. Today, religious rituals are often communal, and they may be logistically and/or financially supported by the local administration.

The impact of the atheistic ideology that was inculcated by central policies for decades, however, should not be underestimated. Secular education and Soviet ideology that aimed to create a new "socialist man" free of religious prejudice raised the question of faith among the rural population—some of whom either do not believe in the supernatural or are at least doubtful. Although only 2.1% of my informants (out of 136 randomly asked at all four sites) said during the interviews that they do not believe in the supernatural, do not perform rituals, do not attend ceremonies, and do not go to religious specialists, my continuous communication with local people and participant observation of communal and family rituals indicated that sacred knowledge has been partly lost and not passed down to new generations. The desire to reconstruct this lost knowledge often leads to borrowing traditions from other locations. Knowledge of these "borrowed" traditions may be transmitted by migrating religious specialists and lay people, or through publications. These publications include newspapers and other popular materials for general audiences as well as specialized literature on religion and ethnography read by the better educated.

The lasting inculcation of atheism enforced by the Soviet state is one of the reasons for the significant variations in local people's knowledge of religious canons and practices. Another results from the onrush of religious revival, which often introduces religious practices native to other districts. On the other hand, some local religious traditions preserved by a few elderly people remain unknown to the broader population.

Many informants in Oka admit that in performing daily libations to the mountains, they do not name all of them, as they do not know all thirteen of the holy ones. They refer to "all thirteen holy mountains." Some mention the most important one and then add "and all other holy mountains ...." It is very important not to aggrieve a mountain by not naming it during the libation or by naming it wrongly. Some people prefer not to perform the ritual, being unsure that they know the entire list of mountains or necessary details of the rite sufficiently well.

16.7 We Will All Have a Patch the Size of a Sleeping Felt

Among local people, there is a belief that arguing over land brings misfortune and death. This belief is widespread in all the research communities. As mentioned, 270 informants were interviewed on the issue. Only 11 of them (three of whom were not native to the communities) said that they did not previously know about the taboo. The other 259 were familiar with the concept. Elderly informants, however, repeatedly stated that the tradition is getting weaker today.

Some recurring reasons people offered to explain why it is "bad" to argue over land follow here: "The land is one whole" [thus should not be divided]. "It can belong to nobody." "A man did not create land and so cannot own it." "The land does not belong to us, we belong to it." Several informants explained that land "is the mother," that "the mother earth (Buryat, *gazar ekhe or delkhei ekhe*) feeds all of us," that "what it provides is always enough for all. It always will be able to support all of us well with even the smallest patch and even in foul weather." These informants emphasized the close relation between land—generous, rich, and prosperous—and the community. The land was often presented as a decision-maker and actor. "The land will remind one [if he/she seized land in conflict]: 'it is not yours, don't take it, everybody is allotted a patch to be buried in'"—one of the interviewees explained.

Other informants explained the taboo through a need to respect deities, the spirit master of the land or the masters of locales: "Each locale has its own master. He will get angry and punish [if someone argues]. Everything has its master, …. One has to ask a master [to allocate him a patch of land] even for a grave."

According to local belief, one's burial plot is the only place that might be thought of as one's own. Its size is defined exactly by the size of "his sleeping felt"—or as people commonly put it in a modern way "one by two" [meters]. Many informants suggested that land apart from the grave cannot be private or personal, some said that it belongs to "deities," and others said that it belongs to "nature," or it is simply "common." "I heard from my childhood that a human being is not eternal; the land is eternal (Buryat proverb *Delkhe munkhe, khun munkhe beshe*). It is *serte* (Buryat, taboo, sin) [to argue over the land]"— an informant explained to me when asked why people should not argue over land. Tradition tells that those who dare to get involved in conflict over land will either die or lose their close relatives. "A whole family may be destroyed for a patch of land"—other local informants repeatedly explained to me.

Nevertheless, few interviewees revealed with embarrassment some cases of people clashing over land. They are stories from members of the communities concerning their own and their neighbors' tensions over hay land boundaries or livestock grazing. "He has cut hay off half of my patch for several years. I pay [taxes] while he takes [hay]. I told him: 'Pay then.' This year I'll cut first and exactly by the boundary. I need nothing extra."

Retelling stories of land conflict, the informants always punctuated their narrations with comments such as:

- They all left us [read "died"] [those who argued over land].
- The deities will punish [them], definitely.

- They quarreled, nothing happened but it may come later.
- It is the tradition [not to quarrel]. Two meters is enough for a person. Whoever argued over the land, as I have noticed, always had a bad outcome.
- It is for sure something bad will happen. There were examples. People are afraid [to argue over the land].

Examples of situations that could have provoked a conflict over land, but never did, were also discussed by the interviewees. Following their belief, some parties with conflicting interests chose to concede rather than get involved in a quarrel, which they believe would threaten their life or their descendants:

- I planted potatoes, the neighbor was not pleased; I kept silent, did not say a word. Since then I never planted a potato there again.
- "We should not argue"—my dad always said, if someone took hay off the patch.
- When everybody argued over the land at the meeting [in the 1990s] I stood up and said loudly: "The land is rich; it always gives enough for all. What we all will get is one by two."
- It is a sin [to argue]. I do not need more than my sleeping felt. It is a sin [also] to move somebody's livestock from a pasture. That is why I have chosen to use the land far away. The costs [of using it] are great, but I'd rather stay away and have nobody come to me.

Community opinion is also important in deciding whether to get involved in a conflict. One of my informants expressed her apprehension over being blamed for somebody's death—a death she might bring on by arguing over her former hay patch:

> We were allotted this land ... in the 1990s, but the recent head of the *sovkhoz* (Russian—Soviet farm) gave it to another man. We have no haymaking land anymore. I am thinking of cutting hay off before him next year but it is a sin for us to argue over land. Well, I will pray thoroughly before that and then go. On the other hand, I would cut it but I am frightened that if somebody dies people will say: "She argued over land."

Involvement in a conflict is often compared with "booking a place for a grave." If someone dares to argue for his/her rights over a patch of land, he/she would be told: "*Ukhekheie gazar bulialdakhymta?*" (rude local Buryat: Are you arguing over land to die?).

Some informants could not explain why arguing over land is taboo. They simply stated: "It is serte" (Buryat, taboo, sin). On the one hand, unlike an ethnographer, a member of a local community would hardly dare to ask the question "why" if he/she is told that something is taboo. It might sound like an attempt to challenge the belief. On the other hand, this quite common inability to provide information on people/land relations beyond the existence of taboo, together with repeated cases of breaking the rule, raises the question as to whether the spiritual connection between people and land has weakened.

Some local views reflect a transformation of the local belief to include both the notion of ownership and the concept suggesting that defense of one's property is the owner's right. The influence of the newly introduced concept of property let some informants to differentiate when the old belief applies. In their view, anyone with legitimate rights to land can defend his/her rights with no risk to himself or his relatives' lives or well-being. These quotations from the interviews well demonstrate that the newly introduced notion of one's land gets intervolved in the complex textile of local values:

- It is sin [to argue over land], but if somebody has taken your land, you can defend it. It is sin to take somebody's land.
- We should not argue over land and also a house or a job but if it is your land, of course you will defend your land.
- I remember our neighbors moved the fence towards us. Our granny said that we should not argue. Today I would not give my land away.

It is important to stress that more stories resulting from conflicts over land were confided in the community that is distinguished by a shortage of land, especially unequal land distribution and de facto private use of large grazing plots. Perhaps, it is not a coincidence that the only attempt to openly challenge the rule not to argue over land was mentioned in this community. An elderly informant who sounded very religious during the interview was the target of disapproving gossip in the community. According to the gossip, the informant had argued over land; when her opponent replied to her with the usual for such a situation "we all will have one by two," she answered that she would die but would be buried on her several dozens of hectares. In contrast to apprehension, uneasiness, regrets, or attempts to explain the circumstances pushing actors toward the uneasy argument over land, which always accompanied local talks about breaking the rule, this case revealed local person's antagonism toward the belief.

Speculating on the reasons for increasing conflicts over land at this community, a *lama* from a local *datsan* sharply pointed out the central issue of injustice in land allocation as the reason for de facto enclosure (Fig. 16.3) and exclusion:

> Unfair allotment of the collective's land is the reason. Everybody wished to receive land near the village [but many did not][6]; as a result, the grass there [near the village] gets used up quickly. So the owners started to fence their plots to preserve the grass for their own livestock.

The unequal distribution of land during the post-socialist privatization process stipulated by the centrally defined policies and reinforced by the corrupted local governance significantly enhanced social inequality in the community creating a stratum of rural poor and landless or almost landless. The inner attempt to oppose the injustice of unequal distribution of land is one factor pushing people to breaking the rule set by the belief. Few informants from the community with the extremely unequal land distribution emphasized the necessity for those who were deprived of their shares during the privatization or whose shares were determined to be unfairly small to argue in defense of their livelihood. "If you argue, you will get it [meaning—you will die]. But these days [people] will argue anyway. Because it is not enough [land], somebody got [a] better [patch], another did not get any. Some people are in tears—they did not get any [land] at all," said one of them.

[6]For details of land distribution in the community, please, see Intigrinova, 2010, pp. 183–184.

Fig. 16.3 Local landscape fenced into individual plots, 2006

16.8 Piety, Doubts, and the Perception of Risks

It also should be noted here that individualization of land use as a general trend in post-socialist policies in Russia has resulted in a number of outside businesses securing certain patches of land in the research area. Mostly this involves land rich in minerals or attractive for tourists. Local interaction with such projects also plays an important role in the complexity of factors influencing sustainability of local spiritual relations with land. They not only clash with local interests over land but they also create conditions challenging local beliefs. One case I observed illustrates this point particularly well.

The case in question concerns a conflict over a sacred site containing unique mineral springs. Access to the springs used to be free for anyone able to make the trip deep into the mountains. The spirit master and two mistresses of the location had always been the only authority from whom local people asked permission to be there; to these spirits, they paid tribute with prayers and libations.

In 2006, a city businessman leased forest land near the springs to develop a tourist resort and attempted to bring things at the springs "in order," according to his own understanding. This understanding did not accept the old log cabins built by local collective farms and populations, their dogs wandering around the territory and horses allowed to heal their battered legs in muddy mineral substances near the resort.

The businessman's attempt to clean the territory of garbage was not in its turn appreciated by the population. They were more concerned with the fact that the businessman's clients, arriving by helicopter, ignored local rules. This conflict of views challenged the local perception of the relationship between deities and people. A young local woman expressed her concern about the situation at the springs in the following way:

> When I was a girl, we could do nothing there. We were prohibited by our parents to laugh and speak aloud. Even brushing our hair, we had to pick up every single hair that fell out and put them between logs in the cabins. Our parents always said that the master spirit would punish us if we misbehave. But now they bathe naked, use bath foam in the mineral baths, scream, laugh and misbehave in many other ways. Nothing happens. Has he [the master] gone blind and deaf?

Her doubts in the bond between people's behavior and deities' benevolence, which underpin local perception of risk and determine the local ways of risk mitigation, allowed the young person to question her adhesion to the local norms of behavior.

The above description of local beliefs demonstrates that breaking religious rules or not following the advice of religious specialists is perceived locally as being risky, possibly resulting in various misfortunes or even fatal outcomes. General well-being and the outcome of various economic activities are perceived to depend on the benevolence of the deities. This loyalty in its turn depends on people's behavior, both personal and collective. Appealing to the spiritual world for protection is one of the local ways to mitigate risks. Rituals performed daily and on special occasions fill the life of local people with constant interaction with the supernatural that "watches" them, protecting or punishing. Every rite, according to local understanding, decreases personal or production risks while improving the actor's fortune.

Another way to maintain personal relations with the macrocosm in balance is to follow religious rules and taboos. The relations between the human microcosm and the spiritual macrocosm determine successes or failures of a believer and his/her well-being. For example, personal health in local perception is interdependent with personal behavior. In medicine based on Buddhist tradition, one-quarter of illnesses are presumptively explained by personal misconduct (Gerasimova, 1989, p. 182).

A ritual or following religious prescriptions are considered to be a risk aversion strategy by an actor who accepts the concept of the universe described above. Following rules prescribed by the religious beliefs and performing the rituals required of them are seen by those accepting the spiritual connections between people and the universe as a safeguard of their livelihood stability. However, for an actor who does not perceive the cosmos this way, ritual does not represent a means of risk management, nor does failure to follow a religious belief represent a risk. When asked about religious issues, the daughter of a deceased state farm director who was often blamed by local informants for privatizing the farm "into one pair of hands" replied: "We rely only on ourselves." It remains a question if those who doubt the deities' verity will choose to follow the behavioral pattern prescribed by a religious belief or prefer to rely on themselves.

16.9 Concluding Remarks

Spiritual relations with chthonic deities have underpinned the norms of local people's behavior for centuries. Although significantly weakened during the Soviet period, these relations have been maintained in the research area. The weakening of the spiritual relations is reflected in the increased diversification of local actors' views on how the universe is constructed and, thus, on how one should build relations with it. Such increase was formed by the clash of Soviet ideology and secular education on the one side and enduring traditional relations with the environment on the other.

Soviet antireligious ideology and repressions pushed religion from the community life to the domestic sphere of local households (Hann, 2006, p. 3), but did not eliminate it. Schools and collective/state farms had to locally impose Soviet ideology and ensure people's behavioral compliance. But as stated above, in the late Soviet period, these institutions, staffed with local people, did not actually pursue this goal. The remoteness of the study area contributed to the preservation of the local people's religious perception of the universe. The Soviet ideology itself, although anti-religious, did not contradict the main local notion of land as a totality that does not belong to a human being.

The post-Soviet policies oriented toward land privatization and development of the market economy in general and land market in particular encouraged property consciousness and fragmentation of land. Gradually, more and more people in the research area are getting involved in the actual process of land privatization and titling, even though the notion of private land imposed by the centrally defined policy and partitioning of land contradict the local religious understanding of land as a totality.

This chapter presented several examples of changes in local beliefs. Some new details are introduced through theological developments by the clergy. Certain religious beliefs weaken if real-life events contradict them. Others are reinterpreted, incorporating new notions introduced by state policies. As mentioned above, some local people incorporate their perception of social justice, which dictates them to stand for owners' rights to a patch. For others, the injustice of the post-Soviet land distribution, which has been extremely unequal and unfair, raises the question of whether the deities will be benignant to the poor and landless. I emphasize these changes and challenges to raise the question of the resilience of the traditional understanding of land and people relations. I wonder whether with further increase in the scale of the private landownership and land commoditization the spiritual relations between land and people will weaken even further.

Under modern conditions, when land privatization is imposed by state policies as a precondition to viable production, the only application of local beliefs that land cannot belong to people is expressed in people's attitude toward their right to own land. Receiving land from the state, different actors understand their relations with it in different ways. Some perceive their rights as allowing them to dwell on and use the patch. Others accept their owners' rights to pass their patches to heirs and to exclude others requesting neighbors even not to walk through it. At the time of the field research, people in the area still felt uncomfortable talking about land sale. None of my informants admitted that they might sell land; only few precariously noted that they might consider buying a patch.

Peoples' spiritual relations with the totality of land in the research area assume their well-being as a result of the balance in these relations, in which people belong to earth and should care for it but also will be cared for by the earth. Such a notion seems incompatible with private property, unless such rights to land are seen as a formality.

Acknowledgments I would like to express my gratitude to the people of Tunka and Oka for welcoming me to their communities and sharing their life experiences. I sincerely thank Professor Katherine Homewood and Professor Philip Burnham at the Department of Anthropology, University College London (UCL), for their generous supervision of my research and the editors of this volume for critical review of the earlier version of the chapter. I also gratefully acknowledge financial support by the Dorothy Hodgkin Postgraduate Award (2005–2008), the Parkes Foundation Small Grants Fund (2005), Chevening scholarship award (2002–2003), and field research grant (2003) at UCL Department of Anthropology.

References

Abaev, N. (Ed.). (1992). *Environmental traditions in cultures of the people of Inner Asia*. Nauka. [Published in Russian] Абаев, Н. (ред.) (1992). Экологические традиции в культуре Центральной Азии. Наука, Новосибирск.

Abaeva, L. L. (2004). History of Buddhism spread in Buryatia. In L. L. Abaeva, & N. L. Zhukovskaya (Eds.), *Buryats* (pp. 397–415). Institute of Ethnology and Anthropology/N.N. Mikluhko-Maklaia. [Published in Russian] Абаева, Л.Л. (2004). История распространения буддизма в Бурятии. Абаева, Л.Л., Жуковская, Н.Л. (ред.): Буряты. Институт этнологии и антропологии им. Н.Н. Миклухо-Маклая, 397–415, Москва.

Batoeva, D. B., Galdanova, G. R., Nikolaeva, D. A., & Skrynnikova, T. D. (2002). *Rituals in traditional culture of the Buryats*. Vostochnaia literatura. [Published in Russia] Батоева, Д.В., Галданова Г.Р., Николаева, Д.А., Скрынникова, Т.Д. (2002). Обряды в традиционной культуре бурят. Восточная литература, Москва.

Galdanova, G. R. (1987). *Pre-lamaist beliefs of the Buryats*. Nauka. [Published in Russian] Галданова, Г.Р. (1987). Доламаистические верования бурят. Наука, Новосибирск.

Galdanova, G. R., Gerasimova, K. M., Dashiev, D. B., & Mitupov, G. T. S. (1983). *Lamaism in Buryatia from the 18th to the early 20th century*. Nauka. [Published in Russian] Галданова, Г.Р., Герасимова, К.М., Дашиев, Д.Б., Митупов, Г.Ц. (1983). Ламаизм в Бурятии 18—начало 20 века. Наука, Новосибирск.

Gerasimova, K. M. (1981). Syncretism of the Dalkha Cult. In K. M. Gerasimova (Ed.), *Buddhism and traditional beliefs of the people of Inner Asia* (pp. 7–45). Nauka. [Published in Russian] Герасимова, К.М. (1981). Синкретизм культа Далха. Герасимова К.М. (ред.): Буддизм и традиционные верования народов Центральной Азии. Наука, 7–45, Новосибирск.

Gerasimova, K. M. (1989). Worshiping masters of land and location. In K. M. Gerasimova (Ed.), *Traditional beliefs of Tibet people in the lamaistic cults* (pp. 189–218). Nauka. [Published in Russia] Герасимова, К.М. (1989). Покровители людей из разряда «хозяев» божеств земли и местности. В Герасимова К.М.: Традиционные верования тибетцев в культовой системе ламаизма. Наука, 189–218, Новосибирск.

Grandin, B. (1986). *Wealth ranking in smallholder communities: A field manual*. Intermediate Technology Publications.

Hann, C. (2006). Introduction: Faith, power and civility after socialism. In C. H. Hann & the 'Civil Religion' Group (Eds.), *The postsocialist religious question: Faith and power in Central Asia and East-Central Europe* (pp. 1–26). Lit, Münster.

Intigrinova, T. (2010). Social inequality and risk mitigation in the era of private land: Siberian pastoralists and land use change. *Pastoralism, 1*(2), 178–197. https://doi.org/10.3362/2041-7136.2010.011

Intigrinova, T. (2015). Production efficiency and beyond: Local determinants in Russian privatisation. *Europe-Asia Studies, 67*(9), 1394–1415. https://doi.org/10.1080/09668136.2015.1085494

Jordan, P. (2004). Ethnic survival and the Siberian Khanty: On-going transformations in seasonal mobility and traditional culture. *Nomadic Peoples, 8*(1), 17–42.

Kasten, E. (Ed.). (2002). *People and the land pathways to reform in post-Soviet Siberia*. Dietrich Reimer Verlag.

Menes, G. (1986). About semantics of the theonym Ul'gen. In *Exploring the historic ethnography of Mongol peoples*. Buryat Branch of the USSR Academy of Science. [Published in Russian] Мэнэс, Г. (1986). О семантике теонима Ульген. Исследования по исторической этнографии монгольских народов. Из-во БФ СО АН СССР, Улан-Удэ.

Mikhailov, T. M. (1980). Ethnic characteristics of the shamanist pantheon of the Buryats (the 19th and 20th century). In T. M. Mikhailov (Ed.), *Buryats livelihood nowadays and in the past* (pp. 73–99). USSR Academy of Science, Siberian Branch, Institute of Social Sciences. [Published in Russia] Михайлов, Т.М. (1980). Этническая характеристика шаманского пантеона бурят (19–20вв). Т.М. Михайлов (ред.): Быт бурят в настоящем и прошлом. Академия наук СССР, Сибирское отделение, 73–99, Улан-Удэ.

Pavlinskaya, L. (2002). *Travelers of the blue mountains: The fate of the traditional culture of the people of Eastern Sayans in the context of its interrelation with the modernity*. Evropeiskiy Dom, St. Petersburg. [Published in Russian] Павлинская, Л. (2002). Кочевники голубых гор: судьба традиционной культуры народов Восточных Саян в контексте взаимодействия с современностью. Европейский Дом, Санкт-Петербург.

Peters, P. E. (1998). The erosion of commons and the emergence of property: Problems for social analysis. In R. C. Hunt & A. Gilman (Eds.), *Property in economic context* (pp. 351–378). University Press of America.

Potapov, L. P. (1946). The cult of mountains in Altai. *Sovetskaia Etnographia, 2*, 145–160. [Published in Russian] Потапов, Л.П. (1946). Культ гор на Алтае. Советская этнография. Sovetskaia Etnographia, 2, 145–160.

Ritter, K. (1894). *The earth science of Asia: The geography of the countries located in Asian Russia and nearby*. Eastern Siberia, lake Baikal, Transbaikalia and Gobi Steppe. St. Petersburg, Vol. 6, Part 1. [Published in Russian] Риттер, К. (1894). Землеведение Азии: география стран входящих в состав Азиатской России или пограничных с нею, Восточная Сибирь, озеро Байкал и приграничные страны, Забайкалье и степь Гоби. Санкт-Петербург. Выпуск 6, Часть 1.

Ziker, J. P. (2003). "Horseradish is no sweeter than turnips." Entitlement and sustainability in the Taimyr Autonomous Region, Northern Russia. In: C. Hann & the 'Property Relations' Group (Eds.), *The postsocialist agrarian question: Property relations and the rural condition* (pp. 361–390). Lit Verlag.

Part IV
Case Studies on the Inner Soil of Individual Scientists

Chapter 17
"The Salt of the Earth": Bernard Palissy (1510–1590) Between Soil Science and Religion

Christian Feller, Jean-Paul Aeschlimann, and Nikola Patzel

> ***Vous estes le sel de la terre**: or si le sel perd sa saveur, avec quoi le salera on? Il ne vaut plus rien, sinon pour estre jetté dehors, & estre foulé des hommes ("You are the salt of the earth. But if the salt loses its saltiness, how can it be made salty again? It is no longer good for anything, except to be thrown out and trampled underfoot." (author's translation) Olivétan Bible, first protestant Bible in ancient French published in 1538; Pierre Robert Olivétan (1506–1538) was a cousin of Jean Calvin.)*
>
> Matthieu (Matthew) 5.13

17.1 Introduction

Since the end of the nineteenth century, Palissy has been considered in France as one of the nation's great men (De Corgnac, ~1880; Desfontaines, ~1880; Du Chatenet, ~1880; Figuier, 1868; Jonveaux, 1874) because of his rediscovery of the white enamel craft and for being a forerunner in fields of natural sciences, including both soil and agricultural sciences, that became mainstream sciences much later.

Two authors of the present chapter have already devoted several articles to Palissy (Feller et al., 2006, 2014) concentrating in particular on his scientific contributions to

C. Feller (✉)
Institut de Recherche pour le Développement (IRD), UMR Eco & Sols, INRA-IRD-SupAgro, Montpellier, France
e-mail: christian.feller@ird.fr

J.-P. Aeschlimann
Agropolis Museum, Montpellier, France

N. Patzel
Freelance Scientist, Office for Soil Communications, Überlingen, Germany

N. Patzel et al. (eds.), *Cultural Understanding of Soils*,
https://doi.org/10.1007/978-3-031-13169-1_17

their respective fields of research, pedology, and agriculture. The existing literature has primarily addressed single aspects of Palissy's work and formed, to a great extent, a corpus of projections by authors rather than striving to depict a historical figure of science for their own scientific theory and personal ideology.

A new edition (2010) of Palissy's C*omplete Works* (*Œuvres complètes*) was issued a decade ago (Fragonard et al., 2010) along with an exceptional body of notes and comments, which consider the activities of Palissy in the framework of the sixteenth century. Based on that, this chapter represents an attempt to (i) assess the ambiguous importance of religion in Palissy's writings and (ii) to determine if there was a core value in a career that appears to have been split into lots of different topics. Indeed, elaborating a comprehensive system of nature could well have been the final aim pursued during his entire life's work, an idea that has been suggested by some authors without being subjected to an in-depth analysis, except for Lamy (2018), who addressed Palissy's attitude between the *Bible* and Nature.

17.2 On the Biography of Palissy[1]

During most of the twentieth century, French school children were routinely told that Palissy had been one of the nation's great men (Fig. 17.1). Lacking firewood to finalize a crucial experiment while trying to rediscover the secrets of making white enamel, he indeed had gone so far as to sacrifice his own furniture and wooden floor to maintain the needed heat in his potter's oven. In teaching this story, Palissy was made a real hero of the scientific mind, to whom the progress of technology was worth big sacrifices. Thus, the image of the ceramist burning his personal belongings in an oven before the eyes of his appalled family has actually illustrated many former French school textbooks (Fig. 17.2).

Palissy succeeded in creating a variety of enamels that mimic nature (Fig. 17.3), which was a phenomenal success and still has many followers to the present day (Amico, 1996; Viennet & Starosta, 2010). What the school textbooks did not mention, however, was that this great French scientist was also one of the first to adhere to the Protestant reformation. As a disciple of Calvin (Huguenot), he barely escaped death on two occasions, once in 1563 while imprisoned in Bordeaux and again in 1572 when he avoided the Saint Bartholomew's massacres in Paris thanks to the protection offered by Catherine de Medicis. He nevertheless died in 1590 in the Bastille prison because he was never prepared to recant his faith. Two hundred years after his death, the French revolution was the first strong wave aiming at secularization of most European states and establishing a "cult of rationality" (Aulard, 1892), which would be more important in political and daily affairs than God.

However, considered to be a great artist and scholar during his lifetime, Palissy was protected by the monarchy and the Catholic nobility due to the royal privilege

[1] See Poirier (2008) for additional details.

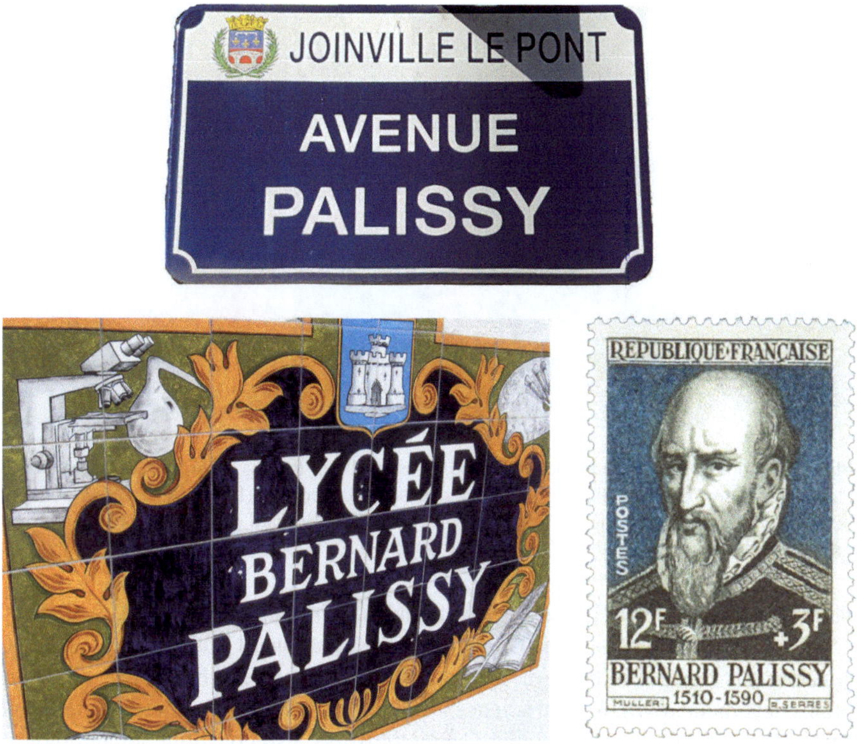

Fig. 17.1 Images that show Bernard Palissy's popularity in France: Street and high school names and postage stamps. (**a**) https://commons.wikimedia.org/wiki/File:Joinville_av._Palissy_plaque.jpg. (**b**) https://lyceepalissygien.fr/accueil/presentation-contact/. (**c**) https://www.phil-ouest.com/Timbre.php?Nom_timbre=B_Palissy_1957

of "Inventeur des Rustiques Figulines du Roy" ("Inventor of the King's Rustic Figurines") that had been bestowed upon him. Every noble in the French kingdom had to acquire some Palissy crockery or dreamed to let him construct a private grotto.

He was very proud of being self-educated, allegedly unable to understand Latin and Greek. Ostensibly ignorant of the ancient scholars, he therefore apparently only trusted his own field observations. As a really exceptional scrutinizer of the environment, he developed a comprehensive knowledge in natural sciences after 1530 during an extensive voyage across France that lasted several years. From 1575 to 1584, Palissy delivered regular scientific conferences in the frame of what he called "my little academy" (Fragonard et al., 2010, *Admirable Discourse,* p. 251), which were attended by the Parisian high society (Fig. 17.4). His experiences and theories were published in three books, that is, *Rustic Grotto* (RG from here on) (Palissy, 1563a),

Fig. 17.2 Popular images in France (left) and Japan (right) of the legend of Bernard Palissy burning his furniture. (**a**) Left. Poster. C. Feller's personal collection. (**b**) Right. Creative common license. https://commons.wikimedia.org/wiki/Category:Bernard_Palissy?uselang=fr#/media/File:Bernard_Palissy,_inventor_of_enamelled_pottery_LCCN2002700144.jpg

Veritable Recipe (VR) (Palissy, 1563b), and *Admirable Discourse* (AD) (Palissy, 1580),[2] later assembled into the *Complete Works* (Fragonard et al., 2010). While RG concentrates on the findings of the ceramist and engineer, the second and third books deal mainly with natural sciences: water, stone, earth, salt, metal, garden, and agriculture. On the other hand, the *Complete Works* also constitute an autobiography, which provides the essence of our knowledge on the life of the author.

Based on these writings, science historians have generally rated Palissy as a tremendous scientist, a forerunner in disciplines like geology, paleontology, mineralogy, crystallography (Cap, 1844 p. i–xxxix, Orcel, 1961 p. 1–17), agriculture (Grandeau, 1879 p. 32–37), pedology (Boulaine, 1989 p. 28–30, 1992 p. 134–135, Feller et al., 2006, 2014), grottos, gardens (Lamy, 2018), fountains and water cycle,

---

[2] Full French titles of the three books (later assembled in the *Complete Works*. See Fragonard et al. 2010):

Grotte rustique (1563) *Architecture et ordonnance de la grotte rustique de Monseigneur le Duc de Montmorency, Pair & Connestable de France.*

Recepte véritable (1563) *Par laquelle tous les hommes de la France pourront apprendre à multiplier et augmenter leurs Thresors. Item, ceux qui jamais eu cognoissance des lettres, pourront apprendre une Philosophie nécessaire à tous les habitans de la terre. Item, en ce livre est contenu le dessein d'un jardin autant délectable et d'utile invention, qu'il en fut onques veu. Item, le dessein et ordonnance d'une Ville de forteresse, la plus imprenable qu'homme ouyt jamais parlé*

Discours admirables de la nature des eaux et fonteines, tant naturelles qu'artificielle, des metaux, des sels & salines, des pierres, des terres, du feu & des emaux (1580). *Avec plusieurs autres excellents secrets des choses naturelles. Plus un traité de la Marne, fort utile & nécessaire, pour ceux qui se mellent de l'agriculture. Le tout dressé par dialogues, esquels sont introduits la théorique & la practique.*

Fig. 17.3 An example of a ceramic dish by Bernard Palissy. Collection of the castle of Écouen (France). Creative commons license: https://commons.wikimedia.org/wiki/Category:Works_by_Bernard_Palissy#/media/File:%C3%89couen_(95),_ch%C3%A2teau,_2e_%C3%A9tage,_salle_de_c%C3%A9ramique_fran%C3%A7aise,_plat_rustique,_Bernard_Palissy,_vers_1565.jpg

and chemistry (Hoefer, 1843 p. 72–98). His opinions on alchemy were largely discussed by Céard (2010).

17.3 God and Nature in the *Complete Works*

Several terms are frequently mentioned in the *Complete Works* (Fragonard et al., 2010). Table 17.1 provides a summary of the total number of occasions the keywords "earth," "stone," "water," "salt," and "metal" for the physical aspects, and "nature" and "religion" for the theological ones appear in the text.

17.3.1 Occurrence of Some Key Words

The occurrence of these seven terms was assessed by simply totaling the number of pages in which each of them appeared in the whole book. The sums obtained are given separately for each of the three volumes, which together represent a total of

Fig. 17.4 Palissy lecturing at his "little academy." Creative commons license: https://commons.wikimedia.org/wiki/Category:Bernard_Palissy?uselang=fr#/media/File:Bernard_Palissy._Wood_engraving_by_C._Maurano_after_E._Morin_Wellcome_V0004434.jpg

Table 17.1 Occurrence of seven keywords in the *Complete Works*

Book title (number of pages)	E Earth (or earth)	W Water	S Stone	N Nature	S Salt	R Religion	M Metal
Rustic Grotto (33)	5	2	8	16	0	3	0
Veritable Recipe (147)	56	44	40	42	28	64	5
Admirable Discourse (299)	171	179	112	94	88	46	60
Complete Works (479)	232	225	160	152	116	113	65

Issued by Fragonard et al. (2010) rated as the number of pages on which each of them—or an equivalent (Examples of "equivalent": Religion (God, Lord, *Bible,* psalm, gospel, evangelist, etc.), nature (natural), earth (it generally means soil or soil material), stone (pebble, rock), water (humid, liquid), salt (different kinds of chemical and organic salts, saltiness, salty, saline), metal (different kinds of metals: gold, iron, mercury, antimony, etc.))—is mentioned at least once

479 pages, retaining Palissy's own wording only, that is, excluding all the editors' notes and comments.

"Earth" has the highest page incidence rate of all chosen words, and "salt" showed about half as often as "earth" in Palissy's work. The word "earth" is essentially used to designate the ground/soil material under our feet. Earth, in the sense of planet, is rarely used unless in biblical quotes, such as "Heaven and Earth." As a first approximation, we can therefore assume the word "earth" as an equivalent of "soil."

Palissy devotes many pages to soil and agriculture and even forestry. Thus, Palissy is the first to mention and describe the soil auger in order to search for marl (which is deep underground) in order to be able to extract it for use as "fertilizer" (Feller et al., 2006). Likewise, for fertilization, he discusses the role, quality, technology, and management of manure as fertilizer extensively (Feller et al., 2014). In this chapter, it is not the soil itself that is discussed, but one of its attributes: "the salt of the earth," all the "salts," according to Palissy, coming from the soil (see this chapter, Sect. 17.4.1, sentence in bold).

"Earth" and "water" are present almost every second page, followed by "stone" and "nature," whereas both "salt" and "religion" show up every fourth page. Despite the fact that alchemy was highly prominent in the scientific world of that period "metal" is by far the least frequent, occurring mainly (on 39/65 occasions) in AD, and especially in the *Treaty of the Metals and Alchemy* (43 p.) and the *Treaty of Potable Gold* (10 p.). One should nevertheless bear in mind that ceramicists had much in common with the material side of European alchemy: both tried to transmute primordial substances from earth by following a recipe in order to get new incorruptible substances.

For a book dealing with natural history, the high frequency of "earth," "water," and "stone" is not really surprising. However, in regard to the terms "salt" and "religion," both occur with considerable frequency in the *Complete Works*. The author probably tried herewith to describe nature as an organized system based on the main agents God and nature, water, and salt.

17.3.2 God Has Created, Nature Is Acting

Palissy held that the whole universe is obviously God's creation; therefore, natural laws are also part of that creation. He frequently reproduced sentences from Psalm 104 or paraphrased citations from Genesis on the splendors of creation[3]:

> … God made the sky and he made the earth. He also made every stone and not a single one has been made ever since… everything had been made at the beginning of the world's creation (VR, p. 132).

The original biblical passage for *Genesis* verse 1 is:

> "In the beginning, God created the heavens and the earth."[4] There is nothing about "stone," etc.

On the other hand, Palissy evidently developed an in-depth perception of the main cycles and transformations of the elements (which would nowadays be called biogeochemical) on the earth's surface:

[3] The various sentences excerpted from Palissy's original work in ancient French are only given here in a modern English translation.

[4] In https://www.biblegateway.com/passage/?search=Genesis+1&version=ESV. Genesis 1 English Standard Version (ESV).

> God did not make nature for remaining inactive, but ordered it to work, produce, engender, consume and destroy (AD, p. 329).

This means that Palissy did not see God the Father permanently maintaining and vivifying or even continuously creating creation. Rather, Palissy adopted the view that the God of the Old Testament created everything once and then retired from His creation and let it be. This is of highest importance, because it has been the precondition of the modern unfolding scientific paradigm that God is out of the game, and nature is acting on its own.

Many of his pages reiterate this idea of a nature that does not remain idle (VR, p. 132, 150; DA, p. 286, 423, etc.). Palissy also wanted to state very clearly that the human being is there to help nature because:

> God wills us to work at helping nature (AD, p. 294)… Everything including sowing wheat and plowing should be nothing else but helping nature (AD, p. 297).

In this, Palissy follows the tradition of Aristotle (fourth century BCE) in his Politeia (VII, 17,1337a): "for the deficiencies of nature are what art and education seek to fill up." (Translated by Benjamin Jowett, 1885.)

In light of the last two quotations, Palissy informs readers that the main task for humans should consist of helping nature through their work and thinking (what he called "philosophy"). This is also why the parable of talents from the Bible is often mentioned (as in Lestringant, 2010) from the very beginning of his writings as shown by the dedication of *Veritable Recipe* to the Marshal de Montmorency:

> … while examining what was His will, I found in His last Testament that He ordered His heirs to multiply the talents He had bestowed on them… (VR, p. 89).

Besides, the formulation "Last Testament" reinforces the idea that God said nothing more to men, which inspired the New Testament.

Helping nature also led Palissy to explain:

> And this is why you have to put manure into the earth… (VR, p. 132).

Therefore, to comply with God's will, Palissy felt compelled to help nature. To do so, he undertook to provide a general explanation of nature's functioning system based on water and salts. At the same time, this gave him a marvelous occasion to show his own talents in terms of knowledge in the fields of geometry and topography (VR, p. 92), as well as of the various soils.

17.4 Nature, Water, and Salts

In each of his writings, Palissy tried to answer various questions associated with the origin, paths, properties, and practical applications of water, be it in artificial fountains, field waters, and their pathways in the soil, rivers, and seas, reasons for the presence of salt in seawater, metal formation, and agricultural practices.

17.4.1 Common and "Congelative"[5] Water

With reference to Genesis, Palissy first claims that water itself was also created right at the beginning of the world:

> … you have to firmly believe that all the waters that exist, will exist and have existed have been created together at the inception of the world (AD, p. 286, 328).

He then went on to demonstrate that all the waters on the earth's surface originate from rain and not from the sea:

> … after I had for a long time and very closely observed the causes for the sources of the natural fountains, and the place from which they could be released, I finally concluded that they proceeded from, and were only engendered by the rains (AD, p. 281).

While describing the general cycle of water, he explains that any rainwater is derived from the clouds, which themselves arise from the continental and marine waters (AD, p. 287):

> Rainwater which falls in winter, rises in summer and falls again in winter. The reverberation of the sun and the dryness of the blowing winds against the earth raise for large quantities of water: which being gathered in the air and formed in clouds are dispersed on one side or another ... and the winds pushing these vapors, the waters fall on all parts of the Earth.

And concerning the fact that it is desalinated water which is evaporated, Palissy used the example of the manufacture of salt in the salt marshes.

As for the origin of sea salt, Palissy answers:

> … **it is rather to believe that the salt of the sea comes from the earth**, being transported there by the waters of the rivers which flow into it as well as by the tumultuous streams which violently hit against the rocks and salty soils (AD, p. 286).

This modern explanation of the hydrological cycle has been recognized as exceptional by Palissy's commentators: in 1843 by Cap (p. xxi) and in 1961 by Orcel (p. 11–12).

According to Palissy, however, water is not just "common" (or "exaltative"[7]), because there is another, invisible water, mixed with the former and called "congelative" (or "germinative," "generative," or even "vegetative"), to which highly peculiar properties are ascribed. Detailed descriptions of these properties can be found in almost every section of the *Complete Works*, for instance, with regard to the formation of crystal, diamond and other diaphanous stones[6]:

[5] The term "congelative" and its substantive (congelation) were explained in note 37 of the *Complete Works* (Fragonard et al., 2010, p. 331). During the sixteenth century, the process of congelation not only referred to the actual freezing faculty but simultaneously designated the amalgamating capacity due either to ice or to heat (Legrand, 2010). Similarly, "generative is the property of generating"; "exaltative/exaltative" is that of rising, in Palissy's works synonymous with sublimate or vaporize (in the chemical sense); "essencive" means that is essential, constitutive, inherent in something; "salsitif/salsitive" is the property of pricking the tongue (the taste of salt).

[6] It is sometimes difficult to distinguish the order of divine creation from the "work of nature" in Palissy's writings. This is the case here for the creation of crystal or different forms of water.

> … the water from which the crystal is being formed is of a kind different from the common waters: and if it is of a different kind we can be sure that there are two waters, one is hence exaltative and the other essencive, congelative, and generative, both being mixed together in such a way that it is impossible to distinguish between them unless one of them is frozen (AD, p. 356–357).

This teaching of "double substances," one rather physical and visible in the common sense and one rather "vital" and "form giving" yet invisible, later became a constituent of the influential philosophy of "vitalism" (Cimino & Duchesneau, 1997).

Palissy understood the system of nature as based on the recognition that every living organism (plant or animal) and inert material (crystal, stone, manufactured object) first existed in a more or less aqueous stage or at least that an aqueous principle went through it. Each of them had hence received at the same time this common water plus that second type of water called generative and congelative. The latter allows any living being or inert material to (i) grow thanks to its generative power and (ii) acquire a certain rigidity thanks to its congelative ("freezing and crystallizing") power. That means realizing and maintaining structures is the task of this second type of water. As an example, in Palissy's mind, the rigidity of dry straw, which has lost its common water, is still due to the congelative water. By absorbing water, animals and humans incorporate at the same time this congelative water, which hardens their bones and gives them a certain rigidity; a similar process occurs in all the plants. This second form of water was said to be generative as it was responsible for transferring the semen among living organisms and the growing principle among the minerals: generating real forms. The idea of a germinative/generative/structuring water was not developed by Palissy himself but was taken up from Paracelsus, who in his *Treaty of Natural Waters* considered water as the matrix of all the creatures (Céard, 2010, p. 161). Paracelsus also stated that stones are the fruits of water and already differentiated between common and special (growing) water. Yet, the acting principle providing all these qualities was seen as being salt.

17.4.2 Salt and Its Virtues

Palissy asserted that there are an enormous number of salts in nature, almost as many as there are different entities:

> … there is such a big number of them that it is impossible for a single man to name them all… To those who want to know how many different species of salts there are, I would answer that there are as many of them as there are different species of savours and scents… (AD, p. 395–396).

The section titled "Diverse Salts" gives a long list of various kinds of "salts"; but Palissy's concept of salt does not systematically refer to what we nowadays call mineral salts or organic salts, as described below.

The author also explained that salt is the element responsible for the congelative capacity of all the items in nature as well as for the generative capacity of living organisms. Clearly spoken, his "salt" was what gave the "generative water" its special quality, originating from soil and rock. Palissy further considers that:

> … the salt is a robust, tangible body (…), preservative and generative of everything and everybody, like timber and all species of plants and minerals. It is an unknown and invisible body, like a spirit and yet taking place and sustaining the item in which it is enclosed ... (AD, p. 406).

Here, we find another link to the Aristotelian "form" versus "matter" philosophy: basically stating that there exists an immaterial force, a kind of "inner nature" named in his writing "Categories" (e.g., 2 b28–30) "secondary substances" that are the general properties that give form to individual matter. For Palissy, it is invisible salt, which gives visible structure to all matter. In his eyes, the salt is body and principle together, here in a perspective that is close to scientific observations:

> The salt whitens everything; the salt hardens everything; it preserves everything; it gives savour to everything; it is a mastic that binds and fills everything… (AD, p. 399–400).

It follows that water is the vector that transports the salts through nature, allowing nature to operate. And by analogy with water, which is the agent transferring and distributing the salt, according to Palissy, there are also two kinds of salts in everything, the "common" and the "congelative."

17.4.3 The Package Water-Salt

Palissy was convinced that with his so-called congelative-generative salt, he had discovered the fifth major element besides Empedocles' (and later Plato's and Aristotle's) traditional four that constituted the whole universe, that is, earth, water, air, and fire. He kept mentioning this fifth element in many sentences and very much insisted on the point that this salt was not yet known to mankind. In the *Treaty of the Metals and Alchemy,* which already foreshadows the existence of this fifth element, he wrote:

> … the matters of the metals are divine semen. I dare say that much divine that they are unknown to mankind and even invisible… (AD, p. 328).

This is a rare statement that shows that in the core of matter, Palissy supposed a divine inside.

This fifth element played a role particularly with regard to the formation of crystal, diamond, and other stones:

> … all metallic matters are liquid, fluid and diaphanous, and unknown amongst the common waters up until their congelation, and as for the fifth element … (AD, p. 355–356).

Similarly, for marl and its fertilizing properties, the same arguments were utilized again:

> … because I told you before that there was one, which the philosophers always ignored, and this fifth element is a generative water (…) when the seeds are thrown on the earth provided with marl, they feed on the generative and congelative water I am calling the fifth element… (AD, p. 512–513).

And again as concerns the topic of agriculture:

> … I always told you, speaking of the fifth element (…) that this is the one which sustains straw and hay, as well as all the species of trees and plants, and even the men and the beasts… (AD, p. 514).

And finally regarding manufactured items:

> … one should not believe that a single thing may turn out to be useful and serve as mirror or spectacles if not by the admirable virtue of the fifth element which binds all the other matters with itself… (AD, p. 515).

For Palissy, this fifth element is immanent in nature and obviously "the" basic and core substance (AD, p. 353, note 1), which asserts the preeminence of salts over the whole of creation. The importance assigned by Palissy to the fifth element is clearly demonstrated by the place it occupies in the section titled "*Excerpt from the Sentences*":

> Despite the fact that the philosophers have concluded that there were only four elements, there is a fifth in the absence of which not a single thing could say I am… (AD, p. 353).

Palissy stands here—in contrast to his pretention—in a long tradition of the "fifth element," called by Aristotle (Greek) *pémptē ousíā*, later better known by its Latin translation *quinta essentia* (the fifth being). Conceptualized by Aristotle as a kind of cosmic ether, in alchemy, it became a synonym for the "stone of the sages" or the "incorruptible gold," the "crystal" or just "lapis" (stone), the "hardened water," the "eternal water," and formulations of the highest value and ultimate aim of alchemical transmutation of matter (or of the human soul) that holds all opposites together. The "salt" being a very important factor of alchemy and science of his time and before, most prominent with Paracelsus, Palissy now merged it with the idea of a fifth element to become the innermost kernel of natural structures. Following its attributes, that "fifth" element in its appearances as special water and special salt can be called "divine." Whether Palissy saw in it a kind of "divine semen" that had flown out of God at the time of creation, or whether he saw matter (soil) itself as having given birth to the basic structures and drivers for the unfolding of the material world, cannot be answered here.

One can remember that the transference of alchemical ideas to chemistry was an important process leading to chemistry becoming a science that one finds acting "in vivo" with Palissy and the material side of his fifth element. This was even still echoed in nineteenth century, for example, with agricultural chemist Adolph Stöckhardt saying (1851, p. 1, authors' translation):

> Following a dim legend, somewhere in nature should be a wonderful jewel hidden or to be distilled out, which would be able to fulfill these two wishes; it was called Stone of the Sages. ... During a millennium, this treasure was sought-after, but not found. ... To a certain degree, chemistry really owns these powers….

17.5 The Discourse of the Commentators

Apart from a few exceptions, Palissy's work has been highly praised and depicted by naturalists of the following centuries. A development line from Palissy's theory of salt to the theory of plant mineral nutrition formulated by Sprengel (1826) and Liebig might be reconstructed (1840; cf. Feller & Aeschlimann, 2014).

17.5.1 Palissy's Salt and the Mineral Theory

A famous sentence often quoted to illustrate the importance of Palissy as a forerunner of the mineral theory of plant nutrition is:

> You've got to accept the idea that when you bring manure into the field, you're giving back to the earth some of what's been taken from it (VR, p. 112) (Authors' English translation).

Sprengel was the first to formulate the "law of the minimum" of plant nutrition in 1828, referring to mineral elements. Liebig (1840) then strongly advanced a so-called mineral theory of plant nutrition, which was based on the most recent scientific advances of the end of the eighteenth century—beginning of the nineteenth century. At that time, it had been shown that plants were able to synthesize their own dry substance utilizing solely elements from the air (carbon dioxide or CO_2) and the mineral salts present in both soil and water. Although Liebig also had vitalist ideas in his worldview (see Chap. 4), his notion of "salt" was a modern chemical one. This mineral approach drastically opposed the organic (humic) theory propagated by Thaer (1809), according to which most plant dry matter originated from a soluble fraction of the soil pertaining to the organic kingdom (Feller et al., 2003). Sprengel's and Liebig's new mineral theory was extremely important because it changed agriculture and opened up the new era of chemical products as major land fertilizers.

Precisely under these circumstances, individuals such as Grandeau (1879) suddenly discovered the importance of Palissy as a scientist (Boulaine & Feller, 1988). By reinterpreting Palissy's original writings, Grandeau made him a forerunner of Liebig three centuries ahead of the mineral theory. To this end, Grandeau (1879, p. 34) extracted only parts of Palissy's sentences and attributed the later meaning of mineral compounds to Palissy's salt concept, thereby transgressing the limit of scientific ethics:

> … under salt, Palissy evidently understood mineral substance …

The distinction between the terms mineral and organic, however, only showed up at some stage during the nineteenth century,[7] and all quotations given this far clearly show that Palissy did not think of salt as being uniquely mineral. Nevertheless, all science historians up to the present day followed Grandeau's interpretation and insisted on the idea of Palissy being an awesome forerunner in the field of agronomy (Feller & Aeschlimann, 2014).

Grandeau's (1879, p. 32–33) dithyrambic description reads as follows:

> Towards the mid-16th century, Bernard Palissy, a man of genius, (…) one has to go back to such masterpieces to find the first precise notion of the role played by the mineral substances in vegetation and the first rational explanation for the actual cause of the fertilizing properties of manure.[8] One cannot refrain from deeply admiring the sagacity of this great

[7] See for instance Histoire de la chimie: https://fr.wikipedia.org/wiki/Histoire_de_la_chimie

[8] The part of this sentence concerning the "actual cause of the fertilizing properties of manure" refers to Palissy's quotation given at the beginning of Sect. 17.5.1.

mind when reading fragments of the *Treaty of Diverse Salts and of Agriculture*, published 1563, which could have been perfectly written by modern-day scientists.

The quotation shows what psychology calls a "projection" (of one's own ideas or intentions to another person), and the function of this projection was to support the importance of their own convictions. There are many similar transpositions of the meaning of the term salt among commentators of other disciplines striving to see Palissy as a genial forerunner (Feller & Aeschlimann, 2014). This search for great forerunners (at best from one's own nation) has frequently been a topic for amateur science historians as emphasized by Fragonard et al. (2010) in the new edition of the *Complete Works*. Lamy (2018) urges research workers to first reinstate former scholars in the frame of their scientific, social, and political context as Kirsop (1961) and Thompson (1954) did in their reappraisal of Palissy's legacy.

17.5.2 What the Commentators Have Omitted

None of the many scientific readers of Palissy (Cap, 1844, Grandeau, 1879, Orcel, 1961, and others) ever evoked the multiple references to religion in his writings: this shows a filtered perception. In so doing, both historians and naturalists have deliberately neglected to deal with one of the most prominent components of his work; several hypotheses may be envisaged to explain this omission. To start with, the occidental hard science culture has decided that taking spiritual dimensions, and even worse religious aspects, into account would be inappropriate, except perhaps with regard to the human and social sciences. This attitude seemed to legitimize a selective perspective. On the other hand, Palissy's resolutely scientific argumentation may well have been perceived hitherto by the commentators as being very much disconnected from his religious engagement.

17.6 Palissy's Distance to God

God's omnipresence in Palissy's writings could lead one to believe that he was indeed a supporter of natural theology (Bodenmann, 2004). This religious conception intends to demonstrate God's existence based on natural and rational arguments and thereby clearly differs from the supernatural theology, which relies on the revelation and the Holy Scriptures. This natural theology once played a major role in the perception of science and the acceptance of its findings; it has nowadays turned up as a historical reference for the so-called creationist movement, one of the books of the English philosopher William Paley (1804)[9] in particular.

[9] Using purely scientific arguments, William Paley (1743–1805) undertook to demonstrate that a creator was perforce responsible for the beauty and complexity of the universe. Charles Darwin had a copy of *Natural Theology* on board the Beagle and to a large extent his theory of evolution may be seen as a rebuttal of Paley's conception.

Quite in contrast with natural theology, Palissy made no attempt to prove God's existence. To him, the Divine omniscience is plainly evident, and he felt commissioned by God to elucidate the laws of the universe. Many of his sentences bear testimony to this deep conviction:

> And if it actually pleased God to bestow me with His gifts....

Furthermore, he never calls on God to support any demonstration and cannot therefore be placed in the ambit of the natural theology. Fragonard (2010, p. 41) also insists on the point that Palissy was not a theologian and hence should not be discussed from a religious perspective. Otherwise Palissy proudly declares:

> ... I have no other book than those of heaven and earth (AD, p. 427).

This clearly shows that his theoretical understanding of the universe was based on two fundamental realms he considered of equal importance: (i) heaven, that is, the Bible, the book directly inspired by God and telling about past creation, and (ii) the earth, the book of the physical, both living and mineral nature at present.

This distance that Palissy is able to put between God and nature appears to be a very modern attitude for his time. The crucial point is that Palissy did remove the creator from his creation—a classical stance of early enlightenment. Then he partly filled the immaterial gap with the special "salt" which later was more easily reduced to mere matter like mineral plant nutrients. Paradoxically, by acknowledging God's creation again and again, Palissy reduced Him into this role and effectively split off the Christian God image from natural sciences. Then, the divine is not inside creation and is not the spirit of life in it, and it is only the primordial cause of nature. Thus, Bourg (2018, p. 82–83), quoting Francis Bacon's preface of the *Instauratio magna scientiarium* (1620), concluded that the idea appears that "the world is no longer [only] what is contemplated but what is transformed." In other words, what we can inform we can transform through our activities. Palissy often repeated that "man must help nature," and this perception is one of the features of his modernity (see Sect. 17.3.2).

17.7 Final Considerations

17.7.1 Palissy Was a Great Scientist

Despite the fact that Palissy's complete opus is virtually littered with references to the *Bible*,[10] his approach was of scientific and/or technical order, including a concealed spiritual dimension. The quality of his thinking is largely acknowledged up to the present day even if a great deal of his hypotheses and interpretations clearly date back to his century (*Albineana,* 2010). Palissy nevertheless ought to be

[10] This was also a necessary strategy at this time to somewhat protect his writing against suspicion of heresy.

recognized as an outstanding observer of nature. Taking the example of Palissy's projected garden (RA), Lamy (2018) emphasized that:

> He highly magnified human actions and did not make God the unique actor of the transformation of nature ….

And we can say: following Palissy, God was no longer acting in nature at all.

Lamy (2018) goes on emphasizing that Palissy's protestant understanding of the biblical message implied in particular the necessity for humans to support nature utilizing scientific expert knowledge, based on empiricism and experiments.

Palissy's observations were indeed not only extremely precise; they were also frequently replicated or even confronted with others made under different environmental conditions, and he systematically only dealt with what he had recorded personally. This determined to a large extent his resolutely practical attitude. He wanted to operate in total independence, gifted by God, but acting independently from Him, and was very proud of that. Yet, he was also generating theoretical concepts arising from his own observations, which could account for the functioning of nature.

17.7.2 Soil/Earth and Salt, the Crucial Fifth Element

As shown above from sentences written by Palissy himself, he understood that all the salts of the earth came from the soil. Therefore, undergirding salt, there are soil/earth and water: soil/earth as the origin of "salt" and water as the vector for salt to infuse and circulate salt through the world. However, in Palissy's writings, there seems to be a confusion on this fifth element, if it is "salt" itself or "water" distributing the "salt." A certain polyvalence and exchangeability of terms referring to a central meaning was quite common in alchemical literature. We may interpret here that for Palissy when the fifth element "special salt" was in water, then the water itself became the fifth element, because it carried salt inside, becoming in a way identical with it.

Palissy repeatedly asserted that God created nature but did not leave it inert; it was made to interact with all the various living and inanimate compartments on every part of the earth. Furthermore, in his eyes, it was human's duty (assigned by God, man being his representative on earth) to assist nature, either in discovering its laws and utilizing them or in laboring usefully. The driving force that makes the world go round, however, was the fifth element, which was at the same time principle and matter, composed of common salt/water intimately mixed with, and non-differentiable from congelative-generative salt/water. The vector of soluble salts was equally congelative-generative and non-differentiable from common salt, which conveyed the growing and/or reproductive principles as well as the characters shaping form and rigidity. Ultimately, the origin of all the salts was to be found in the soil. This fifth element very much represents the heart of Palissy's conception; it showed up 1563 times in VR, was probably developed over the course of the many conferences he delivered from 1575 to 1584 in his so-called little Parisian academy, and assumed its definitive dimension as of 1580 (AD).

17.7.3 How Objective Were Palissy's Reviewers

Palissy's writings have been commented on by naturalists and engineers, historians, and literary critics, all of whom agreed in considering him as one of France's great men. Scientific authorities in particular have been eager to emphasize Palissy as a wide-ranging forerunner as if he alone had discovered all the evidence.

Addressing exactly that yearning of former analysts, sixteenth century specialists participating in a recent congress devoted to Palissy (*Albineana,* 2010) felt it necessary to state that:

> ... it is perhaps more important to retrace the history of the gradual progression of the ideas (i.e., their popularization) than to detect the pure invention (which anyway does not really exist) or the history of geniuses....

The main reason why eighteenth- and nineteenth-century scientists lacked objectivity in regard to Palissy is most likely related to the imperious need arising in all state nations to focus on models and examples of the past apt to galvanize the patriotic, nationalist, and scientific senses of their citizens. At that time, national colors, anthems, heroes, myths, etc. were therefore selected, composed, detected, or invented to exalt the exceptional qualities of defined ethnic groups (Thiesse, 1999, p. 13–14). To successfully construct a mythic scientific giant likely to be unanimously praised at the French national level, Grandeau (1879) and others certainly decided to conceal the fact that Palissy had been a Huguenot.

Most scientific commentators have analyzed Palissy's discourse without considering his constant references to God and his religion. This point is far from harmless and clearly reflects the perception of the average scientist in terms of the kind of statements they are willing to transmit. Specifically, the propensity to make Palissy all the more genial leads scientists to erase every little bit of connection with religion. In a somewhat unconscious way, reviewers make themselves censors of information regarding Palissy's faith, and they remain thereby far from adopting an objective approach to historical analysis. This rough action may be in line with Palissy's own (unconscious or conscious) tendency to remove God from present nature, leaving an open field for human genius, science, and technology. But they miss the supposed ambiguity in his concept of salt, being a hybrid between a physical and an implicitly spiritual concept and by that opening a possibility to overcome the later established Cartesian split between spirit and matter, that is, the complete de-spiritualization of nature in the philosophy of science by René Descartes (1596–1650) and others. Unperceived by his interpreters, Palissy was not only a forerunner of scientific rigors, but his work circumambulated his "salt," an element in nature that has material and immaterial properties at the same time and coming from the basic underground matter (soil and rock).

17.7.4 The Salt of the Earth

The Bible is very present throughout the writings of Palissy. This is not surprising given the period in which he lived and the religious debates occurring at that time; it was also appropriate for the Calvinist protestant he was. However, in the tension between science and religion, he always maintained a certain distance from God and did not at all integrate the Christian God image, or refer to it in any way, when elaborating his universal theory of the salts and especially their status as the "fifth element" of the earth. In short, the scientist Palissy takes over responsibility once it is acknowledged that God has created the universe—and then was supposed to have retired from His creation! This is a landmark in the development of science and its self-understanding. *Cum grano salis* (with a grain of salt), because in the salt of the earth, Palissy saw a divine semen, a central agent for every form and life.

By keeping God out of nature after His original creation (see Sect. 17.3.2), Palissy did not adhere to natural theology nor was he playing the preacher or the propagator of his faith. Maybe Palissy used the paradox to present his theory on the one hand as a forerunner of the famous *Cartesian split* that is the existential duality and separation of spirit (God removed from His creation, only human spirit remaining there) and matter (the object of science, driven by mere causality and not by finality/intentionality). But on the other hand, we may hypothesize that Palissy seemed to be rather unconsciously striving to overcome this emerging Cartesian split by hypothesizing the salt—situated within soil—to be the central material and simultaneously spiritual element in nature and by that attributing some properties of divinity to nature itself.

First of all, Palissy deliberately intended to behave as a scientist and as such did not even quote Matthew 5:13 (see epigraph of this chapter), an oversight which his own commentators did not raise! Maybe Palissy wanted to keep a maximum distance between his concept of salt and the Christian faith he quoted around it. However, one cannot help but think that his *Complete Works* would undoubtedly deserve to be placed under the additional heading of "**The Salt of the Earth**"—but not in the biblical meaning. Palissy's writing often mentioned God to be the creator of the world, but in fact, he did not understand God to be the acting *spiritus rector* (steering mental force) in His own creation. This function was transferred to the salt that became the crucial matter and steering principle at the same time. Salt, being a kind of essence coming out of soil, is a principle of generating life, growth, and structured order at the same time.

By that, in Palissy's view, nature itself contained and issued the force that makes the world go round, by that making the actions of God, the Earth's creator, obsolete in nature—and simultaneously introduced a natural-spiritual element in his worldview, the "salt." Finally, the human task in Palissy's worldview is not to help God rule His creation, but to help nature move it.

References

Albineana. (1992, new ed. 2010). *Actes du colloque Bernard Palissy (1510–1590). L'écrivain, le réformé, le céramiste* [Proceedings of the Bernard Palissy (1510–1590) Symposium. The writer, the protestant, the ceramist]. Cahiers d'Aubigné 4. Honoré Champion, Niort, France, 239 p. + Table.

Amico, L. N. (1996). *À la recherche du Paradis terrestre, Bernard Palissy et ses continuateurs* (256 p.). Flammarion. [The search for the Paradise on earth. Bernard Palissy and his followers]

Aulard, F.-A. (1892). *Le culte de la raison et le culte de l'Être suprême (1793–1794), essai historique*. Paris. [The cult of reason and the cult of the supreme Being (1793–1794), historical essay]

Bodenmann, S. (2004). William Paley et la théologie naturelle – Interpréter le vivant à la fin du XVIIIe siècle. *Variations herméneutiques, 20*,15–29. (Zurich Open Repository and Archive, University of Zurich, ZORA URL: https://doi.org/10.5167/uzh-109611). [William Paley and the natural theology – interpreting the living at the end of the 18th century]

Boulaine, J. (1989). *Histoire des pédologues et de la science des sols*. (pp. 29–30). Inra. [History of the pedologists and of the science of the soils]

Boulaine, J. (1992). *Histoire de l'agronomie en France* (pp. 134–135). Lavoisier. [History of the agronomy in France]

Boulaine, J., & Feller, C. (1988). L'œuvre agronomique de Louis Grandeau (1834–1911). Professeur à l'école Forestière de Nancy. *La Revue Lorraine Populaire, 82*, 193–194. [The agronomic work of Louis Grandeau (1834–1911). Professor at the Nancy school of Forestry]

Bourg, D. (2018). *Une nouvelle Terre* [A new earth] (215 p). Desclée de Brouwer.

Cap, P. A. (1844). Notice historique sur la vie et les ouvrages de Bernard Palissy. In *Œuvres complètes de Bernard Palissy, édition conforme aux textes originaux imprimés du vivant de l'auteur avec des Notes et une Notice historique par Paul-Antoine Cap* (pp. i–xxxix) Dubochet et Cie. [Historical account on the life and books of Bernard Palissy]

Céard, J. (1992, new ed. 2010). Bernard Palissy et l'alchimie. In *Albineana* 4, Cahiers d'Aubigné 4. Bernard Palissy (1510–1590). L'écrivain, le réformé, le céramiste, pp. 155–166. Honoré Champion, Niort, France. [Bernard Palissy and the alchemy]

Cimino, G., & Duchesneau F. (1997). Vitalisms from Haller to the Cell Theory. *Proceedings of the Zaragoza Symposium*, XIXth Int. Congr. of Hist. of Sc., 22–29 Aug. 1993. Firenze.

De Corgnac, E. (~1880). *Les hommes utiles et les bienfaiteurs de l'humanité* (pp. 202–214). Eugène Ardant et Cie. [Useful men and benefactors of the humanity]

Desfontaines, C. (~1880). *Les grands hommes racontés aux enfants. Gutenberg, Bernard Palissy, A.-L. Bréguet* (pp. 61–106). Libr. Hachette. [The great men told to children. Gutenberg, Bernard Palissy, A.-L. Bréguet]

Du Chatenet, E. (~1880). *Les artisans célèbres* (pp. 54–77). Eugène Ardant et Cie. [Famous craftsmen]

Feller, C., & Aeschlimann, J. P. (2014). Soil and salts in Palissy's (1510–1590) view: Was he the pioneer of the mineral theory of plant nutrition? In J. Churchman & E. Landa (Eds.), *The soil underfoot. Infinite possibilities for a finite resource* (pp. 289–302). CRC Press, 421 p.

Feller, C., Blanchart, E., & Yaalon, D. H. (2006). Some major scientists (Palissy, Buffon, Thaer, Darwin and Müller) have described soil profiles and developed soil survey techniques before 1883. In B. P. Warkentin & D. H. Yaalon (Eds.), *Down to earth: A soil science history* (pp. 85–105). Elsevier.

Feller, C., Thuriès, L., Manlay, R., Robin, P., & Frossard, E. (2003). The principles of rational agriculture by A.D. Thaer (1752–1828). An approach of the sustainability of cropping systems at the beginning of the 19th century. *J. Plant Nutr. Soil Sci., 166*, 687–698.

Figuier, L. (1868). *Vie des savants illustres de la Renaissance avec l'appréciation sommaire de leurs travaux* (pp. 156–212). Librairie internationale. [Life of famous Renaissance scientists with a short appreciation of their works]

Fragonard, M. M. (2010). Introduction. In Bernard Palissy. Œuvres complètes. Édition présentée et annotée par K. Cameron K. Céard, J., M.-M. Fragonard, M.-D. Legrand, F. Lestringant, G. Schrenk. Sous la direction de M.-M. Fragonard. *Deuxième édition revue et augmentée* (pp. 11–46). Honoré Champion. [Introduction. In: Bernard Palissy. Complete Works]

Fragonard, M. M. et al. (2010). Bernard Palissy. Œuvres complètes. Édition présentée et annotée par K. Cameron K. Céard, J., M.-M. Fragonard, M.-D. Legrand, F. Lestringant, G. Schrenk. Sous la direction de M.-M. Fragonard. *Deuxième édition revue et augmentée* (636 p.). Honoré Champion. [Bernard Palissy. Complete Works]

Grandeau, L. (1879). *Chimie et physiologie appliquée à l'agriculture et à la sylviculture. 1. La nutrition de la plante*. Berger-Levrault et Cie, xvi + 624 p. [Chemistry and physiology as applied to agriculture and forestry. 1. Nutrition of the plant]

Hoefer, F. (1843). *Histoire de la chimie depuis les temps les plus reculés jusqu'à notre époque*. Tome 2, Section Palissy, 72–98. Hachette. [History of chemistry from the most ancient to our time]

Jonveaux, E. (1874). *Histoire de trois potiers célèbres* (pp. 40–143). Hachette. [History of three famous potters]

Kirsop, W. (1961). The Legend of Bernard Palissy. *Ambix*, *9*,3, 136–154, (cité par Keith Cameron p. 45). https://doi.org/10.1179/amb.1961.9.3.136

Lamy, J. (2018). *Lire la Bible, explorer la nature. Bernard Palissy et le geste expérimental*. Revue d'histoire du protestantisme, 3–4, pp. 375–393 [Reading the Bible, exploring the nature. Bernard Palissy and the experimental process]

Legrand M. D. (1992, new ed. 2010). Notes sur le lexique de Bernard Palissy. In *Albineana* 4, Cahiers d'Aubigné 4. Bernard Palissy (1510–1590). L'écrivain, le réformé, le céramiste. 145–154. Diffusion Librairie Honoré Champion, Niort, France, 239 p. [Notes on the lexicon of Bernard Palissy]

Lestringant F. (1992, new ed. 2010). L'Éden et les ténèbres extérieures. De la *Recepte véritable* aux *Discours admirables*. In *Albineana* 4, Cahiers d'Aubigné, 4. Bernard Palissy (1510–1590). L'écrivain, le réformé, le céramiste. pp. 119–130. Diffusion Librairie Honoré Champion, Niort, France, 239 p. [The Eden and the outside darkness. From the *veritable Recipe to the admirable Discourses*]

Liebig, J. (1840). *Die Chemie in ihrer Anwendung auf Agrikultur und Physiologie*. Vieweg u. S. [Chemistry in its application to agriculture and physiology]

Orcel, J. (1961). Préface. In *Œuvres complètes de Bernard Palissy, édition conforme aux textes originaux imprimés du vivant de l'auteur avec des Notes et une Notice historique par Paul-Antoine Cap*. Nouveau tirage augmenté d'un Avant-Propos de Jean Orcel. Librairie Scientifique et Technique Albert Blanchard, pp. 1–17. [Foreword. In: Complete Works of Bernard Palissy, edition consistent with the original texts as printed at the time of the author, with notes and an historical account]

Paley W. (1804). *Théologie naturelle ou preuves de l'existence et des attributs de la divinité tirée des apparences de la nature*. Traduction française de Charles Pictet, réimpression par Hachette Livre – BnF, 39 p. [Natural Theology or Evidences of the Existence and Attributes of the Deity collected from the Appearances of Nature]

Palissy B. (1563a). *Grotte Rustique* [Rustic Grotto (GR)]. See reference Fragonard et al., 2010.

Palissy B. (1563b). *Recepte Véritable* [Veritable Recipe (VR)]. See reference Fragonard et al., 2010 and note 3.

Palissy B. (1580). *Discours admirables* [Admirable Discourse (AD)]. See reference Fragonard et al., 2010 and note 3.

Palissy B. (2010). *Oeuvres Complètes* [Bernard Palissy. Complete works]. See reference Fragonard et al., 2010.

Poirier, J. P. (2008). *Bernard Palissy. Le secret des émaux* (296 p.). Pygmalion/Flammarion. [Bernard Palissy. The secret of enamels]

Sprengel, C. (1828). Von den Substanzen der Ackerkrume und des Untergrundes, insbesondere, wie solche durch die chemische Analyse entdeckt und von einander geschieden werden können; in welchen Fällen sie dem Pflanzenwachsthume förderlich oder hinderlich sind und welche Zersetzungen sie im Boden erleiden. *Journal für Technische und Ökonomische Chemie, 1828, 2*, 423–474, *3*, 42–99, *3*, 313–351, *3*, 397–421. [About substances of the topsoil and subsoil, in particular how they can be detected and separated by chemical analysis; in what cases they are beneficial or detrimental to plant growth and what decomposition they undergo in the soil]

Stöckhardt, A. J. (1851–1853). *Chemische Feldpredigten für Landwirte*. Georg Wigands Verlag, Leipzig, 2 Bände. [Chemical field sermons for farmers]

Thaer, A. D. (1809). *Grundsätze der rationnellen Landwirtschaft* (4 Bände (1809–1812)). Realschulbuch Ed. [Principles of rational agriculture]

Thiesse, A. M. (1999). *La création des identités nationales. Europe XVIII*[e]*-XX*[e] *siècles* (269 p.). Le Seuil ed. [The creation of the national identities. Europe 18th–20th centuries]

Thompson, H. R. (1954). The geographical and geological observations of Bernard Palissy the potter. *Ann. Sci., 10*(2), 149–165.

Viennet, C., & Starosta, P. (2010). *Bernard Palissy et ses suiveurs du XVI*[e] *siècle à nos jours. Hymne à la nature* (300 p.). Éd. Faton. [Bernard Palissy and his followers from the 16th century to our days. Hymn to nature]

Chapter 18
Darwin and the Earthworm: From God the Father to Mother Earth

Christian Feller, Laurence Feller Girod, and Nikola Patzel

Preliminary Remarks

This chapter from authors with different backgrounds and competences[1] is not strictly written in a classical scientific or historical perspective. One could say that this is one possible psychological interpretation of the eminent scientist named Darwin, linking especially his work on earthworms with his family and cultural background.

This book entitled *Cultural understanding of soils* calls, among other aspects, on what Patzel named "inner soil" (Chap. 22 of this work and references there). This refers to a mental framework of perception of the soil and nature, which is deeply rooted in an individual's psychic and cultural background, in conscious or unconscious references to mythological, religious, and even political ideas and patterns of imagination in the society to which each of us belongs. The psychological dimension of the inner soil is deeply installed in the psyche and influences our thinking; this goes to the point that one does not even realize it, but it can exert a hidden influence on the individual thoughts and valuations, the actions, and what one is ready to experience or not. This is what Freud and different schools of depth psychology call the unconscious.

[1] CF is a soil scientist, LFG a psychiatrist and NP both an environmental (soil) scientist and analytical psychologist.

C. Feller (✉)
Institut de Recherche pour le Développement (IRD), UMR Eco & Sols, INRA-IRD-SupAgro, Montpellier, France
e-mail: christian.feller@ird.fr

L. Feller Girod
Ministère de la Santé et de la Prévention, Centre Hospitalier Le Mas Careiron, Uzès, France

N. Patzel
Freelance Scientist, Office for Soil Communications, Überlingen, Germany
e-mail: nikola.patzel@soilcom.com

N. Patzel et al. (eds.), *Cultural Understanding of Soils*,
https://doi.org/10.1007/978-3-031-13169-1_18

This chapter is a kind of "thought experiment," which aims to invite Darwin on Freud's couch and question how Darwin's unconscious could have influenced not only his life in general but also his scientific work, including that concerning the soil. Obviously, Darwin never spoke to Freud or Jung about his dreams! But one might think that both would have been very interested in having Darwin as a client or in collegial exchange. Darwin was not a stranger to Freud since many books written by Freud mention Darwin and refer to his theories. And Jung named Darwin exemplarily to be an extravert thinker based on the outer facts, in contrast to introvert thinkers like Kant or Nietzsche (Jung, 1921).

The reader may wonder, however, how this "thought experiment" came to the mind of the authors regarding soil. We will mention here two elements that were at the origin of the interest of the authors for this psychological portrait of Darwin.

First, Darwin is generally recognized for his theory of evolution. His work on the soil is unknown to the general public, even if it is starting to become famous now among soil scientists through his last scientific publication in 1881, *The Formation of Vegetable Mould Through the Action of Worms with Observations on Their Habits* (Fig. 18.1). This book was published a year before his death with the earthworm as its "hero" and is the beginning of soil ecology. What is less well known is that

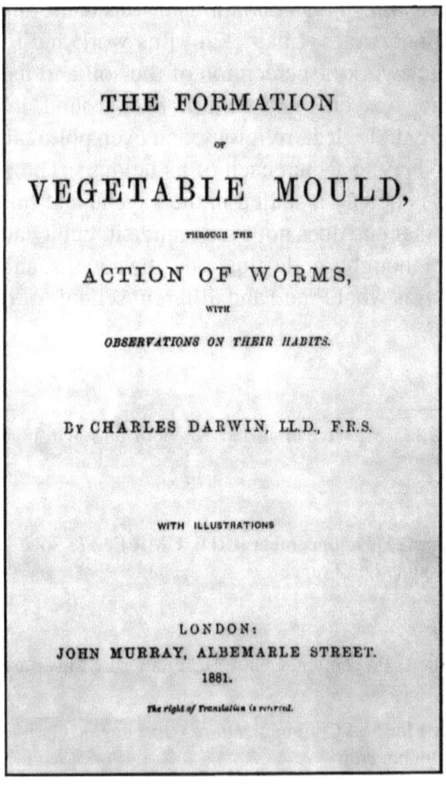
THE FORMATION

OF

VEGETABLE MOULD,

THROUGH THE

ACTION OF WORMS,

WITH

OBSERVATIONS ON THEIR HABITS.

BY CHARLES DARWIN, LL.D., F.R.S.

WITH ILLUSTRATIONS

LONDON:

JOHN MURRAY, ALBEMARLE STREET.

1881.

The right of Translation is reserved.

Fig. 18.1 The cover of Darwin's book devoted to Earthworms. (https://en.wikipedia.org/wiki/The_Formation_of_Vegetable_Mould_Through_the_Action_of_Worms)

Darwin started to work (1837) and publish (1838 and 1840) on the earthworm as soon as he returned from his voyage on the Beagle in 1836.

The second point is that Darwin fell ill and remained so to various degrees, for the remainder of his life only a few years after his return from his trip around the world on the Beagle. The symptoms were severe digestive difficulties combined with circulatory and breathing problems. He himself spoke of this disease almost daily in his correspondence. The origin of this disease has still not been elucidated, but various authors (Bowlby, 1990; Colp, 1977) have attributed a strong psychosomatic dimension to it. It has also been noted that this disease slightly subsided when Darwin finished publishing his main works on the theory of evolution.

Therefore, we look a bit at the psychological aspects of Darwin's background, including his family system and religious elements. Thus, the authors' interest and questioning for a psychological analysis of Darwin with reference to the soil came from this double observation: (i) Darwin's scientific life almost begins (1837) and ends (1881) with the earthworm, and (ii) his illness, which occupied him almost daily, is probably psychosomatic. Would that have any meaning regarding the existence of an inner soil in Darwin's unconscious?

Introduction

Charles Darwin's early work, in part, oscillates between God and the earth. God because the young Darwin, just before embarking on the Beagle at 21 years old, was destined to ecclesiastical training for a career in the Anglican church. God because the voyage on the Beagle, which was dedicated to scientific expeditions around the world and on which Darwin was a natural scientist, would lead to the Theory of Evolution. This theory replaced God to explain change in and the diversity of species in the past and present.

Earth is very present in Darwin's work. Darwin was passionate about geology. His geological observations collected during his voyage on the Beagle were quickly recognized as original and essential, and Charles Lyell, the most famous natural scientist of the time, would become one of Darwin's three great friends along with Joseph Hooker and Thomas Huxley.

If geology was one of Darwin's privileged areas, earth (soil) was another. Darwin devoted very long periods, even years, to the observation of the soil through a character who was dear to him, the earthworm.

In this chapter, we pose the general assumption that the development of a scientific work can have, beyond the scientific approach itself and the passion for the subject by the author, origins that can be registered in a psychological dimension of the scientist. This psychic dimension has conscious and unconscious levels. We will illustrate this approach with the example of Darwin's works on the soil.

To discuss this question, it is first necessary to re-situate Charles Darwin's personal history. This includes his sensitivity, his relationships with his family, his illness (Sects. 18.1 and 18.2), and some of his writings on the soil and earthworms (Sect. 18.3). Section 18.4 will then address the inner dimension of Darwin in relation to his scientific work.

18.1 The Lineage of Charles Darwin

Charles's paternal line was part of an intellectual and scientific bourgeoisie, while his maternal line was that of a famous family of ceramic manufacturers. It all starts at the level of Charles' two grandfathers: Erasmus Darwin on the paternal side and Josiah Wedgwood on the maternal side. These two were great friends and great liberals of the late eighteenth century, for example, campaigning for the abolition of slavery.

18.1.1 The Paternal Line

Erasmus Darwin (1731–1802) was a strong personality who was a doctor, naturalist, inventor, and poet. He and his book *Zoönomia* (Zoonomie or Laws of Organic Life, 1794) are considered, like the French J.-B. de Lamarck and his *Philosophie zoologique* (1809), one of the great precursors of the theory of evolution. Erasmus had 12 children, including Robert Waring Darwin (1766–1848), Charles' father.

Robert was also a family doctor and savvy businessman who tended wealthy English families. He had a great reputation as a doctor, also caring for the poor free of charge (Bowlby, 1990, p. 51). He was a member of the Royal Society. One of his brothers, Charles (1758–1778), was a future doctor but died very young from blood poisoning following a dissection.

Darwin's father Robert was said to be impressive (Fig. 18.2). In his autobiography, Charles Darwin wrote of his father (Darwin, 1986, p. 11):

> He was about 6 ft. 2 inches in height [about 1.90 m], with broad shoulders, and very corpulent, so that he was the largest man whom I ever saw. When he last weighed himself, he was 24 stone [over 150 kg], but afterwards increased much in weight.

Robert married Susannah Wedgwood, the daughter of Josiah Wedgwood, the great friend of his father Erasmus. They had six children, including Charles Robert Darwin (1809–1882). Darwin will therefore bear, in his first name, that of his prematurely deceased uncle and, in the second, that of his father.

18.1.2 The Maternal Line

Josiah Wedgwood (1730–1795) was a British industrialist who was born into a family of potters. He founded Josiah Wedgwood and Sons, aka Wedgwood, a porcelain and earthenware company that still exists today and which contributed to the beginnings of the British industrial revolution. A member of the Royal Society, he was also an inventor of new manufacturing and marketing techniques. He had eight children, in fifth position Susannah, who would marry Robert Darwin.

Fig. 18.2 Portrait of Robert Waring Darwin [1766–1848]. (https://commons.wikimedia.org/wiki/File:Portrait_of_Robert_Waring_Darwin_(1766_-_1848)_Wellcome_L0001575.jpg)

Susannah was therefore the mother of Charles, and Josiah II was one of his uncles, his favorite uncle who will eventually become his father-in-law. Both sides of the family were members of the Unitarian church (Desmond & Moore, p. 12). Summing up, one could say that the paternal branch was "medical" and the maternal branch "earthly" through ceramics.

18.2 Charles Darwin, Some More Biographical Elements

To better understand Darwin's personality, including certain psychological aspects, it is necessary to recall here some biographical elements from the life of Charles Darwin. For more details on Darwin's life, see the works by Bowlby (1990), Desmond and Moore (1992), and Tort (2000).

18.2.1 Youth and Adolescence

Charles's mother Susannah (1765–1817) died when he was 8 years old. In his autobiography, Darwin (1986, p. 6) said he had almost no memory of his mother. He was raised by his father and two of his sisters.

Young Charles was quickly fascinated by natural history and collections (biological, among others) (Darwin, 1986, p. 7) but was not a particularly good student. This created some despair for his father, who one day made this extraordinary statement, which was difficult for Charles to hear:

> You care for nothing but shooting, dogs and rat-catching, and you will be a disgrace to yourself and all your family (Darwin, 1986, p. 10).

After high school, his father wanted Charles to study medicine like his brother Erasmus. Very quickly, Charles realized that he could not stand the sight of blood and wished to abandon this line of training. Failing to become a doctor, his father then destined Charles for a clerical career. Young Charles began theological studies that he found quite satisfying, particularly the approach of William Paley, author of *Natural Theology* (1804). This approach appealed to a scientific discourse, arguing that it is science itself that proves that chance cannot be envisaged to explain the beauty, diversity, and complexity of nature, which necessitated a "creator." But the course that most fascinated Charles was that of the Reverend Henslow. Charles would surely have become a clergyman except that one day in May 1831, while on a hunting vacation at Maer's (Fig. 18.3) with his uncle Josiah II, he received a letter from Henslow indicating that a naturalist was being sought for a two-year long scientific expedition on a ship named the Beagle. Charles was 21 years old, and this project excited him. But Charles needed his father's consent, which he categorically refused to give. However, Robert left a small door open, saying:

> If you find a common sense man who advises you to go, I will agree.

Charles immediately thought of his favorite uncle "Jos." Indeed, Josiah II supported his nephew and countered Robert's negative arguments to make positive ones in favor of this trip for young Charles. Josiah's arguments won the support of Charles's father (Darwin, 1985, pp. 144–145).

Therefore, without Josiah II (maternal line), Charles would never have traveled with the Beagle.

He embarked on December 27, 1831, and did not return until October 2, 1836, a trip around the world in 5 years instead of the two initially planned.

Fig. 18.3 Maer Hall, the house of Darwin's uncle Josiah II. (Taken from the churchyard Maer_ Hall_ geograph.org.uk_-_207996. Wikimedia commons. wikimedia.org)

18.2.2 The Trip on the "Beagle"

We will mention here only a few facts, which are necessary to understand additional discussions at the end of this chapter. Among the works carried by Darwin on the Beagle were *The Natural Theology* of Paley (1804) and *The Lost Paradise* of Milton (1667).

The first general point to note is scientific. The collection of scientific information by young Charles was prodigious and very diverse (Earth sciences, animal and plant biology, and anthropology with observations on the populations encountered). Throughout the journey, Charles sent whole cases of collected materials and commentaries to his old friend Henslow (Bowlby, 1990, pp. 186–190) who distributed them to leading English scholars.

The second point concerns Charles' health. Apart from being seasick from time to time, Charles must have been in exceptional health when you consider that he traveled thousands of kilometers on foot or horseback and climbed major mountains to make his observations and collect samples. In South America alone, he went on eight major expeditions totaling approximately 4800 km with ascents in the Andes to elevations of 3600–3700 m (Bowlby, 1990, pp. 179–184). Soon after he returned to England, Charles became a completely different man, constantly suffering from health problems.

The third point is that Charles, a Christian believer at the beginning of this extraordinary trip, returned agnostic. William Paley was no longer his scientific base, and paradise was really lost!

18.2.3 Darwin's Health Problems

Charles was still in good health when he returned to England in 1836: he went for walks and visited his family and his friends. But from 1837 until approximately 1872, he was very often ill with digestive disorders, nausea, dizziness, insomnia, rashes, nervous tics, "black flies," and general weakness. It was also in 1837 that Charles annotated on the paper of his notebooks a first branched sketch of the formation of the Genres, his first step toward the theory of evolution (Tort 2000, p. 61). Beginning in 1842 (33 years old), Charles was no longer able to take long walks and did not like to go out much. He tried various hydrothermal treatments and cures, but none provided long-term relief, and no doctor was able to cure him, not even his father. In one of his rare letters, Robert (Darwin's father) wrote:

> ... You know that I never write anything besides answering questions about medicine and therefore as you are not a patient, I must conclude (Bowlby, 1990, p. 82).

With such a sentence, Charles could only conclude that he must be sick in order to communicate with his own father!

In Charles' correspondence, which is enormous (published in five volumes, Francis Darwin (Darwin, 1887, 1888; Darwin & Seward, 1903), the theme of disease comes up very often, and he regularly laments the time lost because of it (Darwin, 1986, p. 71).

Charles' disease has been the subject of numerous writings, including Colp (1977) and especially Bowlby (1990). Numerous hypotheses have been offered to explain what afflicted him (Bowlby, 1990, pp. 457–466), in particular, Chagas disease (a trypanosomiasis) that Darwin could have contracted during his trip around the world. Current medical science has concluded that Chagas is unlikely for different reasons: the main being that someone with Chagas would have died in the 1800s, which was not the case for Charles. The most widely accepted assumption today is that Darwin was prone to psychosomatic disorders of multifactorial origin (Bowlby, 1990, p. 465).

Charles himself almost recognized that his disorders were of a psychosomatic nature. Here are some of his quotes about it. From 1838:

> I find the noddle and the stomach are antagonist powers ... What thought has to do with digesting roast beef, I cannot say, but they are brother faculties (Bowlby, 1990, p. 215).

Indeed, this quote is a poignant description of psychosomatic symptoms.

Similarly, in 1863, during a tense discussion with his friend Lyell:

> "A good severe fit of Eczema would do me good" ... "[without my eczema I felt] languid and bedeviled" (Bowlby, 1990, pp. 372–373).

In the same psychological perspective, one can wonder about the role his chosen wife (Emma, his cousin, whom he married in 1839) played in his illness. Indeed, even before their marriage, Emma showed much compassion for, and perhaps enabling of, Charles' bouts with mysterious illnesses. In 1838 she wrote:

> ... feel sure that nothing could make me so happy as to feel that I could be of any use or comfort to my own dear Charles when he is not well ... So don't be ill any more my dear Charley till I can be with you to nurse you and save you from brothers … .

In fact, Emma was a great "nurse" for Charles. At the slightest annoyance or stress, Darwin was sick; he was legendary for such behavior.

Thus, a week before his wedding (January 1839), he wrote to Emma with humor:

> I wish the awful day was over. I am not very tranquil when I think of the procession: it is very awesome (Bowlby, 1990, p. 229);

or, as he was ill, he did not go to his father's funeral (November 1848), no more than to that of his master Henslow (Bowlby, 1990, p. 364), and not to that of his daughter Annie (April 1851) whom he adored (Bowlby, 1990, p. 296), or of his sister Susan in October 1866 (Bowlby, 1990, p. 384). Or even during events assuming confrontations such as the famous debate about the *Origin of Species* on June 30, 1860, known as "Battle of Oxford", during which the Darwinists were represented by Thomas Huxley (nicknamed the "Darwin bull-dog") and the "fixists" (the creationists of this time) were defended by the Oxford bishop Samuel Wilberforce, a debate which was to the advantage of Darwinists (Tort, 2000, pp. 80–81; Bowlby, 1990, pp. 354–359).

The symptoms of the disease disappeared very markedly from 1871 to 1872, the years when *The Descent of Man* (Darwin, 1871) and *The Expression of Emotions* (Darwin, 1872) were published, which to some extent completes the saga of the theory of evolution when it integrates humans into the whole of the theory as a species like any other. By 1872, the "birth" of the theory of evolution was finished, along with its accompanying anxieties. Darwin was then able to devote a lot of time to his friend the earthworm.

18.2.4 Darwin on His Paternal Family

In his autobiography (Darwin, 1986, p. 24), Darwin suggests that intellectually he owes little to his paternal family. On his paternal grandfather:

> At this time [around 1825–1826] I greatly admired the *Zoönomia*; but on reading it a second time after an interval of ten or fifteen years, I was much disappointed, the proportion of speculation being so large to the facts given.

While Erasmus has been considered a precursor to the theory of evolution, it was not until the third edition of the *Origin of Species* that Darwin referred to his predecessors in a "Historical Record." And he only mentions the contribution of his eminent grandfather in a footnote! (Bowlby, 1990, p. 363).

While Darwin regularly displayed great admiration for his father Robert as a doctor, he was very critical of his scientific thought:

> My father's mind was not scientific... I do not think that I gained much from him intellectually… (Darwin, 1986, p. 19).

It was the same when he talked about his brothers and sisters (Darwin, 1986, p. 19–20). He finally concluded (Darwin, 1986, p. 20):

> I am inclined to agree with Francis Galton in believing that education and the environment produce only a small effect on the mind of any one, and that most of our qualities are innate.

Finally, numerous family testimonies show how different the paternal and maternal homes were: sad and heavy in Dr. Robert Darwin's house (Bowlby, 1990, p. 58), cheerful, playful and culturally enriching in Josiah II's (Darwin, 1986, pp. 27–28; Bowlby, 1990, pp. 91, 230–231, 264–265).

From these biographical elements, we can infer a few essential points concerning Darwin from a psychological perspective:

First, despite the external admiration displayed by Charles Darwin for his "imposing and authoritarian" father (as was his grandfather too), it seems that Charles unconsciously rejects this paternal ancestry. Despite the medical and scientific paternal family genealogy a father-son bond was missing, which created a somewhat melancholic family atmosphere. This was expressed in different ways: by writing on various occasions that he did not owe much to this ancestry, by giving up on all the careers that his father chose for him (medical or clerical studies), and, finally, by showing that no doctor, not even his father, could cure his illness.

Then these psychosomatic disorders seemed to frame the whole period (1837–1872) during which Darwin constructed and put on paper the theory of evolution. Darwin was subject to numerous and violent psychological stresses. The main reason for this stress was supposedly crucial conflicts. These conflicts were a hidden conflict with the paternal side of his family—and, coinciding, a conflict with the male-dominated religious-cultural background of his time. Both conflicts were (1) too dangerous to become fully aware to his conscious mind and (2) were supposedly in part rooted in conflicts between unconsciously effective currents in the spirit of his time. As a result, Darwin (3) internalized the fights appearing on the familiar and social level so that they became part of his own embodied personality, and (4) as a visible result, he externalized them in somatic (symbolically transformed) forms as various sickly manifestations. And (5) when the new spirit inside Darwin that was linked with his maternal side won important battles in his inner conflict, these victories appearing in the publication of his main works on the theory of evolution, his psychosomatic illness was markedly reduced. Point 3 of the somatization of conflicts suggest that instead of facing them directly Darwin reliably became immediately ill whenever he was called to a debate. Part of the complex was Darwin's religious problem. What happened when the former religious beliefs no longer held true for him? The fact that Darwin neurotically avoided funerals (which were of course very religious rituals at that time) may hint at (6) how horrible some final questions were to him, but (7) he was happily compensated on the other side by the digging earthworm!

The image of the *Pater familias*, associated with the imposing physique and personality of his father (Fig. 18.2), is not unlike that of "God the Father," that authoritarian and not always benevolent "Almighty God" that drove Adam and Eve out of Paradise. In Charles' case, his father's imposing and authoritarian presence (and all that was symbolized by him) likely led to an experience of physical and psychological suffering and the desire to move away from this environment. First, he fled to his world tour, and then his conflicts fled into his diseases.

In contrast, the image of the maternal line is that of welcome, pleasure, gentleness, and freedom. It is also that of understanding between the generations with an uncle (Josiah II) who discussed things with Charles and defended him against God the Father. It was Uncle Jos who convinced young Charles's father to let him board the Beagle. Originally an uncle, Jos also became Charles's father-in-law when Charles married Emma. This maternal line is comfort, even in the times of Charles' own illness. While Charles' father (who we associate with "God the Father") was powerless to cure Charles' problems, Emma "the nurse" (who we may associate with "Mother Earth") brings immense comfort to her husband Charles. This maternal and mothering lineage is also that of the earth. Through ceramics, the Wedgwoods are "people of the earth."

In addition to the paternal and maternal ancestries, Charles seems to have repressed the loss of his mother at a young age to death. All the authors who have written about Charles' illness believe that his mother's death and unresolved grief may have played an important role in the psychological development of young Charles (Bowlby, 1990, pp. 463–465). The loss of the original female made him suffer, and its return brought him relief.

18.3 Darwin, the Soil and the Earthworm

Charles' relationship to the earth was noted at the beginning of this text. Beyond geology, Charles was a great observer of the soil, in particular of one of its inhabitants, the earthworm. He became interested in earthworms in 1837 (Darwin, 1838, 1840), and his last scientific work, *The Formation of Vegetable Mould Through the Action of Worms with Observations on Their Habits* (Darwin 1881), concerned soil and earthworms (Fig. 18.1). A French edition followed in 1882. The most recent French edition (Darwin, 2001) corresponds to a new translation with a long introduction by Patrick Tort, an edition to which we will refer later in this text. For scientific analyses of Darwin's work on earthworms, see Feller et al. (2003, 2006) and Browne et al. (2003). The focus of this chapter is more the place of this work in Charles' life and how he treated the character "earthworm" rather than the detailed scientific content of these works.

18.3.1 The Common Life of Darwin and the Earthworm

Charles first met the earthworm in 1837. He was 28 years old, just back from his trip on the Beagle. Already famous for his shipments of samples collected during the trip and his comments, Charles was immediately welcomed to the prestigious London Geological Society where he was asked to give scientific communications. And what subject did he propose to his audience for his fourth intervention?

> The formation of vegetable mould by earthworms.

Some of the members were angry, hoping for something grander than a worm speech (Desmond & Moore, 1992).

After returning from his trip on the Beagle, Charles visited Uncle Josiah II. While they were walking in a meadow, Josiah told his nephew that he saw the layer of ash and debris from marl and bricks that had covered the soil disappear from the surface of the meadow in just a few years. The two accomplices dug a trench (Fig. 18.4) in the underlying soil and found the debris 7–10 cm deep. They deduced that there were earthworms that, through their droppings which were composed of deep earth, covered the debris that was on the surface!

Charles immediately drew an exceptional conclusion from this observation: in 15 years, earthworms were able to mix the first 10 centimeters of the soil of England. That meant that all the soil in the vegetal layers of gardens, fields, and meadows passed through the intestines of earthworms in just a few years. In other words, earthworms constantly changed this layer of soil that is so important for agriculture. Moreover, Darwin believed that this layer should not be called "vegetable mould" but rather "animal mould."

Throughout his life (1837–1880), alone or with his sons and through thousands of experiments, Charles studied the action of earthworms on the soil, their behavior, and their importance to humanity. The earthworm was well established in Darwin's brain (Fig. 18.5). Beyond the role of earthworms in the formation of soils and in the improvement of their fertility, Darwin was interested in many other aspects of earthworms (Feller et al., 2003, 2006; Brown et al., 2003) such as: their sensory (sight, hearing, smell, touch, taste) and mental qualities. To study these aspects, he made thousands of observations in the field or through experiments. For intelligence, he imagined how the worms were able to manage their food that is to bury the leaves that they consume in their galleries. He concluded that:

> ... worms, although ranked low on the animal organization scale, have some degree of intelligence;

- their role in the erosion-sedimentation cycle, which made him think that the earthworm, at the level of planet Earth, participates in the shaping of landscapes and the formation of sediments,
- their importance in the protection of archeological remains (Fig. 18.6), because without them, these would be exposed and therefore looted.

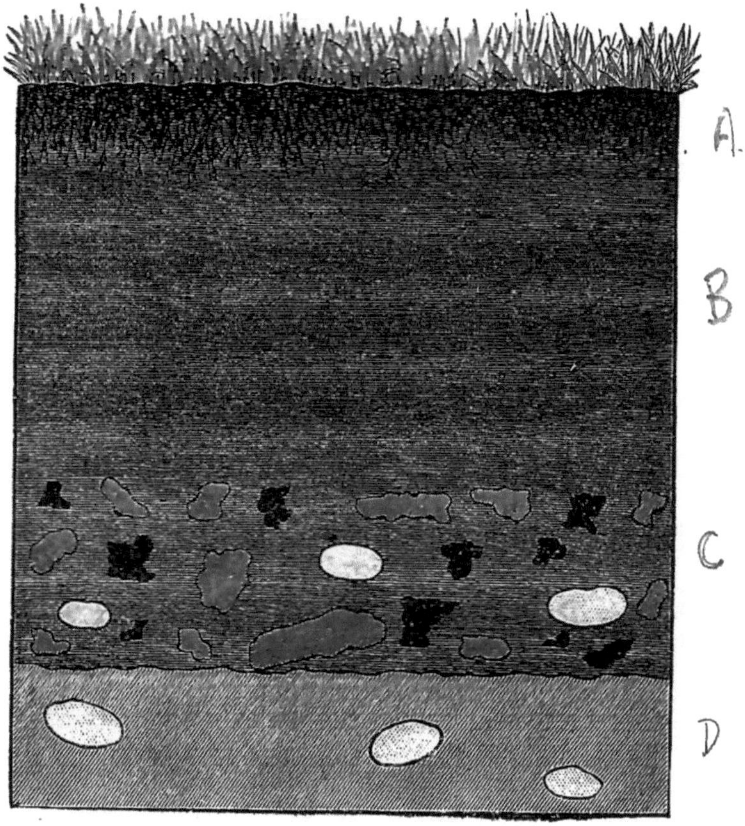

Fig. 18.4 Section of the vegetable mold in a field, drained and reclaimed 15 years previously. (**a**) turf; (**b**) vegetable mold without any stones; (**c**) mold with fragments of burnt marl, coal-cinders, and quartz pebbles; (**d**) subsoil of black, peaty sand with quartz pebbles. (Illustration scanned from Darwin (1882, p. 110))

All these observations lead Charles to consider this insignificant animal as very important to humanity and practically as a "friend of man." A synthesis of this work of about 40 years was published in 1881 (Darwin, 1881). It was his last scientific publication and came out 6 months before his death, one could say "before [the earthworms] join"!

With this work on the soil and the earthworm, Darwin is recognized today as the founder of soil ecology and of the concepts of biomantles and bioturbation. Dokuchaev disputed the generalization made by Darwin on the sole effect of earthworms on soil bioturbation (Dokuchaev, 1967, p. 337), but in general, the majority of the quantitative data Darwin published are still considered valid today (Feller et al., 2003).

Charles illustrated his book with various perfectly represented soil profiles (Fig. 18.4): a soil profile already illustrated its first publication in 1838, and Darwin was probably the first to publish such diagrams.

Fig. 18.5 The earthworm was well established in Darwin's brain. (The worm turns—cartoon of Charles Darwin, 1881–1882, Popular Science Monthly Volume 20). https://upload.wikimedia.org/wikipedia/commons/0/00/PSM_V20_D414_The_worm_turns.jpg)

The worm is therefore a friend of the soil scientist. One could say that with Charles, the earthworm changed status: it became the friend of Man (Blanchart et al., 2005). The penultimate sentence of the book shows Charles' feeling about the earthworm (Darwin, 2001):

> ... It is doubtful that there are many other animals that have played as important a role in world history as these creatures of a lower organization.

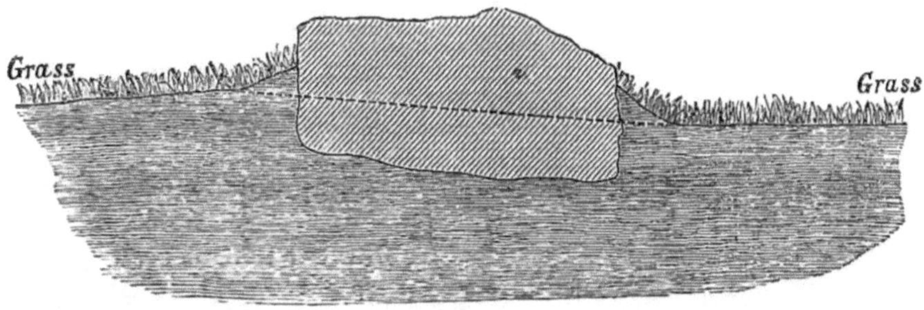

Fig. 18.6 Section through one of the fallen Druidical stones at Stonehenge, showing how much it had sunk in the ground. (Illustration scanned from Darwin (1882, p. 128))

18.3.2 The Success of the Worms in 1881 and the Pleasure of Darwin

The work, published on October 10, 1881, was a great publishing success at the time (Graff, 1983, pp. 5–18; Feller et al., 2001, Darwin, 2001), with 3500 copies sold in one month and 8500 three years later, more than the sale of *The Origin of Species*! It was translated into many languages, including the French edition in 1882 (Darwin, 1882). When the book was published, a cartoonist from the humorous newspaper *Punch* considered the worm as the ancestor of Man (symbolized by Darwin) in the cycle of evolution (Fig. 18.7), in a way a rapprochement between the two books, the *Worms* and the *Origin of Species*, a point to which we will return later.

Charles expressed his pleasure many times regarding both his research work and the publication of his work on *Worms*. In various letters to Julius V. Carus (Darwin, 1888), he mentions:

> "... As far as I can judge it will be a curious little book" (September 21, 1880), or "The subject was for me a real hobby, and it can be treated with ridiculous details" (April 1881) and "It is a subject of little importance, and I do not know if it will interest some readers. But that interested me" (May 1, 1881).

In November 1881, in a letter to Mr. Thiselton Dyer.

> My book was received with almost laughable enthusiasm ...

From these few quotes, we can deduce the following points:

(i) Charles considered his work as "a curious little book" with "subject of little importance" and one that might not interest readers, but the subject was interesting to him. When Darwin wrote these words, perhaps he remembered the lack of enthusiasm that his communication had been received with in 1837 at the Geological Society of London. But it is also as if he did not situate this book in line with his other works ("curious book"), and probably, it did not matter if the book sold or not and interested the world or not. What was essential was the interest that Charles himself had in it.

Fig. 18.7 Man is but a worm. (Caricature published in Punch's Almanach for 1882. https://upload.wikimedia.org/wikipedia/commons/b/b5/Caricature_of_Charles_Darwin_Wellcome_L0003921.jpg)

(ii) Darwin associated the reception of the book with an "almost laughable enthusiasm" and said this book was "a real hobby." The fact that Darwin speaks of "hobby" when writing a book is quite extraordinary, because generally, for all his other works, he described the work of writing as real suffering:

> What a magnificent occupation natural history is, if there was only to observe, and never to write! (Darwin, 1888, vol. 2, p. 381)

Darwin's remarks therefore placed this book outside the subjects of his usual works, showing a real pleasure in writing, with humorous and joyful remarks about the reception of the work. It is also likely that this interest in earthworms, and the pleasure it brought to Darwin, was born with his beloved uncle Josiah II of the maternal line. We are very far from the throes of *The Origin of Species*. Let us not forget that the Darwinian adventure of the earthworm is completely, literally, and figuratively, for him and his own family (his wife and children), in the maternal line, "Mother Earth"! And with the earthworm, Darwin felt free, not being an actor in a war as with his other works. He enjoyed the fool's freedom of a man who playfully downplays how essential his statements are, thus protecting himself, this time, this one time, from criticism and attacks. What a good trick, what a mischievous joy!

18.4 God, Darwin, the Serpent, and the Earthworm

18.4.1 God and Darwin

As already mentioned in the introduction to this chapter, the theory of evolution also had as its hidden objective a struggle against the idea of a divine creation of nature. In other words, it challenged the work of God. In a letter to his son George in 1873, Charles mentions Voltaire who:

> … had discovered that direct attacks on Christianity ... have little permanent effect [and that] only slow, silent attacks of bias are good[2];

To fight against the Judeo-Christian God image was also to stand on the side of Adam and Eve, who disobeyed God by eating the fruit of the tree of good and evil, which was also the tree of knowledge. With their expulsion from the Garden of Eden, "thanks" to the serpent, Adam and Eve (in Hebrew "adamah" means soil that covers the earth and "eva" means life) lost eternity but entered the cycle of life and therefore death.

There is the God of heaven and that on earth. Charles, as a respectful son, did not disobey his father, whose goodness he often proclaimed (Robert provided for Charles' needs until Charles earned his own living by selling his works). Publicly, Charles acknowledged Robert's superiority; he did not openly disobey his father. However, he also did not obey Robert. Charles did not become a doctor or a clergyman, and he enlisted Josiah II's help to allow him to embark on the Beagle. Neither his father nor Erasmus, his grandfather, were scientific benchmarks for him. Charles did not want to owe anything to either his Father in heaven or to his father on earth.

It seems that Charles, who did not want to consider himself indebted to his own "God the father" and the rest of his paternal family for his intellectual inheritance, somehow moved away from it. He could identify with Adam who left the kingdom of God the Father and, as a consequence, joined that of Mother Earth. Here, we

[2] Darwin Correspondence Project, "Letter no. 9105," accessed on February 16, 2020, https://www.darwinproject.ac.uk/letter/DCP-LETT-9105.xml. Also published in *The Correspondence of Charles Darwin*, vol. 21

propose to associate the Mother Earth archetype (Jung, 1939a) in Darwin's unconscious as represented by the "maternal line," with soil ("vegetable mould") and the clay of ceramics. It could be inferred that Darwin, at the family as well as the cultural and concealed religious level, likely preferred the earth (Mother Earth) to heaven (God the Father).

This preference can also be found, but there consciously, in his scientific work. His famous book *On The Origin of Species* did not invoke a divine creator to understand the origin of the diversity of living things, including Man (the mankind) on planet Earth. According to Darwin, the diversity of living beings is a non-divine creation.

Thus, we may say that Darwin moved away from God the Father, both on the level of his personal life with a marked preference for his maternal line and in his worldview and scientific work with the theory of evolution, which aimed to weaken or replace the role of the God of heaven in the creation of the living including that of Man. The release of Man from Heaven was linked with the serpent/snake, and Darwin's hero and friend on Earth was the earthworm, with its appearance and symbolism being close to the snake and also in its meaning of the serpent in the myth losing heavenly existence to gain the earthly snake.

18.4.2 The Snake and the Earthworm

The earthworm and the snake are symbolically close to each other, because of their similar appearance and living space. In German, for example, the same word (Gewürm) can apply to all sorts of snakes, worms, and insect larvae. Eskinazi (1992) introduces her "History of Snakes" with the associations that can often be made between snakes and other crawling animals. These include the orvet, or "the earthworm, harmless earthworm, tireless worker swallowing, rejecting the earth, contributing in its own way to its renewal" (Eskinazi, 1992, p. 55).

Many cultural areas know the symbol of the snake in their creation myths, including the Aborigines of Australia; Massim of Papaousia, New Guinea; Dogon of Mali; Egypt with the god Atum sometimes shown in the shape of a snake, among the Hindus, in China, etc. (Sadaka, 2019, pp. 10–14). Sadaka (2019) described the symbol of the snake in connection with creation, destruction and rebirth, knowledge and wisdom, fertility and immortality, even with a cosmic force in life. Sadaka (2019) mainly considered the snake as a chthonian[3] symbol, in other words, related to the earth. We can hypothetically apply elements of the symbolism of the snake to that of the earthworm.

The Greek symbolic underground was "well populated with serpents" (Ogden, 2013, p. 248). In the positive aspect, the snake represented the *Agathos Daimon* and

[3] According to the *Universalis Encyclopedia*, the "chtonian" adjective relates to the earth, the catacombs, the underground world.

its female form *Agathe Tyche*: a helpful spirit from below that also dwelled in the fields, associated with the cornucopia. A second central symbolism of the snake was linked with the healing god Asklepios and his daughter Hygieia: that is, the healing chthonic God/Goddess in the form of the snake, or in human shape with the snake as consort (Ogden, 2013, p. 277 and passim.).

In Greek religion, the association of the snake with Goddesses was very present. One version of the symbol was the snake entwined around a vessel that means receptivity and creativity united. Another common symbol was the Goddess holding snakes in her hands. Thus, following Neumann, the symbolism of the snake incorporates female characteristics, showing an aspect of the creative mother Goddess, as well as male characteristics, showing the phallic fecundating consort of the Goddess. Not surprisingly, the "snake of the earth" was connected with the mystery of fertility (Neumann 1997, p. 143–145). In his article "Serpent de mères" ("Mother's Snakes"), Maillard (1992) also described the serpent, not only as a phallic symbol but also as one very often associated with the Mother Goddess. The snake knows the secrets of life and fertility. Maillard reminded us that Demeter, the Greek goddess of the earth and fertility, is often represented with an ear of wheat in one hand and a snake in the other. Zeus fecundated her in the shape of a snake, and there are images of Demeter (or Persephone) cultivating a garden of phalloi (see Chap. 2).

In Genesis, God said to the serpent:

> You walk on your belly and you will eat earth every day of your life (*Bible*, 1998, Genesis 3:14).

In a way, the serpent is therefore relegated to the rank of earthworm for eternity, because snakes have been decreed by God to feed on earth, just like worms. According to religious speculations around the serpent of paradise, it was associated with the first Man Adam, being his chthonic *daimon* who makes him disobey God (Jung, 1939b). Jewish legends even name the snake the first wife of Adam, namely, Lilith, with Lilith interpreted as the devil's daughter, the devil being a son of God (Jung, 1952). This shows that in some Jewish beliefs, the snake was not just an "animal" of God's creation, it represented a split-off part, or an opponent to God himself, and by that having its own divine quality, too. Therefore, in Christian dualism, the devil was often called "the big serpent." This interpretation goes very deep, but for the purpose here, we can just state that the snake in the Christian realm was associated with the unconventional feminine, this one being traced back to a divine part that has been split off from the official god image.

But even though the symbolism of the snake is mainly negative in Christianity, according to Hannah (2006): "The serpent that comes from the earth is much more positive (than the Christian serpent), for it comes with the healing of nature." Its more general symbolism is ambivalent, showing the chthonic world of "dark" drivers as well as healing force and wisdom (Jung, 1952). In European symbolic alchemy, the *humidum radicale* (root-like humidity) as well as the *spiritus vitae* (spirit of life) were addressed with the symbol of the snake, said to be the healing agent and a driver of development in life (Jung, 1944).

Now we may hypothesize: The big story behind the Darwinian move to change the worldview of people aims at the redemption of the serpent—it was the serpent sitting on the "tree of knowledge" that seduced Eve and Adam to taste its fruits. It was the serpent that was condemned by Jahwe to "eat the earth" and humans to eat the fruits of the earth. Thus, the serpent was possibly the appropriate strong symbol in Darwin's unconscious too, to seduce him to the tree of knowledge and to eat the (also) sweet fruits of the earth.

The earthworm can be interpreted as the "little snake." That means the earthworm symbolized not the "big snake" that was seen by the terrified antimodernist popes of his time behind the theories of Darwin, but the earthworm is the snake in a form closer to humans that does not cause fear. Thus, by giving tender attention and good words to the earthworm, Darwin in a way addressed the religious split and conflict between "God and the Serpent" in an indirect way. In the shape of the earthworm, Darwin gave back to the symbolism of the snake its creative and friendly side. In Charles, the serpent-worm who asserts itself against "God the Father" is not to be cursed but, on the contrary, to be recognized as a benefactor of humanity. And in a subtle way, one may see here a shift from the heavenly creation to the earthy creativity.

Based on this interpretation, the worm, this tiny little but strong and friendly response to the unconscious serpent, has become a hero. The kingdom of the worm is "this world"—the soil, the earth, where it works "miracles" for humanity: (i) the worm makes earth useful to man to the point that this earth should be called "animal" rather than "vegetable" mold; (ii) the worm plows the land, even in the absence of the peasant; (iii) this land that, thanks to the worm, sees its fertility grow and thus nourishes humanity better (fertility); and (iv) the landscapes and their beauty owe a lot to the worm.

As Phillips (1999, p. 47) said of *Darwin's worms:*

> … worms that created the earth.

and:

> Everything happens as if the earth was reborn again and again through the body of the worms. Darwin replaced a myth of creation with a myth of maintenance (p. 60).

18.4.3 There Is the Heaven, There Is the Earth ... Two Works That Make It One

The *Origin of species* (Darwin, 1859) and *The Worms* (Darwin, 1881) are both eminent scientific works and recognized as such, but they may also be charged with a strong unconscious load that Charles obviously ignored since he did not go on Freud's couch, that is, did not confront what happened in his inner world! Let us recall what Freud said of science in 1916 in the *Introduction to Psychoanalysis*

(Freud, 1916, pp. 266–267). This quote is often referred to as "the three narcissistic wounds inflicted on man":

> Over the centuries, science has inflicted two serious denials on the naive selfishness of mankind. The first time was when it showed that the earth, far from being the center of the universe [Copernicus]... The second denial was inflicted on humanity by biological research, when it reduced to nothing the claims of man to a privileged place in the order of creation, by establishing his descendants of the animal kingdom and by showing the indestructibility of its animal nature [Darwin]... A third denial will be inflicted on human megalomania by psychological research these days which proposes to show to the "self" that he is not only master in his own house, that he is reduced to being content with information rare and fragmentary about what is happening, outside of his consciousness, in his psychic life. Psychoanalysts are neither the first nor the only ones who launched this call for modesty and meditation, but it is to them that the mission seems to fall to extend this way of seeing with the greatest zeal and to produce its support of materials borrowed from experience and accessible to all. Hence the general outcry against our science, the forgetting of all the rules of academic politeness, the unleashing of an opposition that shakes all the shackles of an impartial logic.

If we give a little credit to the third denial, to this "ego" who is no longer master at home due to the existence of the unconscious, we can also consider applying this approach to the field of scientific production, which does not detract from it. Charles Darwin is a fine example.

In *On The Origin of Species*, the hidden character that Charles consciously and unconsciously confronts through the theory of evolution is the God his culture believed in. He is not the first to confront the Christian God image. In France, Lamarck had attempted it but had been strongly attacked by Cuvier (Continenza, 2018, pp. 17–36), and then, a little later in England (at the time of Charles), Grant and Chambers fought the "fixist" Owen (Tort, 2000, p. 67), who was also called "The British Cuvier" (Continenza, 2018, p. 33). Confronting the church is therefore a difficult fight, which explains Charles' two decades of waiting before daring to publish his theory. According to Charles' strategy, "it is better to make slanting, slow and silent assaults" (see Sect. 18.4.1 and Footnote 2). As noted by the Lacanian psychoanalyst Jean-Pierre Winter (2001), this world of the father can only be bearable, provided that:

> ... the father is rejected out of the earthly world, that he long live or survive in eternal and infinite spaces and respond to the poet's prayer [Jacques Prévert, 1945]:
>
> Our Father who is in heaven.
> Stay there!
> And we will stay on our earth.
> Who is sometimes so pretty.

But what Charles did not perceive is that there was a coincidence and association of the fight at the cultural level and the fight at the personal level, which was the fight against his own father, "Almighty God the Father." His inner struggles (visible in his externalized symptoms of illness) were unavoidable on Darwin's way to his considerable contribution to human thought in general and scientific thought in particular, by explaining the history and evolution of the living on planet Earth without involving any divine force. On a personal level, there was a kind of reconciliation, as his father ended up admiring his son (Bowlby, 1990, p. 283), the son who so much felt like he was not appreciated by his father and who, in fact, ultimately refused all of the life choices offered by Robert. On the cultural level, the fight between "creationists" and "Darwinists" that started with Darwin's theory on the origin of species still exists in some societies today. And what happened on a transpersonal level with the snake, we don't speculate here…

We could also mention here a third fight, unconscious or not, and that is the fight against the English society of the second half of the nineteenth century. The rich and well-to-do of this world diverted Charles' theory of evolution to make "social Darwinism," a theory which, through the principle of "struggle for life" and the "law of the strongest," justifies the exploitation of man by man in this colonial and Victorian era (Tort, 2000, p. 131–135). We just want to point out that the animal species that Darwin gave most of his attention was not the lion, but the earthworm. For social Darwinists, the lion would have been the better role model. No wonder they did not care about the earthworm, in spite of the fact that Darwin showed that insignificant animals like the earthworm "have played such an important role in the history of the World," also demonstrating, by the way, that little people can play a major role for humanity.

But all this came at a cost, which was the loss of Charles' health. It was necessary to wait for the end of the publications linked to the theory of evolution, around 1872, before the sickly symptoms subsided. The new paradigm had won the battle inside Darwin. And Darwin found a second way instead of fighting, that is, gently undermining (if undermining ever can be "gentle") the worldview of his time with the help of the harmless yet powerful "little snake," that is, the earthworm. The snake seduced Eve and Adam, the earthworm seduced Charles Darwin and gave him joy and satisfaction, a compensation for all the hard struggles between the "big animals" in his inner soil that undermined his health. No wonder Darwin made the earthworm his hero.

The worm is the one who, through its burials and incessant activity of "digestion" of death, brings life out. This worm stands in contrast to the Christian God image: as interpreted by Phillips (1999, see above), one could consider the worm as "the creator of the earth." Continuing this interpretation, in Charles' final work, the serpent prevailed over God and the earth prevailed over heaven. It was on the one side a painful fight and on the other side an easy and pleasant victory because it was a question of saving Mother Earth, which was in Darwin's new paradigm effectively more important for humanity than the Garden of Eden of God the Father. It was also a display of Charles Darwin's preference for the maternal line—this earthly line—which was associated in Charles' mind with moments that were pleasant and/or essential for him (journey on the Beagle, marriage to his cousin, and ... the earthworms).

For Darwin, it was necessary to give supremacy to the earth over heaven, and this was done with the *Worms*! Finally, while the work on the *Worms* is perhaps Darwin's book that seems to have the least to do with the *Origin of Species*, maybe it is only the hidden and unconscious continuation of the same struggle against his father and

God the Father for the benefit of his mother and Mother Earth, in a way two works that make it one.

With *The formation of Mould by the Action of Worms*, we have a fine example of an "external" soil science associated with the eminently effective "inner" soil of Darwin's unconscious.

References

Bible (1998). *La Bible de Jérusalem.* Nouvelle édition revue et corrigée. Paris, Desclée de Brouwer, 2169 p. [The Jerusalem's Bible.]

Blanchart, E., Brown, G. G., Chernyanskii, S. S., Deleporte, P., Feller, C., & Goulet, F. (2005). Perception et popularité des vers de terre avant et après Darwin. In C. Feller et J.P. Legros D (eds), N° Spécial « Histoires d'Hommes et de Concepts en Science du Sol ». *Étude et Gestion des Sols, 12*, 145–152. [Perception and popularity of earthworms before and after Darwin.].

Bowlby, J. (1990). *A new biography* (xiv + 511 p). Hutchinson.

Brown, G. G., Feller, C., Deleporte, P., & Chernyanskii, S. S. (2003). With Darwin, earthworms turn intelligent and become human friends. *Pedobiologia, 47*, 924–933.

Cahiers Jungiens de Psychanalyse (1992). *Le Serpent.* N° 72 (1er trimestre 1992). Commission paritaire, Paris, 88 p. [The Serpent.]

Colp, R., Jr. (1977). *To be an invalid. The illness of Charles Darwin* (285 p). The University of Chicago Press.

Continenza, B. (2018). *Darwin. L'arbre de vie* (217 p.). Belin/Humensis. [The tree of life.]

Darwin, Ch. (1838). On the formation of mould. *Proceedings of the Geological Society of London, 2*, 274–276.

Darwin, Ch. (1840). On the formation of mould. *Transactions of the Geological Society London II, Ser, 5*(III), 505–509.

Darwin, Ch. (1859). *On the origin of species by means of natural selection, or the preservation of favoured races in the struggle for life.* John Murray.

Darwin, Ch. (1871). *The descent of man, and selection in relation to sex* (Vol. 2). John Murray.

Darwin, Ch. (1872). *The expression of emotions in man and animals.* John Murray.

Darwin, Ch. (1881). *The formation of vegetable mould through the action of worms with some observations on their habits.* John Murray.

Darwin, Ch. (1882). *Rôle des vers de terre dans la formation de la terre végétale*, Traduit de l'anglais par M. Levêque. Préface de M. Edmond Perrier. C. Reinwald Libr.-Ed., Paris, xxviii + 264 p. [Role of earthworms in the formation of vegetable mould.]

Darwin, Ch. (1985). *Darwin (1809–1882). Autobiographie. La vie d'un naturaliste à l'époque victorienne.* Texte original restitué, présenté avec annexes et notes par la petite-fille de Charles Darwin, Nora Barlow. Belin, Paris, 175 p. [Autobiography. The life of a naturalist in the Victorian era.]

Darwin, Ch. (1986). *Charles Darwin. Autobiographies.* Edited by Michael Neve and Sharon Messenger. With an Introduction by Michael Neve. Penguin Books, xxvi + 98 p. [Autobiography. The life of a naturalist in the Victorian era. Based on the original text, presented with appendices and notes by Charles Darwin's granddaughter, Nora Barlow.]

Darwin, Ch. (2001). *La formation de la terre végétale par l'action des vers avec des observations sur leurs habitudes.* Précédée de Patrick Tort, « Un regard vers la terre ». Éditions Syllepse, Paris, 196 p. [The formation of vegetable mould through the action of worms with some observations on their habits. Preceded by Patrick Tort, "A look at the earth".]

Darwin, E. (1794, 1796). *Zoönomia; or the laws of organic life* (2 vol.). J. Johnson.

Darwin, F. (1887). The life and letters of Charles Darwin, including an autobiographical chapter (3 vol). John Murray.

Darwin, F. (1888). *La vie et la correspondance de Charles Darwin. Avec un chapitre autobiographique*. Traduit de l'anglais par Henry C. de Varigny, Paris, C. Reinwald Libraire-Editeur, 2 vol., 1888, 701, 794 p. [The life and correspondence of Charles Darwin. With an autobiographical chapter.]

Darwin, F., & Seward, A. C. (1903). *More letters of Charles Darwin* (2 vol.). John Murray.

Desmond, A., & Moore, J. (1992). *Darwin* (xix + 808 p.). Pengui Books.

Dokuchaev, V. V. (1967). *Russian Chernozem. Selected Works of V.V. Dokuchaev, Vol. 1. Translated from Russian, Israel Program for Scientific Translation* (ix + 419 p.). US Department of Agriculture and National Science Foundation.

Eskinazi, M.-C. (1992). Histoire de serpent. In *Cahiers 1973 de Psychanalyse, Le Serpent*. 1992. N° 72 (1er trimester 1992). Commission paritaire, Paris, 88 p. [Snake/serpent story.]

Feller, C., Brown, G.G. and Blanchart, E. (2001). Darwin et le biofonctionnement des sols. *Étude et Gestion des Sols, 7,* 395–402. [Darwin and the biofunctioning of soils.]

Feller, C., Brown, G. G., Blanchart, E., Deleporte, P., & Chernyanskii, S. S. (2003). Charles Darwin, earthworms and the natural sciences: Various lessons from past to future. *Agriculture, Ecosystems and Environment, 99*, 29–49.

Feller, C., Blanchart, E., & Yaalon, D. H. (2006). Chapter 4, some major scientists (Palissy, Buffon, Thaer, Darwin and Müller) have described soil profiles and developped soil survey techniques before 1883. In B. P. Warkentin & D. H. Yaalon (Eds.), *Down to earth: A soil science history* (pp. 85–105).

Freud, S. (1916). *Introduction à la psychanalyse*, IIe partie, chap. 18, trad. S. Jankélévitch, Payot, coll. « Petite Bibliothèque «, 1975, p. 266–267. [Introduction to psychoanalysis.]

Graff, O. (1983). Darwin on earthworms – The contemporary background and what the critics thought. In J. E. Satchell (Ed.), *Earthworm ecology: From Darwin to Vermiculture*. Chapman and Hall.

Hannah, B. (2006). *The Archetypal Symbolism of Animals. Lectures given at the C.G. Jung Institute, Zurich, 1954–1958*. Chiron Publications.

Jung, C. G. (1953-). The Collected Works of C. G. Jung. Edited by Herbert Read, Michael Fordham, Gerhard Adler, 20 vol. London (Routledge). [*The English and German editions adopt similar paragraph numbers*].

Jung, C. G. (1921 1st ed., 1960). *Psychologische Typen*. Cited following Collected Works, vol. 6: § 632. [Psychological Types.]

Jung, C. G. (1939a 1st ed., 1954). *Die psychologischen Aspekte des Mutterarchetypus*. Cited following Collected Works vol. 9/1, § 148–198. [The psychological aspects of the mother archetype.]

Jung, C. G. (1939b 1st ed.; 1954). *Aion. Beiträge zur Symbolik des Selbst*. Cited following Collected Works vol. 9/2, § 385. [Aion. Researches into the phenomenology of the self.]

Jung, C. G. (1944 1st ed.; 1952). *Psychologie und Alchemie*. Cited following Collected Works vol. 12, § 527 f. [Psychology and Alchemy.]

Jung, C. G. (1952 1st ed. 1967). *Antwort auf Hiob*. Cited following Collected Works vol. 11, § 619. [Answer to Ijob.]

Lamarck, J.-B. (1809). *Philosophie zoologique*. Dentu. [zoological philosophy.]

Maillard, C. (1992). Serpents de mère. In *Cahiers Jungiens de Psychanalyse, Le Serpent*. 1992. N° 72 (1er trimestre 1992). Commission paritaire, Paris, 88 p. [Mother/Sea snakes.]

Milton, J. (1667 original edition). 1995: Paradise lost. A poem written in ten books. P. Parker, Creed Church neer Aldgate. New Frenche edition in 1995, « Le Paradis perdu ». Trad. de l'anglais par François-René de Chateaubriand. Édition de Robert Ellrodt. Gallimard, Paris, 434 p.

Neumann E. (1997, 1st ed. 1974). *Die Große Mutter. Eine Phänomenologie der weiblichen Gestaltungen des Unbewussten*. Zürich. [The Great Mother. A phenomenology of the female formations from the unconscious.]

Ogden, D. (2013). *Drakōn*. Dragon Myth and Serpent Cult in the Greek and Roman Worlds.

Paley, W. (1804). *Théologie naturelle ou preuves de l'existence et des attributs de la divinité, tirées des apparences de la nature*. Trad. de l'anglais par Ch. Pictet. Impr. De la Bibliothèque britan-

nique, Genève, xvi + 391 p. [Natural Theology: or, Evidences of the Existence and Attributes of the Deity; Collected from the Appearances of Nature.]
Phillips, A. (1999). *Darwin's worms* (48 p). Faber & Faber Ltd.
Prévert, J. (1945). *Paroles* (224 p.). Éditions du Point du Jour. [Paroles.]
Sadaka, J. (original ed. 2014, 2019). *Le Serpent. Symboles, mythes et caractères* (135 p.). Mon Petit Editeur. [The serpent. Symbols, myths and characters].
Tort, P. (2000). *Darwin et la science de l'évolution* (159 p.). Gallimard. [Darwin and the science of evolution].
Winter, J.-P. (2001). *Choisir la psychanalyse* (202 p.). France Loisirs. [Choose psychoanalysis.]

Part V
Soil Relation and Care

Chapter 19
Taking Soil Care Seriously: A Proposition

Anna Krzywoszynska

19.1 Introduction

How to conceptualize and encourage connections between people and soils that would create good outcomes for both soils and societies has been an important debate in soil scholarship (e.g., Bouma, 2015; Grunwald et al., 2017; Salazar et al., 2020). The term "care" has rarely been debated in the soil science literature (except Yaalon, 1996), although in practice, it is often associated with soil protection and conservation efforts. For example, the Food and Agriculture Organization of the United Nations motto for World Soil Day 2017 was "Caring for the Planet starts from the Ground," and the Horizon 2020 Soil Health and Food Mission Board's Roadmap, titled "Caring for soil is caring for life," seeks to embed soil care in societal values and practices (Soil Health and Food Mission Board, 2020).

Care also reoccurs in the naming conventions of contemporary soil research and action programs, such as Landcare in Australia and European Union (EU) funded projects SoilCare and RECARE (Preventing and Remediating degradation of soils in Europe through Land Care). Inspired by this frequent use of the care concept in relation to efforts to create good futures for soils and societies, in this chapter, I engage with understandings of care developed within feminist, eco-feminist, and science and technology studies literatures. I propose that the concept of care as proposed by these literatures is relevant to some of the key issues relevant to soil research, namely, how we come to know the needs of soils, what does it mean to be responsible for soils, and how to cultivate attention to soils. In this chapter, I use the term "soil research" to refer to soil-related inquiry as carried out within any discipline (not only soil science), as well as within nonacademic organizations such as think tanks, governance bodies, practitioner groups (e.g., farmer-led research), or

A. Krzywoszynska (✉)
Department of Anthropology, University of Oulu, Oulu, Finland
e-mail: anna.krzywoszynska@oulu.fi

N. Patzel et al. (eds.), *Cultural Understanding of Soils*,
https://doi.org/10.1007/978-3-031-13169-1_19

artistic practice. The aim of this chapter is to demonstrate the practical relevance of the concept of care to these arenas.

19.2 What Is Care?

In a broad sense, care can be understood as the ethic and practice of "looking after" something or someone, characterized by both an attentive interest and a concern for the well-being of the object or subject of care (Nassauer, 2011). Care has been a long-standing object of inquiry in feminist studies. Their approaches have critiqued understandings of social relations as interactions between contained and autonomous individuals and have instead stressed interdependence and relationality as the inescapable condition of human life; "the guiding thought of the ethic of care is that people need each other in order to lead a good life and that they can only exist as individuals through and via caring relationships with others" (Sevenhuijsen, 2003: 183). Fisher and Tronto described care as "… a species' activity that includes everything that we do to maintain, continue, and repair our 'world' so that we can live in it as well as possible. That world includes our bodies, our selves, and our environment, all of which we seek to interweave in a complex, life-sustaining web" (1991: 40). This systemic and practice-oriented view of care seeks to push thinking about care beyond the primary focus on individual sentiment or emotion and encourage the recognition that care is also a practice, a series of activities (see also Mol, 2008). As a result, care is also seen as a political issue, one that is relevant to the understanding and shaping of desirable societal relations (Tronto, 1993).

Building on this, care as an ethic and as a practice has been extended to thinking about eco-social relations, with eco-feminist and science and technology studies. Arguing that interdependence and relationality are an ontological condition of all being, Puig de la Bellacasa proposed an understanding of care as a relational activity between human and nonhuman entities: "... for interdependent beings in more than human entanglements, there has to be some form of care going on somewhere in the substrate of their world for living to be possible" (Puig de la Bellacasa, 2017: 5). Care thus becomes a way to think about the practical and ethical implications of building relations with more-than-human entities to ensure a livable future (Puig de la Bellacasa, 2017). Discussing soils in particular, Puig de la Bellacasa works with the food web model to argue for a relational perspective on care as circulating between soils and humans: "A care approach would look not only at how soils and other resources produce output or provide services to humans; it would look also at how humans are providing for the soil community in order to maintain, continue and repair this living web" (Puig de la Bellacasa, 2015, p. 703). Crucially, care ethic philosophers are careful to stress that care does not automatically result in good outcomes and that it can be turned into an instrument of oppression: care can be used as a form of control. As a result, caring itself must involve "an ongoing critical engagement with the terms of its own production and practice" (van Dooren, 2014, p. 292). Furthermore, care is always relational; it unfolds in relations, and its "goodness" can only be judged from within these relations; there is no absolute moral or practical blueprint for what constitutes good care. As a result, practicing care requires an attentiveness to our own participation in its becoming and the effects it produces, effects which are always contextual (Puig de la Bellacasa, 2012).

In the following sections, I discuss three issues that are of key importance to research efforts, which seek to create careful relations with soils. Firstly, caring means taking the needs of others as the basis for action. How do we come to know what the needs of soils are and how is this informed by how we understand soil relationalities? Secondly, caring means taking up the responsibility for the delivery of care. What does it mean to be responsible for soil care and what issues arise when we consider how the responsibility for soil care is distributed in society? Thirdly, care requires attentiveness to soils as the object of care, to relations that structure the care, and to care processes. How can we help to cultivate such attentiveness to soils? In the following, I first explore the relationship between problem framing and the understandings of what constitutes soil needs. I then discuss the question of responsibility for soil care and investigate the relationship between framing soil needs and "tasking" specific societal groups with soil care-related actions. In the final section, I explore the importance of attentiveness and highlight how attentiveness can change and transform soil need framings and so transform practices of soil care.

19.3 Needs: How Do We Come to Know the Needs of Soil?

The question of how to care for the needs of soil is intrinsically linked with how we conceptualize soils and relations between soils and societies[1]. Soils are relational; they come into being and function as dynamic assemblages between different materials and organisms (including humans), which co-constitute one another. Bring to mind a tree in an orchard. Follow in your mind's eye the tip of the root, with its complex associations of fungal hyphae, bacterial colonies, roots of other trees, with its immersion in water, air, and mineral media. Here comes the grower, compacting the soil with a tractor, spreading compost made from organic waste from a nearby town. Such a diversity of actors, human and nonhuman, all of whom can be seen as participating in soil functions and processes. Where does a tree root end and a soil begin? Where does a farmer end and a soil begin? How do we recognize "needs" in this system of relations, and how to do this in a way that creates good outcomes for soils, their associated ecosystems, and for humans?

There is not one single conceptual model of what a soil is and of what constitutes a desirable state of soil and its functions. The way that various materials, biophysical processes, and societal relations are interpreted and bundled up into the category "soil," and the ways these are evaluated and valued, is historically, geographically, and culturally specific. The variety of existing and historical conceptions of soils and soil-human relations has been explored in ethnopedology (Barrera-Bassols et al., 2006; Pauli et al., 2016; Williams, 2006) and in historical studies of soil science (Hartemink, 2016). Various scientific ways of defining soils can therefore be

[1] Indeed, even starting with a distinction between "soils" and "societies" here establishes a particular conceptualization of soil relationalities.

seen, like ethnopedological definitions, as societally constructed ways of making sense of soils (see, e.g., Richelle et al., 2018).

Caring means taking the concerns and needs of others, including nonhuman others (Puig de la Bellacasa, 2017), as the basis for action (Tronto, 1993, p. 105). In the light of the historical and geographical situatedness of our conceptual models of soils and considering their relational nature, deciding what constitutes a soil's need involves the process of framing. Frames identify and give meaning to a situation by defining what a situation is, what forms of knowledge about it are relevant, and what other situations it is related to (Halffman, 2019). They are a way of reducing the complexity of the world by identifying what matters. Because different people or groups hold different perspectives and values, or indeed different ontologies (Mol, 2002), framing processes are often a site of power struggle, and multiple frames can exist simultaneously.

How soils are framed within knowledge cultures, political discourses, and other arenas of life has consequences for how soil needs are conceptualized, experienced, and acted on. In relation to framing processes, two trajectories can be identified: a "closing down" and an "opening up" (Stirling, 2008). The objective of closing down is to strengthen a frame through, for example, "defining the 'right' questions, finding 'priority' issues, identifying 'salient' knowledge, recruiting 'appropriate' protagonists, adopting 'effective' methods, highlighting 'likely' outcomes" (Stirling, 2008, p. 279). This makes it possible to clearly indicate "possible courses of action (…) which appear as preferable under the particular framing conditions" (Stirling, 2008, p. 279). The linked objective is often to achieve the dominance of a particular frame, making it "a matter of fact": something that has well-defined boundaries and properties and the existence of which is seemingly "natural" or objective and so undisputable in relation to values or experiences (Latour, 2004). The "opening up" trajectory in contrast invites critical reflection on existing frames and proposes alternatives. Its objective is to pose alternative questions, explore neglected issues, include marginalized perspectives, triangulate contending knowledge, test sensitivities to different methods, consider ignored uncertainties, examine different possibilities, and highlight new options (Stirling, 2008: 280). When frames are opened up, "matters of fact" transform into "matters of concern": their properties and relations are no longer clear or beyond dispute, and the workings of the frames are made visible (e.g., power relations or value commitments which underpin them; Latour, 2004). The objective of "opening up" is to highlight the plurality of problem frames by exploring previously ignored attachments or conceptualizations and potentially to create a dialogue between them.

Struggles and debates over different framings of what constitutes good soil quality are a pertinent example. Being "of good quality" is often framed by different societal actors as a soil need—a state that is desirable for soils to have. However, what constitutes good soil quality, and how soil quality is to be measured, continues to be debated. Baveye (2020) argued that a singular, universal "soil quality" definition is not achievable; instead, researchers need to start with the different desired land uses or soil functions in defining multiple desirable "soil qualities." Such an approach shifts from attempts to achieve a soil quality "matter of fact," a strong undisputable frame that hides its underpinning values and seeks universality, to a multiplication of soil qualities as "matters of concern," taking into account their various and potentially conflicting value commitments to particular land uses and therefore to particular visions of eco-social relations.[2] Opening up "matters of fact" to uncover the many "matters of concern" serves to highlight the value commitments, which underpin an understanding of soil needs (see e.g. Setälä et al., 2014; Bartke & Schwarze, 2015; Valujeva et al., 2016: on trade-offs between soil functions).

Going further, problem framings themselves can be generative of care if they attend to the needs of those humans and nonhumans who contribute to the maintenance, continuation, and repair of the world. This move, from "matters of concern" to "matters of care," Puig de la Bellacasa argues, "involves not only detecting what is there, what is given in the thing we are studying, but also to think about what is not included in it and about what this thing could become—for instance if other participants were gathered by/in it" (Puig de la Bellacasa, 2011, p. 96).

Scoones' (2015) study of soil quality framings in the Ethiopian highlands is illustrative of such an approach. In development policies, these highland soils were framed as having poor quality. This poor soil quality was in turn framed as undermining peasant livelihoods. A top-down, large-scale, engineering-oriented approach to soil conservation was adopted, combined with a push for inorganic fertilizer as part of a "Green Revolution" package (Scoones, 2015, p. 21). The bottom-up perspective of the farmers, which Scoones (2015) engaged with, however, offered a different problem framing, in which different soils, their fertilities, gender dynamics, and peasant livelihoods were connected in complex ways. For the peasant families, keystone soils were those in the *darkoa* gardens; there, soil fertility was carefully managed, primarily by women's labor, through additions of domestic organic matter and complex cropping strategies. These small areas were, however, not included in development policies, which only accounted for the larger *shoka* fields, which were used for male-dominated, marketable monocultural crop production. Cultivation of *shoka* soils was a more risky and costly undertaking from a livelihoods' perspective. Policy support for soil quality improvements in these areas was not successful in creating long-term positive relationships between people and soils, which would lead to lasting soil quality and livelihood improvements. Reading Scoones (2015) study as opening up soil quality from a care perspective illustrates the importance of actively seeking the voices of those who are silenced or excluded from problem framings, such as *darkoa* gardens and women workers. It was they who in fact supported livable socio-ecological relations in the region, and it was their entangled needs that ought to have been the focus of soil quality and livelihood enhancement strategies.

[2] On the relationship between soil quality conceptualizations and values related to eco-social relations see Engel-Di Mauro (2014).

19.4 Responsibility: What Does It Mean to Be Responsible for Soil Care?

Thinking with care as a practice as well as an ethic calls up the question of responsibility; "to care implies more than simply a passing interest or fancy but instead the acceptance of some form of burden" (Tronto, 1993, p. 103). To care requires taking responsibility for the object or subject of care and practically responding to the need (taking care of), as well as organizing, delivering, and evaluating care (caregiving and care-receiving). How can we understand responsibility in relation to caring for soils? Who "owns" the issue of soil care, and whose role it is to act in caring ways toward soils? In the academic literature, environmental and resource governance is often seen as a focal point for efforts to embed soil care. In these approaches, responsibility is situated with the nation states and other governance bodies who are expected to embrace caring for soils as a key value. While some frameworks envisage the state as regulating land use in line with criteria based on soil properties, for example (Schulte et al., 2014), most soil governance approaches advocate bringing in a wide range of societal actors through multilevel, adaptive, networked, reflexive, and deliberative modes of governance (for a full discussion, see Weigelt et al., 2014). In these approaches, the soil science community is often ascribed a particular responsibility for educating others about the importance of soils and for the provision of expert advice to governance actors at all levels (Keesstra et al., 2016; Montanarella & Vargas, 2012; Bouma, 2015).

Soil scientists can thus be seen to participate in "responsibilization" (Shamir, 2008), in that they seek to create a new form of moral agency (concern for soil), which would lead to voluntarily action (care for soil). A number of authors argue that a promising pathway for this is through linking responsibility for soils with already articulated governance objectives, such as reducing carbon emissions (Minasny et al., 2017), ensuring food security (Panagos et al., 2016), sustainable development (Keesstra et al., 2018), securing of a functional environment (McBratney et al., 2014), or valuing nature appropriately within capitalism (Davies, 2017).

Taking a care perspective can add further dimensions to these governance approaches, helping to nuance responsibilization efforts. Critics argue that governance approaches are rooted in enlightenment ideas in which responsibility is linked with the idea of free, autonomous, and rational individuals (Pellizzoni, 2004) or organizations (Shamir, 2008) who have full knowledge of the situation, are able to anticipate consequences of their actions, and as a result can be held liable. Responsibility based on liability demands "a state that clearly knows what to ask for, and how to apply controls and sanctions" (Pellizzoni, 2004, p. 550). However, the uncertainty inherent in environmental issues, such as those pertaining to soils, and the waning of state authority in environmental governance make liability a weak instrument for ensuring responsible conduct (Pellizzoni, 2004). For example, in a study of the regionally devolved natural resource management governance in Australia, Wallington and Lawrence (2008) found that while farmers and regional communities were made responsible for identifying problems and solutions, their responsibility was framed as a duty to achieve pre-specified biophysical outcomes (indicators of natural resource quality). As a result, stakeholder behavior became limited to "reporting back" and avoiding blame. In agreement with Pellizzoni (2004), Wallington and Lawrence (2008) argued that in the governance of natural resource management, a different notion of responsibility is needed, based not on liability or accountability but also on future-oriented responsiveness: "a willingness to understand and confront the other's commitments and concerns with ours, to look for a possible terrain of sharing" (Pellizzoni, 2004, p. 227). Such an approach to responsibility would enable governance actors to articulate and debate their own objectives without being limited by the procedural demands of reporting against externally imposed metrics. These insights are very relevant to current debates in soil governance, which often focus on the appropriate role for soil health and/or quality indicators (e.g., Juerges & Hansjürgens, 2018).

In relation to care, Tronto (1993) argued that responsibility needs to be seen as a political issue, in that how responsibility for care (and so the labor involved in care) is societally distributed reflects and reinforces power inequalities and so should in itself be an object of public debate. To distribute the responsibility for care in a just way requires that "individuals and groups be frankly assessed in terms of the extent to which they are permitted to be care demanders and required to be care providers" (Tronto, 1993, p. 169). Because care is typically undervalued as well as costly, some people and groups seek to avoid care and occupy positions of "privileged irresponsibility," while others are unequally burdened with the work of repairing the damage.

In relation to soils, while the benefits of soil use typically accrue to private individuals or groups (e.g., food production, mineral extraction), the costs of soil degradation are typically borne by the society at large (e.g., water purification, soil remediation, health impacts; Tironi & Rodríguez-Giralt, 2017). A care perspective on responsibility encourages a critical and careful attention to how responsibility is framed in governance approaches through the asking of questions such as: what or who has responsibility toward (who or what benefits), what constitutes responsible conduct (what does good care look like), and who is to be responsible for what (how is care labor to be distributed). Furthermore, the care perspective stresses interdependencies, which governance approaches with their focus on rational individuals may neglect to consider; it sees all societal actors (including bio-physical actors, as per Puig de la Bellacasa, 2017) as both contributors to and recipients of care. Responsibility for caring is thus not conceptualized as a dyadic relation between individuals involved in a give-and-take; rather, care is "a generalized condition that circulates through the stuff and substance of the world" (Puig de la Bellacasa, 2017, p. 122). This shifts the question of responsibility from individuals who are expected to care in spite of adverse conditions, to collectives or broader socio-material arrangements, which can enable or preclude care.

Farmers, as individuals who act directly on soils, are often ascribed the responsibility for soil care in governance approaches through a combination of push and pull policies (Juerges & Hansjürgens, 2018). A care ethic approach complicates the idea of farmers' responsibility for soil care as an exercise of autonomous agency and draws attention to the interdependencies and relations they themselves operate within, beyond a simple recognition of the conflicting demands between resource utilization and preservation (c. f. Prager et al., 2011). Further, recognizing that care is both a material practice and a disposition shifts the focus from individual-focused push and pull tactics to the broader relations, which prevent or enable care. For example, farmers may be structurally excluded from participating in cycles of care when their food-growing is practiced in a context of uncertain short-term tenancies (e.g., Mahadevan, 2008), or in the wake of colonial cultural destruction and land appropriation (Green, 2020). Land justice in the form of secure land tenure and land rights for indigenous communities as well as resisting contemporary land grabs are thus a crucial element of soil care (Blaikie & Brookfield, 1987; Williams & Holt-Giménez, 2017; Hurlbert et al., 2019).

Further, a broader understanding of responsibility for soil care in food growing would lead us to ask: what systemic structures drive harmful practices? And what is the shape of socio-ecological relations that drive soil care? This requires a reflection on the functioning of the food systems within which farmers participate to, for example, identify actors who may be benefitting from soil damage while occupying positions of "privileged irresponsibility" in respect to soil care, such as corporations or retailers (Falkner & Kalfagianni, 2009), or to identify structural incentives to soil harm, such as the neoliberal deregulation of food trade (McMichael, 2011). An inspiring model for soil care as a systemic project is food sovereignty, a movement that states that the right of people to healthy and culturally appropriate food produced through ecologically sustainable methods necessarily demands social relations free of oppression and exploitation (Wittman, 2009). Such a framing rearticulates responsibility for soil care as emergent and inextricable from collective struggles for healthy and just socio-ecological relations, in food production (Lang, 2009) and beyond (Country et al., 2015). Overall, the care perspective on the question of responsibility for soils calls for a shift from a focus on individual agents acting in relation to clear reportable objectives and broadens the attention to relations that embed or prevent soil care as culturally and politically systemic.

19.5 Attentiveness: How to Cultivate the Art of Becoming Attentive to Soils?

In many ways, attentiveness is the most important aspect of care both as a practice and as an ethic. Attentiveness is the first step in identifying a need in someone or something other than oneself; it is also necessary throughout the care cycle to assess the quality of care and to learn about how the care may be improved. Inattentiveness to soils and a lack of soil education and awareness are much lamented among soil scholars (Davies, 2017; Kareiva, 2014; Bouma et al., 2012; Lobry de Bruyn et al., 2017). This lack of attentiveness to soils can be seen as both an individual and a collective issue, and cultivating attentiveness to soils thus requires action on both of these levels as well as on their interconnections.

For individuals, attention to soils is challenged by the limits of embodied perception. Processes within soils operate not only on scales, which are difficult to relate to through the body, but partly also on timescales, which challenge intuitive understandings of environmental processes. These challenges of scale and temporality mean that processes of soil change can be difficult to observe. Scientific forms of observations that make evidence of soil harm or soil degradation visible do not easily correspond with embodied forms of attention practiced by nonscientists and can be indeed actively resisted as relevant forms of evidence (see, e.g., Brockett, 2016: on contestations around soil carbon). As a result, there are some efforts in the soil research community to educate the broader public or specific groups (e.g., farmers) to pay attention to soils through the use of support tools and procedures, which focus on particular aspects of soils, such as structure (Ball et al., 2018), or earthworm presence (Stroud, 2019), hoping that attention to these indicators will generate and help direct practices of soil care.

Such attention-directing tools are significant beyond their particular uses in, for example, land-management decisions, in that their use creates situations from which an ethic of soil care can arise. Relational perspectives see ethics as emergent from within situations, rather than as imposed onto them from the outside through formalized moral frameworks (Levinas, 1969). The use of soil assessment tools can help to create situations of phenomenological recognition of soil as worthy of our noninstrumental ethical concern (Smith, 2011) and as needful of our assistance in some way, giving rise to the ethical impulse to care for soil (O'Brien, 2020). Furthermore, attentive engagements with particular aspects of soils can direct attention to the interconnectedness of the biotic and abiotic elements of soils and of the impacts of particular human activities on these. Attentiveness can unfold through a care network model (Krzywoszynska, 2019), in which an initial attentiveness to one aspect of soils leads to a further attentiveness to other interconnected aspects. For example, farmers' attentiveness to soil structure can lead to an attentiveness to soil biota and result in changes to land management practices so that the needs of soil biota are respected (Krzywoszynska, 2019). Attentiveness can thus have a transformative effect on human-soil relations, leading, for example, to a questioning of models of land use, which neglect the needs of soil organisms.

Attentiveness or inattentiveness to soils is also a societal structural issue, in that how soils are or are not perceived is informed by shared understandings and valuations. Many of the practices, which underpin modern societies, are built on the invisibility of soils as living ecosystems. For example, modern farming practices based on the combination of mechanical power and chemical nutrition have both resulted in and been enabled by a systemic loss of attention to soils as ecosystems (Krzywoszynska, 2020), while the rise of the modern city has similarly depended on a "backgrounding" of soils, which were sealed and transformed into media for covering infrastructure (Meulemans, 2020). Attentiveness is thus shaped both by an embodied capacity of noticing and responding to soils and collective framings of what counts as soils and about soils. In relation to perceptions of soil harm or soil damage, Schneider et al. (2010) found that among Swiss farmers, soil erosion was perceived as a normal part of farming; as a result, even easily observable spectacular erosion events were not strong drivers for the adoption of soil care practices. In contrast, among farmers practicing conservation agriculture and other farming systems based on minimal soil disturbance, attention to soil health as a perceptible reality and a shared professional value is emerging (O'Brien, 2020; Roesch-McNally et al., 2018).

As a result, instigating soil care means changing what and who is paid attention to.[3] Attentiveness may mean giving voice to those entities, such as soil microorganisms, who live in and make the soil and whose interests or needs are not usually recognized (Puig de la Bellacasa, 2015), and to those groups and individuals whose connections to soils are sidelined or disregarded, often as a result of historical injustices (Shiva, 2008). Further, taking a care perspective encourages questions not only around what the objects of attention in relation to soils are but also what temporalities are considered relevant. The lack of broad long-term soil monitoring and site-specific soil experimental work is today seen as a major problem in the soil science community, impeding the observation of long-term dynamics crucial to soil care (Richter et al., 2007). Similarly, the lack of secure long-term land access can be seen to limit the potential for long-term attentiveness to soils, which informs care as a form of skill (Krzywoszynska, 2016). Paying attention to temporalities of soil renewal can indeed suggest a need to revisit the fast temporalities of industrial food production (Puig de la Bellacasa, 2015). Crucially, the care perspective encourages the broadening of soil attention beyond the narrow field of soil experts. Practices of soil care are always localized and dynamic, developing in relation to different and changing bio-geo-physical conditions as well as different forms of land use and conflicting interests. Furthermore, care by its nature never ends; it is an ongoing activity that is constantly revised in relation to new knowledge and new needs (Mol, 2008). As a result, in optimal care situations, attention circulates at all scales, informing care practices in a virtuous cycle of learning. This attention includes attention to the care processes itself to ensure that care does not become another form of oppression or domination. A key challenge for the future of soil research lies in designing and supporting such circulation of soil attention without foreclosing the question of what matters in relation to soils.

19.6 Conclusions: Caring for Soils as Caring for Life

Soils need our care today, and we need soils to care for us now and into the future. This chapter argued that care theories as developed in the feminist, eco-feminist, and science and technology literatures can usefully inform research and governance efforts, which aim to create better relations between soils and societies. Firstly, care and the care ethic are relevant to the staging of soil-related situations and interventions; the very question of what constitutes "a soil" and what constitutes "soil

[3] For a fuller discussion of the potential of and limits to attentiveness in creating soil care, see Krzywoszynska (2019).

needs" can become an object of attention and of care. This will help to uncover the relationality of "soil needs," prevent the foreclosing of crucial questions around what the object of care should be and what activities are relevant, and help to include less powerful voices, both human and nonhuman. Secondly, the care ethic is relevant to the question of what it means to be responsible for soils, and how responsibility for soils is societally distributed. The care perspective challenges assumptions about autonomous agency embedded in soil governance efforts, which seek to instill a sense of responsibility for soils among different societal actors. By stressing interdependence, the care perspective encourages a shift in soil governance, away from a focus on individual agents acting in relation to clear objectives and toward attending to broader socio-material relations, which can support or prevent soil care. Thirdly, care is both sparked into being and propelled by attentiveness, which is both an individual and collective practice. Cultivating soil attentiveness requires the training of individual embodied perception, not only in order to inform people about soils from, for example, a scientific perspective, but also to create opportunities for the emergence of the impulse to care for soils, recognizing them as worthy of ethical concern. It also requires the creation of new collective framings, which link attention to soils with societally valued ways of living off and with the land.

Seen from the perspective of care, humans and soils can be seen to participate in unequally reciprocal cycles of care, which together enable livable worlds (Puig de la Bellacasa, 2015). Linking soils and humans as participants in a wider life-sustaining web of socio-ecological relations means moving beyond utilitarian approaches to framing soils, primarily in relation to how they may be best optimized to serve the needs of (particular) humans. Rather, in this care framing, the needs of soils are recognized as legitimate in and of themselves, and responding to those needs is seen as contributing to the soils' own processes of maintenance and repair and, thus, sustaining both soil ecosystems and their wider, life-enabling webs of relations.

Humans are situated as attentive participants whose efforts need to contribute to the maintenance and repair of this life-giving web. The care perspective thus encourages a rethinking of human needs, which have to be carefully coordinated in relation to activities and needs of all other entities and processes, biotic and abiotic, which enable the soils to exist and function. In other words, the care ethic perspective on soils aims to think with the question—what does caring for soils demand *of us*? Care perspectives can thus contribute to the important intellectual and political project of conceptualizing soils as independent of but entangled with human activities and of conceptualizing human well-being as intrinsically bound with the well-being of soils.

References

Ball, B. C., Hargreaves, P. R., & Watson, C. A. (2018). A framework of connections between soil and people can help improve sustainability of the food system and soil functions. *Ambio, 47*(3), 269–283.

Barrera-Bassols, N., Zinck, A. J., & Van Ranst, E. (2006). Symbolism, knowledge and management of soil and land resources in indigenous communities: Ethnopedology at global, regional and local scales. *Catena, 65*, 118–137.

Bartke, S., & Schwarze, R. (2015). No perfect tools: Trade-offs of sustainability principles and user requirements in designing support tools for land-use decisions between greenfields and brownfields. *Journal of Environmental Management, 153*, 11–24.
Baveye, P. C. (2020). Bypass and hyperbole in soil research: Worrisome practices critically reviewed through examples. *European Journal of Soil Science*, 1–20. https://doi.org/10.1111/ejss.12941
Blaikie, P., & Brookfield, H. (1987). *Land degradation and society*. Methuen.
Bouma, J. (2015). Reaching out from the soil-box in pursuit of soil security. *Soil Science and Plant Nutrition, 61*, 556–565.
Bouma, J., Broll, G., Crane, T. A., Dewitte, O., Gardi, C., Schulte, R. P. O., & Towers, W. (2012). Soil information in support of policy making and awareness raising. *Current Opinion in Environmental Sustainability, 4*, 552–558.
Brockett, B. (2016). *An interdisciplinary approach to mapping soil carbon*. PhD thesis, Lancaster University.
Country, B., Wright, S., Suchet-Pearson, S., Lloyd, K., Burarrwanga, L., Ganambarr, R., Ganambarr-Stubbs, M., Ganambarr, B., & Maymuru, D. (2015). Working with and learning from Country: Decentring human authority. *Cultural Geographies, 22*(2), 269–283.
Davies, J. (2017). The business case for soil. *Nature, 543*, 309–311.
Engel-Di Mauro, S. (2014). *Ecology, soils, and the left: An ecosocial approach*. Palgrave Macmillan.
Falkner, R., & Kalfagianni, A. (2009). *Corporate power in global agrifood governance*. MIT Press.
Fisher, B., & Tronto, J. C. (1991). Towards a feminist theory of care. In E. Abel & M. Nelson (Eds.), *Cricles of care: Work and identity in women's lives* (pp. 35–63). State University of New York Press.
Green, L. (2020). *Rock | Water | Life: Ecology and humanities for a decolonial South Africa*. Duke University Press.
Grunwald, S., Clingensmith, C. M., Gavilan, C. P., Mizuta, K., Wilcox, R. K. K., Pinheiro, É. F., Ceddia, M. B., & Ross, C. W. (2017). *Integrating new perspectives to address global soil security: Ideas from integral ecology* (Global soil security) (pp. 319–329). Springer.
Halffman, W. (2019). Frames: Beyond facts versus values. In E. Turnhout, W. Tuinstra, & W. Halffman (Eds.), *Environmental expertise: Connecting science, policy and society*. Cambridge University Press.
Hartemink, A. E. (2016). The definition of soil since the early 1800s. *Advances in Agronomy, 137*, 73–126.
Hurlbert, M., Krishnaswamy, J., Davin, E., Johnson, F. X., Mena, C. F., Morton, J., Myeong, S., Viner, D., Warner, K., Wreford, A., Zakieldeen, S., & Zommers, Z. (2019). Risk management and decision making in relation to sustainable development. In *Climate change and land: An IPCC special report on climate change, desertification, land degradation, sustainable land management, food security, and greenhouse gas fluxes in terrestrial ecosystems* [P. R. Shukla, J. Skea, E. Calvo Buendia, V. Masson-Delmotte, H.-O. Pörtner, D.C. Roberts, P. Zhai, R. Slade, S. Connors, R. van Diemen, M. Ferrat, E. Haughey, S. Luz, S. Neogi, M. Pathak, J. Petzold, J. Portugal Pereira, P. Vyas, E. Huntley, K. Kissick, M. Belkacemi, J. Malley, (eds.)]. In press.
Juerges, N., & Hansjürgens, B. (2018). Soil governance in the transition towards a sustainable bioeconomy – A review. *Journal of Cleaner Production, 170*, 1628–1639.
Kareiva, P. (2014). Foreword. In G. J. Churchman & E. R. Landa (Eds.), *The soil underfoot: Infinite possibilities for a finite resource* (pp. xi–xii). CRC Press.
Keesstra, S. D., Bouma, J., Wallinga, J., Tittonell, P., Smith, P., Cerdà, A., Montanarella, L., Quinton, J. N., Pachepsky, Y., van der Putten, W. H., Bardgett, R. D., Moolenaar, S., Mol, G., Jansen, B., & Fresco, L. O. (2016). The significance of soils and soil science towards realization of the United Nations Sustainable Development Goals. *The Soil, 2*, 111–128.
Keesstra, S., Mol, G., De Leeuw, J., Okx, J., Molenaar, C., De Cleen, M., & Visser, S. (2018). Soil-related Sustainable Development Goals: Four concepts to make land degradation neutrality and restoration work. *Land, 7*, 133.

Krzywoszynska, A. (2016). What farmers know: Experiential knowledge and care in vine growing. *Sociologia Ruralis, 56*, 289–310.

Krzywoszynska, A. (2019). Caring for soil life in the anthropocene: The role of attentiveness in more-than-human ethics. *Transactions of the Institute of British Geographers, 44*(4), 66–675.

Krzywoszynska, A. (2020). Nonhuman labor and the making of resources: Making soils a resource through microbial labor. *Environmental Humanities, 12*, 227–249.

Lang, T. (2009). Reshaping the food system for ecological public health. *Journal of Hunger and Environmental Nutrition, 4*, 315–335.

Latour, B. (2004). *Politics of nature*. Harvard University Press.

Levinas, E. (1969). *Totality and infinity. An essay on exteriority*. Duquesne University Press.

Lobry de Bruyn, L., Jenkins, A., & Samson-Liebig, S. (2017). Sharing soil knowledge to improve land management and sustainable soil use. *Soil Science Society of America Journal, 81*, 427–438.

McBratney, A., Field, D. J., & Koch, A. (2014). The dimensions of soil security. *Geoderma, 213*, 203–213.

McMichael, P. (2011). Food system sustainability: Questions of environmental governance in the new world (dis)order. *Global Environmental Change, 21*, 804–812.

Meulemans, G. (2020). Urban pedogeneses: The making of city soils from hard surfacing to the urban soil sciences. *Environmental Humanities, 12*, 250–266.

Minasny, B., Malone, B. P., McBratney, A. B., Angers, D. A., Arrouays, D., Chambers, A., Chaplot, V., Chen, Z.-S., Cheng, K., & Das, B. S. (2017). Soil carbon 4 per mille. *Geoderma, 292*, 59–86.

Mol, A. (2002). *The body multiple: Ontology in medical practice*. Duke University Press.

Mol, A. (2008). *The logic of care: Health and the problem of patient choice*. Routledge.

Montanarella, L., & Vargas, R. (2012). Global governance of soil resources as a necessary condition for sustainable development. *Current Opinion in Environmental Sustainability, 4*, 559–564.

Nassauer, J. I. (2011). Care and stewardship: From home to planet. *Landscape and Urban Planning, 100*, 321–323.

O'Brien, A. T. (2020). Ethical acknowledgment of soil ecosystem integrity amid agricultural production in Australia. *Environmental Humanities, 12*, 267–284.

Panagos, P., Imeson, A., Meusburger, K., Borrelli, P., Poesen, J., & Alewell, C. (2016). Soil conservation in Europe: Wish or reality? *Land Degradation & Development, 27*, 1547–1551.

Pauli, N., Abbott, L. K., Negrete-Yankelevich, S., & Andrés, P. (2016). Farmers' knowledge and use of soil fauna in agriculture: A worldwide review. *Ecology and Society, 21*(3), 19.

Pellizzoni, L. (2004). Responsibility and environmental governance. *Environmental Politics, 13*, 541–565.

Prager, K., Schuler, J., Helming, K., Zander, P., Ratinger, T., & Hagedorn, K. (2011). Soil degradation, farming practices, institutions and policy responses: An analytical framework. *Land Degradation & Development, 22*, 32–46.

Puig de la Bellacasa, M. (2011). Matters of care in technoscience: Assembling neglected things. *Social Studies of Science, 41*, 85–106.

Puig de la Bellacasa, M. (2012). 'Nothing comes without its world': Thinking with care. *The Sociological Review, 60*, 197–216.

Puig de la Bellacasa, M. (2015). Making time for soil: Technoscientific futurity and the pace of care. *Social Studies of Science, 45*, 691–716.

Puig de la Bellacasa, M. (2017). *Matters of care: Speculative ethics in more than human worlds*. University of Minnesota Press.

Richter, D., Hofmockel, M., Callaham, M. A., Powlson, D. S., & Smith, P. (2007). Long-term soil experiments: Keys to managing earth's rapidly changing ecosystems. *Soil Science Society of America Journal, 71*, 266–279.

Richelle, L., Visser, M., Bock, L., Walpole, P., Mialhe, F., Colinet, G., & Dendoncker, N. (2018). Looking for a dialogue between farmers and scientific soil knowledge: Learnings from an ethno-geomorphopedological study in a Philippine's upland village. *Agroecology and Sustainable Food Systems, 42*(1), 2–27.

Roesch-McNally, G., Arbuckle, J. G., & Tyndall, J. C. (2018). Soil as social-ecological feedback: Examining the "ethic" of soil stewardship among corn belt farmers. *Rural Sociology, 83*, 145–173.

Salazar, J. F., Granjou, C., Kearnes, M., Krzywoszynska, A., & Tironi, M. (2020). *Thinking with Soils: Material politics and social theory*. Bloomsbury.

Schneider, F., Ledermann, T., Fry, P., & Rist, S. (2010). Soil conservation in Swiss agriculture—Approaching abstract and symbolic meanings in farmers' life-worlds. *Land Use Policy, 27*, 332–339.

Schulte, R. P. O., Creamer, R. E., Donnellan, T., Farrelly, N., Fealy, R., O'Donoghue, C., & O'Huallachain, D. (2014). Functional land management: A framework for managing soil-based ecosystem services for the sustainable intensification of agriculture. *Environmental Science and Policy, 38*, 45–58.

Scoones, I. (2015). Transforming soils: Transdisciplinary perspectives and pathways to sustainability. *Current Opinion in Environmental Sustainability, 15*, 20–24.

Setälä, H., Bardgett, R. D., Birkhofer, K., Brady, M., Byrne, L., de Ruiter, P. C., de Vries, F. T., Gardi, C., Hedlund, K., Hemerik, L., Hotes, S., Liiri, M., Mortimer, S. R., Pavao-Zuckerman, M., Pouyat, R., Tsiafouli, M., & van der Putten, W. H. (2014). Urban and agricultural soils: Conflicts and trade-offs in the optimization of ecosystem services. *Urban Ecosystem, 17*, 239–253.

Sevenhuijsen, S. (2003). The place of care: The relevance of the feminist ethic of care for social policy. *Feminist Theory, 4*, 179–197.

Shamir, R. (2008). The age of responsibilization: On market-embedded morality. *Economy and Society, 37*, 1–19.

Shiva, V. (2008). *Soil not oil: Environmental justice in a time of climate crisis*. North Atlantic Books.

Smith, M. (2011). Dis (appearance): Earth, ethics and apparently (in) significant others. *Australian Humanities Review, 50*, 23–44.

Soil Health and Food Mission Board. (2020). *Caring for soil is caring for life*. Ensure 75% of soils are healthy by 2030 for healthy food, people, nature and climate: Interim report of the Mission Board for Soil Health and Food. Luxembourg, Netherlands.

Stirling, A. (2008). "Opening up" and "closing down" power, participation, and pluralism in the social appraisal of technology. *Science, Technology & Human Values, 33*, 262–294.

Stroud, J. L. (2019). Soil health pilot study in England: Outcomes from an on-farm earthworm survey. *PLoS One, 14*, e0203909.

Tironi, M., & Rodríguez-Giralt, I. (2017). Healing, knowing, enduring: Care and politics in damaged worlds. *The Sociological Review, 65*, 89–109.

Tronto, J. C. (1993). *Moral boundaries: A political argument for an ethic of care*. Routledge.

Valujeva, K., O'Sullivan, L., Gutzler, C., Fealy, R., & Schulte, R. P. O. (2016). The challenge of managing soil functions at multiple scales: An optimisation study of the synergistic and antagonistic trade-offs between soil functions in Ireland. *Land Use Policy, 58*, 335–347.

van Dooren, T. (2014). Care. *Environmental Humanities, 5*, 291–294.

Wallington, T. J., & Lawrence, G. (2008). Making democracy matter: Responsibility and effective environmental governance in regional Australia. *Journal of Rural Studies, 24*, 277–290.

Weigelt, J., Müller, A., Beckh, C., Srigiri, S. R., & Töpfer, K. (2014). Pathways towards sustainable soil and land governance: Discussing the contribution of the Global Soil Week. In J. Weigelt, A. Müller, C. Beckh, & K. Töpfer (Eds.), *Soils in the nexus: A crucial resource for water, energy and food security* (pp. 139–160). Oekom.

Williams, B. J. (2006). Aztec soil knowledge: Classes, management and ecology. In B. P. Warkentin (Ed.), *Footprints in the soil: People and ideas in soil history* (pp. 17–42). Elsevier.

Williams, J. M., & Holt-Giménez, E. (2017). *Land justice: Re-imagining land, food, and the commons*. Food First Books.

Wittman, H. (2009). Reworking the metabolic rift: La Vía Campesina, agrarian citizenship, and food sovereignty. *The Journal of Peasant Studies, 36*, 805–826.

Yaalon, D. H. (1996). Soil science in transition: Soil awareness and soil care research strategies. *Soil Science, 161*, 3–8.

Chapter 20
Soil Care, Culture, and Eco-Identities

Sabine Grunwald and Kay R. Kastner-Wilcox

20.1 Introduction

The concepts of soil health, soil quality, and soil security share the common aim to preserve and improve the world's soil resources (Mizuta et al., 2018). These notions mirror the values and morals of cultures associated with soil/land/nature. For example, Janzen et al. (2011) and Robinson et al. (2013) asserted that soils have supported human needs and sociocultural aims through regulation of the global climate and providing critical services for global food webs, water, energy and nutrient cycles, and biodiversity. However, such a view of soil as a service provider for people inherently assumes an anthropocentric perspective that has been prevalent throughout modernity[1]

[1] **[Footnotes 1 to 3 refer to Europe, the USA and the "Global West"].** *Modernism* is characterized by realism-naturalism, objectivism, autonomy of humans, individualism, and liberal capitalism during the time of the Enlightenment, "The Age of Reason" (about eighteenth to twentieth century) encapsulating sciences, economy, technical fields, arts, and philosophy. In contrast, *premodernism* is characterized by realism-supernaturalism, mysticism, and/or faith and religion, and humans are subject to Gods divine will, collective altruism, and feudalism during medieval times (about 400–1400 CE; Hicks (2011). Modernity brought about the loss of belief in religion, the expansion of markets, commodification of things through capitalism, rise of science and technology, growth of mass culture and its influence, most markedly the invasion of bureaucracy into private life, and changing beliefs about relationships between man and woman (Butler, 2010).

S. Grunwald (✉) · K. R. Kastner-Wilcox
Soil, Water, and Ecosystem Sciences Department, University of Florida,
Gainesville, FL, USA
e-mail: sabgru@ufl.edu

N. Patzel et al. (eds.), *Cultural Understanding of Soils*,
https://doi.org/10.1007/978-3-031-13169-1_20

and postmodernity[2] spanning from about the fifteenth century to present. Globally, cultures have left their mark on the environment with profound soil carbon losses estimated at 133 Pg C over the past 12,000 years, geographically centered in cropping regions and grazing lands under human management (Sanderman et al., 2017). History has shown that the prosperity of cultures depends on soils, specifically carbon-rich soils, to provide crops, food, and fiber. Fertile soils are one of the fundamental ecological resources for human survival, but soil degradation has contributed to some cultural-ecological suicides, as exemplified by the collapse of the Mayan and some Polynesian cultures (Diamond, 2011). Multiple contemporary examples demonstrate the inseparability between soil, food, culture, economy, and politics. For instance, global climate change, droughts, and severe soil degradation all contributed to the deterioration of Syria's economic and political downturn, which spawned a devastating civil war beginning in 2011 and severely impacted Syrian culture (Gleick, 2014). Though soil degradation and cultural collapse stories have been interlinked, they are often complex and involve a variety of anthropogenic factors that can be studied through archaeological records (see Brevik et al., 2018).

Although soils have been classified, mapped, quantified, objectified, commodified, and managed globally as a valuable resource to serve a variety of human-centered needs (e.g., crop, food, and fiber production; Grunwald, 2006, 2010; Field et al., 2018), soils have been metaphorically raped, mutilated, demeaned, devalued, and even killed, leaving hostile barren land. According to Amundson et al. (2003), a sizable fraction of soils is in danger of profound loss or even complete extinction to fulfill anthropocentric needs such as urbanization and agriculture. Soils' abuse by a variety of cultures through centuries stands in contrast to steadily increasing scientific knowledge about soils, its processes, and functions. Grunwald et al. (2017a) pointed out that the pure recognition that soil is degraded is at risk of degradation due to a specific use or is unhealthy in one way or another will not necessarily invoke action to counteract and secure soils suggesting *cognitive dissonance* between knowledge and action. Grunwald et al.'s (2017a) critique suggested that simple assessment of soils, soil-ecosystem services, functions, and goods from a pure science perspective does not necessarily translate into pro-soil politics, decision-making, and adaptation of agricultural management. Similarly, Amundson (2017) argued that soil knowledge and facts produced by empirical and deterministic science (including probabilistic models) need to be linked with core values and beliefs that people hold to sustain both soils and cultures. The underpinnings of an

[2] *Postmodernism* is a broad movement that developed in the mid-to-late twentieth century across philosophy, the arts, literature, and the intellectual world in general (Hicks, 2011). Postmodernists aimed at liberating ethical and political doctrines, as exemplified by Jacques Derrida, Louis Althusser, Michel Foucault, Jean-Francois Lyotard, and others. Postmodernism is characterized by deconstruction of previously trusted political, philosophical, artistic, linguistic, or other views. Postmodernists have radically attacked the objectivist claims of science and theories that explain the world. The critical postmodern techniques were especially liberating in the realms of ethical and social problems, for example, deconstructing patriarchy, authority, and oppressive power structures freeing women and cultural minorities. Postmodern discourse also involved cultural identities and deconstruction of self and other (Butler, 2002).

integrating lens that combines cultural, political, and science perspectives are found in morals and ethics. Both inform how people relate to soil/land/nature and manage, use, conserve, or preserve soils.

The significance of soils as a life-support system for food and fiber production has been duly recognized (Yaalon, 2000). Yet the effects that global human economies have produced on soils are profound. Globally, 35.9 Pg year^{-1} of soil was eroded in 2012 with an overall increase in global soil erosion driven by cropland expansion (2001–2012). The greatest increases in erosion were predicted to occur in sub-Saharan Africa, South America, and Southeast Asia. The least developed profit-oriented cultures have been found to experience the highest estimates of soil erosion rates (Borrelli et al., 2017). Significant cultural shifts of massive proportion are called for toward higher levels of consciousness at which people are not only aware that all sentient beings depend on the physiosphere and biosphere, including soils, but are also willing to act accordingly (Esbjörn-Hargens & Zimmerman, 2009). From a philosophical perspective, Wilber (2000a) articulated mutual effects between matter (soils, nature) and the depth (development) of consciousness that enacts (i.e., brings forth) cultural, economic, and political practices, action, and management of the environment that are either nourishing, neutral, or out of balance.

Although environmental ethics and moral philosophy are thousands of years old, both have been revisited in times of more recent ecological crisis (Burnor & Raley, 2011; Schmidtz & Willott, 2012). These environmental ethics and morals were born out of perspectives rooted in modernism and a stronghold of objectivism, rationalism, control of the environment, and the scientific method. Modernism differs from postmodernism, which is characterized by nihilistic, cultural subjective, and relativistic stances toward soil/land/nature/environment. Our thesis is that deficiency in care and compassion for soil/land/nature/environment in individualized postmodern cultures has sabotaged soil health and soil security. We posit that the emergent cultural metamodernism[3]—characterized by hope, harmony, authenticity, and affect[4]—will give rise to new forms of ethics and morals of soil/land/nature/environment due

[3] *Metamodernism* is a proposed reactive cultural, philosophical, and aesthetic response to postmodernism. Metamodernism blends aspects of both modernism and postmodernism, with the latter characterized by relativism, nihilism, deconstruction, and the rejection of grand narratives. In contrast, metamodernism is characterized by hope, romanticism, sincerity, authenticity, affect, feeling tones, and the potential for universal truths and grand narratives (Akker et al., 2017). According to Akker and Vermeulen (2017), metamodernism emerged in the 2000s. Freinacht (2017) pointed out that metamodern refers to a major developmental stage of society in response to capitalism and liberal democracy.

[4] *Affect* is emotion of subjectively experienced feelings, such as happiness, sadness, fear, or anger (Colman, 2015). According to Damasio (2000), *feelings* are inwardly directed and private, whereas *emotions* are outwardly directed and public. The full and lasting impact of feelings requires consciousness, because only along with the advent of a sense of self do feelings become known to the individual having them. There are three stages of affective processing: (1) a state of emotion (which can be triggered and executed unconsciously), (2) a state of feeling (which can be represented nonconsciously), and (3) a state of feeling made conscious.

to the ability to integrate objective and subjective, rational/logic, and affective approaches toward the environment through re-awakened care and compassion.

The purpose of this chapter is to explore soil ethics in the context of cultural epochs (modernity and postmodernity) and socially constructed cultural identities (subjectivism, relativism, pluralism, and objectivism). In general, cultural identities are associated with specific moral viewpoints toward soil/land/nature or environmental ethics. We discuss these ethical attitudes that manifest in the form of ecological (eco)-identities, which express relationships toward soil/land/nature/environment. Findings are critically discussed in the context of ecopsychology that links psychology, ecology, and ethics. Lastly, a vision for a soil ethics of the future situated within the cultural context of metamodernity is presented. This soil ethics emphasizes soil care as a core element.

20.2 Cultural Identities and Moral Cultural Perspectives of Soil/Land/Nature

Different kinds of human moral agents have been identified that adjudicate between conflicting demands of morality in regard to soil/land/nature acknowledging people's needs (e.g., food, self-aggrandizement, aesthetics, and economic profit). Ethical views operate concurrently in individuals and cultures and cover the spectrum from subjectivism, relativism, pluralism, and absolutism (objectivism) (Thiroux & Krasemann, 2016). *Cultural identity* expresses the sense of belonging toward a culture and entails shared sets of beliefs, values, relationships, religion, ethics, and more (Tyson, 2015)

Subjectivism assumes that moral principles hold only for particular people but may differ from one person to another. In short, each individual's moral standard depends on what that person believes to be right in any given moment. In essence, subjectivism negates the function of morality itself (Burnor & Raley, 2011). According to Hicks (2011), postmodernism is grounded in social subjectivism, social construction of reality, and collectivist egalitarian ethics. Postmodernism advocates for deconstruction and the collapse of metanarratives that have given people meaning, identity, and purpose, while it rejects any absolutist, universal truths about human beings, the environment, and the cosmos (Lyotard, 2003). This attitude is problematic from a soil-centric perspective because in postmodern cultures, there is no moral incentive nor underlying moral reasoning to care for or preserve soils and the environment, which may turn easily into amoral self-centered or corrupt behavior (e.g., exploiting soils and the environment).

Cultural relativism asserts that the same moral principles hold true for members of a particular subculture or social group. These principles are determined by what the majority of that subculture accepts as morally correct. However, these principles may differ from one another; for example, morals in an indigenous culture and cultures of the West or South show distinct differences often clashing within and among diverse multicultural nation states (Burnor & Raley, 2011). Relativism embraces polycentric power structures and individualism prominently found in Western culture, while anti-institutionalism is favored to give individuals unlimited choices of beliefs and values (Baghramian & Carter, 2015). According to Ferrer (2002), relativism maintains that "everything is the same as everything else, and that no distinctions can be therefore made between what is good or bad, right or wrong; true or false is both blatantly self-contradictory and morally repulsive" (p. 211). One subculture may conspicuously believe that the Earth is flat, others may claim that the Earth is Gaia (a living organism that is sacred), while a science subculture holds the view that the Earth is a biogeochemical system consisting of discrete elements. Relativism is problematic in that it does not discern truth claims and rejects any moral ideals or standards (e.g., thresholds for the application of mineral fertilizers to soils or preservation of nature reserves) and thus opens the door to any moral view as long as there is a collective of people believing in it. Such tribalism becomes especially damaging to the environment when leaders in legislatures, executives, or politicians embrace a relativistic worldview that is not caring in regard to soil/land/nature.

Burnor and Raley (2011) pointed out that the greatest trouble with relativism is that in a complex and globalized world with diverse intermixed cultures in which each person belongs to more than one distinct cultural group with diverse values and beliefs, relativism easily slips into subjectivism. Values such as freedom, equality (e.g., equal value for all soils including highly fertile Mollisols and low-carbon unproductive Inceptisols), equity (e.g., impartial and fair access to soil and the land), or environmental justice (e.g., fracking for profit that destroys soils) are subjective from a relative cultural perspective.

"Anything goes" relativism is spontaneous, often irrational, and in opposition to some other view without justification (Wals, 2010). In contrast, *pluralistic relativism*, as proposed by Wong (2006), posits that different cultures may have different but equally true moral standards (relativism). However, most cultures share common core fundamental values and principles, for example, human rights or preservation of soils to sustain life on Earth (objectivism). Pluralism is emancipatory in the sense that it gives voice to different interpretations of soils by different cultures, while it insists that some truths are unshakable (e.g., soils produce food and need to be secured for human survival).

Cultural objectivism asserts that the same moral principles hold universally, for every individual, across time, and every culture on the planet (Burnor & Raley, 2011). According to Thiroux and Krasemann (2016), objectivism takes a conservative stance claiming that there is only one right way, no matter what, ignoring the diversity of moral options. Such ethics and focus on empirical sciences, the scientific method, reason, and logic have been prominent in modernism, the Enlightenment period from the eighteenth to twentieth century, and somewhat in contemporary cultures that value objectification and quantification as superior to the felt lived subjective experience (Hicks, 2011). Grunwald et al. (2014) asserted that objectifying soil reduces them to material viewed as a resource to be mapped and quantified (e.g., soil taxonomic classes, pedon descriptions, or spectral readings of soils). Such a view purports that humans and soil organisms are concrete pieces in a planetary cycle in which constituent biogeochemical elements are quantifiable, manageable, and deterministically controllable and predictable (Jacobson et al., 2000). Likewise, the commodification of soil/land through property rights and ownership fits the objectivism worldview (Rose, 2012). Thiroux and Krasemann (2016) pointed out that an absolutistic scientific view makes the claim for absolute "laws" in nature. Objectivism is also found in cultures where rule-based ethics that enforces environmental laws and regulations are authoritatively imposed onto people.

The absolute objectivist view is prominent in contemporary soil science culture around the globe. It has been critiqued as reductionistic lacking consideration of subjective and phenomenological aspects (e.g., the felt sense and sacredness of soils) and a more integrative view balancing objective and subjective ways of knowing (Grunwald et al., 2017b; Mickey et al., 2017; Varela et al., 2016; Wilber, 2000a). These more integrative worldviews suggest that digital soil maps and models, quantification of the genetic code of all soil organisms, and more objective data of the microstructure in all soils will not be able to address global soil, ecological, and food crises due to the lack of relationality and connectivity between people and soil/land/nature.

20.3 Ecological Identities

Cultural identities pertaining to soil/land/nature and the environment in general express themselves as *ecological (eco)-identities* that mirror human developmental levels of consciousness. These identities and their underlying environmental ethics were presented by Esbjörn-Hargens and Zimmerman (2009) and include Eco-Guardian, Eco-Warrior, Eco-Manager, Eco-Strategist, Eco-Radical, Eco-Holist, Eco-Integralist, and Eco-Sage (Table 20.1). The eco-identities, part of *integral ecology*, are situated within *integral theory* that includes subjective and objective as well as individual interior and collective perspectives (Wilber, 2000a, b). Grunwald et al. (2017a, b) applied integral ecology and integral theory to develop a meta soil model for soil security, which also includes psycho-ecological levels of development.

Eco-identities are supposedly based on a human developmental model assuming that self[5] (in individuals) and cultural selves develop along a specific trajectory that expresses their level of consciousness from a narrow self/selves toward more expansive ones: ego-, ethno-, socio-, world-, planet-, to kosmo-centric self/selves (Esbjörn-Hargens & Zimmerman, 2009). Following this theory, eco-levels are ecological stages of development of eco-perspectives along a hierarchical ladder model

[5] *Self* has been defined as a person's distinctive individuality, identity, essential nature, or collection of personal characteristics. In Western psychology, the self plays a critical role in providing self-coherence, self-expression, and self-identification (Colman, 2015). In sociology, the self is considered a socially constructed mental image that arises through the process of socialization (Cooley, 2013).

Table 20.1 Ecological identities and associated moral values toward the environment

Ecological identities	Moral values toward the environment	Eco-self description	Environmental ethics	View toward/for soils
Eco-Guardian[a]	Animistic	Romantic ethos: self-focused on creating safety and satisfying basic needs. Performs rituals to maintain control and power. Create safety and satisfy basic needs. Views nature as ensouled/ spirited. Examples: paganism, pantheism, nature worship, indigenous practices, shamanism, neopaganism, neo-shamanism, deep ecologists. Cultural epoch: pre-modernism; metamodernism.	Respects and fears nature. Mystery of nature. Belief in spiritual beings. Impulsive, opportunistic/ self-protective.	Soils are sacred, holy, have a soul or are spirit.
Eco-Warrior	Egocentric	Heroic ethos: self-protective and self-serving. Leads through power and strength. Wants to conquer nature and/or cultures. Wants immediate reward. Examples: ecoterrorism, mining, fracking, rock climbing, and hunting. Cultural epoch: modernism.	Interest in "me" and "mine" Individual needs are articulated and heard. Views others as competitors for goods, space, and dominance. Exploits nature as macho/ warrior over nature.	Soils are radically exploited for human profit.

(continued)

Table 20.1 (continued)

Ecological identities	Moral values toward the environment	Eco-self description	Environmental ethics	View toward/for soils
Eco-Manager	Ethnocentric	Stewardship ethos: self-conformist who is rule-oriented and concerned with group membership. "us versus them" thinking. Maintain order and follow the law (law of the state or divine order). Obedience. Dedicated to managing nature from a religious or secular perspective. Examples: US Environmental Protection Agency, wildlife management, environmental legislation. Cultural epoch: modernism.	Stewardship approach to the environment: passes laws and established institutions that serve to act as stewards over nature. Anthropocentric: Dominion of humans over nature. Aim to protect one's group and community values rather than protecting the Earth.	Soils are managed and viewed as a resource to support food production and soil ecosystem functions and services.
Eco-Strategist	Sociocentric	Rational ethos: appreciates scientific empiricism (nature/world is measurable and predictable). Self-identity competes for wealth, influence, and social standing. Conserves resources for consumption. Wants to manage nature strategically for one's or the group's sake. Opportunistic, independent, and achievement oriented. Examples: natural capitalism, biogeochemistry, digital soil mapping, pedometrics, science of ecology, industrial agriculture. Cultural epoch: modernism.	Exploits nature as a result of greed and a focus on short-term profits. Rational approach to the environment (rationalism). Deontological ethics. Utilitarian perspectives.	Soils are degraded and used for human gain but sustained to secure human needs.

(continued)

Table 20.1 (continued)

Ecological identities	Moral values toward the environment	Eco-self description	Environmental ethics	View toward/for soils
Eco-Radical	World-centric	Equality ethos: sensitive to other's experiences. Values personal experience and feelings. Promotes eco-social justice; teamwork; social responsibility and sensitivity. Exposes shortcomings of modernity. Embraces postmodern perspective. Examples: ecopsychology, deep ecology, ecofeminism, environmental justice, green politics, social construction of nature. Cultural epoch: postmodernism.	Pluralistic ethics: ability to consider multiple perspectives. Promotes sharing of resources without discrimination in regard to gender, class, race, or other. Socially engaged "green" activism.	Soils, humans, and all organisms have equal rights and value. Therefore, soils need to be preserved.
Eco-Holist	Planet-centric	Holistic ethos: Holarchical worldview. Self-autonomous individual who is conformable amidst complexity, recognizing that individuals occupy multiple contexts. Multi-perspectival: appreciates paradoxes and contradictory perspectives and values. Examples: Gaia hypothesis, Aldo Leopold's land ethics, process ecology, biodynamic agriculture, ecological postmodernism. Cultural epoch: postmodernism.	Maps the complexity of interrelationships within and between ecosystems. Holistic understanding of complex ecosystems and interactions between nature and people (both are attributed with value).	Biocentric view that aims to secure soil health from a holistic perspective. Soil are viewed as living natural bodies.

(continued)

Table 20.1 (continued)

Ecological identities	Moral values toward the environment	Eco-self description	Environmental ethics	View toward/for soils
Eco-Integralist	Planet-centric	Inclusive ethos: Integration of immanence and transcendence; integration of thinking and feelings. Honors and integrates multiple perspectives. Seeks and values paradox and multiple perspectives; biocentric perspective. Recognition of the Buddhist emptiness of all phenomena; luminosity of the planet. Values interiority: heart-based experiences allow to feel the suffering of other people and the planet. Examples: integral soil security, Buddhists ecodharma, integral community development; environmental philosophy, eco activism. Cultural epoch: participatory turn; nonconceptual (post-metaphysical); metamodernism.	View themselves as part of maps that represent nature/ecosystems. Deep acceptance of all life forms and nature without the need to change or manipulate them. Appreciation of nature. Care for nature. Acknowledgment that the environment is getting ecologically worse *and* better at the same time.	Care and compassion for soils and the environment evoke deep appreciation that inspire the integration of soil, organisms, humans at a deep feeling level.

(continued)

Table 20.1 (continued)

Ecological identities	Moral values toward the environment	Eco-self description	Environmental ethics	View toward/for soils
Eco-Sage[a]	Kosmo-centric	Unity ethos: Ego-aware-self individual who has direct perception of every organism and nature in general being a manifestation of the Divine integrating multimodal and multidimensional elements across contexts in the service of humanity. Capacity to witness themselves in the present moment. Integrates all other eco-selves. Transpersonal state of consciousness: one with nature; cares and protects nature because it is part of one's being ("being Earth"; "being the universe"); unitive perspective. Examples: Nondual ecology, transcendentalism, nondual spiritual activism, shamanism. Cultural epoch: participatory turn; nonrepresentational (post-metaphysical); metamodernism.	Experiences the world as an immanent expression of timeless Spirit. Deep commitment to all sentient beings. Care for nature and All. Identifies with the Totality of life and All. Transcended self; kosmic view that all are One (non-dual). Participation in the mystery of kosmic undetermined unfolding.	Soils and people are one and the same. And because of such nondual realization soils are spiritualized and protected as one would protect one's own life or one's child.

Modified after Esbjörn-Hargens and Zimmerman (2009)

Ecological identities are hierarchically arranged in form of a ladder model from Eco-Guardian to Ego-Sage dependent on levels of human development. The ecological identities are associated with specific characteristics that one holds and determine the predominant moral view toward the environment and soils (environmental ethics)

[a]Wilber (2000c) elaborated on the *pre/trans fallacy*, which is the confusion of pre-rational structures with trans-rational structures simply because both are nonrational. In other words, the *pre/trans fallacy* expresses that transpersonal and mystical states experienced at higher developmental stages *are not the same* as those at lower levels. The claim Wilber makes suggests that these states of consciousness are not the same (e.g., a nondual state of consciousness and a magical experience of a child are not identical). In the context of eco-selves, an Eco-Guardian may be confused with an Eco-Sage and vice versa. The confusion works in both directions, with transpersonal experiences (e.g., non-dual states of consciousness) being interpreted as regression to infantile states (e.g., shamanistic states of consciousness), while experiences due to the incomplete formation of self-boundaries (e.g., psychotic states, magical thinking) are often labeled as transpersonal

with rungs (i.e., the eco-levels). Subjectivist ethics plays a major role at the Eco-Guardian level, while it may also be found at higher eco-levels to some extent. Relativistic ethics can slide easily into subjective ethics specifically when egocentric and narcissistic tendencies are pronounced. Objectivist ethics dominates the Eco-Warrior, Eco-Manager, and Eco-Strategist. Objectivist tendencies are also associated with the Eco-Radical and Eco-Holist and intermix with relativistic ethics to make specific claims that substantiate their view toward soil/land/nature (e.g., environmental justice represents the Eco-Radical, or individuals or cultures that contaminate and destroy soils represent the Eco-Warrior). The Eco-Holist is most prominently found among soil scientists that overemphasize the exterior view of soil, for example, to classify, map, assess, and measure soils using laboratory, field, and remote sensing methods rooted in a reductionistic-deterministic view of the world. Pluralistic ethics can be associated with the Eco-Integralist and Eco-Sage with a capacity of expanded awareness of the complexity of cultural, soil/land/nature, economic, and political perspectives that are weighted against each other. Both of these eco-selves (or eco-identities) are able to adjudicate ethical conflicts through handling of paradoxical situations (e.g., develop integrative adaptation and mitigation strategies for soil health due to global climate change). The Eco-Sage embodies special insight and wisdom through the ability to combine knowledge of soil/land/environment and cultures, spiritual subjective knowing, and felt sense experiences that expand from individual to collective and cosmic levels. It is noteworthy to mention that each of the eco-selves also has a shadow side that brings out the pathology of a specific eco-identity. For example, the shadow side of the Eco-Sage could be a false prophet, spiritual leader who abuses his/her insights for personal power harming the environment or is simply ecologically inauthentic.

Human developmental models have been prominent and assume developmental stages (or phases) from smaller/lower to larger/higher levels (Ardelt & Grunwald, 2018; Kegan, 1982; Wade, 1996). These hierarchical developmental models expand self-identity or levels of consciousness toward larger conceptions (Wilber, 2000c). Eco-identities are assumed to represent specific developmental stages from the Eco-Guardian to the Eco-Sage. Although this stage model explains some ecological thinking, perceptions, and actions/management, it falls short to account fully for all situations and different moral valuations.

Expansive identities at higher eco-identity levels have been associated with environmental ethics (Esbjörn-Hargens & Zimmerman, 2009). Extensionists expand the range of interpersonal virtues to include "larger than human values" based on virtue ethics (from I → You → Others → Cosmos). The extensionist's view entails the identification with all humanity as moral virtue. For example, McFarland et al. (2013) presented empirical evidence and a quantitative instrument, which is rooted in Alfred Adler's concept of *oneness with humanity* to express the innate potential in all humans to connect to something larger than themselves. This instrument is also grounded in Abraham Maslow's concept of *self-actualization,* which embraces identification with and concern for all humanity. Similarly, transpersonal constructs that go beyond ordinary personal self-conceptions toward the transpersonal realm evoke deep care for the environment (Friedman, 1983). Ecological collectives (communities, cultures, nations, or the planet viewed as *Gaia*) that are viewed as larger than the individual self are deeply grounded in moral imperatives to care, nurture, and protect soil/land/nature and sentient beings (Callicott, 2017). These larger eco-conceptions are often accompanied by compassion for all sentient beings, their well-being, and reduction of harm and suffering for human and nonhuman beings alike. Morin (1999) calls for planetary solidarity and union as a lived expression of planetary consciousness and interconnectedness. *Integral ecologies* embrace such a "larger than self, larger than human" point of view to respond to pressing soil, water, climate, and ecological crisis in general (Mickey et al., 2017).

The critique of hierarchical development models, such as the hierarchical eco-identities model, entails the lack of empirical evidence to evidence stages that often extend over long periods of time. Although longitudinal research in human development is much more costly and timely than cross-sectional research, there is ample evidence in support of hierarchical developmental stage models (e.g., Baker & Mednick, 1984; Kegan, 1982; Lerner et al., 2009). Another obstacle is that hierarchical models are linear and sometimes regressive phases have been observed that point rather to spiral or nonlinear developmental phases. Hierarchical models claim that there is an absolute goal (highest stage). For example, Esbjörn-Hargens and Zimmerman (2009) asserted that the highest psycho-ecological developmental stage is the Eco-Sage that is "one with nature." Such non-dual conceptions have been questioned because they assume that oneness/non-duality is superior when compared to other conceptions, such as "nature is the God/the divine," "nature is the Feminine," or atheistic conceptions of reality (Ferrer, 2002). Rapley (2004) critiqued developmental studies from a postmodern perspective arguing that they are an artifact of modernity's science hubris dominated by the scientific method.

Despite the critique of hierarchical developmental models, they offer a lens to discern eco-identities associated with pre-modernity to modernity (e.g., the Eco-Guardian to Eco-Strategist), and postmodernity to metamodernity (e.g., the Eco-Holist to Eco-Sage). *Ecopsychology* emerged in response to ecological destruction observed in cultures that embraced modernity and postmodernity. This kind of psychology is interested in understanding the relationships between humans and the natural world, specifically relations that are destructive to the environment (Reser, 1995; Roszak, 2001). Metzner (1995) elaborated on the psychopathology of the human-nature relationship that has lost respect for nature and lacks humility and care for the environment. Ecopsychology, which combines psychology, ecology, and ethics, claims to provide the theoretical and practical underpinnings for healing of a dying planet shattered by global climate change and eco-crisis (O'Connor, 1995).

20.4 Ecopsychology

There have been many attempts to explain both human-induced global environmental degradation as well as society's inability or unwillingness to assess and address the risk that environmental degradation poses to us and our decedents. In 1940,

Benton MacKaye, the father of the Appalachian Trail (Appalachian Mountains, USA), argued that the sprawling expansion of metropolitan areas, and especially the outward flow of population, resulted in the degradation of valued natural resources, including psychological resources. Recognizing interrelationships between humans and their environment was instrumental for regional urban planning (MacKaye, 1990). However, society has often shown lack in concern regarding the environment as exemplified by severe environmental degradation in the United States in the 1930s during the dust bowl (Worster, 2004) and in the 1960s (Reser, 1995).

How society is able to make ethical decisions about our environment in the face of increasingly conflicting needs and opinions will require the exploration of the connection between the ecological, spiritual, and psychological crises that have resulted from society's separation from the "more-than-human world" upon which we rely. These "wicked" problems are deeply rooted and will not be solved without the significant alteration of both thought and action, requiring the redirection from individualistic thinking and practice to a collective level of cultural and social engagement. This challenge requires the confrontation of society's anthropocentric, theological, psychological, philosophical, political, and economic biases (Fisher, 2013).

The environmental movement in the United States, which began on the first Earth Day in April 1970, reflected a major increase in public awareness of and concern about environmental problems (Freeman, 2002). This movement has now grown to become the largest, most solidly organized political cause in human history. The environmental movement, like every political campaign, has a psychological dimension—the need to persuade people to alter their environmentally destructive behavior, debating values, and questioning motivations (Brown, 1995). The psychological dimension is extremely important, given that "the pivotal psychology reality of our time" is the survival of the human species (Macy, 1995, p. 241). The question to be answered in this regard is whether ecology and psychology are complementary disciplines.

Psychology is the science of behavior and mental processes (Myers & Dewall, 2015), and the core issue for psychology has been: Who am I? Where is the "me?" "Where does the "me" begin and where does the "me" end? In the past, this has been considered an arbitrary distinction in the field. The paradigm is that the divider, or cut, between self ("me") and the natural world is wholly arbitrary—it may be as close as the individual's heart, or the skin, or as far out as the distant stars or deepest oceans. The distance is not, in itself, the issue. The acknowledgment of the boundary between "me" and "not me" is far less important than the recognition of the uncertainty of the boundary. The uncertainty itself allows for the consideration of the actual merging of ecology and psychology (Hillman, 1996). During a conference in 1990 titled, "Psychology as if the Whole Earth Mattered," a group of ecopsychologists concluded "if the self is expanded to include the natural world, behavior leading to the destruction of this world will be experienced as self-destruction" (Roszak, 1995, p. 12).

At this same conference, Walter Christie observed:

> The illusion of separateness we create in order to utter the words "I am" is part of the problem of the modern world. We have always been far more a part of great patterns on the globe than our fearful egos can tolerate knowing ... To preserve nature is to preserve the matrix through which we can experience our souls and the soul of planet earth (Roszak, 1995, p. 12).

Scull (2008, p. 68) defines ecopsychology as "an explicitly moral psychology" that explores this issue and searches for the roots of the personal and social problems that have led to this disconnection between people and nature. In the epilogue of *The Voice of the Earth*, Roszak (2001) sets out eight principles that provide a framework for ecopsychology. One of those principles addressed a sense of "ethical responsibility" felt as deeply as the responsibility experienced toward other people, weaving both into social relations and politic discourse (Reser, 1995). Land ethics states that "ethics ... [are] actually a process of ecological evolution" and "its sequences may be described in ecological as well as in philosophical terms" (Leopold, 1949, p. 108).

Ecopsychology offers an ethical lens to the study of soil health, soil quality, and soil security with the additional provision of psychological sensitivity, ecological expertise, and environmental activism. This rich mixture may lead to a new, more philosophically grounded form of scientific communication providing for more effective environmental policymaking (Brown, 1995).

20.5 An Emergent Environmental Ethics of Care Founded in Ecopsychology and Metamodernism

To summarize, we stated that both modern and postmodern cultures were associated with soil degradation, soil erosion, and soil carbon losses. Modernism is characterized by objectivism, rationalism, beliefs in the superiority of science and technology, and control and domination of the environment, while it has favored logic over affects and feelings marginalizing care and compassion. Thus, modernity valued industrialization, profit, individualism, and technological developments more than an intimate relation to soil/land/nature. Postmodernity has been a backlash to the shortcomings of modernity aiming to deconstruct its notions and replace it with cultural subjectivism, relativism, and a collective attitude of "whatever, I don't care" or "I care only for my peer group." Such a relativistic cultural stance refuses to show compassion and care for soil/land/nature, because it rejects belief in a shared narrative of soil health or soil security.

Modern and postmodern cultures prevalent in the West and North have resisted adopting an ethics of care for the environment as evidenced by persistent ecological crisis in various forms and soil insecurity at the global scale. Both cultural epochs—modernism and postmodernism—have shown strong tendencies to externalize and project people's material needs, wants, ideas, and beliefs onto others and communities or critique and deconstruct them. Along with these tendencies came disinterest in affects, felt sense experience, spiritual connection to nature, romanticizing the beauty of nature, and participatory intuitive ways of relating to soil/land/nature has marginalized care and compassion for the totality of the environment. The objectification and commodification of nature prevalent in modern and postmodern cultures have outpaced subjective inner forms of knowing that undergird care and compassion for soil/land/nature. Patzel (2010) described such inner forms of knowing as archetypal images and emotional sensing related to soils, for example, Mother Earth.

The emergent cultural epoch of metamodernity has the potential to evoke a new ethics of care and compassion fostering connection with soil/land/nature (Rowson & Pascal, 2021). According to Akker et al.'s (2017) assertions, the main characteristics of metamodernism entail hope, romanticism, sincerity, authenticity, affect and feeling, the potential for universal truths, and grand narratives. Freinacht (2017) pointed out that metamodern refers to a major developmental stage of society in response to the shadow sides of capitalism and liberal democracy. Freinacht stated that metamodern society is characterized as "a listening society" that is sensitive to the inner dimension of human beings. Freinacht (2015) proposed the following metamodern characteristics: (1) increased awareness of emotions; (2) belief in technology to achieve desirable outcomes (e.g., green technologies that conserve and preserve the environment rather than technologies like fracking that exploit the environment for human profit); (3) understanding of human and social developmental hierarchies; (4) reconstruction of political, social, and other structures that must follow postmodern deconstruction; and (5) integrative thinking of inclusive "both-and" rather than dualistic "either-or." Metamodern metaphysics provides an emergent global vision that builds on harmony and love rather than modern and postmodern claims that have been rather destructive toward the environment (Stein, 2018; Severan, 2021). Corsa (2018) asserted that the expected metamodern turn allows one to embrace polyphonic grand narratives effectively to address the global environmental crisis through a new global metamodern ethics. Metamodern (grand) narratives, for example, social and racial equality or environmental justice, give meaning to people because they recognize humanity as a whole. The envisioned metamodern ethics is pluralistic, acknowledges the diversity of cultures, and is globally respectful and responsive through expansive horizons that include the global commons, soil/nature-humans, and life on Earth. Such metanarratives honor sentient beings' fullness and wholeness. Vermeulen and van den Akker (2010) stated that a structure of feeling, a sensibility, an emotive awareness, and belonging to collective humanity are part of metamodernism. Ecopsychology provides the framework for an environmental ethics rooted in care and compassion for people and the environment. The commonality of an environmental ethics motivated by emotions/feelings of care and compassion for "a larger whole" in ecopsychology and metamodern culture are striking. Both—ecopsychology and metamodern culture—are not only poised to overcome the shadow sides of modernity and postmodernity but also provide an optimistic future to care for the well-being of humans and soil/environmental health. This vision for a soil ethics of care is dependent on the presupposition that the development of human consciousness takes a leap allowing movement from Eco-Strategist toward Eco-Holist up to Eco-Sage.

A metamodern culture that cares equally for people and the environment—including soils—aims to provide the foundation to minimize soil degradation and preserve soil health. This kind of care ethics is aligned with the concept of soil/land degradation neutrality, soil carbon neutrality, and soil security, the latter aimed to support soil functions and services as well as enhance food security. In that sense, soil care ethics is suggested to be the missing piece in the puzzle to secure healthy soils. Knowledge of the extent of soil degradation or indicators that express the biophysical-chemical health of soils are not sufficient to motivate a culture to care, protect, and secure soils. Soil care and compassion for soil/nature are sensibilities that everybody is inherently capable of expressing. All, including soil scientists, can be Eco-Holists or Eco-Sages. It does not matter where you stand now ("do not feel bad that you are not a Sage yet") but start now to change and strive for more personal wholeness and wisdom.

References

Akker, R. V. D., & Vermeulen, T. (2017). Periodizing the 2000s, or, the emergence of metamodernism. In R. V. D. Akker, A. Gibbons, & T. Vermeulen (Eds.), *Metamodernism: Historicity, affect, and depth after postmodernism* (pp. 1–20). Rowman & Littlefield.

Akker, R. V. D., Gibbons, A., & Vermeulen, T. (Eds.). (2017). *Metamodernism: History, affect, and depth after postmodernism*. Rowman & Littlefield.

Amundson, R. (2017). Whose security is important? Communicating environmental risk about soil to a diverse audience. In D. J. Field, C. L. S. Morgan, & A. B. McBratney (Eds.), *Global soil security* (pp. 383–387). Springer.

Amundson, R., Guo, Y., & Gong, P. (2003). Soil diversity and land use in the United States. *Ecosystems, 6*(5), 470–482.

Ardelt, M., & Grunwald, S. (2018). The importance of self-reflection and awareness for human development in hard times. *Research in Human Development Journal*, 1–13. https://doi.org/10.1080/15427609.2018.1489098

Baghramian, M., & Carter, J. A. (2015). Relativism. In E. N. Zalta (Ed.), *The Stanford encyclopedia of philosophy* (2018th ed.). Metaphysics Research Lab, Stanford University. https://plato.stanford.edu/archives/win2018/entries/relativism/

Baker, R. L., & Mednick, B. R. (1984). *Influences on human development: A longitudinal perspective*. Kluwer-Nijhoff.

Borrelli, P., Robinson, D. A., Fleischer, L. R., Lugato, E., Ballabio, C., Alewell, C., Meusburger, K., Modugno, S., Schütt, B., Ferro, V., Bagarello, V., Oost, K. V., Montanarella, L., & Panagos, P. (2017). An assessment of the global impact of 21st century land use change on soil erosion. *Nature Communications, 8*(1), 1–13. https://doi.org/10.1038/s41467-017-02142-7

Brevik, E. C., Homburg, J. A., & Sandor, J. A. (2018). Soils, climate, and ancient civilizations. In W. R. Horwarth & Y. Kuzyakov (Eds.), *Developments in soil science: Climate change impacts on soil processes and ecosystem properties* (Vol. 35, pp. 1–28). Elsevier.

Brown, L. (1995). Ecopsychology and the environmental revolution: An environmental forward. In *Ecopsychology: Restoring the earth, healing the mind* (1st ed., pp. xiii–xvi). The Sierra Club.

Burnor, R., & Raley, Y. (2011). *Ethical choices: An introduction to moral philosophy with cases* (2nd ed.). Oxford University Press.

Butler, C. (2002). *Postmodernism*. Oxford University Press.

Butler, C. (2010). *Modernism*. Oxford University Press.
Callicott, J. B. (2017). How ecological collectives are morally considerable. In S. M. Gardiner & A. Thompson (Eds.), *The Oxford handbook of environmental ethics* (pp. 113–124). Oxford University Press.
Colman, A. M. (2015). *Oxford dictionary of psychology* (4th ed.). Oxford University Press.
Cooley, C. H. (2013). *Human nature and the social order* (Rev .ed. (original ed. 1922)). HardPress.
Corsa, A. J. (2018). Grand narratives, metamodernism, and global ethics. *Cosmos and History: The Journal of Natural and Social Philosophy, 14*(3), 241–272.
Damasio, A. (2000). *The feeling of what happens: Body and emotion in the making of consciousness*. Mariner Books.
Diamond, J. (2011). *Collapse: How societies choose to fail or succeed*. Penguin Books.
Esbjörn-Hargens, P., & Zimmerman, M. E. (2009). *Integral ecology: Uniting multiple perspectives on the natural world*. Integral Books.
Ferrer, J. N. (2002). *Revisioning transpersonal theory: A participatory vision of human spirituality*. State University of New York Press.
Field, D., Morgan, C. L. S., & McBratney, A. B. (Eds.). (2018). *Global soil security*. Springer.
Fisher, A. (2013). Ecopsychology at the crossroads: Contesting the nature of a field. *Ecopsychology, 5*(3), 167–176. https://doi.org/10.1089/eco.2013.0031
Freeman, A. M., III. (2002). Environmental policy since Earth Day I: What have we gained? *Journal of Economic Perspectives, 16*(1), 125–146. https://doi.org/10.1257/0895330027148
Freinacht, H. (2015). *5 things that make you metamodern* [Blog]. Metamoderna. https://metamoderna.org/5-things-that-make-you-metamodern/
Freinacht, H. (2017). *The listening society: A metamodern guide to politics*. Metamoderna ApS.
Friedman, H. L. (1983). The self-expansiveness level form: A conceptualization and measurement of a transpersonal construct. *Journal of Transpersonal Psychology, 15*(1), 37–50.
Gleick, P. H. (2014). Water, drought, climate change, and conflict in Syria. *Weather, Climate, and Society, 6*(3), 331–340. https://doi.org/10.1175/WCAS-D-13-00059.1
Grunwald, S. (Ed.). (2006). *Environmental soil-landscape modeling: Geographic information technologies and pedometrics*. CRC Press.
Grunwald, S. (2010). Current state of digital soil mapping and what is next. In J. L. Boettinger, D. W. Howell, A. C. Moore, A. E. Hartemink, & S. Kienast-Brown (Eds.), *Digital soil mapping* (pp. 3–12). Springer. http://link.springer.com/chapter/10.1007/978-90-481-8863-5_1
Grunwald, S., Cao, B., Xiong, X., Ross, C. W., Patarasuk, R., Hempel, J., West, L. T., Andrews, S. S., Wills, S., & Loecke, T. D. (2014). *Part II—Integration of data to work towards a Meta Soil Carbon Model in the U.S* (D. Arrouays, N. McKenzie, J. Hempel, A. C. Richer de Forges, & A. B. McBratney, Eds., pp. 239–244). Taylor and Francis.
Grunwald, S., Clingensmith, C. M., Gavilan, C. P., Mizuta, K., Kastner-Wilcox, R. K., Pinheiro, E. F. M., Ceddia, M. B., & Ross, C. W. (2017a). Integrating new perspectives to address global soil security: Ideas from integral ecology. In D. J. Field, C. L. S. Morgan, & A. B. McBratney (Eds.), *Global soil security* (pp. 319–330). Springer Nature.
Grunwald, S., Mizuta, K., Ceddia, M. B., Pinheiro, E. F. M., Kastner-Wilcox, R. K., Gavilan, C. P., Ross, C. W., & Clingensmith, C. M. (2017b). The meta soil model: An integrative multi-model framework for soil security. In D. J. Field, C. L. S. Morgan, & A. B. McBratney (Eds.), *Global soil security* (pp. 305–318). Springer Nature.
Hicks, S. R. C. (2011). *Explaining postmodernism: Skepticism and socialism from Rousseau to Foucault*. Ockhams's Razor.
Hillman, J. (1996). A psyche the size of the earth: A psychological forward. In *Ecopsychology: Restoring the earth, healing the body* (pp. xvii–xxiii). Sierra Club Books.
Jacobson, M., Charlson, R. J., Rodhe, H., & Orians, G. H. (2000). *Earth system science. Biogeochemical cycles to global changes* (Vol. 72). Academic.
Janzen, H. H., Fixen, P. E., Franzluebbers, A. J., Hattey, J., Izaurralde, R. C., Ketterings, Q. M., Lobb, D. A., & Schlesinger, W. H. (2011). Global prospects rooted in soil science. *Soil Science Society of America Journal, 75*(1), 1–8. https://doi.org/10.2136/sssaj2009.0216

Kegan, R. (1982). *The evolving self: Problem and process in human development.* Harvard University Press.

Leopold, A. (1949). The land ethic. In F. O. Ndubisi (Ed.), *The ecological design and planning reader* (pp. 108–121). Island Press. http://www.cfr.washington.edu/classes.esrm.458/Leopold.%201949.pdf

Lerner, R. M., Schwartz, S. J., & Phelps, E. (2009). Problematics of time and timing in the longitudinal study of human development: Theoretical and methodological issues. *Human Development, 52*(1), 44–68. https://doi.org/10.1159/000189215

Lyotard, J.-F. (2003). *Postmodern fables* (G. van den Abbeele, Trans.). University of Minnesota Press.

MacKaye, B. (1990). *The new exploration: A philosophy of regional planning.* University of Illinois Press.

Macy, J. (1995). Working through environmental despair. In *Ecopsychology: Restoring the earth, healing the mind* (pp. 240–262). Sierra Club Books.

McFarland, S., Brown, D., & Webb, M. (2013). Identification with all humanity as a moral concept and psychological construct. *Current Directions in Psychological Science, 22*(3), 194–198. https://doi.org/10.1177/0963721412471346

Metzner, R. (1995). The psychopathology of the human-nature relationship. In R. Roszak, M. E. Gomes, & A. D. Kanner (Eds.), *Ecopsychology: Restoring the earth, healing the mind* (pp. 55–67). Sierra Club Books.

Mickey, S., Kelly, S., & Robbert, A. (Eds.). (2017). *The variety of integral ecologies: Nature, culture, and knowledge of the planetary era.* State University of New York Press.

Mizuta, K., Grunwald, S., & Phillips, M. A. (2018). New soil index development and integration with econometric theory. *Soil Science Society of America Journal, 82*, 1017–1032. https://doi.org/10.2136/sssaj2017.11.0378

Morin, E. (1999). *Homeland earth: A manifesto for the new millennium* (S. M. Kelly & R. LaPointe, Trans.). Hampton Press.

Myers, D. G., & Dewall, C. N. (2015). *Psychology* (11th ed.). Worth Publishing Ltd.

O'Connor, T. (1995). Therapy for a dying planet. In T. Roszak, M. E. Gomes, & A. D. Brown (Eds.), *Ecopsychology: Restoring the earth, healing the mind* (pp. 149–155). Sierra Club Books.

Patzel, N. (2010). The soil scientist's hidden beloved: Archetypal images and emotions in the scientist's relationship with soil. In E. R. Landa & C. Feller (Eds.), *Soil and culture* (pp. 205–226). Springer.

Rapley, J. (2004). Developmental studies and the post-development critique. *Progress in Development Studies, 4*(4), 350–354.

Reser, J. P. (1995). Whither environmental psychology? The transpersonal ecopsychology crossroads. *Journal of Environmental Psychology, 15*(3), 235–257. https://doi.org/10.1016/0272-4944(95)90006-3

Robinson, D. A., Hockley, N., Cooper, D. M., Emmett, B. A., Keith, A. M., Lebron, I., Reynolds, B., Tipping, E., Tye, A. M., Watts, C. W., Whalley, W. R., Black, H. I. J., Warren, G. P., & Robinson, J. S. (2013). Natural capital and ecosystem services, developing an appropriate soils framework as a basis for valuation. *Soil Biology and Biochemistry, 57*, 1023–1033. https://doi.org/10.1016/j.soilbio.2012.09.008

Rose, C. M. (2012). Liberty, property, environmentalism. In D. Schmidtz & E. Willott (Eds.), *Environmental ethics: What really matters, what really works* (pp. 420–437). Oxford University Press.

Roszak, T. (1995). Where psyche meets Gaia. In *Ecopsychology: Restoring the earth healing the mind* (pp. 1–17). Sierra Club Books.

Roszak, T. (2001). *The voice of the earth: An exploration of ecopsychology (Second).* Phanes Press.

Rowson, J., & Pascal, L. (Eds.). (2021). *Metamodernity: Dispatches from a time between worlds—Crisis and emergence of metamodernity.* Perspectiva Press.

Sanderman, J., Hengl, T., & Fiske, G. J. (2017). Soil carbon debt of 12,000 years of human land use. *Proceedings of the National Academy of Sciences, 114*(36), 9575–9580.

Schmidtz, D., & Willott, E. (2012). *Environmental ethics: What really matters, what really works*. Oxford University Press.
Scull, J. (2008). View of ecopsychology: Where does it fit in psychology in 2009? *The Trumpeter, 24*(3), 68–85.
Severan, A. (2021). *Metamodernism and the return of transcendence*. Palimpsest Press.
Stein, Z. (2018). Love in a time between worlds: On the metamodern "return" to a metaphysics of eros. *Integral Review, 14*(1), 187–221.
Thiroux, J., & Krasemann, K. W. (2016). *Ethics: Theory and practice* (11th ed.). Pearson.
Tyson, L. (2015). *Critical theory today* (3rd ed.). Routledge.
Varela, F. J., Thompson, E., & Rosch, E. (2016). *The embodied mind: Cognitive science and human experience* (Revised Edition). MIT Press.
Vermeulen, T., & van den Akker, R. (2010). Notes on metamodernism. *Journal of Aesthetics & Culture, 2*, 1–14.
Wade, J. (1996). *Changes of mind: A holonomic theory of the evolution of consciousness*. State University of New York Press.
Wals, A. E. J. (2010). Between knowing what is right and knowing that is it wrong to tell others what is right: On relativism, uncertainty and democracy in environmental and sustainability education. *Environmental Education Research, 16*(1), 143–151. https://doi.org/10.1080/13504620903504099
Wilber, K. (2000a). *A theory of everything: An integral vision for business, politics, science and spirituality*. Shambhala.
Wilber, K. (2000b). *Sex, ecology, spirituality: The spirit of evolution*. Shambhala.
Wilber, K. (2000c). *Integral psychology: Consciousness, spirit, psychology, therapy*. Shambhala.
Wong, D. B. (2006). *Natural moralities: A defense of pluralistic relativism*. Oxford University Press.
Worster, D. (2004). *Dust bowl: The southern plains in the 1930s*. Oxford University Press.
Yaalon, D. H. (2000). Down to earth. *Nature, 407*(6802), 301. https://doi.org/10.1038/35030260

Chapter 21
Take Care of Soils: Toward a Pluralistic Integral Soil Ethics

Sabine Grunwald

21.1 Introduction

Environmental ethics reveals the intricate interconnection between nature/Earth and beliefs, values, and cultural awareness. Ethics to protect cute panda bears, pristine wildlife preserves in the wilderness of Alaska, or beautiful exotic butterflies have received much attention, while an explicit comprehensive soil ethics has not been developed. Rather, soil as an invisible and voiceless part of nature has been generalized under land ethics (Esbjörn-Hargens & Zimmerman, 2009) or environmental ethics (Schmidtz & Willott, 2012). Thompson (2011, 2017) is one of the few who addressed soil ethics; however, his ethic was narrowly focused on agricultural stewardship, good farming, and food production as related to soils. Such a narrow account does not do justice to all soils on planet Earth, specifically nonfarmed soils, soils of ecological significance (e.g., tundra soils in the Arctic), urban soils, or natural soils in nature preserves or protected areas. This chapter aims to speak on behalf of all soils on Earth and present a pluralistic account of ethical strands applied explicitly to soils. The *Pluralistic Integral Soil Ethics* (PISE) synthesizes multiple ethical pillars into a coherent soil ethics that informs the use, management, preservation, conservation, security, health, quality, and sacredness of soils.

From a historical perspective, one of the most important ethics applied by humanity to nature was the "Natural Law" with questions on "Natural Right or Rights."

S. Grunwald (✉)
Soil, Water, and Ecosystem Sciences Department, University of Florida, Gainesville, FL, USA
e-mail: sabgru@ufl.edu

N. Patzel et al. (eds.), *Cultural Understanding of Soils*,
https://doi.org/10.1007/978-3-031-13169-1_21

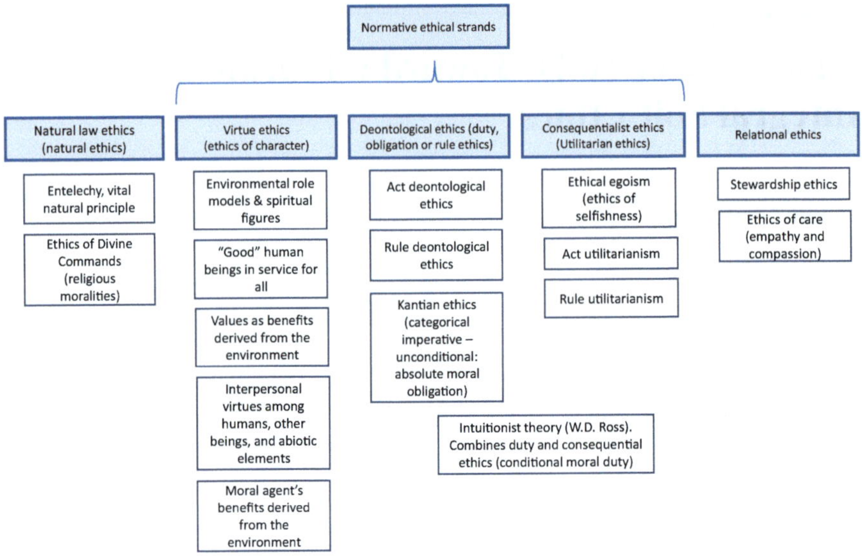

Fig. 21.1 Overview of ethical perspectives applied to soils/nature/environment

Nowadays, *normative ethics* address questions about "Who or what matters—that is, which things ought we to care about?" and "How do things matter—that is, how ought we to take them into consideration?" (Sandler, 2017, p. 223). There are three major normative ethical strands that address morality and various hybrid ones. The first, *virtue ethics*, focuses on the character of those who are acting, while both *consequentialist ethics* and *deontological ethics* focus on the status of the action, rule, or disposition itself (Fig. 21.1). These ethical strands will be discussed before proposing the PISE approach.

Kohak (2000) argued that the human approach to nature has often been in the spirit of "cowboy ethics (ethical relativism),"[1] with an interest mainly in our own individual desires and sublimely unconcerned about consequences, such as local and planetary soil degradation, leaving behind serious damage to our immediate environment. This kind of "wild west ethics" in regard to anthropocentric amplified global climate change and its effects on soil and food security are one of the most pressing planetary problems of our time (Cook et al., 2013; Kohak, 2000). In

[1] Kohak (2000) used the term "cowboy ethics" (or "wild west ethics"), which refers to an ethics in analogy to life in the Wild West of America. It expresses a time when ethics (e.g., virtue, deontological or other ethics) were replaced by the pistol or shotgun implying that the cowboy who shoots best wins or has the power to control the outcome. Cowboy ethics applied to soils (or in broadest sense nature, the environment) refers to impulsive ethics of those in power (ownership) to decide how to manage and use soils and determine soils' quality and health. Thus, this ethics is relativistic. In this approach, soil as a common good with all of its functions and services for the greater public good, is superseded by individuals (e.g., landowners or owner of a farm corporation) decision-making about soil health and quality.

contrast, White (1967) argued that Christian theology taught that humans are meant to control nature and that this was the root of much of Western civilization's mistreatment of nature. However, from a contemporary ethical pluralistic perspective, there is not a single unified view in regard to soil and nature ethics (Hinman, 2013). Therefore, I will first critically discuss the different ethical perspectives of soil, nature, and the Earth before presenting a proposal for pillars of a novel integrative soil ethics.

21.2 Ethical Perspectives of Soil, Nature, Earth

21.2.1 Soil Natural Ethics: Natural Law and Natural Rights

Natural Law Theory claims that it is a moral obligation to maintain and promote those natural values toward which nature aims. Natural Law was instilled by Aristotle's thought of *entelechy*, which is the vital principle that realizes what is otherwise merely potential (entelechy, "having the aim in itself") (Fig. 21.1). Natural functions express the purpose of nature, that is, they ultimately aim at achieving some end or goal that has natural value. According to this theory, we have a moral obligation to care for ourselves and others and the Earth, including soil, by upholding life, protecting the environmental, and not harming other species (preservation of nature). This ethics calls for cooperation and harmonization with nature because nature is attributed intrinsic value. Natural Law theorists view the cosmic order as rational and accessible through human reason. Catholic thinkers like Thomas Aquinas reasoned for Natural Law ethics because the rationality and goodness of the natural order are explained as created by a good God who endowed humans with reason so that people understand the natural order. Since God is good, it explains the goodness of creation as a whole (Burnor & Raley, 2011). This Natural Law perspective extends intrinsic (inherent or essential) value to all organisms (biocentric view), even if they are considered as nonconscious organisms (e.g., soil fungi or plants) (Varner, 2012). The biocentric view considers the following: (1) humans as members of life on Earth with equal value among humans and nonhumans, (2) Earth viewed as an interconnected complex web with function of one element (e.g., climate) interconnected with the function of another element (e.g., soil carbon), and (3) each organism pursuing its own good in its own way (Taylor, 2012).

The general principle of Natural Law Theory is considered absolute. Although the motivation to expect the greatest personal satisfaction and fulfilment in life for being moral is compelling, this ethical view is problematic. First, what reason is there to believe that the natural is inherently good? Nature has good and bad sides, including evil expressions. Second, what does "natural" mean? Natural is undefinable and vague. Are human technologies unnatural and not part of nature? Natural rights in this theory are considered as universal, for example, the right to life, right to health, or right to liberty, whereby all natural rights must be held equally for all. However, this poses questions in case of conflict as to which of the natural rights to uphold as primary (e.g., the right to protect soil, pandas, or human rights).

Stone (2012) asked whether the natural environment (e.g., soil, water, trees) should be given legal rights. The common environmental law makes natural objects rightless because nature ethics is often overpowered by anthropocentric ethics. The latter ethics assume that natural objects are senseless objects without consciousness. The moral perspective objectifies natural things like soils, hillslopes, or valleys so that they can be manipulated at will with psychic distance between human and nature. This kind of contemporary environmental ethic stands in contrast to myths, spiritualities, and religions (e.g., pantheism, Buddhism, or Taoism). Some of these spiritualities regard the Earth and all sentient beings as one organism of which mankind is a part. The attribution of inherent worth to nature evokes an ethic of respect for nature (Taylor, 2012), which also imbues virtue ethics.

21.2.2 Virtue Ethics and Valuation of Soils, Nature, Earth

Virtue ethics ("being good") is based on either character virtues or relationships and is one of the normative ethics (Traer, 2007) (Fig. 21.1). According to Hursthouse and Pettigrove (2016), the main characteristics of virtue ethics are moral character and discernment between virtues (i.e., an excellent trait of character) and vices (i.e., immoral or wicked character). Environmental virtue ethics assert how one ought to interact with soil/nature/environment and focuses on the character dispositions "to do the right thing" and "act for the right reasons." Sandler (2017) pointed out that it is uncontested in environmental ethics that the kind of people we are (i.e., one's identity) is crucial to how we engage with the natural environment. *Values* reflect what is morally acceptable in terms of a culture, whereas *virtues* reflect individual human characteristics. The Meta Soil Model introduced by Grunwald et al. (2017a) stressed the tight coupling between the "Who" (i.e., the subject/person; identify of a person), "What" (i.e., soil/nature/environment), "How" (i.e., the approach or worldview), "Why" (i.e., values attributed to soil/environment), and "for What/Whom" (i.e., purpose or need in collective view) that cannot be separated and are intimately tied to virtue ethics.

Environmental virtues are manifold and entail (Sandler, 2012): (1) environmental role models and spiritual figures, (2) "good" human beings who act to maintain and promote the well-being and flourishing of larger ecological communities, (3) value as benefits derived from the environment, (4) extension from standard interpersonal virtues among humans to include nonhuman beings (spirituality) and abiotic elements, and (5) moral agent's benefit derived from the environment, such as food from soil, aesthetic value of soil, and physical value to build structures. The following sections describe each of the environmental virtues in detail.

1. ***Environmental role models and spiritual figures*:** Environmental role models like Rachel Carson and Aldo Leopold advocated pro-nature narratives, which inspired conservation management and environmental protection in North America (Esbjörn-Hargens & Zimmerman, 2009). Idealized role models may be beings of the unknown world, such as *Pṛthvī* (Earth goddess) an idealized deity in Buddhist mythology representing Mother Earth, source of life, and fertility. She serves in her role as *Bhūmi* (Sanskrit, soil, ground or foundation) as the womb of the Earth that is free from deceit and full of truth (Shaw, 2015). Buddhist beliefs inspired ecodharma, in which people express deep care for sentient beings, the Earth, and the whole cosmos (Loy, 2019).
2. ***"Good" human beings who act to maintain and promote the well-being and flourishing of larger ecological communities*:** Environmental virtuous good human beings are social beings that perceive themselves as members of the human community (e.g., Chico Mendes, Al Gore or David Suzuki) or even broader as member of sentient beings that support stewardship of the land (Sandler, 2012). Virtuous soil heroes are often less well known to the public and include the ordinary organic farmer or permaculturist. Broader conceptions of soil, land, beings, and the cosmos as super-ecologies are rooted in the transcendence of the natural world, thereby connecting with the divine, sacred, or some cosmic spiritual dimension (Kripal, 2014).
3. ***Value as benefits derived from the environment*:** Environmental benefits equated with virtues depend upon the prioritization or ranking of environmental values to discern what is virtuous or vicious. Moral excellence (i.e., goodness and righteousness of a person) in a given culture is associated with knowledge and understanding of soils and covers a broad range of *intrinsic* and *extrinsic* motivational principles (Grunwald et al., 2017a, b). *Intrinsic motivations* arise from within the individual as an internal truth or transpersonal knowing. Grunwald et al. (2017b) referred to this knowing from within as cognizance that is the awareness and perceptions held by individuals. Intrinsic valuation of nature and all life on Earth is referred to as biocentrism (or ecocentrism). For example, the biocentric notion asserts that if living beings, including earthworms, are attributed intrinsic value, it follows that it is conducive to allow them to pursue their own goods unconstrained by human interference (e.g., rejecting technological and managerial interferences, such as agricultural management and ploughing imposed on soils). Such view prioritizes the protection of all sentient beings, for example, soil fungi or bacteria, as an ideal (Sandler, 2012). This kind of ecocentrism advocates holism to value nature and equality for all beings understood as interconnected parts of a whole. From a practical point of view, the notion of equality of all life is untenable—who is more valuable, a prey or predator, soil microorganisms, or human community? The shadow side is that the notion of eco-/biocentric superiority maximizes ecological welfare and may violate human rights and well-being because other life forms are attributed equal value to human life. This kind of rivalry toward life has been equated to environmental fascism and from a practical standpoint is impossible to achieve (Zimmerman, 1995).

In contrast, *extrinsic motivations* for protecting nature arise from outside the individual, such as environmental change (e.g., land use change or urban sprawl), pollution of soils, or soil carbon losses due to global climate change. For example, environmental injustices and environmental racism are observed in the case of corporate, developer, or governmental abuse of the soil environment to disenfranchise the poor who live off the land (Shrader-Frechette, 2012). Sociocultural factors, such as gender (Buckingham & Kulcur, 2009) as well as race and class (Mohai & Bryant, 1992), have been involved in environmental injustices.

4. ***Extension from standard interpersonal virtues among humans to include non-human beings (spirituality) and abiotic elements.*** From a philosophical perspective, Thompson (2017) argued that the ethics of soil are not only based on externalized motivations (e.g., potential effects of agriculture on soil and water) but also on religion, spirituality, and folklore, which point to the spiritual or intrinsic values underlying how cultures relate to soil. Historically, in medieval cultures, it was evident that nature is God's creation, thus, to destroy nature and sentient beings was considered heresy; though in practice, this belief was often not realized with few exceptions (e.g., Saint Francis of Assisi). In the Judeo-Christian tradition, one value was the *utilitarian* (or *extrinsic*) value in which a being derived from serving some higher (divine) goal and the other its own (or *intrinsic*) value simply because man is God's creature and God is love. The latter implying that life is a value for itself (Kohak, 2000).

   Virtue ethics are associated with *valuation of soils*, for example, in the form of soil ecosystem services (Adhikari & Hartemink, 2016). *Ecofeminism* builds on the valuation of nature and the feminine that have been both oppressed and dominated and must be freed to deeply connect with nature/soils/Earth and the feminine (Warren, 1997). Ecofeminism brings together feminism and the green movement. According to MacKinnon and Fiala (2018), ecofeminists argued that environmental problems are caused by a male-centered view of nature grounded in dominion over nature. In contrast, women value caring, relationships, and cooperation with nature whereby mother-child mirrors the human-Mother Nature (Gaia) relationship. Soil/Earth are considered feminine and sacred in ecofeminism that has been marginalized in Western cultures.

   Deep Ecology is a holistic virtue-based ecology that does not separate humans from the natural environment (Drengson & Inoue, 1995). This ecology recognizes the fundamental interconnectivity, interdependence, and intrinsic values of all living beings as part of the web of life. The acknowledgement of value inherent in the living nature, of which soil and humans are part, stems from the deep ecological awareness that nature and the self are one; thus, at its core, it is spiritual awareness. According to Deep Ecology, the individual feels connected to the cosmos as a whole and is evoked to act virtuously by respecting nature. The virtues of Deep Ecology stand in sharp contrast to *Shallow Ecology*, which is considered anthropocentric (Capra, 1995). For Deep Ecology, the contemporary environmental crisis is ultimately a spiritual crisis, and thus, globally, both are pertinent contemporary issues due to the overpowering of anthropocentric and egocentric values over biocentric values (Mickey et al., 2017).

5. ***Moral agent's benefit derived from the environment, such as food from soil, aesthetic value of soil, physical value to build structures.*** Moral agent's benefits derived from the environment, such as food from soil, aesthetic value of soil, or physical value of soil to build structures, to justify dispositions by preserving and cultivating these opportunities and goods, and also to enjoy them (Sandler, 2012). The virtue of a meaningful relationship with the natural environment is only possible for those who are receptive and open-minded toward soil/nature/environment. This view suggests that only those who are able to embody soil/nature/Earth see value in preserving them. Disembodied and dissociated individualized urban cultures have shown a propensity to disidentify and separate, not only from their own individual bodies but also from natural bodies (Caldwell, 2014; Stanley, 2016; Turner et al., 2004).

Human separation from nature was recognized earlier by representatives of the Frankfurt School of Critical Theory who argued that through narrow positivistic conceptions of rationality focused on science, power, and technological control to solve all human problems, nature is no longer considered fearsome, mysterious, and uncontrollable. Instead positivism and consumerism have disenchanted nature and, thus, disrupted culture's intimate relationship with the Earth (Brennan & Lo, 2015). Specifically, cultures of the North and West have marginalized subjective direct ways of knowing nature on behalf of technological advancements that have infused sciences. For example, in soil science, the use of remote and proximal sensors and geospatial technologies to quantify soils from space and the laboratory has reached new peaks and separate people from the soil that is sensed (Grunwald et al., 2015). This trend contrasts subjective sensual experiences of soils in nature and has led to distancing between soil and people, specifically in technology and information-oriented cultures. Gottlieb (2019) argued that it is a personal individual connection between people and soil/Earth that evokes fulfilment and belonging, which serve as pillar for an ecological democracy. Importantly, the aim in such deliberate democracy is not about equal individual freedom (humans and other beings alike), but a good, shared, harmonious life in community.

Virtue ethics are grounded in liberative personal feelings (i.e., motivation from within) and intimate relationships (e.g., compassion for nature and empathic relating with life), while *principle-based ethics* asks one to detach or separate from one's personal feelings to make a moral decision (externalized ethics; Burnor & Raley, 2011). Thus, the latter emphasizes impartiality and obeyance of prescriptive moral principles but evokes less motivation to act morally (e.g., to care for nature and preserve soil) when compared to virtue ethics. Kantian ethic, a principle-based ethic, asserts that ideally the moral person is motivated by duty as a moral obligation that is unconditional and applies to all people in all situations (categorical imperative), which has been questioned to truly motivate action because of a lack of personal desire (MacKinnon & Fiala, 2018; Rohlf, 2018).

The drawback of virtue ethics is that a vicious character motivation (e.g., narcissistic aggrandizement and egoic profit orientation) may overpower principle-based ethics (e.g., protective environmental regulations). For example, Bomberg (2017) pointed out that US President Donald Trump's anti-environmental crusade eroding environmental policies and regulations in the United States was motivated by economic profit rather than care for the environment. To reverse limits on carbon emissions from power plants by Trump's order to the US Environmental Protection Agency and denial of the global climate change phenomenon not only amplifies climate change but also impacts soils. This kind of vicious virtue is an example of psychological egoism that overpowers principle-based ethics and care for the greater good (MacKinnon & Fiala, 2018; Tollefson, 2017). Davies (2017) stressed the risks of soils' cultural undervaluation as common good and pointed out the lack of international legislation that governs and protects soils directly despite its inherent value as carbon sink. According to Thiroux and Krasemann (2016), one question that has been raised is how virtues are determined and who is really a "good virtuous person." Contemporary philosopher A. MacIntyre argued that virtues depend at least partly on the beliefs and practices that constitute a culture (cited in MacKinnon & Fiala, 2018). Prima facie *value* ("at first glance") expresses the phenomena that a value, which seems neutral or positive at first impression (e.g., soil health), can be overcome by other interests or values in a community or culture (e.g., economic profit) (MacKinnon & Fiala, 2018).

21.2.3 Consequentialist Ethics and Soils

Consequentialist ethics entails ethical egoism, act utilitarianism, and rule utilitarianism (Thiroux & Krasemann, 2016) (Fig. 21.1). *Ethical egoism* asserts that the morally right act, for any particular situation, is the act that will produce the greatest amount of *utility* (usefulness) or the least amount of disutility for oneself. This kind of ethics is selfish or egocentric (Burnor & Raley, 2011). In contrast, in *utilitarianism*, an act is considered moral if it is useful in bringing about a desirable or good end for the greater common good or for the benefit of a majority ("bring pleasure and happiness to everyone") (Thiroux & Krasemann, 2016). Utilitarianism is exemplified in the concept of soil security, which is defined as "the freedom from risks of losing (1) a specific or group of soil function(s), (2) goods and services that soils provide to benefit humans and—in its broadest sense—(3) sustainability of life on Earth" (Grunwald et al., 2017a, p. 306). Ethical egoism rejects the notion of moral equality that is the belief that all persons and cultures have equal moral worth and deserve equal treatment. The implications are that the sole underlying motivation of egoism is personal advantage, control, and self-fulfillment (Burnor & Raley, 2011). The consequences of ethical egoism are often power imbalances between an oppressor and oppressed. The masculinization of agriculture oppressing women farmers and creation of monocultures incentivized by biotechnology industries in India with devastating effects for soils (Shiva, 1999), corporations and shareholder policy (Korten, 1998), and consequences of such corporate culture in Brazil's mining disaster (Garcia et al., 2017) exemplify ethical egoism. The consequences of egoism featuring profit gains rather than soil security and human health are exemplified by the recent glyphosate[2] controversy with contrasting claims in regard to human toxicity (carcinogenic risk) by multibillion dollar industry conglomerates, environmental advocates (Benbrook, 2019), and empirical studies (Zhang et al., 2019). Specifically, enculturated egoism and socially sanctioned egoism may pose greater threats to emerging trends, such as Digital Agriculture (Shepherd et al., 2018) and Digital Soil Mapping/Pedometrics (McBratney et al., 2018). The digital divide of cultures in the West-North and South-East may amplify socio-ethical issues that impart the utility of profit orientation through technological advancements onto the use and management of soils.

Consequentialist ethics differs widely and exhibits ambiguities based on what is considered a "good" consequence (e.g., pleasure, happiness, individual satisfaction, welfare of others, soil quality) (Alexander & Moore, 2015; Traer, 2007). From a biocentric perspective, the maximization of naturalness (i.e., excluding agricultural, forest, and other kinds of management of soils) as an intrinsic value of *soilness* stands in contrast to anthropocentric perspectives. The latter emphasize consequential utilities (e.g., maximize the crop yield in Alfisols, maximize soil security, or optimize soil carbon sequestration).

Among the consequentialist ethics, *act utilitarianism* posits that "everyone should perform that act which will bring about the greatest amount of Good over Bad for everyone affected by the act," while *rule utilitarianism* argues that "everyone should always establish and follow that rule or those rules that will bring about the greatest good for all concerned" (Thiroux & Krasemann, 2016, p. 41 and p. 43) (Fig. 21.1). Act utilitarianism suffers from the need for each person to re-assess each situation and possible consequences, and thus, it is unable to provide any cultural relevant universal morals (MacKinnon & Fiala, 2018). This need is difficult to meet by people, and cosmopolitan-industrialized cultures that lack knowledge about soils have no relationship to soil (e.g., urban dweller, disembodied people, or office workers) for whom soil lacks visibility and sensual experience, is valued as less important than economic security and personal well-being, and perceives no stake in the well-being of soil (soil $\neq$ common good). Bouma et al. (2012) pointed out that awareness and affinity of mankind to their soils have been lacking in contemporary cultures in Europe. These authors argued that traditional approaches of soil information databases and atlases and nonlegal binding directives (e.g., the European Union Thematic Strategy for Soil Protection) have been ineffective to address the consequences of management that impart on soil sustainability. Instead, citizen groups and local awareness raising efforts to bring people and soil together were suggested to enhance the capacity to build relationships *with soil, for soil.* In contrast, the recent COVID-19 pandemic and its impacts on work chains of consumers and agricultural producers may raise the awareness of the consequences of interrupted food production and quality, which is dependent on soil health and management.

[2] Glyphosate is a broad-spectrum herbicide (e.g., Roundup).

Soil science has been instrumental in assessing possible consequences of soil management using empirical and modeling studies. For example, assessment of global soil carbon losses (i.e., the effects or consequences) due to global climate warming (i.e., the cause or stressor) suggests that the losses are contingent on the size of the initial soil carbon stocks and that effects are larger in high-latitude areas. This study estimated that a business-as-usual climate scenario would lead to a loss of 55 ± 50 Pg of carbon from the topsoil by 2050, suggesting that rising temperatures stimulate net loss of carbon to the atmosphere (Crowther et al., 2016). Consequentialist ethics applied to a gamut of scenarios of soil carbon losses/gains due to global climate change reveal which imagined future would provide the best and worse consequences in terms of the soil carbon sequestration function.

Despite the ability to assess consequences using the scientific method, the limitations of consequentialist ethics are manifold. In general, a difficulty with consequentialist theories is the assessment of future outcomes for oneself and for others, with an acceptable level of certainty (e.g., family, neighborhood, nation, or the whole globe; Thiroux & Krasemann, 2016). Another major detriment of act utilitarianism is that people, particularly if they do not view themselves as a stakeholder of the environment, lack the cultural awareness and environmental literacy to assess the complexity of soil-environmental issues, and are under the influence of mass media that may debunk scientific studies as pseudoscience (Lipps, 1999). A cultural engrained anti-science mentality tends to distort science-based findings.

Rule utilitarianism suffers from identifying universal sets of rules that cover multiple situations (Burnor & Raley, 2011). For example, to assess soil carbon sequestration and water holding capacity considering different soils, crops, climate, geology, topography, cultural patterns, and management practices is highly complex. To develop universal sets of moral rules for the multiverse of soils and cultures to address soil health, soil security or soil care seems extremely challenging because these soil constructs are ill defined (Mizuta et al., 2018). Utilitarian ethics often employ a "cost-benefit" approach in which the "end justifies the means" to maximize utility (Thiroux & Krasemann, 2016). While soil scientists are fully aligned with consequentialist ethics (i.e., "to study what happens to soils under specific conditions or change"), this kind of approach in a moral decision context makes less sense. For example, one may argue from a utilitarian perspective that soil health is best achieved globally by reducing the number of people on the planet. Such a thought is considered morally absurd because what maximizes desirable utility is not necessarily good for everyone, specifically refuting the basic human right to live. Marginalized people (e.g., women, immigrants, poor people, and indigenous people) are especially vulnerable in patriarchal cultures to suffer consequences (e.g., forced migration due to climate change and soil degradation). The chasm between soil science, which has embraced the cause-effect (or consequence) paradigm, stands in contrast to the consequentialist morals from a cultural perspective with unforeseen and somewhat amoral consequences for some cultures.

Kohak (2000) pointed out that the North-South divide of consequences imposed on soils (e.g., global climate change) will be quite unequal, with industrialized countries buffered through the use of technologies and wealth, while developing countries, specifically on the African continent, will more severely suffer from the disruption of cultures and economies. Gibbs and Salmon (2015) estimated that 1–6 billion ha of soils globally (dependent on the assessment method) are degraded with most degradation centered in Third World cultures on the African continent and in South-East Asia. According to Kohak (2000), some cultures have become global raptors, exemplified by the inequities in terms of soil resource usage with about 20% of humanity inhabiting the overconsuming world that use about 80% of all global resources and generate corresponding waste. Specifically, native populations in the Third World now desperately reach out as globalization threatens to destroy their soils that provide a means of balanced survival. The consequences of technological, energy and digital divides combined with the population explosion have amplified the West-East and North-South separation with culturally uneven distributions of consequences for the utility of soil quality and health. These soil-centric utilities are in competition with anthropocentric utilities (e.g., to attain the greatest happiness or well-being). However, it is unclear how preferences are negotiated in regard to competing consequences (e.g., is it more important "to combat climate change and soil degradation which impacts food security" or "people to be free and self-actualize"?). It is also disputable whether quantity or quality is more important to maximize utility (MacKinnon & Fiala, 2018). For instance, is it more moral to secure one field and soil (e.g., conservation management) or secure a larger soil region, but allow some degradation through agricultural management (e.g., grow corn with some nutrient-enrichment of waterways)? Environmental consequences for oneself are often perceived differently than for collectives. Schultz (2001) found a clear distinction in attitude between egoistic, altruistic, and biospheric environmental concerns. Opotow and Weiss (2000) showed that moral denial in environmental conflicts is due to self-involvement (i.e., unwillingness to make sacrifices to foster the greater good or others well-being) and outcome severity (e.g., concealment of environmental harmful outcomes, double standards, reduction in moral standards, dehumanization of other stakeholders).

According to act utilitarianism, one ought to reject other alternatives in order to maximize utility (Burnor & Raley, 2011). In the case of maximizing utility, suppose I pay $10 to help sequester carbon in soil on a field, while on a farm or region, I would morally be asked to pay more ($100, $1000, etc.) to do more good for the sake of others and the planet to make up global soil carbon losses. However, many people are less inclined to act that way. For example, Chaikaew et al. (2017) found that people in a large watershed in southeastern United States were only willing to pay <US $2/household/year for ecosystem services, including soil carbon sequestration and climate adaptation, which is less money than a fast food meal.

21.2.4 Deontological Ethics and Soils

In contrast to consequentialist ethics, *deontological ethics* rejects consequences as a basis of morality and instead focuses upon duties (or principles) and intentions ("what we ought to do") (Burnor & Raley, 2011) (Fig. 21.1). *Act nonconsequentialist* theories refer to morals grounded in what one believes, feels, or intuits to be the right action to take. This ethic is highly individualistic and grounded in personal experience and intuition. Argument in support of this kind of act deontology is that morals do not require any reason or logic because direct perception provides a way of knowing what is "right or wrong." Act deontological ethics are often grounded in spiritual/religious beliefs. It is assumed that supernatural beings or some spiritual force communicates to humans how they should act morally (Thiroux & Krasemann, 2016). According to the Pew Research Center (2015), two-third of Americans identify as Christians (mainly Evangelicals and Catholics). Worldwide, more than eight in ten individuals identified with a religious group with 31.5% Christians, 23.2% Muslims, 15% Hindus, and 7.1% Buddhists, and the rest classified themselves as other or unaffiliated (Pew Research Center, 2012). These numbers suggest that religion and spirituality play a major part in many people's lives and inform their moral attitude toward the environment.

Rule deontological (nonconsequentialist) theories argue that rules are the only basis for morality (Thiroux & Krasemann, 2016). For example, *Divine Command Theory* asserts that if one believes there is a God(s)/goddess(es), the divine provides moral commands to people. If individuals obey the sacred commands, irrespective of the consequences that might ensue, they are "good" moral people. The intersectionality between individual spiritual/religious beliefs and cultures enforces socially sanctioned behaviors. An argument against this kind of deontology is that there is no proof or evidence that this kind of moral is certain and correct. In addition, rule deontological moral is immune to rational critique because it applies only to the individual or cultural group who is assumed to have direct access to "higher spiritual power" (Thiroux & Krasemann, 2016). The drawback of spiritual/religious inspired morals is that hierarchies of superiority are established by claiming to hold "higher moral standards," while others' standards are inferior. This often leads to righteousness arguing for superiority based on spiritual/religious insights (Ferrer, 2002, 2017). Globally, there are trends in regard to the erosion of dogmatic organized religions (Fernandes, 2003), the rise of religious fundamentalist movements (Doktor, 2003), and an increase in individualized personal spiritualities (Moore, 2014).

Spirituality involves holding one's deepest moral values and seeks to understand one's place in the environment and the cosmos, while religion refers to organized and institutionalized religion according to Taylor (2010). In Taylor's view, spiritualities or religions that consider nature to be sacred are imbued with intrinsic value worth of reverent care. In general, nature religion touches on religious perceptions and practices that are characterized by a deep reverence for nature. The view of sacredness of nature is found in nature mysticism and earth religions, such as animism and pantheism. Importantly, the connection to nature (including viewing soils as sacred) expressed through spiritual (peak) experiences brings forth communion and healing. For example, Adams (2010) provided an account of nature's participatory psyche viewed through the nondual transpersonal lens in support for an experiential ethical imperative to care and protect nature. Bidwell (2015) presented an embodied spiritual experience in nature that was brought forth through a Buddhist-Christian hybridized identity. Environmental spiritual/religious expressions are found among indigenous cultures, for example, the Chipko Andolan (Whyte & Cuomo, 2017), nondual Eastern traditions (e.g., Buddhism; Kaza, 2019), as well as American Transcendentalists, for example, Henry David Thoreau (Taylor, 2010). These traditions share in common that insights and wisdom come through direct experience of nature. Nature and the Earth are believed to be sacred evoking ethics of care rooted in interconnectedness of humans, nature, and the cosmos (Taylor, 2017).

Another prominent rule of deontological ethics is *duty ethics,* developed by Immanuel Kant (Thiroux & Krasemann, 2016). This ethic established morality by goodwill and reasoning forming absolute universal moral rules. Kant's categorical imperative holds unconditionally for everyone and every situation without exception and involves the principle of ends ("persons should be treated with respect and worth in themselves") but never simply as means ("as valuable only as ways of obtaining something else that we do value"; Burnor & Raley, 2011). The motivation for moral action according to moral rules, permission, and codes within deontology that are socially constructed are based on the avoidance of social sanctions and penalties (Alexander & Moore, 2015). Limitations of duty ethics are that moral dilemmas arise (e.g., protect soils versus exploit them for profits) when absolute rules that oppose each other are not resolved. Kant provided the reversibility criterion (Golden Rule concept) implying that one should not destroy nature, because it may backfire on the one aiming to destroy nature. A major criticism of rule deontology is that ignoring consequences is risky, specifically when applied to environmental and soil degradation (Thiroux & Krasemann, 2016). Problematic with Kant's categorical imperative is that it assumes a person's freedom and inherent rationality, which Kant calls a person's autonomy to act morally. However, from a social constructivist perspective, considering that intersectionality of race, gender, spirituality/religion, economy, and other social constructs inform people's beliefs that are often subjective and individual refutes Kant's ethic. W.D. Ross's *intuitionist theory* offers a reinterpretation of absolute deontological ethics that can resolve moral conflicts through the prima facie approach (i.e., ranking of duty preferences) by consideration of consequences (Fig. 21.1). This ethic is called *conditional moral duty* based on fundamental binding moral duties (fidelity, reparation, gratitude, justice, beneficence, self-improvement, and nonmaleficence). Such conditional moral duty supports a stance of care and gratitude toward nature (Burnor & Raley, 2011; Thiroux & Krasemann, 2016).

Rule-based ethics have inspired soil policies with mixed results (Bouma, 2005; Bouma et al., 2012). For example, despite the significant increases in soil degradation in the European Union (EU) and acknowledgment of the importance of soil health, there is not to date an EU soil-specific legislation with binding measures (Paleari, 2017). The United Nations Sustainable Developmental Goals focused on land degradation neutrality are an integral part of law, policy, and management at the national level that have not been ratified in the form of legally binding universal soil protection and multilateral agreements yet (Boer & Hannam, 2015). Bunzl (2012) pointed out the governance gap at global, national, and regional levels about soil and environmental protection are overshadowed by more engagement from nongovernmental organizations and small activist groups than binding legislation. Global institutionalized efforts, for example, the Global Soil Partnership (GPS) and the Intergovernmental Technical Panel on Soils (ITPC), have no national mandate to enforce the implementation of soil safeguard.

21.2.5 Stewardship Ethics and Soils

Stewardship ethics are grounded in beliefs that soil is a fundamental source of life, and thus, it is society's responsibility to care for the planet. Sauer and Nelson (2011) eloquently stated that "We are, each of us, people of the soil" (p. 5) recognizing that *soil stewardship* is an ethical issue that soil science cannot solve alone. The human-nature relationship rooted in the Western Judeo-Christian notion of despotism toward nature has been blamed for environmental degradation according to White (1967) and Passmore (1980). Despots, masters, or shepherds of nature view themselves apart from nature, positioned as superior and being above all other life. This dualistic notion assigns humans the right to manipulate nature from the outside, rather than to cooperate with nature acknowledging a nondual perspective (Plumwood, 2012).

Passmore (1980), in *Man's Responsibility for Nature*, discerned two interlocking views in dialectic opposition—stewardship as humanities care for nature (ethic of preservation for nature's own sake) and the myth of the garden (nature to be dominated by humankind with human masters that can exploit nature). This dialectic chasm contrasts the love of nature against productionism, nature extremists against the production technologists. In softer forms, stewardship of preservation of soils (nature essentialism) can be discerned from stewardship of conservation of soils (dominion over nature) (Thompson, 2017).

Agricultural stewardship has tied moral duty to care for soil and nature due to (1) religious duty to protect and nurture God's creation in the Judeo-Christian tradition ("the farmer as shepherd of nature"), (ii) productionist duty to keep soils healthy and fertile to grow food ("good farming"), (iii) preservationist motivation tied to intrinsic value of soils and the land ("farmer avoids the bad consequences of soil degradation, now and in the future, because soils cannot be restored to fertility if completely degraded"), (iv) folklore of stewardship as a virtue, a character trait ("the good farmer"), and (v) romanticized stewardship ("the farmer is the custodian of the land for all—rural, urban, and natural land"; Thompson, 2017).

One major argument countering agricultural stewardship ethics is that in contemporary cultures, specifically in industrialized cultures, urbanization trends are prevalent with less people involved in agriculture. Hence, such ethics is impartial and relies on those who manage the land on behalf of those who live in cities and are separated from soil/land. More than 55.3% (2018) lived in urban areas globally and 82% (2018) lived in urban areas in the United States, suggesting that the number of farmers and people in rural communities relative to urban people are small (The World Bank Data, 2019). According to Adamopoulos and Restuccia (2014), in the richest countries, farms larger than 20 ha operate 70% of land, while in the poorest countries, 70% of land is operated by farms smaller than 5 ha. Subsidence farmers in smallholder farm settings, for example, in East African cultures, have been marginalized, and farmers often farm for survival and livelihood rather than as shepherds for all of humanity (Tittonell et al., 2010). Thompson (2017) questioned agricultural stewardship ethics because duties to nature seldom prevail over productionist ethics, specifically in industrialized cultures in which agribusiness are run for profit.

A critique of stewardship ethics is that stewards have perceived themselves as owners of land/soil under the maxim of controlling a resource "it's my land and I can do with it what I want" (Little, 1986, p. 316). Little (1986) asserted that the unsuccessful efforts to codify land stewardship ethics into policy to protect land/soil rather than to exploit them has hampered its success. Another limitation is that sustainability is rooted solely in scientific assessment (Singh et al., 2012); however, it leaves people and cultures out of the picture, objectifying them as human agents.

21.2.6 Ethics of Care and Soils

Ethics of care (*relational ethics*) are a relatively new ethics that is focused on the feminine or care perspective instead of the masculine justice perspective (Burnor & Raley, 2011). Women's views on ethics are different, not inferior, but equal to men's ethics. The latter emphasize rights, justice, living by rules, being independent, and competition, while women's ethical view is grounded in working to maintain close relationships, generosity, harmony, and reconciliation (Thiroux & Krasemann, 2016). According to Gilligan (1993), women tend to focus on care and relationships exemplified by the ideal care model of mother/child relation, while men tend to focus on justice, universal rights, and moral principles. Ethics of care emphasize interpersonal relationships, human interdependence, relational autonomy, impartiality, particularism, and emotions, especially healthy relation-building emotions (e.g., sympathy, empathy) in support of morally healthy relationships. Noddings (2003) asserted that caring relationships between carer and cared-for requires the latter to recognize oneself as recipient of care and mutuality ("good relationships"). Modern and ancient cultures have shown pronounced patriarchal social structures with more masculine than feminine expressions that have been perceived as uncaring and oppressive (Fernandes, 2003). Burnor and Raley (2011) remarked that care ethics is radical because it aims at degenderized morality in which the feminine (care ethics) and masculine (justice ethics) are in balance and shared by both men and women. Feminine care ethic has been instrumentalized in the *Gaia theory* developed by James Lovelock and Lynn Margulis (Kelly, 2017; Taylor, 2010), feminine earth spirituality (Low & Tremayne, 2001), and feminism and environment (Brennan & Lo, 2015).

Objections of care ethics entail the nature/nurture problem, which asks whether one's genderized perspectives are the products of society or whether they reflect innate biological differences between men and women. *Nature* represents those beliefs and personal traits that are innate and unchanging, regardless of outside influences, whereas *nurture* refers to those beliefs and traits that are the products of one's environment, particularly cultural expectations and social norms (Burnor & Raley, 2011). The largest problem is that care in close family relationships is tender and intimate, whereas care extended to communities, sentient beings, soil, nature, the Earth, or future generations becomes more abstract, distant, and impersonal. Caring as morally normative ethics views all kinds of relationships as fundamentally good, trusting, open, and respectful (Burnor & Raley, 2011). Though only the full integration of cognitive empathy, emotional empathy, and empathic concern enacts deep care and subsequent action (e.g., to preserve nature and soils). *Cognitive empathy* describes how a person thinks and views other's perspectives, while *emotional empathy* is focused on feeling what the other is feeling in one's body and allows deep relating at a visceral level to the suffering of another person. The third is *empathic concern,* which lies at the heart of compassion. Compassion is at the core of relieving suffering from self, others, and the world by evoking compassionate behavior (Neff, 2003). Purely cognitive empathy provides factual understanding but has no sympathetic feelings, while emotional empathy is an embodied experience. They both rely on each other to support action, but empathic concern is a necessity to bring forth compassion that then evokes prosocial actions (Mikulincer & Shaver, 2010). Thus, from a psychological perspective, *soil care* arises from a compassionate lived emotive relationship between carer (person/community) and cared-for (soil/land/nature). In this sense, care is reciprocal similar to a symbiotic relating between carer and cared-for, in contrast to altruistic care that is selfless in nature. This conception of soil care stands in contrast to earlier definitions that are system oriented. According to soil scientist Yaloon (1996), soil care is "the activity of selecting and implementing, locally and regionally, a system of soil and land use management suitable for maintaining and improving soil usefulness and quality for any selected purpose" (p. 3). Puig de la Bellacasa (2015) touched on the challenges to move beyond the predominant objectified product (e.g., soils, crops) and techno-scientific orientations in contemporary cultures toward soil care. From a care ethical perspective, this involves a reframing of the commodification of soils as a receptacle for crops, a soil map or model, toward soil as a living community that is valued and nurtured. Importantly, soil care rooted in empathy and compassion is associated with secure attachment and the ability to engage in intimate relationships (Mikulincer & Shaver, 2012) or cultivation of deliberate compassion and care practice (e.g., mindfulness meditation and loving-kindness practices; Goleman & Davidson,

2017). Burnor and Raley (2011) emphasized that care ethics treat relationships, rather than individuals, as morally basic; thus, care ethics depend upon the mutuality of relationships (e.g., cultures that appreciate and love soils/nature and this joy is reflected back to the lover).

21.3 Pluralistic Integral Soil Ethics

The diversity of ethical views in relation to soil/nature/environment and moral standards is striking, whereby theories often stand in opposition to each other. Brennan (2012) pointed out that pluralism offers an avenue to reconcile various rival ethics, even if none of them is universally applicable across the globe in different cultures. Ethical pluralism resists the temptation to rely on only one favorite ethical approach to a moral dilemma, for example, soil degradation or loss in soil carbon. Instead, multiple ethical perspectives are employed in dialogue and negotiation to assess better or worse solutions in specific circumstances. The assumption in ethical pluralism is that each ethics contains part of the truth but none of which contain the whole truth (Hinman, 2013).

The proposed *Pluralistic Integral Soil Ethics* (PISE) acknowledges different but equally true moral standards and also attributes fundamental core values/principles shared by the human collective. The pluralistic soil ethical framework considers three core pillars (Fig. 21.2), which can be complemented by additional ethical strands. The first proposed pillar in PISE attributes values to soils and considers people's moral standards toward soils (virtue ethics). Soil values are identified across the spectrum from sacredness–naturalness–care–preservation–conservation–security/sustainability/stewardship–health/quality/fertility–production/management–valueless. Soils deserve veneration when considered to be holy or divine carrying transpersonal cosmic value. *Naturalness* refers to soilness as the intrinsic value of "being soil" as a living and breathing organism with its own agency and purpose. *Care for soils* is grounded in compassion and intimate and mutually fulfilling relationships between people-soil/land/nature. Equal value is attributed to soils, life, and humans. Both *preservation* and *conservation of soils* aim to protect soils but with different justification. The preservation of soils is grounded in nature essentialism and adopts a biocentric perspective, while conservation of soils adopts an anthropocentric stance. Soil conservation is rooted in the belief that men control soil/land/nature to derive benefits for people (e.g., soil ecosystem services or agricultural food production). The *security of soils* refers to the freedom from risks of losing functionality, goods and services, and sustainability of life on Earth. Security considers vulnerabilities, threats (e.g., global climate and land use change), and resilience of soil as an ecosystem. *Soil sustainability* is associated with "good" stewardship for future generations. *Soil quality*, *soil health*, and *soil fertility* are tightly coupled to agricultural production systems. These three soil values also stress the importance to keep soils in the "green zone" so that they do not harm (contaminate) people (and perhaps other organisms). *Production* objectifies and

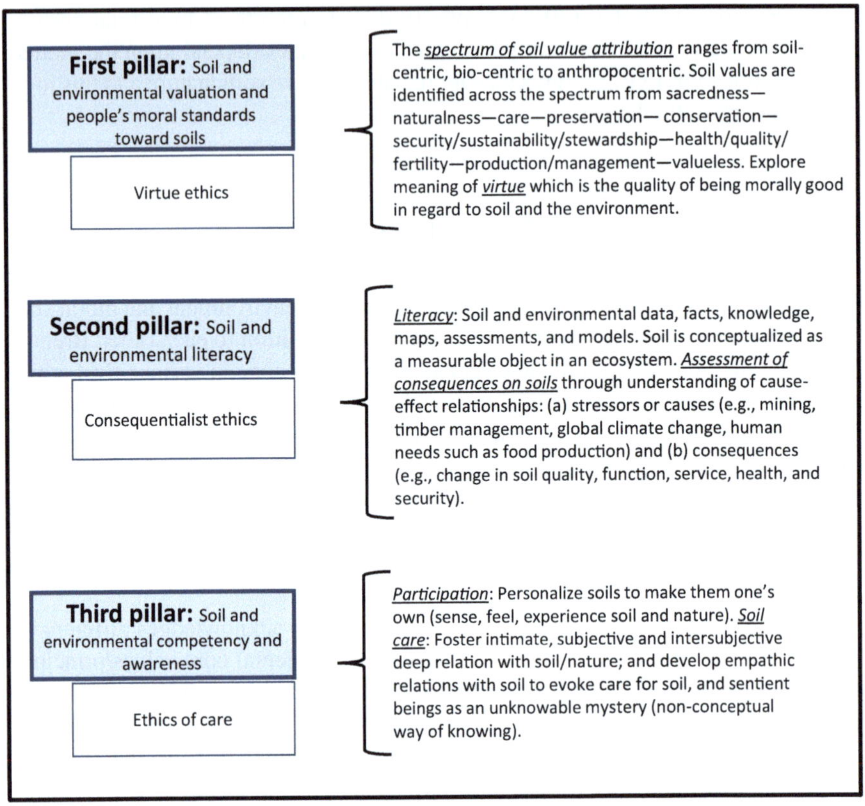

Fig. 21.2 Three core pillars of the proposed Pluralistic Integral Soil Ethics (PISE)

commodifies soils as a resource to be used and managed to produce food, fiber, and goods to support people (extrinsic motivation), while *exploitation of soils* is associated with destruction without attribution of value to soil/land/nature. In summary, the value spectrum of soil ranges from soil-centric, biocentric, to anthropocentric value attribution.

The second pillar of PISE entails soil and environmental literacy (soil data, facts, knowledge, maps, and models) to assess possible consequences to soil based on egoistic ethics, act utilitarianism, and/or rule utilitarianism morality (consequential ethics). The understanding of soils as systems is of critical importance to assess cause-effect relationships and identify possible threats that would change soils in one way or another. Rationality is undergirding this pillar that addresses issues of sustainable soil management and survival of humanity with finite soil resources (stewardship ethics). Consequentialist ethics probes into the consequences (or impacts) onto soils due to (a) behavior, for example, conservation management [act utilitarianism], (b) adoption of rules and regulations how to manage soils [rule utilitarianism], and/or (c) measurable impacts, for example, soil erosion or soil carbon losses [egoistic ethics and stewardship ethics]. The second pillar is depersonalized and aims to objectify soils and how they change and evolve. Ideally, such assessment would also need to consider the full value spectrum from sacredness to production and exploitation of soils. However, a majority of soil knowledge is tied to production agriculture supporting strongly an anthropogenic view. More research to address soil care, naturalness, and sacredness of soils from spiritual, artistic, cultural, psychological, feminine, and social perspectives would be desirable.

The third pillar of PISE is focused to personalize soils to make them one's own. This pillar aims to enact (bring forth) soil awareness as a lived, subjective experience that awakens the mystical, unknowable paradoxical beingness of soils. Ethics that are abstract, conceptual, and distant from one's center of life evoke no motivation to engage and relate. In contrast, closeness to soil/land/nature opens spaciousness to new ways of being human that extend into the transpersonal realm. Intuitive and intimate relating to soil and the land is found especially among indigenous people and in rural communities in India or the African continent. Thus, the third pillar entails environmental competency, spirituality/religion, and curious participation in learning experiences with soil/land/nature. These may involve the exploration of humanness and soilness as intersubjective forms of beingness. A participatory approach engages people in the environment to bridge the separateness between people and soil/nature. What touches people's hearts personally evokes ethics of care through deep empathic relations with soil, for soil, and sentient beings as an unknowable mystery.

References

Adamopoulos, T., & Restuccia, D. (2014). The size distribution of farms and international productivity differences. *American Economic Review, 104*(6), 1667–1697. https://doi.org/10.1257/aer.104.6.1667

Adams, W. W. (2010). Nature's participatory psyche: A study of consciousness in the shared earth community. *The Humanistic Psychologist, 38*(1), 15–39. https://doi.org/10.1080/08873261003635708

Adhikari, K., & Hartemink, A. E. (2016). Linking soils to ecosystem services—A global review. *Geoderma, 262*, 101–111. https://doi.org/10.1016/j.geoderma.2015.08.009

Alexander, L., & Moore, M. (2015). Deontological ethics. *Stanford Encyclopedia of Philosophie*. https://stanford.library.sydney.edu.au/archives/win2015/entries/ethics-deontological/

Benbrook, C. M. (2019). How did the US EPA and IARC reach diametrically opposed conclusions on the genotoxicity of glyphosate-based herbicides? *Environmental Sciences Europe, 31*(1), 2. https://doi.org/10.1186/s12302-018-0184-7

Bidwell, D. R. (2015). Enacting the spiritual self: Buddhist-Christian identity as participatory action. *Spiritus: A Journal of Christian Spirituality, 15*(1), 105–112. https://doi.org/10.1353/scs.2015.0014

Boer, B., & Hannam, I. (2015). Developing a global soil regime. *International Journal of Rural Law and Policy, 2015*(1), 1–13.

Bomberg, E. (2017). Environmental politics in the trump era: An early assessment. *Environmental Politics, 26*(5), 956–963. https://doi.org/10.1080/09644016.2017.1332543

Bouma, J. (2005). Soil scientists in a changing world. In *Advances in agronomy* (Vol. 88, pp. 67–96). Academic. https://doi.org/10.1016/S0065-2113(05)88003-4

Bouma, J., Broll, G., Crane, T. A., Dewitte, O., Gardi, C., Schulte, R. P. O., & Towers, W. (2012). Soil information in support of policy making and awareness raising. *Current Opinion in Environmental Sustainability, 4*(5), 552–558. https://doi.org/10.1016/j.cosust.2012.07.001

Brennan, A. (2012). Moral pluralism and the environment. In D. Schmidtz & Willott (Eds.), *Environmental ethics: What really matters, what really works* (pp. 357–369). Oxford University Press.

Brennan, A., & Lo, Y. S. (2015). Environmental ethics. *Stanford Encyclopedia of Philosophy*. https://plato.stanford.edu/entries/ethics-environmental/

Buckingham, S., & Kulcur, R. (2009). Gendered geographies of environmental injustice. *Antipode, 41*(4), 659–683. https://doi.org/10.1111/j.1467-8330.2009.00693.x

Bunzl, J. M. (2012). Discovering an integral civic consciousness in a global age—Global problems, global governance, and denial. *Journal of Integral Theory and Practice, 7*(1), 105–123.

Burnor, R., & Raley, Y. (2011). *Ethical choices: An introduction to moral philosophy with cases* (2nd ed.). Oxford University Press.

Caldwell, C. (2014). Mindfulness and bodyfulness: A new paradigm. *The Journal of Contemplative Inquiry, 1*(1), 69–88. https://journal.contemplativeinquiry.org/index.php/joci/article/view/6

Capra, F. (1995). Deep ecology—A new paradigm. In G. Sessions (Ed.), *Deep ecology for the twenty-first century* (pp. 19–25). Shambhala.

Chaikaew, P., Hodges, A. W., & Grunwald, S. (2017). Estimating the value of ecosystem services in a mixed-use watershed: A choice experiment approach. *Ecosystem Services, 23*, 228–237. https://doi.org/10.1016/j.ecoser.2016.12.015

Cook, J., Nuccitelli, D., Green, S. A., Richardson, M., Winkler, B., Painting, R., Way, R., Jacobs, P., & Skuce, A. (2013). Quantifying the consensus on anthropogenic global warming in the scientific literature. *Environmental Research Letters, 8*(2), 024024. https://doi.org/10.1088/1748-9326/8/2/024024

Crowther, T. W., Todd-Brown, K. E. O., Rowe, C. W., Wieder, W. R., Carey, J. C., Machmuller, M. B., Snoek, B. L., Fang, S., Zhou, G., Allison, S. D., Blair, J. M., Bridgham, S. D., Burton, A. J., Carrillo, Y., Reich, P. B., Clark, J. S., Classen, A. T., Dijkstra, F. A., Elberling, B., et al. (2016). Quantifying global soil carbon losses in response to warming. *Nature, 540*(7631), 104–108. https://doi.org/10.1038/nature20150

Davies, J. (2017). The business case for soil. *Nature News, 543*(7645), 309–311. https://doi.org/10.1038/543309a

Doktor, T. (2003). Religious fundamentalism. In R. L. Piedmont & D. O. Moberg (Eds.), *Research in the social scientific study of religion* (pp. 204–213). Bril.

Drengson, A., & Inoue, Y. (Eds.). (1995). *The deep ecology movement: An introductory anthology*. North Atlantic Books.

Esbjörn-Hargens, S., & Zimmerman, M. E. (2009). *Integral ecology: Uniting multiple perspectives on the natural world*. Integral Books.

Fernandes, L. (2003). *Transforming feminist practice: Non-violence, social justice, and the possibilities of a spiritualized feminism*. Aunt Lute Books.

Ferrer, J. N. (2002). *Revisioning transpersonal theory: A participatory vision of human spirituality*. State University of New York Press.

Ferrer, J. N. (2017). *Participation and the mystery: Transpersonal essays in psychology, education, and religion*. State University of New York Press.

Garcia, L. C., Ribeiro, D. B., Roque, F. de O, Ochoa-Quintero, J. M., & Laurance, W. F. (2017). Brazil's worst mining disaster: Corporations must be compelled to pay the actual environmental costs. *Ecological Applications, 27*(1), 5–9. https://doi.org/10.1002/eap.1461

Gibbs, H. K., & Salmon, J. M. (2015). Mapping the world's degraded lands. *Applied Geography, 57*, 12–21. https://doi.org/10.1016/j.apgeog.2014.11.024

Gilligan, C. (1993). *In a different voice: Psychological theory and women's development*. Harvard University Press.

Goleman, D., & Davidson, R. J. (2017). *Altered traits: Science reveals how meditation changes your mind, brain, and body*. Penguin Random House.

Gottlieb, R. S. (2019). *Morality and the environmental crisis*. Cambridge University Press.

Grunwald, S., Clingensmith, C. M., Gavilan, C. P., Mizuta, K., Kastner-Wilcox, R. K., Pinheiro, E. F. M., Ceddia, M. B., & Ross, C. W. (2017a). Integrating new perspectives to address global soil security: Ideas from integral ecology. In D. J. Field, C. L. S. Morgan, & A. B. McBratney (Eds.), *Global soil security* (pp. 319–330). Springer.

Grunwald, S., Mizuta, K., Ceddia, M. B., Pinheiro, E. F. M., Kastner-Wilcox, R. K., Gavilan, C. P., Ross, C. W., & Clingensmith, C. M. (2017b). The meta soil model: An integrative multi-model framework for soil security. In D. J. Field, C. L. S. Morgan, & A. B. McBratney (Eds.), *Global soil security* (pp. 305–318). Springer.

Grunwald, S., Vasques, G. M., & Rivero, R. G. (2015). Fusion of soil and remote sensing data to model soil properties. *Advances in Agronomy, 131*, 1–109. https://doi.org/10.1016/bs.agron.2014.12.004

Hinman, L. M. (2013). *Ethics: A pluralistic approach to moral theory* (5th ed.). Wadsworth.

Hursthouse, R., & Pettigrove, G. (2016). Virtue ethics. *The Stanford Encyclopedia of Philosophy*. https://plato.stanford.edu/entries/ethics-virtue/

Kaza, S. (2019). *Green Buddhism: Practice and compassionate action in uncertain times*. Shambhala.

Kelly, S. (2017). Integral ecology and Edgar Morin's paradigm of complexity. In S. Mickey, S. Kelly, & A. Robbert (Eds.), *The variety of integral ecologies* (pp. 81–101). Suny Press.

Kohak, E. (2000). *The green halo: A bird's-eye view of ecological ethics*. Open Court.

Korten, D. C. (1998). Do corporations rule the world? And does it matter? *Organization & Environment, 11*(4), 389–398. https://doi.org/10.1177/0921810698114001

Kripal, J. J. (2014). *Comparing religions*. Wiley Blackwell.

Lipps, J. H. (1999). Beyond reason: Science in the mass media. In J. W. Schopf (Ed.), *Evolution! Facts and fallacies* (pp. 71–90). Academic.

Little, C. E. (1986). Has the land ethic failed in America: An essay on the legacy of Aldo Leopold symposium "Stewardship of land and natural resources: I. The philosophy of resource control". *University of Illinois Law Review, 2*, 313–318.

Low, A., & Tremayne, S. (Eds.). (2001). *Sacred custodians of the earth: Women, spirituality and the environment*. Berghahn Books.

Loy, D. R. (2019). *Ecodharma: Buddhist teachings for the ecological crisis*. Wisdom.

MacKinnon, B., & Fiala, A. (2018). *Ethics: Theory and contemporary issues* (9th ed.). Cengage Learning.

McBratney, A. B., Minasny, B., & Stockman, U. (Eds.). (2018). *Pedometrics* (1st ed.). Springer.

Mickey, S., Kelly, S., & Robbert, A. (2017). Introduction: The history and future of integral ecologies. In S. Mickey, S. Kelly, & A. Robbert (Eds.), *The variety of integral ecologies—Nature, culture, and knowledge in the planetary era* (pp. 1–30). State University of New York Press.

Mikulincer, M., & Shaver, P. R. (Eds.). (2010). *Prosocial motives, emotions, and behavior*. American Psychological Association.

Mikulincer, M., & Shaver, P. R. (2012). Adult attachment orientations and relationship processes. *Journal of Family Theory and Review, 4*(4), 259–274. https://doi.org/10.1111/j.1756-2589.2012.00142.x

Mizuta, K., Grunwald, S., & Phillips, M. A. (2018). New soil index development and integration with econometric theory. *Soil Science Society of America Journal, 82*, 1017–1032. https://doi.org/10.2136/sssaj2017.11.0378

Mohai, P., & Bryant, B. (1992). Environmental injustice: Weighing race and class as factors in the distribution of environmental hazards. *University of Colorado Law Review, 4*, 921–932.

Moore, T. (2014). *A religion of one's own: A guide to creating a personal spirituality in a secular world*. Gotham Books Publ. by the Penguin Group.

Neff, K. D. (2003). The development and validation of a scale to measure self-compassion. *Self and Identity, 2*(3), 223–250. https://doi.org/10.1080/15298860309027

Noddings, N. (2003). *Caring: A feminine approach to ethics and moral education*. University of California Press.

Opotow, S., & Weiss, L. (2000). New ways of thinking about environmentalism: Denial and the process of moral exclusion in environmental conflict. *Journal of Social Issues, 56*(3), 475–490. https://doi.org/10.1111/0022-4537.00179

Paleari, S. (2017). Is the European Union protecting soil? A critical analysis of community environmental policy and law. *Land Use Policy, 64*, 163–173. https://doi.org/10.1016/j.landusepol.2017.02.007

Passmore, J. (1980). *Man's responsibility for nature: Ecological problems and Western tradition*. Duckworth.

Pew Research Center. (2012). *The global religious landscape*. Religion & Public Life. https://www.pewforum.org/2012/12/18/global-religious-landscape-exec/

Pew Research Center. (2015). Religion in America: U.S. religious data, demographics and statistics. *Religious Landscape Study of 35,000 Americans from 50 States in the U.S.* https://www.pewforum.org/religious-landscape-study/

Plumwood, V. (2012). Finding our place in nature/dominating nature: Being prey. In D. Schmidtz & E. Willott (Eds.), *Environmental ethics: What really matters, what really works* (pp. 266–271). Oxford University Press.

Puig de la Bellacasa, M. (2015). Making time for soil: Technoscientific futurity and the pace of care. *Social Studies of Science, 45*(5), 691–716. https://doi.org/10.1177/0306312715599851

Rohlf, M. (2018). Immanuel Kant. *Stanford Encyclopedia of Philosophy*. https://plato.stanford.edu/entries/kant/#CatImp

Sandler, R. (2012). Environmental virtue ethics. In D. Schmidtz & E. Willott (Eds.), *Environmental ethics—What really matters, what really works* (pp. 286–291). Oxford University Press.

Sandler, R. (2017). Environmental virtue ethics: Value, normativity, and right action. In S. M. Gardiner & A. Thompson (Eds.), *The Oxford handbook of environmental ethics* (pp. 223–233). Oxford University Press.

Sauer, T. J., & Nelson, M. P. (2011). Science, ethics, and the historical roots of our ecological crisis: Was White right? In T. J. Sauer, J. M. Norman, & V. K. Sivakumar (Eds.), *Sustaining soil productivity in response to global climate change* (pp. 3–11). Wiley.

Schmidtz, D., & Willott, E. (2012). *Environmental ethics: What really matters, what really works*. Oxford University Press.

Schultz, P. W. (2001). The structure of environmental concern: Concern for self, other people, and the biosphere. *Journal of Environmental Psychology, 21*(4), 327–339. https://doi.org/10.1006/jevp.2001.0227

Shaw, M. (2015). *Buddhist goddesses of India*. Princeton University Press.

Shepherd, M., Turner, J. A., Small, B., & Wheeler, D. (2018). Priorities for science to overcome hurdles thwarting the full promise of the 'digital agriculture' revolution. *Journal of the Science of Food and Agriculture*. https://doi.org/10.1002/jsfa.9346

Shiva, V. (1999). Monocultures, monopolies, myths and the masculinization of agriculture. *Development, 42*(2), 35–38. https://doi.org/10.1057/palgrave.development.1110033

Shrader-Frechette, K. (2012). Environmental justice: Creating equality, reclaiming democracy. In D. Schmidtz & E. Willott (Eds.), *Environmental ethics: What really matters, what really works* (pp. 204–216). Oxford University Press.

Singh, R. K., Murty, H. R., Gupta, S. K., & Dikshit, A. K. (2012). An overview of sustainability assessment methodologies. *Ecological Indicators, 15*(1), 281–299. https://doi.org/10.1016/j.ecolind.2011.01.007

Stanley, S. (2016). *Relational and body-centered practices for healing trauma: Lifting the burdens of the past*. Routledge.

Stone, C. D. (2012). Should trees have standing? Toward legal rights for natural objects. In D. Schmidtz & E. Willot (Eds.), *Environmental ethics: What really matters, what really works* (pp. 85–89). Oxford University Press.

Taylor, B. (2010). *Dark green religion: Nature, spirituality and the planetary future*. University of California Press.

Taylor, B. (2017). The sacred, reverence for life, and environmental ethics in America. In S. M. Gardiner & A. Thompson (Eds.), *The Oxford handbook of environmental ethics* (pp. 248–264). Oxford University Press.

Taylor, P. W. (2012). The ethics of respect for nature. In D. Schmidtz & E. Willott (Eds.), *Environmental ethics: What really matters, what really works* (pp. 102–113). Oxford University Press.

The World Bank Data. (2019). *Urban population (% of total population) globally*. https://data.worldbank.org/indicator/sp.urb.totl.in.zs

Thiroux, J., & Krasemann, K. W. (2016). *Ethics: Theory and practice* (11th ed.). Pearson.

Thompson, P. B. (2011). The ethics of soil. In T. J. Sauer, J. M. Norman, & M. V. K. Sivakumar (Eds.), *Sustaining soil productivity in response to global climate change* (pp. 31–42). Wiley-Blackwell. https://doi.org/10.1002/9780470960257.ch3

Thompson, P. B. (2017). *The spirit of the soil: Agriculture and environmental ethics*. Routledge.

Tittonell, P., Muriuki, A., Shepherd, K. D., Mugendi, D., Kaizzi, K. C., Okeyo, J., Verchot, L., Coe, R., & Vanlauwe, B. (2010). The diversity of rural livelihoods and their influence on soil fertility in agricultural systems of East Africa—A typology of smallholder farms. *Agricultural Systems, 103*(2), 83–97. https://doi.org/10.1016/j.agsy.2009.10.001

Tollefson, J. (2017). How Trump plans to wipe out Obama-era climate rules. *Nature News*. https://doi.org/10.1038/nature.2017.21726

Traer, R. (2007). *Doing ethics in a diverse world*. Routledge.

Turner, W. R., Nakamura, T., & Dinetti, M. (2004). Global urbanization and the separation of humans from nature. *Bioscience, 54*(6), 585–590. https://doi.org/10.1641/0006-3568(2004)054[0585:GUATSO]2.0.CO;2

Varner, G. (2012). Biocentric individualism. In D. Schmidtz & E. Willott (Eds.), *Environmental ethics: What really matters, what really works* (pp. 90–101). Oxford University Press.

Warren, K. J. (Ed.). (1997). *Ecofeminism: Women, culture, nature*. Indiana University Press.

White, L. (1967). The historical roots of our ecologic crisis. *Science, 155*(3767), 1203–1207. https://www.jstor.org/stable/1720120

Whyte, K. P., & Cuomo, C. J. (2017). Ethics of caring in environmental ethics: Indigenous and feminist philosophies. In S. M. Gardiner & A. Thompson (Eds.), *The Oxford handbook of environmental ethics* (pp. 234–247). Oxford University Press.

Yaloon, D. H. (1996). Soil science in transition: Soil awareness and soil care research strategies. *Soil Science, 161*(1), 3–8.

Zhang, L., Rana, I., Shaffer, R. M., Taioli, E., & Sheppard, L. (2019). Exposure to glyphosate-based herbicides and risk for non-Hodgkin lymphoma: A meta-analysis and supporting evidence. *Mutation Research, Reviews in Mutation Research, 781*, 186–206. https://doi.org/10.1016/j.mrrev.2019.02.001

Zimmerman, M. E. (1995). The threat of ecofascism. *Social Theory and Practice, 21*(2), 207–238. https://doi.org/10.5840/soctheorpract199521210

Chapter 22
Inner and Outer Soil

Nikola Patzel

22.1 Humans Perceive Humus

When looking at soil, consider what is in front of your eyes and what is behind the eyes. This is a psychological-cultural essay on soil.

A human being smells fungi and organic soil. A person methodically observes features of a soil or ponders over a research question. No way to deny that we are living beings and no machines who perceive soil or brood over phenomena and questions. Researchers are living bodies, and at the same time, their subject of cognition can be approximately characterized as human psyche: containing self-awareness and consciousness, subconscious processes, and a psycho-spiritual world within and around that is mainly unconscious, albeit affecting the human ego permanently with a varying strength.

These seemingly trivial statements signify that mind matters for soil perception and that every representation and interpretation of soil that a human "sees" is a psychic one; we just don't know what the psyche or spirit really *is*, what the I that posits a non-I *is*, and what the soil *is*, or what somebody (who is that?) means by the word "is." But don't mind, at least we can say: human bodies are no robots, and the working hypothesis that there is a largely autonomous psycho-spiritual dimension to every human is quite useful and satisfying and of course applies to natural scientists or farmers that are at work as well. This dimension makes up our inner world. The inner world includes the inner nature, and the inner nature includes the inner soil. So we can note: our inner soil is not the inner representation of the outer soil alone—there is something to it that is independent from the content of consciousness and sensory input from outside. "*An approach to research that would keep soul in mind*

N. Patzel (✉)
Freelance Scientist, Office for Soil Communications, Überlingen, Germany
e-mail: nikola.patzel@soilcom.com

N. Patzel et al. (eds.), *Cultural Understanding of Soils*,
https://doi.org/10.1007/978-3-031-13169-1_22

requires a method that not only includes the researcher in the process, but also takes into account the deep unconscious of the researcher" (Romanyshyn, 2007, p. 305).

I invite you to empathize with the following examples: What happens in your imagination and feeling when your awareness turns to a broken crumb with a bowing enchytraeus (a little whitish worm)? What comes to your mind when looking at a pasture soil profile with wild carrot roots, at open soil at the edge of the hollow way, or at farmland with a hidden catena (sequence of soils) that expresses in the height of the maize? Which emotions associate with your impression when gazing from a vantage point at a whole landscape of our earth under the enlightening sun—or in the mystifying light of the full moon? Or when writing a soil policy paper or other stuff at your desktop: Is your inner world active to give inner images about connections and possible destinations, or giving flashbacks of the past or hunches of a future? Maybe try for some moments more consciously turning your awareness that has been seemingly focused on what is in front of your eyes to what is behind the own eyes. Swiss Psychologist Carl Gustav Jung (1875–1961) addressed these two directions of awareness coining the terms "extraversion" and "introversion": looking outward or looking inward, taking situational orientation from outside or from inside—areas of perceptual experience and impression that can be discerned more or less (Fig. 22.1).

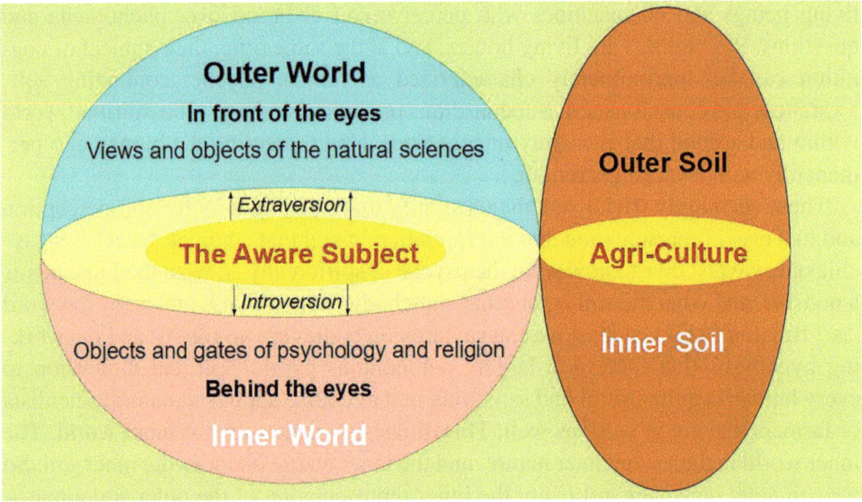

Fig. 22.1 Schematic representation of the position of human consciousness between its external and its internal environment. "I" interact with the outer and inner world; both may be one in essence. This view can also be applied to culture-forming social processes. (Graph: author)

22.2 Factors of Inner Soil Formation

To soil scientists and farmers who are familiar with their physical ground, the *outer soil* appears to be well familiar, even though most of farmland nature—think of soil microbiota—remains unknown and hardly graspable by hand or mind. Rock and water, air, and energy together with the whole terrestrial tree of life over time are commonly considered being outer soil forming factors. On the other side, when looking for factors of *inner soil* formation that are behind the eyes and largely beyond the scope of humans outer senses, the "anatomy of the psyche" (Edinger 1985) and the mostly unknown expanses of psycho-spiritual realms that influence human individuals and groups might be considered.

Look: Humans all have a very similar genome and anatomy, for example, almost each individual disposing of 10 fingers. All human eyes are basically similar; our hearts and brains share common basic structures. Thus, it need not surprise that also the human structures of consciousness and the appearances of the psycho-spiritual sphere of experience show many trans-cultural phenomenological and structural similarities all over the world; this is not meant as a reductive biologistic argument, but as a comparison. These similarities in human's inner world can be called archetypal patterns.[1] Imagine historically grown strands of cultural traditions and development in their interplay with psycho-spiritual experiences. They have been going along with different human nature and soil perceptions, understandings, and relation, like they are unfolded in many case studies in this book and summarized in the general conclusions of it (Chap. 26).

How to face the contents and impulses of inner soil? For a scientist, a low-threshold entry door to that might be to look at other people. In 2010, I wrote in a paper called "The Soil Scientist's Hidden Beloved: Archetypal Images and Emotions in the Scientist's Relationship with Soil" for the IUSS book of Landa & Feller on *Soil and Culture* (Patzel, 2010c, p. 205): "*From the outset, the practice of modern soil science has been a double-grounded process, being simultaneously in touch with outer observations and inner images. Scientists have studied soils with their senses and their means of consciousness, while also being fascinated, driven and guided by other factors of their inner life. These internal factors are emotional inner images and ideas, and they often act unconsciously. Some keywords of emotional inner images, expressed by pioneers and leaders of soil science, are: Mother Earth, vital force, the Stone of the Sages, the encompassing whole and total control. They can be found in the writings of Liebig, Sprengel, Fallou, Dokuchaev, and Jenny, as well as contemporary soil scientists, and I call them their inner soil.*"[2] The empirical basis regarding the inner soil of soil science and its cultural context is broadened in

---

[1] On the development of this theory, see Carl Gustav Jung, collected works *vol. 8, § 317–325, 417 f., 420, 440; vol. 9/1*, § 1–86, 153, 155, 259; *vol. 9/2, § 13; vol. 18/1*, § 523, 531.

[2] A concept of "inner soil" was first presented at the soil department of the University of Stuttgart-Hohenheim (Germany) in 2003 and first published in Patzel and Lindenthal 2009 and Patzel 2010a, b, c.

Chap. 4; here, in this chapter, some theoretical considerations are done—not in the form of a literature review, but rather essayistically scientific.

When scientists follow regarding soil emotive patterns of imagination like Mother Earth and like the basic elements of soil and life, and are addressing vital forces and health in nature and so on, they are likely in interaction with factors of their inner soil, not solely with produces of observation and consciousness. In monographs, papers, and interviews, some scientists tell about their inner motivation, for example, using metaphors that have a symbolic content that hints at mythical patterns of imagination or spiritual experiences.

One may discuss controversially how independent, or not, the scientific findings and theories of leading soil scientists were and are from their sociocultural context and worldview and from their inner soil. There are questions also of methodological nature: how to trace effects of mythical and deeply rooted ideological patterns in the mental landscape of an individual, and how to interpret symbolic images inside? Inner soil can personally be experienced as motivating and fascinating, spiritual, and numinous, but in most cases, it can only be traced indirectly by observations and the comparative cultural study of what is said about soil and done with it.

Maybe even more difficult regarding the inner soil is to perceive it in action at present. For example, what are inner collective drivers of what people call "digitalization of agriculture" with "big data" and sensor-fed automatic decision-making and robotics in soil management? What is effective in those who believe that rebuilding humus in agriculture can save the earth from the started and ongoing climate change? Which background context might be included when speaking about pest control, about soil health, or an organic whole? What about the bio-economy paradigm? And from the Neolithic Revolution to the present day, "the peasant" has become a primary symbol of human self-understanding in nature. This is until today the projection target for a large spectrum of social ideas, for example, from cultural conservatism to disruptive progress, or from nature's gardener to a resource manager within global value chains. Do new developments in science and technology imply that their inner soil changes as well?

Concerning the last question, I suggest at least considering the possibility that quite ancient mythical patterns in the human psyche may join in and play along even in the most recent waves of development in science, technology, and agricultural policy. It can, for example, be interesting to assess contemporary moves on structures of the Demeter myth (fosters relation-oriented and some spiritual approaches), or whether it expresses the myth of Prometheus (fosters scientific "enlightenment" and technology). Both mythical patterns were early documented in ancient Greece. How do they frame the role of humans in nature (Patzel, 2016, 2017)?

Myths need not be considered ahistorical or eternal. The history of religion and basic worldviews show that people's inner world changes faster than their physical evolution. Goethe's word in his Faust (1908/1831, p. 79), "*Formation, transformation, the eternal mind's eternal recreation,*"[3] applies to inner processes, too. But however, within change, there appears to be some continuity and recurrence as well,

[3] "Gestaltung, Umgestaltung, des ewgen Sinnes ewge Unterhaltung" (Goethe, 1888/1831, verse 6287).

which also means that history can repeat itself in a similar way. This can also be seen again and again worldwide in our present societies, for example, their resource politics, their inner and outer alliance politics, and their wars. In addition, it can be observed that some old religious myths and mythologemes continue to fascinate people today: for example, in the representation of some of them in best sellers such as Tolkien's Lord of the Rings opus, Rowling's Harry Potter series, Riordan's Percy Jackson series and sequels, or movies such as the Star Wars or Game of Thrones series.[4] So it need not come as a complete surprise if some mythical structures are also at work in the area of real human relationship with nature and the way people deal with the soil: inner patterns that did not just fell from the sky or rise from the depths yesterday. The theory of "archetypal patters" cited from Jung is just one of several recent attempts to find a new language to scientifically deal with such phenomena.

Whether the factors of inner soil formation are spiritual in the sense of coming from divine sources beyond, or whether they emerge from living matter like soil life, plants, animals, and humans as an immanent quality, is not being determined by drawing the hypothesis of inner soil, and it is everybody's choice how to integrate it in their own worldview.

22.3 Psychology of Inner Soil

Some different psychological dimensions can be seen involved with inner soil. At an individual subjective level, whether or not a human being has ever experienced mud, a well-structured soil or an earthworm in hands and which emotions, images, and ideas were evoked with such personal experiences can be part of the factors of inner soil formation of a person. Likewise, attitudes and behaviors one's mother or father, grandmother or grandfather showed towards nature, as well as the relationships with academic teachers in soil science, can matter in shaping or letting unfold the inner soil of a person. In the case of having grown up on a farm, what was the experienced reality there? It is legitimate and interesting to reflect on the effects of personal background on ideas and behavior of individual scientists, farmers, extension service workers, etc. Personal experiences and familial and social conditions affect the inner soil, be they archetypal or just individual achievements. This can provide lifelong motivation for dynamic developments—or create strict convictions and taboos that are difficult to change. Such things can be significant for the history of science, as the example of Darwin and the earthworms show (see Chap. 18).

[4] All these examples come from the Anglo-Saxon world and have international followings. In the film sector, examples with no less impact in some world regions, though unknown in the West, could be found, for example, from the very large Indian film industry (or the Chinese, Nigerian and more industries with large national or continental markets).

When looking further at social and collective levels, that is, interpersonal and transpersonal aspects of inner soil, a comparative approach can be helpful to gain an orientation on common patterns. Of course at this point, debates can occur that refer to the problem of transcultural mental (anthropologic, psychological) universals of humanity. However, at first glance, it is sufficient for our purpose here to assume that there are not only differences between humans, but people do also share ideas and inner images. Thus, similar to the uncontested basic commonalities of all human bodies, there are psychic commonalities, too. In depth psychology, this led to the concept of the "collective unconscious" or "objective psyche" that shows "archetypal patterns"—but there are different ways to name that. Jung spoke of "*forms that appear spontaneously and more or less universally, regardless of tradition, in myths, fairy tales, fantasies, dreams, visions and delusions*" (CW 11, § 5). These were "*an innate disposition to parallel imaginations, or [...] universal, identical structures of the psyche [...]. They correspond to the biological term 'pattern of behaviour'*" (CW 5, § 224). These "*structural elements of the human soul" would* correspond to a *"collective basic layer of the soul*" of the human being (Jung CW 9/1, § 262). This theory was based on observations of similar motifs in various fairy tales, myths, and their appearance in art and dreams across different eras, languages, and cultures.[5]

Patterns of behavior in nature relations also strongly rely on the environmental conditions of the living space of human communities, as was just recently confirmed for human hunters and gatherers as well as other species by Barsbai et al. (2021), and the patterns of imagination and behaviors are co-shaped by the sociocultural environments as well. Even though archetypal patterns sometimes appear to have a timeless presence by re-occurring, for example, in dreams and spontaneous ideas of individuals again and again, they do seem to change and evolve, and their manifestation and interpretation involve individual shapes, social processes, and constructions in culture (Fig. 22.2).[6]

22.4 Philosophical Remarks

In philosophical terms, it can be said that the distinction between outer and inner soil is an epistemological and not an ontological one. More simply said: What one *perceives* and deals with scientifically is not by definition identical with what "really" *is*. Perceiving something means that there is *something*, but without knowing for sure *what* is there. It may be less trivial than sometimes assumed by scientists to distinguish which parts of an interpretation of observations and data rely more on the outer soil and which rely more on the inner soil. But one can assume that a personally satisfactory theory in science is one where you have the best fit

[5] For observations and arguments backing these assumptions, see from Jung in the collected works vol. 18/1 the paragraphs 80 f., 92 f., 138, 190, 195, 218, 221 f., 231, 250, 262, 271, 299, 324, 353 f., 358, 366, 368, 385, 402, 406 f., 409, 512, 521–559, 563, 578, 589, 595, 830.

[6] For a recent review research on archetypes, see Roesler 2016.

Fig. 22.2 An example of the "inner soil" in the eye of the beholder. Pagan and Christian iconography seem to be merged here. Book illustration from Christine de Pizan (1406), folio 13v. The text accompanying this illustration speaks of a "deesse ceres" (goddess Ceres). The sheaf doll of the Wheat Mother (Fr. mère du blé) was called "La Cérès" in France until the nineteenth century (Mannhardt, 1884, p. 318 f.).

between your present inner and outer reality, here of your outer and inner soil (Pauli, 1957; von Franz, 1968, 1988).

Such connections are rediscovered from time to time in various practical or theoretical contexts and can be acknowledged without any exclusivity, originality, or priority claim. I want just to mention Swiss Philosopher Paracelsus von Hohenheim (1493/1494–1541): In the sixteenth century, he wrote about an inner sky and earth in humans: The whole cosmos were inside every human being with the specific qualities of its luminaries and elements, acting inside man (Paracelsus, 1965, p. 526 f.). Because of the similarity of the inner and outer world, human imagination was supposed by him to produce effects outside: "Thus the [inner] imagination attracts the [outer] colour and shape" (Paracelsus, 1932a, p. 268, transl. by auth.). All living beings, including humans, had within them "a light of nature in which we may well learn, that is, that we may experience it" (Paracelsus, 1932a, p. 274). "For nature gives a light by which it can be recognized in its own shine" (Paracelsus, 1932b, p. 41).

A common difficulty of natural scientists in considering psychic, even spiritual, content as relevant to their field is that they assume it to be "subjectively valid only" (if valid at all), in contrast to their outer object of scientific research having "objective properties". It is natural to think that, because the inner world is experienced within the subject, that is, the individual person, and is in many modern societies rather not perceived to come from "outside" or having an existence independently of the perceiving subject. However, this is not the only arguably position on that. One may also argue that an inner soil is as objective as an outer soil and that even before maybe considering a fundamental unity of matter and mind, objectivity, and subjectivity is not necessarily to be attributed to "matter" or "psyche and spirit" respectively, only.

Not only highly emotional psycho-spiritual experiences can be perceived being "just (or even utter) reality," but also quite concrete visions and intuitions have led to new scientific findings, and they are often perceived as coming from outside. In German, the word "Einfall" means a sudden thought that literally "falls into my mind" (from outside); the term was formerly used by late middle age mystics and then was generalized to common language. There is evidence that important scientific breakthroughs started with inspirations from the unconscious, in some cases dreams or similar and in others sudden flashes of insight. This way, the mathematician Carl Friedrich Gauss found the law of quadratic reciprocity (Gauss 1900/1805) and his colleague Henri Poincaré discovered the automorphic functions (Poincaré, 1912, pp. 51–62); the pharmacologist Otto Loewi proved the function of acetylcholine as a neurotransmitter (Loewi, 1960, pp. 168–190), and the chemist Dmitri Mendeleev saw the periodic table of the elements (Strathern, 2000, pp. 307–317). An at its time especially sensational and controversial publication was August Kekulé's account of how his dream image of a circular snake helped him to discover the ring structure of the benzene molecule (Wotiz, 1993). These examples hint to the idea that epistemological polarity between an inner and outer world does not necessarily imply a strict separation. When we come to the frontier of awareness of the outer soil, we may face the inner soil, and when we come to the frontier of awareness of the inner soil, we may face the outer soil. There might even be common patterns and essential links between the two.

22.5 Conclusions

The sciences of soils and agriculture are no more devoid from the "inner world" than other sciences; the inner soils of the men and women, their societies, and cultural matrices are involved in their relation with nature and soil. This psychic maybe spiritual world of inner soil contains individually acquired personal elements and structures, and it has "objective" interpersonal and transpersonal, archetypal elements and windows that are wide open. Of course, the concept of "inner soil" presented here is just one of many ways of speaking and how to enhance human self-perception when perceiving and understanding soil and relating to it.[7] Plurality in how one understands soils can be based on the diversity of outer soils as well as in the diversity of inner soils. But however, the inner world with the inner soils, independently of how we name that, can be hypothesized to be basically "one," like the earth with all its diverse soils is "one." Symbols can be bridges. Romanyshyn stated that *"The term 'poetics of research' does not mean that research is poetry. It means (...) that a researcher is attuned to the gap of what is said and what is always left unsaid, that between conscious and unconscious, which is bridged by the symbol"* (2007, p. 220). The inner world works. American psychologist William James wrote early regarding the question of the relation of science and religion (1902, p. 501):

> ... since one of the duties of the science of religions is to keep religion in connection with the rest of science, we shall do well to seek first of all a way of describing the 'more', which psychologists also recognize as real. The subconscious self is nowadays a well-accredited psychological entity; and I believe that in it we have exactly the mediating term required. Apart from all religious considerations, there is actually and literally more life in our total soul that we are at any time aware of. ... [p. 506] Yet the most unseen region in question is not merely ideal, for it produces effects in the world. (...) But that which produces effects within another reality must be termed a reality itself, so I feel as if we had [p. 507] no philosophic excuse for calling the unseen or mystical world unreal.

To approach the inner soils of humans, it can be suggested: (1) to take note of a variety of worldviews and emotive ideas concerning soil; (2) to note that from the outset, modern soil science (as well as any agriculture) has been a double-grounded process, being simultaneously in touch with outer observations and inner patterns of imagination; (3) to consider the possibility that mythical patterns in the human psyche may join in and play along even in the most recent waves of development in science, technology, and agricultural policy—with potentially negative or positive outcomes. After all, one may be facing the decision to make a fourth step in consciously approaching one's own inner soil, or not: how can these drivers be more consciously faced and partly integrated into the life of a soil scientist or another soil stakeholder who is concerned in daily practice with the outer soil? Is the inner soil split off from the official life and work and therefore kept unconscious or placed in a separate tray of ideas? Or is it more integrated into soil science and soil communication, unfolding while cautiously and carefully considering its diverse and ambivalent opportunities? Documented inner soils from different cultural areas may help to find one's own.

---

[7] Approaches with similarities to the one pursued here can be found, for example, in the body of literature on *deep ecology*, in *integral ecology*, in some approaches to *environmental psychology*, and also in views that come out of religious or spiritual traditions.

References

Barsbai, T., Lukas, D., & Pondorfer, A. (2021). Local convergence of behavior across species. *Science*. https://doi.org/10.1126/science.abb7481

de Pizan, C. (1406). (*Epistle of the Goddess* [the proper noun Othéa is assumed to mean the poetic Greek formula Ὦ θεά (O Thea) = oh Goddess]) Épître d'othéa. Bibliothèque nationale de France (Paris), Département des Manuscrits, Division occidentale, Côte: Français 606.

Edinger, E. (1985). *Anatomy of the psyche. Alchemical symbolism in psychotherapy*. Chicago: Open Court.

Gauss, C. F., & H. Olbers (1900). (Letters) *Briefwechsel*. Ed. C. Schilling. Vol. I, Berlin. Letter from Gauss to Olbers from September 3, 1805, pp. 267–270.

Goethe, J. W. von (1908). *Faust II* (A. Latham, Trans.). (Original written in 1831 and published 1888.)

James, W. (1902). *Varieties of religious experience*. New York: Random House.

Jung, C. G. (1957–1990). *Collected works* (R. F. C. Hull, Trans.). Princeton University Press.

Loewi, O. (1960). An autobiographical sketch. *Perspectives in Biology and Medicine, 4*(1), 3–25.

Mannhardt, J. W. (1884). *(Mythological research from the estate)* Mythologische Forschungen aus dem Nachlasse. Ed. H. Patzig. Vol. LI of Quellen und Forschungen zur Sprach- und Culturgeschichte der germanischen Völker.

Paracelsus, Th. B. von Hohenheim (16th cent/1932a). (Fragment of the book on the symbols) Fragment des Buches über die Charaktere. (Ueber die Symbole.). In B. Aschner (Ed.), *(Collected works of Paracelsus, Vol. 4)* Paracelsus. Gesammelte Werke Bd. 4 (pp. 273–276).

Paracelsus, Th. B. von Hohenheim (16th cent/1932b). (Book about the nymphs, sylphs, pygmies, salamanders and the other spirits) Buch über die Nymphen, Sylphen, Pygmäen, Salamander und die übrigen Geister. In B. Aschner (Ed.*), (Collected works of Paracelsus, Vol. 4)* Paracelsus. Gesammelte Werke Bd. 4 (pp. 41–70).

Paracelsus, Th. B. von Hohenheim (16th cent/1965). *Werke. Das Buch Paragranum* (Collected works, vol. 1). Ed. W.-E. Peukert. Vol. 1.

Patzel, N., & Lindenthal, T. (2009). (Dealing with soil in organic agriculture) Der Umgang mit Böden im ökologischen Landbau. In *(German handbook of soil science)* Handbuch der Bodenkunde. 02/09, chapter 1.2.9.

Patzel, N. (2010a). *(The „inner soil". Psycho-cultural orientation and motivation in soil culture, the example of organic farming)* Der „innere Boden". Seelisch-kulturelle Orientierungen und Antriebe in der Bodenkultur am Beispiel des biologischen Landbaus. Der kritische Agrarbericht (pp. 266–774).

Patzel, N. (2010b). European religious cultivation of the soil. In E. Landa & C. Feller (Eds.), *Soil and culture*. Springer (pp. 261–276)

Patzel, N. (2010c). The soil scientist's hidden beloved: archetypal images and emotions in the scientist's relationship with soil. In E. Landa & C. Feller (Eds.), *Soil and culture*. Springer (pp. 205–226).

Patzel, N. (2016). (Christian and non-Christian roots of our contemporary relation to soil and nature) Christliche und ausserchristliche Wurzeln unserer heutigen Boden- und Naturbeziehung. *Kultur und Politik, 3*(2016), 26–27.

Patzel, N. (2017). (Politics and market, Demeter and Prometheus) Politik und Markt – Demeter und Prometheus. *Kultur und Politik, 3*(2016), 24–25.

Pauli, W. (1957, first ed. 1952). *(The influence of archetypal ideas on the formation of scientific theories of Kepler)* Der Einfluß archetypischer Vorstellungen auf die Bildung naturwissenschaftlicher Theorien bei Kepler (pp. 109–194). C. G. Jung, & W. Pauli (Eds.). Naturerklärung und Psyche.

Poincaré, H. (1912). *Science et méthode*. Paris: Flammarion.

Roesler, C. (2016). *(The concept of archetypes by C.G. Jung: Theory, research and application)* Das Archetypenkonzept C.G. Jungs: Theorie, Forschung und Anwendung. Stuttgart: Kohlhammer.

Romanyshyn, R. (2007). *The wounded researcher. Research with soul in mind.* New York: Routledge.

Strathern, P. (2000). *Mendeleyev's dream. The quest for the elements*. Scientific American.

Von Franz, M.-L. (1968). (The unconscious and the sciences) Das Unbewusste und die Wissenschaften. In C. G. Jung, et al. (Eds.), *(Man and his symbols)* Der Mensch und seine Symbole (pp. 304–310). Zurich: Walter.

Von Franz, M.-L. (1988). *(Psyche and matter)* Psyche und Materie. Einsiedeln: Daimon.

Wotiz, J. (Ed.) (1993). *The Kekule Riddle. A Challenge for Chemists and Psychologists*. Clearwater (FL): Cache River Press.

Part VI
Soil Education and Cultural Language of the Soil

Chapter 23
Roots of Soil Perceptions by University and Secondary School Students in Minas Gerais, Brazil

Cristine Carole Muggler, Arthur Stefanelli Gasparini, and Débora Cristina Lucas dos Santos

23.1 Introduction

Soils occupy a special place in the Earth system and in the human experience. They play a very important role in human history. This role, which always existed, became more and more relevant with the transition from the hunter-gatherer lifestyle to a lifestyle ever more dependent on food production and, later on, to complex agricultural industrial societies. With these transitions, the influence and importance of soils have grown more relevant in human societies, as well as knowledge about their characteristics related to their use and management. Replacing hunting and gathering with agricultural production and animal husbandry made our interactions with soils increasingly close.

The knowledge of how to manage soils gradually increased with agricultural development throughout time. Learning how to deal with and improve soil fertility fostered agricultural production, allowing a considerable increase in the number of people who could live in a certain area. In addition to greater population density, the higher agricultural productivity also created the possibility of generating food surpluses, which led to the division and specialization of labor (Diamond, 2002). In this way, agriculture could sustain a portion of society not dedicated to food production and to the surge and growth of cities (Brevik et al., 2018). The rise and spread of cities together with, and thanks to, agriculture has occurred repeatedly in the past thousands of years. Although this phenomenon was not new, industrialization from

C. C. Muggler (✉) · A. S. Gasparini
Soil Department, Federal University of Viçosa, Minas Gerais, Brazil
e-mail: cmuggler@ufv.br; arthur.gasparini@ufv.br

D. C. L. dos Santos
Plant Pathology, Federal University of Viçosa, Viçosa, Brazil

N. Patzel et al. (eds.), *Cultural Understanding of Soils*,
https://doi.org/10.1007/978-3-031-13169-1_23

the eighteenth century and, more recently, the rise of industrial intensive agriculture after the Second World War resulted in huge migration from rural to urban areas. In 2007, for the first time on a global scale, the number of people living in urban areas[1] exceeded those living in rural areas (Ritchie & Roser, 2018).

Since the advent of tillage-intensive agriculture, human activities have accelerated soil loss and degradation to levels several orders of magnitude higher than the rate of pedogenesis. These activities often disturb the soil so intensely and profoundly that they unbalance important cycles and destroy soil functions, producing effects whose consequences can negatively affect future generations (Amundson et al., 2015). Although we often consider soil and water degradation to be recent phenomena, the use and modification of these resources have been impacting the natural environment throughout various civilizations. In many ancient societies, the deleterious effects of soil degradation were major causes of their ruin (Diamond, 2005; Hillel, 1992; Hudson, 1985). Soil degradation, and its catastrophic effects on apparently prosperous human societies, is therefore not an uncommon nor recent phenomenon, requiring that serious measures be continually developed and applied to mitigate them. However, history also shows societies that developed long-term sustainable agricultural systems (Dotterweich, 2013) that can inspire and encourage the creation and adoption of soil conservation measures today.

The move to growing cities, where soils became more and more invisible, under housing and all kinds of urban equipment, expanded the detachment between humans and the soil, apart from traditional indigenous societies. However, despite this increased gap, humankind has never been more dependent on the soil than nowadays. Most of the food, fiber, and water production systems that humans rely on are entirely dependent on several billion hectares of soil (Richter & Markewitz, 2001). Therefore, our dependence on soils in the near future tends to increase because our food will continue to be supplied almost entirely by agricultural lands (Brady & Weil, 2010). The 2020–2021 Coronavirus crisis just corroborated this dependence as it partly disrupted both food production and distribution chains, impacted global food and nutritional security, and raised questions about how the resilience and productivity of local food systems can be improved and the reliance on external supplies reduced (Lal et al., 2020).

More dependence on soils and less engagement and knowledge of soils by most people evoke a contradiction that is at the roots of the need to communicate and educate better and more widely about soils. This has to be done in a broad educational process that goes beyond formal education, including the subjectivities of art and spirituality, where values and attitudes toward soils can be reviewed and rebuilt (Muggler, 2015). The International Year of Soils in 2015 and the International Union of Soil Sciences (IUSS) Soil Decade 2015–2024 represent efforts from the soil scientific communities to cope with this lack of soil knowledge in society. Soil education has been fostered in a multitude of initiatives (Kosaki et al., 2020; Lima

[1] Areas that encompass cities, towns and suburbs, with a high density of human built structures and a predominance of non-agricultural jobs

et al., 2020) reaching a wide range of audiences both in formal and nonformal education.

In Brazil, general soil contents are expected to be taught in basic education schools within the disciplines of sciences and geography, as indicated by the national curriculum basis (Brasil, 2017). Still, soil information is seldom taught by school teachers, who consider soil a difficult subject (Cirino et al., 2015) and rely strongly on what is presented in school books, which is very incomplete and conceptually weak (Santos, 2011). The fact that soil information is approached through formal education in schools, however, does not seem to promote the engagement of students with the soil any further than to inform them about its importance from an agricultural and environmental point of view. From this partial approach, the connections between people and soils, as well as their connections to nature, are not developed or strengthened, resulting in devaluation and being careless with soils and nature. The question that remains is how to sensitize people to soils in a way they feel connected and engaged so they will care for soil? In this context, a better understanding and consideration of the differences in perceptions about soils among people with different life experiences with soils may give clues to improve our (academic) communication with society.

The Soil Education Programme (PES in Portuguese) of the Earth Sciences Museum Alexis Dorofeef in Minas Gerais, Brazil, was created in the year 2000 and since then has developed a wide array of activities on education and communication. Throughout its existence, the program has undergone several changes related to concepts and methodologies driven by practice and experience as well as by the perceptions of the limits of communication of our soil science knowledge. From a basic soil properties approach, the program enlarged to encompass agroecology, arts, and culture. To further enhance it, it was decided to take a closer look at soil perceptions and its roots among its current and potential users.

Within the Soil Education Programme, we observed that students coming from the countryside or from small towns, periurban and urban areas bordering the rural areas, show a different and wider knowledge about soils, probably related to learning experiences other than the formal scholar approach. Thus, differences in perceptions about soils between these individuals and the ones from urban areas exist, probably related to differences in their life history and experiences, including their previous contacts with soils. The understanding of those differences and the use and consideration of them in outreach activities could improve that outreach and make education and communication about soils more effective.

In this context, a study with secondary school and first-year university students was carried out to investigate their perceptions about soils related to their origin and background. The study focused on the characterization of the students' previous experiences with soils and their understandings and perceptions regarding soil use and functions. It also evaluated whether the rural, urban, or "rururban" (rural + urban) origin of the students had any influence on these perceptions. The results and discussion of the objective and subjective, rational, and emotional roots of soil perception among students are presented and discussed in this chapter. The idea that inspired this research is that scholarly instruction, life experiences, and place of belonging of individuals as well as different audiences need to be considered in soil education programs.

23.2 Study Context and Methods

The study was conducted in Viçosa (80,000 inhabitants), Minas Gerais, South-East Brazil. The town houses the Federal University of Viçosa, a traditional academic institution within Agrarian Sciences, with 12,000 students in total. The study was conducted with 100 students, half of whom were undergraduate students aged 18–24 and the other half, 16- to 18-year-old secondary school students.

Their background information, as well as their perceptions about the soil, were obtained through semi-structured interviews (Duarte, 2004). Those interviews were done with an open script that can be changed according to the answers and new aspects that arise in the conversation. This modality of data collection allows researchers to obtain knowledge of the ways in which each of the interviewees perceives and signifies their reality. It also allows researchers to obtain consistent information that lets them describe and understand the logic that governs the relationships that are established within that group of interviewees, which, in general, is more difficult to obtain with other data collection methods (Duarte, 2004). The interviews took place in the educational institutions of the students, outside classrooms, on patios or lawns, and during breaks, when they were randomly invited to participate. The researchers excluded undergraduate students who had already taken, or were taking, classes in soil science. This aimed to minimize the academic influence on the perceptions of the interviewed students, given the focus on the nonscientific cultural perceptions and values sought in this research. The enquiry was designed to be done individually in an open dialogue using elements of Paulo Freire's pedagogy (Freire, 1981, 1996). The script of the semi-structured interview had five parts:

1. Provenance of the student, if urban, rural, or "rururban." The rururban category was used to designate intermediate origins between rural and urban, as well as mixed origins, a quite recurrent situation in the country, and to separate these students from those born and raised in cities. Due to intense rural migration since the 1970s, it is common that people of rural origin live in urban areas and some of their family still live in a rural area. This rururban category also includes areas and life situations that, even though influenced by the urban, manage to maintain a rural essence (Rua, 2005). In these areas, the urban environment is often permeated by typical elements of the rural, such as traditions and customs. These areas are endowed with a hybridism that gives them unique characteristics, with urbanities inserted in the rural space, and ruralities present in the urban space, without one context leading to the elimination of the other (Jacinto et al., 2012).
2. The word "Solo" (soil) was shown on an A4 sheet, and the student was asked to say the first idea that came to her/his mind seeing it.

3. *Solo* or *terra?* This question was focused to discern between "soil" or "land/earth." In Portuguese, these two words are usually used interchangeably to refer to soil, but they carry quite different meanings (semantic connotations). The word "solo" is usually employed in academic and technical circles and is seen as a formal way of referring to "soil." The word "terra," on the other hand, is the popular, common sense word, widely used for soils, but it also means Earth, the planet, as well as a person's place of origin, motherland, and the place where a person makes her/his living or land they own. Besides, it carries many more subjective meanings, related to one's connection to a place. This polysemy is not exclusive of the Portuguese language[2].
4. An open question about the main functions and/or uses of soils.
5. A small dialogue where the researchers investigated the students' previous contacts with soils, through questions such as "Why do you perceive the soil like this" or "Were/are soils present in your life in any/which way?" or "What is your relation to soils?."

Apart from the questions that were direct (1 and 3), the answers obtained for question 2 were organized in a set of generating words, and the answers for questions 4 and 5 were systematically analyzed and grouped into categories to allow a more quantitative analysis. Multiple answers were allowed for the last two questions.

23.3 Results and Discussion

The majority of the interviewed students were from urban origin (49%), followed by rururban (34%) and rural (17%). Among the university students, 64% had an urban origin, and rural/rururban were 18% each. Fifty percent of the secondary school students were rururban, 34% urban, and 16% rural.

Table 23.1 shows the frequency of the words used by the students to express the idea that came to their minds when exposed to the word "solo" (soil). Those ideas and words used by the students were systematized into a set of generating words, since they may be used as educational tools in soil education programs, following the socio-constructivist perspective used by Paulo Freire (Freire, 1987).

The most used word, by far, was "terra." Despite the fact that the word is used as a synonym for soil, its dominance in the responses suggests that people's understanding of the soil is broader than the technical meaning of the word used in academia. It also shows that the technical and scientific communication about soils, as well as school contents, do not really change the way youngsters refer to (and perceive) soil. The fact that "terra" is feminine and "solo" is masculine brings up the

[2] It can also be observed, for instance, in French with the word "sol," which means soil, but which often suggests an "origin," or in German, where words for soil are "Boden" [related with Engl. bottom], "Erde" [related with Engl. earth], and "Grund" [related with Engl. ground]), the latter signifying not only soil, but also "inner foundation," "cause" or "reason" (Feller & Patzel, 2015)

Table 23.1 Set of generating words and number of mentions obtained from the answers given by the students exposed to the word "solo"

Generating words			
Portuguese	English meaning	Number of responses	Comments
Terra	Soil, dirt	58	
Plantar	To plant, to grow	12	
Chão, piso	Ground, floor	7	In the sense of support
Cultivo, lavoura	Cultivation, plantation	6	Also agricultural field
Vida	Life	3	
Alimento	Food	3	Includes feeding
Território	Territory	2	
Quintal, horta	Backyard, vegetable garden	2	
Geografia/geologia	Geography/geology	2	
Decomposição	Decomposition	1	
Substrato	Substrate	1	
Rocha	Rock	1	
Barro	Mud/dirt	1	
Origem	Origin	1	Religious meaning

relation between the male and the female, as was indicated by Patzel (2010). Fertile soil benefits are compared to motherhood, a female quality. In a way, the soil is perceived as being a life provider, a feminine aspect. On the other hand, the science of soils was developed by men, where the relation to the female soil was seen and explained from a male-dominated standpoint (Patzel, 2010). This intuitive perception of female/male can be related to the many factors that build the "inner soil" mentioned by Patzel in Chap. 22, as the unconscious parts of mind, soul, and spirit reflected in symbolic ideas and expressions, religious symbols, myths, and other types of inner images and narratives.

The word "chão" (ground), which can also be used as a synonym for soil in some cases, was cited only by 6% of respondents. The use of the word "terra" may also indicate a more affective connection to soils, given by cultural and familiar, as well as ancestral, roots (Baldwin et al., 2017; Tobasura Acuña et al., 2015).

It is also interesting to see the use of "plantar" and "cultivo," to plant and cultivation, respectively. The latter is the more technical term in relation to the former that is the current and common sense way to refer to this action. We could have joined both, but this difference in *where* and *how* they are used is relevant. Life and origin follow the same pattern, and the origin may also have religious roots.

Aside from the 58 mentions of "terra" (soil), there were many (23) other words assigned to "solo" that are directly related to food production and subsistence, such as "plantar" (to plant), "cultivo" (cultivation/farming), "alimento" (food), "lavoura" (farm or agricultural field), "quintal" (cultivated backyard), and "horta" (vegetable garden), showing that the students made a direct connection between soil and food, emphasizing the productive function of soils. The indirect association between soils and food is also revealed by words such as "vida" (life) and "decomposição" (decomposition), which relate soils to the biotic, living dimension of organic matter.

The mention of soils as a component of the landscape or a geographic/political element, in words such as "chão" (ground), geography/geology, and "território" (territory), appeared a few times (10). Moreover, the word "terra" can also be used as a synonym for land and territory. Thus, the perception of soil as a component of the landscape, or a geopolitical element, is also relevant and related with one's sense of belonging.

The identification and selection of generating words relevant to specific groups may be a useful tool in the teaching-learning process of soils and soil science, where these are meaningful in the student's reality. The connection between the subject to be addressed and their daily lives contributes to its understanding and makes the learning meaningful (Ausubel, 1968). Equally important is its application in the context of extension and communication, to improve the dialogue capacity between the academic and the popular (Freire, 1983). In this context, the preference and choice for using the word "solo" or "terra" should be considered.

When asked about their preference for using the term "solo" or "terra," the majority of the students (60%) preferred to use the word "terra" instead of "solo" to refer to soil. Figure 23.1 shows that there is a very clear distinction between the preferences of students of urban, rururban, and rural origin regarding the use of one term or the other. Students of urban origin, to whom soils are less visible and present in daily life, used the more technical term that does not carry other connotations or

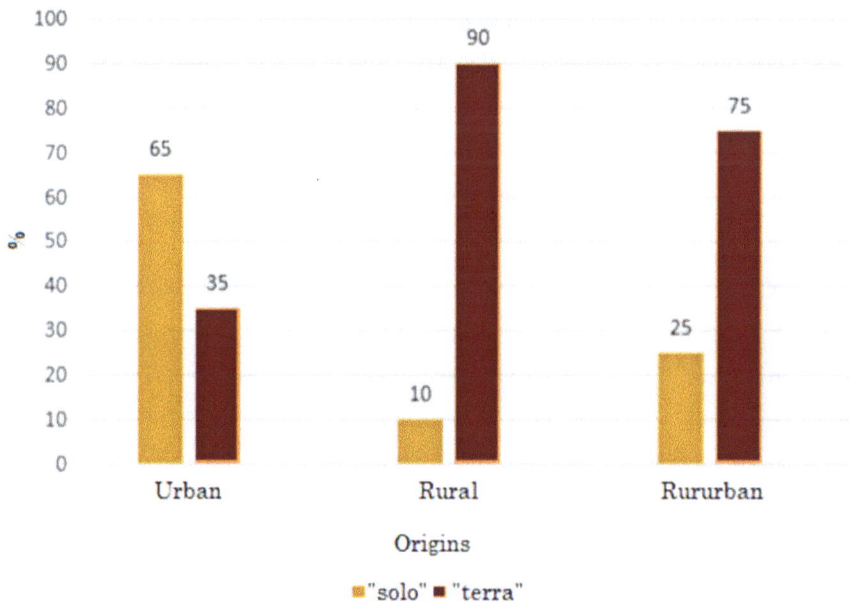

Fig. 23.1 Use of the words "solo" (soil) or "terra" (soil) by students of rural, urban, and rururban origin

symbolic meanings, indicating a weaker connection to soils. Conversely, students of rural origin used the word "terra," with all its symbolic and subjective meanings that embody a sense of belonging, overall (Hagerty et al., 1992). The rururban students also preferred to use "terra," probably related to their rural roots as family ties or origins. This may be interpreted as an indication that the affective connection to soils is stronger than the schooled learning.

These results indicate that local singularities, such as the more rural or urban origin of students, can be fundamental in defining the terms and approach used to dialogue about soils. Moreover, it can increase connection to or detachment from soils. The use of the word "terra," which brings in a stronger connection to soils, despite its other meanings, would be more suitable, as long as it belongs to the language repertoire of the audience an educator is attempting to reach. This debate is relevant, since the effectiveness of an educational process is related to meaningful learning (Ausubel, 1968), where the use of terms and concepts rooted in the student's reality plays an important role (Freire, 1987, 1996). In this context, the use of the word "solo" for students from rural origin can be a distancing factor between the educator and the learner, rather than bringing the teacher closer to the student and the student to the soil.

In the question about soil uses and functions, the students could answer and list as many as they liked. Even so, most of the students mentioned only one or two uses and functions, with 150 answers in total. This shows that the knowledge and understanding of the environmental and social role of soil is very limited among these youngsters, at least as a spontaneous statement. If they would have been asked to answer a closed questionnaire with many possibilities, probably there would have been more answers. All the answers were fit in seven groups: productive, support, water and climate regulation, habitat and biodiversity, archive, nutrient cycling, and others (Fig. 23.2).

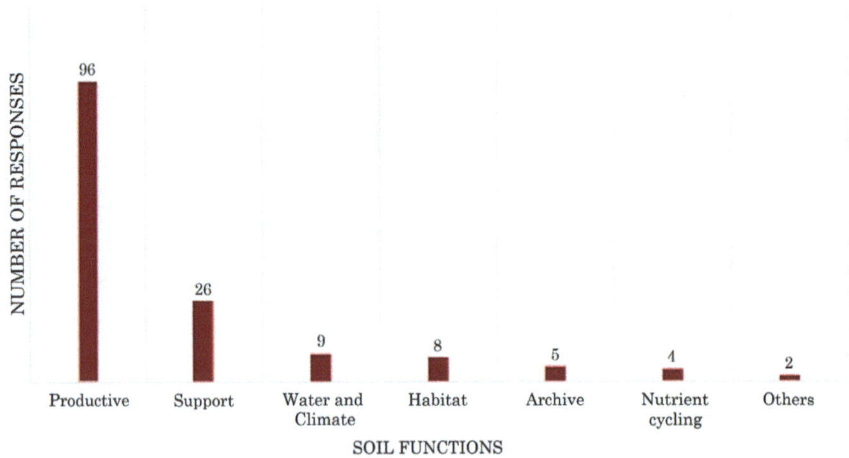

Fig. 23.2 Functions attributed to soils by university and secondary school students

As for the functions attributed to the soil, the large majority of the students, regardless of their origin, indicated the productive one, that is, the perception of the soil as a biomass producer. This common and broad perception of the association between soil and agriculture can be observed throughout historical time (Bridges & Cattizzone, 1996) and will probably continue to be observed in the future, given the continuous and growing demand for agricultural resources by humanity (Davis et al., 2016).

The dominant perception of the productive function of soils, especially of food, associated with an affective connection with the land, indicates that the soil tends to be seen as a provider. This also shows a productivist view of the soil, that is, the perception of soil as a natural resource from which food and other consumer goods are obtained and taken for granted. In this context, several ecosystem services provided by soils, such as storage and filtering of water, are ignored, unknown or seen as secondary by a large part of society. This interpretation may be particularly relevant if the individual soil affective connection is weak. The soil function of support, which is the basis for the construction and standing of buildings, roads, and other infrastructure equipment, was also well recognized.

The functions that go beyond the anthropic use of soils (in this study, represented by the productive and support functions) were less mentioned by the students. This shows that important functions, such as those referring to ecosystem regulation, as well as cultural and spiritual ones (Millennium Ecosystem Assessment, 2005), are little regarded by students from urban areas and a little more by the others. These results indicate that, in general, people have little awareness of the role that soil plays in their lives in less obvious ways, such as its influence on biodiversity and climate regulation (Aparin & Suhacheva, 2002; Muggler et al., 2006).

Despite the striking differences between urban, rururban, and rural areas, the roles that students from these different backgrounds attribute to soils were quite similar (Table 23.2). However, students of rural origin were able to recognize more soil functions than the others.

The most contrasting differences between these groups occurred in terms of nutrient cycling and "archive" functions, which were much more perceived by students of rural origin. The function of nutrient cycling, which is taught in formal basic education and therefore accessible to students of urban origin, can be readily observed in the daily lives of students of rural origin, who, for example, use manure to fertilize the land. Daily observation and experience with soil functioning of those

Table 23.2 Relation between students' origins and the functions assigned to soil by them

	Soil functions					
Students' origin	Production	Support	Water and climate	Habitat and biodiversity	Nutrient cycling	Archive
Urban	96	27	10	8	2	4
Rururban	97	27	9	6	3	6
Rural	94	24	12	12	12	12

The values represent the percentage (%) of students of each origin who mentioned the function. Note that students were allowed more than one answer students makes this knowledge much more significant and present, which seems to explain their greater perception of this phenomenon. The function of archive, memory, or cultural heritage concerns the fact that the soils hold a testimony of past human activities and constitute a means of permanence of archeological artifacts, keeping traces of their own history (Lahmar et al., 2004). People in rural areas are also more exposed to this archive function and can realize that it exists in their relations with soils.

The answers to the open question about the students' relations to soils or how soils came or are in their lives were many and were grouped into eight categories: ludic, school learning, belonging, natural resource, environmental role, ancestry, "pés de barro"[3] (muddy feet), and spirituality (Fig. 23.3). The figure shows that more than half of the students had an initial contact with soils mainly in a playful way (playing with and on the soil) and at school (formal education). The previous contact of people with soils in a playful, pleasant way explains why many times soil perceptions are related to affective memories. In the context of soil education and communication, these memories can be retrieved to bring significance to learning, by connecting the dimensions of experience and learning for each individual.

Even for students who did not have the opportunity to play with soils in childhood, there is a great potential in handling soil materials casually during activities like painting or modeling. This comes from the fact that pleasant feelings regarding this activity may have been rooted in human nature over millennia, since humans settled and learned to deal with soils (Harrison et al., 2012). Since pre-historic times, imprinting rocks with soil pigments was usual, as shown by rupestrian paintings found all over the world.

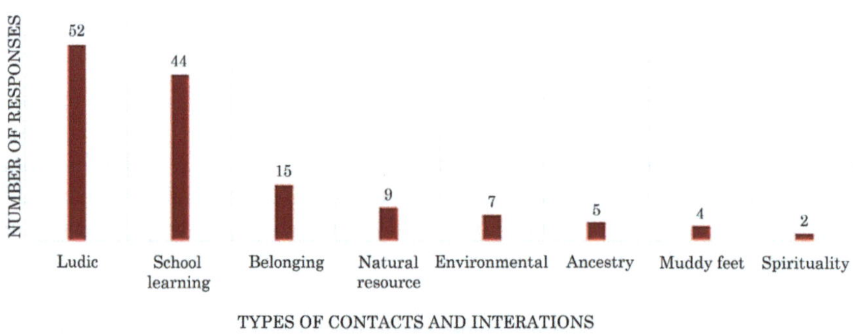

Fig. 23.3 Categories of types of contacts and interactions with soils, mentioned by university and secondary school students

[3] "Pés de barro" literally means muddy, dirty feet. It comes from the fact that students coming from rural areas used to arrive at school with their feet or shoes muddy from the dirty roads. It is less common nowadays since school buses became common after 2005. Before, they used to come to school on foot, and this made them readily identified as rural, which carried a negative stigma, given by social representations of prejudice against the rural as a geographic origin and place (Albuquerque Junior, 2007)

Some of the students mentioned that they were related to soils by the so-called sense of belonging (feeling part of a place and/or a community), ancestry (holding a historical identity or generation ties), and spirituality (relating soils with sacred and religious experiences, both spontaneous and within rituals). All students that mentioned these relations and interactions with soils were of rural or rururban origin. This shows a much deeper, diverse, and complex involvement of these students with soils, compared to students of urban origin. It is worth mentioning that many of the rural communities those students come from have African or indigenous ancestry that is usually not mentioned or considered by the families, due to a common social representation that devalues these peoples.

Although most of the students' experiences with soils were positive or neutral, some students of rural origin reported a negative feeling, identified as "pés de barro" (muddy feet) stigma. This stigma is related to the discrimination suffered by youngsters of rural origin in the Brazilian schools situated in urban areas, as are most of the secondary schools. There is a sociocultural construction that considers people coming from rural areas rustic, primitive, and "less civilized" (Albuquerque Junior, 2007), which results in a negative stereotype (Campos, 2012). This perception is associated with soils because of the mud (soil) that sticks to the feet or shoes of students coming from rural areas, "denouncing" their origin. The result is that these students develop a negative feeling toward the soil, whose presence on their shoes can carry a connotation of non-belonging in the school environment. The association between negative self-image related to soil is not exclusive to the Brazilian reality, as, for instance, terms such as soiled and dirt poor present in American English (Brevik & Hartemink, 2010). Soils are so common and seemingly abundant that they are often treated with disdain, which can be seen in the use of words like "dirty" and "muddled" in the English language (Hillel, 1992), and influence social representations of people coming from rural areas, as observed in our investigation.

The origin of the students has great influence on their mentioned previous contacts and interactions with soils, as shown in Table 23.3. Students of rururban and, especially, rural origin reported previous contact with the soil linked to recreational activities. For students of urban origin, this form of contact proved to be much less

Table 23.3 Relation between students' types of contacts and interactions with soils and their origin

	Urban	Rururban	Rural
Types of contacts and interactions	%		
Ludic	17	41	64
School learning	63	3	11
Belonging	0	26	14
Utilitarian (natural resource)	6	15	2
Environmental role	12	0	0
Ancestry	2	9	2
"Muddy feet"	0	3	5
Spirituality	0	3	2

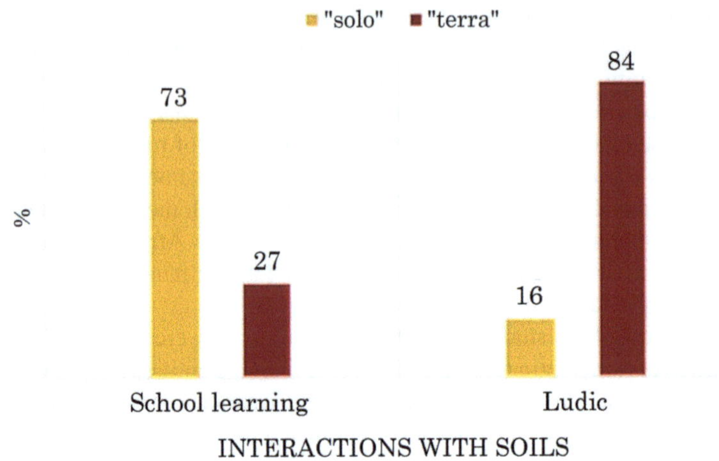

Fig. 23.4 Relation between the students' types of contacts and interactions with soils (ludic or schooled) and their choice of words "solo" or "terra" to refer to soil

common, probably due to the very "impermeable" nature of urban areas. Contact with soils through school instruction, on the other hand, was of greater importance for urban origin students.

The student's previous contact and interaction with soils also influences how they refer to them, as shown in Fig. 23.4. The use of "solo" over "terra" is essentially done by students who had contact with soils at school, mostly of urban origin as reported before. This is due to the fact that the use of the technical term "solo" is preferred in the school environment, encouraging its use by the students. In the same way, ludic contact with soils and use of the word "terra" are intimately associated. A ludic contact or memory with soils incorporates a subjective, affective connection, and therefore the use of the term "terra" is consistent.

23.4 Final Considerations

The Soil Education Programme of the Earth Sciences Museum Alexis Dorofeef started in 2000. After 10 years, we realized that soil education in the region was happening spontaneously, not directly induced by the museum activities, and in unforeseen ways. The Programme was then assumed as a social movement and renamed Movement "*Conhecer e Gostar de Solos*" (Know and Like Soils). This transformation incorporated the fact that the knowledge of soils people carry has diverse and, even unexpected, roots.

At that time in the early twenty-first century, activities on soil education and outreach consisted mainly of soil workshops, itinerant expositions, and projects developed at basic education schools. Among the museum workshops, the most successful has been the paints and painting with soil materials. It uses soils to prepare paints and to paint with them, including wall painting. Those workshops have been taken as an opportunity to approach soils and their importance to life and environment with different groups of people. They start with questions about how people consider and see soils and follow up with considerations about their formation, properties, distribution in the landscape, and their functions. It also emphasizes the fact that the tropical soils of the region offer a wide diversity of colors that wonders and intrigues people.

Interest in the workshops is rooted in the pleasure of ludic nature people experience in handling, modeling, and painting with soil materials, from paper sheets to house walls. Often the dialogues that happen within the soil education activities bring up surprising and unexpected perceptions and knowledge about soils that people have. In order to look closer at the soil knowledge brought by different people and how our understanding of it could make our communication about soils more effective, we investigated the roots of soil perceptions among the museum's current and potential audiences.

The study was done with youngsters coming from different places and origins, from typically rural to urban areas. As shown by the results previously described, we corroborated that students coming from the countryside or from rururban environments (small towns and periurban areas) carry a different knowledge about soils, which is rooted in learning and experiences other than the ones provided by formal school education. These contrasts are related to differences in their life history and experiences as well as their previous contacts and interactions with soils. The results have also shown that students' perception of soils is strongly related to affectionate memories and subjective experiences of ludic nature.

Youngsters of rural and rururban origin showed a closer link to and wider knowledge of soils, while the ones of urban origin knew less and showed a greater detachment from soils. This is not really surprising, since people living in rural areas have permanent contact and familiarity with soils. This still does not necessarily imply a good understanding of soils' functions and functioning. It only shows that youngsters of rural or rururban origin have a stronger connection to soils that is created by family ties, affection, belonging, and spirituality. This is indicated by the recurrence of the use of the word "terra" for soil, which brings forth simultaneously the meanings of *my* land, the place where *I* belong to/come from, and the religious mentioning of soils as godsends, which is rooted in ancestralism and spirituality.

In our experience in rural areas of the region, among family farmers, many of which have indigenous roots or belong to quilombola communities (former African enslaved populations), there is a prevalence of the idea of soil as mother earth, the gift from the creator that has been given to families, which cared for it throughout generations. This also unravels the question of gender consideration and devaluation of the feminine in our modern society and science. As stated by Patzel (2010): "For an individual or a culture which is dominated by male properties and ideals, the task to have an equal, fruitful and sustainable *relationship* with the feminine, is a difficult one, which requires personal and cultural change and engagement by women and men involved with the soil."

Our study, as well as our experience in 20 years of soil education activities, indicated that to be effective and knowledge enriching, soil education and outreach have to consider social and cultural aspects and the ways and words that are meaningful and representative to different people. The use and consideration of those differences could improve and make education and communication about soils more effective. On the other hand, much has to be done to involve and connect youngsters of urban origin to soils. A strategy may be to approach basic needs of people, such as food and water, that are directly related and provided by soils and explore this interconnectivity (Ball et al., 2017), even beyond a strictly scientific perspective.

A persistent challenge in our soil education task is to realize what really makes people engage with soils and their care. The usual answer is that knowledge is a key point of engagement. Then, the next question is: which knowledge? We assume that it is the knowledge rooted in tangible and intangible life experiences of people that goes far beyond scientific knowledge. Thus, for an effective dialogue between researchers and soil science professionals to be made with society, other dimensions (e.g., personal, cultural, spiritual, sense of belonging, and affective ones) must be considered and worked on. The consideration of these dimensions is especially relevant in a constructivist context of environmental or soil education, in which students' previous experiences are reclaimed and valued, giving sense and significance to learning.

It is not possible to teach and engage people with soils without considering their cultural and social reality, as well as their spiritual and ancestral foundation. If soil education efforts are limited to the technical and scientific concepts, many of the meanings that soils have for society are disregarded. This becomes especially important for people with direct or family rooted rural as well as indigenous origin, who have a stronger affective and spiritual connection with soils.

Acknowledgments The authors thank the support and inspiration given by the students we interviewed and to the small farmers we have been interacting with in our research and extension activities. We are deeply grateful to the reviewers Nikola Patzel, Christian Feller, Eric C. Brevik, and Sabine Grunwald whose comments and suggestions were stimulating and contributed greatly to improve the text.

References

Albuquerque Júnior, D. M. (2007). *Preconceito contra a origem geográfica e de lugar: As fronteiras da discórdia*. Cortez (Preconceitos; v. 3), São Paulo, Brazil.

Amundson, R., Berhe, A. A., Hpmansm, J. W., Olson, C., Sztein, A. E., & Sparks, D. L. (2015). Soil and human security in the 21st century. *Science, 348*(6235). https://doi.org/10.1126/science.1261071

Aparin, B., & Suhacheva, E. (2002). Methodology of uninterrupted ecological education and soil science. In *World congress of soil science, 17., Thailand, 2002. Abstracts* (p. 1685). IUSS.

Ausubel, D. P. (1968). *Educational psychology: A cognitive view* (1st ed.). Holt, Rinehart and Winston.

Baldwin, C., Smith, T., & Jacobson, C. (2017). Love of the land: Social-ecological connectivity of rural landholders. *Journal of Rural Studies, 51*, 37–52, ISSN 0743-0167. https://doi.org/10.1016/j.jrurstud.2017.01.012

Ball, B. C., Hargreaves, P. R., & Watson, C. A. (2017). A framework of connections between soil and people can help improve sustainability of the food system and soil functions. *Ambio, 47*, 269–283. https://doi.org/10.1007/s13280-017-0965-z

Brady, N. C., & Weil, R. R. (2010). *Elements of the nature and properties of soils*. Pearson Educational International.

Brasil (2017). *Ministério da Educação. Base nacional comum curricular: educação é a base*. Brasília, 595 p.

Brevik, E. C., & Hartemink, A. E. (2010). History, philosophy, and sociology of soil science. In W. Verheye (Ed.), *Soils, plant growth and crop production, Encyclopedia of Life Support Systems (EOLSS)*. EOLSS Publishers. http://www.eolss.net

Brevik, E. C., Homburg, J. A., & Sandor, J. A. (2018). Soils, climate, and ancient civilizations. In W. Horwath & Y. Kuzyakov (Eds.), *Changing Soil Processes and ecosystem properties in the Anthropocene. Developments in Soil Science*. Elsevier.

Bridges, E. M., & Cattizzone, M. (1996). Soil science in a holistic framework: Discussion of an improved integrated approach. *Geoderma, 71*(3–4), 275–287.

Campos, J. T. (2012). Uma pesquisa pioneira para a compreensão da cultura caipira. *Estudos Avançados (USP. Impresso), 26*(76), 315–335.

Cirino, F. O., Muggler, C. C., & Cardoso, I. M. (2015). Sistematização participativa de cursos de capacitação em solos para professores da educação básica. *Terræ Didatica, 11*(1), 21–32.

Davis, K. F., Gephart, J. A., Emery, K. A., Leach, A. M., Galloway, J. N., & D'Odorico, P. (2016). Meeting future food demand with current agricultural resources. *Global Environmental Change, 39*. https://doi.org/10.1016/j.gloenvcha.2016.05.004

Diamond, J. (2002). Evolution, consequences and future of plant and animal domestication. *Nature, 418*(6898), 700–707.

Diamond, J. (2005). *Collapse: How societies choose to fail or succeed*. Viking.

Dotterweich, M. (2013). The history of human-induced soil erosion: Geomorphic legacies, early descriptions and research, and the development of soil conservation—A global synopsis. *Geomorphology, 201*, https://doi.org/10.1016/j.geomorph.2013.07.021.

Duarte, R. (2004). Entrevistas em pesquisas qualitativas. *Educar em revista, 24*. https://doi.org/10.1590/0104-4060.357

Feller, C., & Patzel, N. (2015). Soil between nature and culture. In S. Northcliffe (Ed.), *Task force: Soil matters – Solutions under foot*. (on behalf of the IUSS), Catena Verlag, GeoEcology Essays.

Freire, P. (1981). *Ação Cultural para a liberdade: e outros escritos* (5th ed.). Paz e Terra.

Freire, P. (1983). *Extensão ou Comunicação* (7th ed.). Paz e Terra.

Freire, P. (1987). *Pedagogia do Oprimido*. Paz e Terra.

Freire, P. (1996). *Pedagogia da Autonomia: saberes necessários à prática educativa*. Paz e Terra.

Hagerty, B. M. K., Lynch-Sauer, J., Patusky, K. L., Bouwsema, M., & Collier, P. (1992). Sense of belonging: A vital mental health concept. *Archives of Psychiatric Nursing, 6*(3), 172–177.

Harrison, R., Strahm, B., & Yi, X. (2012). *Soil education and public awareness. Soils, plant growth and crop production, Encyclopedia of life support systems (EOLSS)* (Developed under the Auspices of the UNESCO). Eolss Publishers.

Hillel, D. (1992). *Out of the Earth: Civilization and the Life of the Soil*. University of California Press.

Hudson, N. W. (1985). A world view of the development of soil conservation. *Agricultural History, 59*(2), 326–339.

Jacinto, J. M., Mendes, C. M., & Perehouskei, N. A. (2012). O rural e o urbano: contribuições para a compreensão da relação do espaço rural e do espaço urbano. *Revista Percurso, 4*(2), 173–191.

Kosaki, T., Lal, R., & Sánchez, L. B. R. (Eds.). (2020). *Soil Sciences Education: Global Concepts and Teaching*. Catena-Schweizerbart.

Lahmar, R., Dosso, M., & Ruellan, A. (2004). *Cadernos de Proposições para o Século XXI: Salvar nossos solos para proteger nossas sociedades*. Instituto Pólis.

Lal, R., Brevik, E. C., Dawson, L., Field, D., Glaser, B., Hartemink, A. E., Hatano, R., Lascelles, B., Monger, C., Scholten, T., Singh, B. R., Spiegel, H., Terribile, F., Basile, A., Zhang, Y., Horn, R., Kosaki, T., & Sánchez, L. B. R. (2020). Managing Soils for Recovering from the COVID-19 Pandemic. *Soil Systems, 4*(3). https://doi.org/10.3390/soilsystems4030046

Lima, M. R., Vezzani, F. M., Silva, V., & Muggler, C. C. (2020). *Iniciativas de Educação em Solos no Brasil* (1st ed.). Sociedade Brasileira de Ciência do Solo.

Millenium Ecosystem Assessment. (2005). *Ecosystems and human well-being*. World Resources Institute.

Muggler, C. C. (2015). Soil and education. In S. Northcliff et al. (Eds.), *Task force: Soil matters – Solutions underfoot* (Geoecology essays) (pp. 147–150). Catena Verlag.

Muggler, C. C., Pinto Sobrinho, F. A., & Machado, V. A. (2006). Educação em solos: princípios, teoria e métodos. *Revista Brasileira de Ciência do Solo., 30*(4), 733–740.

Patzel, N. (2010). The soil scientist's hidden beloved: Archetypal images and emotions in the scientist's relationship with Soil. In E. R. Landa, & C. Feller, (Eds.), *Soil and culture*. Springer Science and Business Media.

Richter, D. D., & Markewitz, D. (2001). *Understanding soil change*. Cambridge University Press.

Ritchie, H., & Roser, M. (2018). *Urbanization*. Published online at OurWorldInData.org. Retrieved from https://ourworldindata.org/urbanization. Accessed on 3 Mar 2020.

Rua, J. (2005). A ressignificação do rural e as relações cidade-campo: uma contribuição geográfica. *Revista da ANPEGE, 2*(2), 45–66.

Santos, A. A. (2011). *Saberes de solos em livros didáticos da educação básica*. MSc dissertation, Universidade Federal de Viçosa.

Tobasura Acuña, I. T., Obando Moncayo, F. H., Moreno Chavez, F. A., Morales Londoño, C. S., & Henao Castaño, A. M. (2015). From soil conservation to land husbandry: An ethical- affective proposal of soil use. *Ambiente & Sociedade, 18*(3), 121–136. https://doi.org/10.1590/1809-4422ASOC802V1832015

Chapter 24
Degrees Pursued by Students in Different Countries to Train for Careers in Soil Science

Eric C. Brevik, Damien Field, Jacqueline Hannam, Maja Krzic, Rainer Horn, Cristine Muggler, Jude Odhiambo, Yoshitaka Uchida, Danny Itkin, Hong-sheng Wu, Liana Pozza, Laura Bertha Reyes-Sánchez, and Thomas Scholten

24.1 Introduction

Knowledge about soil allows us to address a wide range of issues, from food production to environmental concerns to human health and beyond (Bouma, 2015; Brevik & Sauer, 2015; McBratney et al., 2014). Just as soil knowledge is important to be able to address many different issues in modern society, there are different

E. C. Brevik (✉)
College of Agricultural, Life, and Physical Sciences, Southern Illinois University, Carbondale, IL, USA
e-mail: Eric.Brevik@siu.edu

D. Field · L. Pozza
Sydney Institute of Agriculture, School of Life and Environmental Sciences, University of Sydney, Sydney, Australia
e-mail: damien.field@sydney.edu.au; liana.pozza@sydney.edu.au

J. Hannam
School of Water, Energy and Environment, Cranfield University, Cranfield, UK
e-mail: j.a.hannam@cranfield.ac.uk

M. Krzic
Faculty of Forestry/Faculty of Land and Food Systems, University of British Columbia, Vancouver, BC, Canada
e-mail: maja.krzic@ubc.ca

R. Horn
Institute for Plant Nutrition and Soil Science, CAU Kiel, Kiel, Germany
e-mail: rhorn@soils.uni-kiel.de

C. Muggler
Earth Sciences Museum Alexis Dorofeef, Department of Soils, Federal University of Viçosa, Viçosa, Minas Gerais, Brazil
e-mail: cmuggler@ufv.br

N. Patzel et al. (eds.), *Cultural Understanding of Soils*,
https://doi.org/10.1007/978-3-031-13169-1_24

educational paths that can lead to a career in soil science (Brevik et al., 2020; Diochon et al., 2017). These paths can provide insight into how soils are viewed by the institutions that offer soil preparatory degrees, but to this point in history, a formal evaluation of these paths has not been conducted for most countries.

This chapter seeks to explore ways that various academic cultures around the world view soils. When soil and culture projects are undertaken, they often focus on topics such as indigenous knowledge, soil-related art forms, and what other nonscientific approaches to understanding soils can add to our knowledge and understanding of this topic (e.g., Critchley et al., 1994; Huckleberry, 2015; Landa & Feller, 2010). The perspectives of many different cultures around the world are explored in this book, and this is well and fitting because there is much to be learned from these viewpoints. However, cultural projects such as these often fail to consider the scientific perspective of soils, where they fit within various societies, and how that influences academic placement of soil science. The project reported in this chapter was undertaken as one step toward including part of the scientific perspective in this broader cultural discussion, recognizing that science is part of culture too. The goal of this study was to investigate the offerings and focal points of soil science curricula in the academy in 11 different countries around the globe. Eleven countries have been included to evaluate whether or not expert opinion, as determined by where various academic institutions have placed soil science in the curriculum, is rather uniform worldwide, or if different countries and cultures have different perspectives on this issue.

J. Odhiambo
Department of Soil Science, School of Agriculture, University of Venda, Thohoyandou, South Africa
e-mail: Jude.Odhiambo@univen.ac.za

Y. Uchida
Faculty of Agriculture, Hokkaido University, Sapporo, Japan
e-mail: uchiday@chem.agr.hokudai.ac.jp

D. Itkin
Department of Geography and Environmental Studies, University of Haifa, Haifa, Israel
e-mail: itkind@post.bgu.ac.il

H.-s. Wu
Department of Agricultural Resources and Environment, Nanjing University of Information Science and Technology, Nanjing, China
e-mail: wuhsglobe@sina.com

L. B. Reyes-Sánchez
Department of Agricultural Engineering. Higher Studies Faculty of Cuautitlán, National Autonomous University of Mexico, Mexico City, Mexico
e-mail: lbrs@unam.mx

T. Scholten
Department of Geosciences, University of Tübingen, Tübingen, Germany
e-mail: thomas.scholten@uni-tuebingen.de

24.2 Materials and Methods

The procedures followed for each country represented in this study were similar, but not exactly the same. Therefore, this section summarizes the procedures for each country. After information on degree programs was collected, the degrees offered were sorted into one of nine categories: (1) agriculture and agronomy; (2) biological and ecological sciences; (3) crop and soil sciences; (4) environmental sciences; (5) geosciences (geography, geology, and similar); (6) miscellaneous; (7) natural resources, water science, and resource management; (8) plant science, plant and soil science, and horticulture; and (9) soil science and soil and water science. The degree subdivisions were based on the names of the bachelor's[1] degrees offered, with the miscellaneous category covering all degrees that did not fall into one of the other categories. The categories chosen also allow easy comparison to other recent studies (e.g., Diochon et al., 2017; Brevik et al., 2018; Brevik, 2019) that have investigated soil science education issues.

24.2.1 Australia

To evaluate the degrees offered by postsecondary institutions in Australia, the websites of universities that were considered likely to have soil science in their programs were visited. Degree information was evaluated at the unit of study level on their websites. The university was considered to have a bachelor's soil science program if they offered an undergraduate degree that met the standards set by the Australian accrediting body for Certified Practicing Soil Scientists (CPSS), which requires a tertiary qualification with a minimum of four soil-related units of study.

24.2.2 Brazil

Specific soil science training does not exist at the bachelor's degree level in Brazil. Therefore, we searched websites of different universities to determine the type and extent of soil science content taught in specific courses offered by the related degrees. Soil science training in Brazil is considered a post-bachelor's (graduate level) specialization that can be obtained within specific master's and doctorate degrees.

[1] A bachelor's degree is an undergraduate academic degree awarded by a college or university to a person who has completed undergraduate studies (Webster's, 1996)

24.2.3 Canada

We searched the websites of the faculties, schools, and/or departments (from here termed "units") to identify soil science programs and courses offered at postsecondary institutions in Canada. We excluded civil engineering programs because they usually focus only on physical soil properties related to building or subsurface media.

24.2.4 China

Data was gathered in China through a survey that was sent to soil scientists and teachers at ten universities, who were asked to poll appropriate people at their institutions. The survey questions focused on the type(s) of degree(s) they expected someone who was training to work as a soil scientist in China would receive. Respondents were not limited to naming a single degree; they could provide as many degree names as they felt were justified.

24.2.5 Germany

Websites of universities as well as applied universities in Germany were explored, and officials of the German Soil Science Society (DBG, headed by Dr. Christian Ahl and the German authors of this chapter) were asked to obtain an overview about the education system referring to soil sciences. Final degrees are typically BSc with six semesters of coursework and MSc with four additional semesters of study. Applied university diplomas require eight semesters of curricula. Since November 2010, and based on a decision of the ministries of all federal states, the Fachhochschulen were renamed with terms that translate to Applied Universities and had to extend their curricula from six to at least eight semesters and offer corresponding master programs to align with European Union guidelines for the mutual acceptance of degrees. Currently, soil science is offered as both bachelor's and master's programs.

24.2.6 Israel

Websites of postsecondary institutions were explored to ascertain which academic institutes offered soil science degrees and programs in Israel. The curricula of six potential institutes, either known or considered likely to include soil courses, were examined using "soil" as the search keyword. Further search parameters included personal communication with colleagues from the academy and officials at the Ministry of Agriculture and Rural Development. Degree programs were evaluated to determine whether students were able to earn a bachelor's degree in soil science or an equivalent degree. Data collection included both academic programs that offer a full educational track in soil science and alternative tracks that enable graduate students to be active as soils experts (FCEE, 2018; RHSFAFE, 2018). However, it excluded programs that teach various aspects of soil science but offer only a limited number of relevant courses (these programs were found mostly within departments of geosciences and environmental studies).

24.2.7 Japan

The National Institution for Academic Degrees and Quality Enhancement for Higher Education website (www.niad.ac.jp) was utilized to investigate which universities offered soil science degrees and programs in Japan. The search investigated the names of academic degrees in Japan in 2016, and this data covers 97.6% of the universities in Japan (735/753 universities). In addition, colleagues from Japanese universities were consulted concerning the degree offerings at their universities.

24.2.8 Mexico

A search was conducted through the websites of Mexican universities and the Agricultural Technological Baccalaureate Centers to determine which offered degrees that included training in soil science.

24.2.9 South Africa

To identify soil science programs and courses offered at postsecondary institutions in South Africa, websites of university faculties, schools, and/or departments offering programs likely to have a soil science component were explored. The focus was on faculties offering a soil science degree as a qualification as well as related programs offering select soils courses but whose recipients of such degree programs cannot be considered soil scientists.

24.2.10 The United Kingdom

Data collection was conducted using a website search of universities and colleges awarding higher education degrees (Bachelor of Science) in the United Kingdom. Degree courses that offered soil science as a module (either compulsory or optional) with "soil" in the module title were selected. This was used to denote significant soil science content of the module and each module is typically offered for one semester out of a six- or eight-semester degree program.

24.2.11 The United States

To evaluate the degrees offered by postsecondary institutions in the United States, the websites of universities that were considered likely to have soil science programs were visited. Degree information on the websites was evaluated to determine if the university offered a bachelor's level degree in soil science. The university was considered to have a bachelor's level soil science program if they offered an undergraduate degree that met the US Office of Personnel Management (OPM) standards for employment in the 0470–Soil Science series (OPM, 2018), which includes completing at least 15 credit hours (225 classroom hours) in soil science coursework. A more in-depth discussion of what was considered to constitute a soil science degree program in the United States is provided in Brevik (2019).

24.3 Results

24.3.1 Australia

There were 36 universities in Australia that had soil-related units of study in their undergraduate degree programs, but 11 universities had programs with the required number of soil units of study to meet the CPSS criteria. In total, there were 175 units of study on soil and a further 42 units of study clearly identified as having an engineering focus on soil. Bachelor's degrees in "Soil Science" are not articulated in Australia, but students may obtain a major in soil science as part of a cognate degree program.[2] Most soil science subjects were part of agriculture-focused degrees, and these usually had the most in-depth studies of soil. More recently, the concepts of precision agriculture, digital soil mapping, soil security, and application of new technologies, such as remote and proximal sensing techniques, have been incorporated into soil education primarily through agriculture programs. Outside of agriculture, most universities taught soil science as part of their environmental science

[2] A cognate discipline is a similar or closely related subject area

programs. The environmental programs focused on systems thinking and soil chemistry as it relates to contamination and remediation. Plant and soil science programs tended to focus on turf and vineyard management. While concepts of soil management were covered across all programs, they were featured most prominently in those programs majoring in soil science or water science.

24.3.2 Brazil

Bachelor's degrees in soil science do not exist in Brazil. Soil science content is taught within degrees such as Agricultural Engineering, Agronomy, Biology, Cartography and Geodesics, Environmental and Sanitary Engineering, Forestry, Zootechny (Animal Science), Geography, Geology, and other related environmentally oriented courses. The amount of class time during which soil contents are covered varies from 30 to 300 h, with those hours being compulsory and higher in the Agronomy and Forestry curricula. Anyone who holds a bachelor's degree in those fields can work in soil science. However, most professionals employed as a soil scientist seek specific postgraduate training that adds soil sciences specialization. This additional training may occur after someone begins working as a soil scientist, which is common in many jobs, or before beginning work as a soil scientist in the case of academic careers. As for post-bachelor's degrees, there are 30 graduate programs in Soil Science (master's and doctorate) within 16 Agrarian Sciences Centers at different universities where the Soil Science departments are housed.

24.3.3 Canada

A total of 45 postsecondary institutions were identified in Canada that offered soil science courses that prepare students for future work in soil science. Some of those institutions had multiple units (i.e., faculties, departments, programs), resulting in a total of 54 units that provided soil science training. The greatest proportion of those units (39%, or 21 total) were focused on geosciences (geography, geology, and earth sciences), followed by environmental science (20%, or 11 total). Interestingly, only four units (or 7%) were focused on agriculture, and the remaining units were in the areas of science (9%, or 5 total), forestry (7%, or 4 total), biology (7%, or 4 total), and natural resources (4%, or 2 total). Only three units offered a degree titled "soil science," and they represent the only remaining Soil Science Departments in Canada. To grant a bachelor's degree in soil science (or soil-related discipline), Canadian postsecondary institutions typically require five courses per semester (or

15-semester credits[3]) spread over a variety of subjects for a total of about 120 semester credits over four years of study. At the Canadian universities that offer a complete soil science degree, students need to obtain 15–21 of those 120 semester credits from soil science courses.

24.3.4 China

The surveys resulted in returns from 220 individuals with 552 names or categories for degree programs that prepare students to work as a soil scientist in China, an average of 2.5 degrees named per respondent. The most commonly named degree categories were "soil science" and "soil and water science," which were named by 82% of respondents. This was followed by "environmental sciences" and "agriculture and agronomy," named by 61% and 55% of respondents, respectively.

24.3.5 Germany

Soil Science as a subject is offered within agriculture, horticulture, environmental sciences, geography, geology, geoecology, and hydrology programs. In total, 51 universities and applied universities offer soil science education in various curricula. Currently, soil science is offered both as a bachelor's and a master's program at a limited number of universities and applied universities. In most cases, soil science is a compulsory subject in geosciences, agriculture, environment, biology, hydrology, forestry, and contiguous programs with largely different scopes and intensities of soil science specialization. The university bachelor's degree can be finalized after at least six semesters, while the university master's program requires four additional semesters that add a specialization onto the basic soil sciences topics with very detailed and specific topics exploring soil processes and functions. Soil science as an independent final degree in Germany is only offered at a few universities, while in most cases, soil science is one part of the total study. The number of credit hours to be trained as a soil scientist varies greatly between universities and study programs. For the BSc, a minimum of six and an average of 24 credit points[4] for soil science courses can typically be assumed, while for the corresponding master courses, a minimum of four and an average of another 24 credit points are taken. The maximum credit points to be trained as a soil scientist may even reach 120 credit points.

---

[3] A semester hour in Canada equals 15–16 contact hours per semester

[4] 25–30 hours of total workload are required to receive one credit point in the German higher education system

24.3.6 Israel

Only one department at one postsecondary institution in Israel provides a bachelor's degree in soil science (Soil and Water Sciences). However, one other postsecondary institution offers a bachelor's degree in Environmental, Water, and Agricultural Engineering that accredits students in areas of soil science and enables them to become soil specialists. Both units set their own standard for qualifying in soil sciences, that is to say, there is no national standard for being certified in this field. Further qualifications at the master's and PhD levels are offered in each unit, with each university again setting their own standards.

24.3.7 Japan

Japanese universities offer >700 different types of bachelor's degrees, but none of them include the name "soil." Instead, students pursue a career in soil science through their graduate research topics. In Japanese universities, most students must complete their graduate dissertations in their final year, supervised by a professor. Thus, the students are allocated or split into different laboratories toward the end of their undergraduate courses. The "soil science" laboratory is one of the choices. Therefore, if a student wants to become a soil scientist, he/she must enroll in the course that is taught by professors, including the principal investigator, of the "soil science" laboratory. For example, Hokkaido University in Japan offers a Bachelor of Agriculture, and the School of Agriculture at Hokkaido University is further split into seven departments. The department of Bioscience and Chemistry is comprised of ten fields including soil science. Thus, the undergraduate students who are accepted into this department (approximately 35 students per year) may choose to do their dissertations in soil science toward the end of their courses (three to five students per year since the department has ~35 students per year with ten scientific fields). However, the "soil conservation" laboratory belongs to a different department (the Department of Bioresource and Environmental Engineering) within the School of Agriculture at Hokkaido University. Additionally, the Department of Applied Bioscience within the same school offers some dissertation topics on soil microbiology. Thus, there is no single way to become a soil scientist in Japan, but if a student completed his/her graduate dissertation in one of the "soil" laboratories, he/she is considered a soil scientist. Soil science laboratories are mostly located within the school of agriculture at the major universities in Japan (e.g., Kyushu University, Kobe University) or within the school of horticulture (e.g., Chiba University).

Students in Japanese universities take approximately 127 credits[5] to receive a bachelor's degree. Within those 127 credits, approximately ten credits come from their research work in the laboratory in their final year. Another two to four credits are typically taken as soil science coursework (either compulsory or elective credits) in the second or third year of bachelor's studies.

24.3.8 Mexico

Soil science concepts are taught within bachelor's degrees such as agronomy, agricultural engineering, biology, earth sciences, environmental and sanitary engineering, environmental sciences, geography, and environmental and forestry engineering. Only one bachelor's degree is offered explicitly in soil science. Baccalaureate-level training in soil science exists in the general education offerings at the Autonomous University of Chapingo (UACh is its acronym in Spanish) and the National and Autonomous University of Mexico (UNAM is its acronym in Spanish) and within the technical education at the Agricultural Technological Baccalaureate Centers (CBTA is the acronym in Spanish). A specialization in soil science can also be obtained at the master's and doctorate levels. The number of class hours that focus on soil science varies considerably between universities, and there is no official standard of professional certification specifically in soil science studies.

24.3.9 South Africa

A total of 16 postsecondary institutions were identified in South Africa that offered bachelor's degrees preparing students to work as a soil scientist. The single largest category (seven total, or 44%) was associated with plant production (agriculture, agronomy, crop science, or plant science), while the second largest category was soil science (six total, or 38%). All six of those institutions offered a degree titled "Soil Science." The remaining degrees were in the areas of environmental science (two total) and geosciences (geology, total of one). In order to practice as a soil scientist, upon graduation with a relevant bachelor's degree, one has to be registered with the South African Council for Natural Scientific Professions (SACNASP), which is mandated by the Natural Scientific Professions Act 27 of 2003 to register scientists in various categories and fields of practice including soil science. The requirement to be registered as a

[5] One credit hour in Japan is equal to 15 hours of lecture-based instruction over the course of a semester

Professional Natural Scientist in the field of soil science is a recognized four-year bachelor of science degree plus three years appropriate work experience in the field of soil science. An individual who meets the qualification requirements for Professional Natural Scientist while not yet meeting the work experience requirements is registered as a candidate natural scientist, pending attainment of the required work experience. Only a registered person may practice in a consultancy capacity.

In the bachelor's qualification referred to above, at least 70% of the required credits obtained during the first year of study are expected to be in a spectrum of natural scientific disciplines. At least 80% of the required credits obtained during the second and third year of study should include a spectrum of natural scientific disciplines, and for the fourth year, it should be 100%. The majority of these modules should be in the respective field of practice or directly supportive of the field of practice.

24.3.10 The United Kingdom

There were no specific degree programs (undergraduate, Bachelor of Science) that offered soil science as the sole lead program (i.e., as the sole title of the degree), although one institution offered a degree in "Plant and Soil Science." A total of 59 individual degree programs from 29 institutions offered a soil science module or modules (up to four) as part of the degree program. The majority of degrees were in agriculture and agronomy (agriculture) (27%), environmental science (29%), and geosciences (dominated by physical geography) (32%).

24.3.11 The United States

Of the 77 bachelor's level degrees identified in the United States that qualified recipients to work as a soil scientist for the federal government, the majority (43 total, or 56%) were associated with agriculture or plant production (agriculture, agronomy, crop science, or plant science). The next largest category was environmental science (21 total, or 27%). Only three degrees were titled "soil science," and another five had soil as the first word in their title (Soil & Land Resources; Soil and Hydrologic Sciences; Soil and Water Sciences, Soil Science Specialization; Soil and Waste Resources, Soil Science option; and Soil, Environmental and Atmospheric Sciences, Soil Resource Management emphasis).

24.4 Discussion

24.4.1 Australia

Australia was among the countries providing the fewest number of bachelor's (undergraduate) degrees that meet the criteria to qualify as a major in soil science. The number of courses reflects the student demand and the course requirements for each of the Australian institutions. As described in Field et al. (2017), the inclusion of soil science in a diverse range of vocational degrees has resulted in a shift from a disciplinary to multidisciplinary contexts of learning and teaching. This trend includes the increasing involvement of industry in curricula activities (Hartemink et al., 2014) and also affects the presence and quantity of time allocated to the degree in its home courses. Between the early 1990s and 2005, two-thirds of universities surveyed recorded a decrease in enrollments in soil science, which Hartemink et al. (2008) attributed to lack of soil science in secondary education curricula, university reorganization, and relabeling of the discipline. It has been suggested that the limited exposure to soil science in K-12 and the lack of popularity of degree programs have influenced the choice of soil science as a major.

To change students' (and their parents') views regarding soil science and emphasize its importance, more work is needed to connect people to soils and encourage soil science education. Community outreach is being encouraged and supported by Soil Science Australia through the development of a National Soils Education Committee, providing teaching resources, engaging in art programs and in-school programs. There are also projects funded by outside organizations, such as the Teabag Project, which is a crowdsourcing campaign teaching understanding of climate change and soil carbon among primary schools. Connectivity with the public may also be increased by more publication of news articles to get research out to the general population, as much research output is more focused on impact factor rather than increasing public knowledge (Lobry de Bruyn et al., 2017).

24.4.2 Brazil

To understand Brazil's situation, one has to look into its cultural and educational development as well as its economy. Brazil is a very large country, mostly situated in the tropics and rich in natural resources. Tropical ecosystems are resilient and capable of very quick recovery from degradation. In this context, soils are taken for granted and overlooked as living systems that have to be protected. Since the arrival of Europeans in the sixteenth century, Brazil has been a supplier of commodities. This began with the export of wood and other plant goods and minerals, followed by agricultural exportation becoming a huge business. Despite the fact that most of the food consumed in the country is produced by family farms (Instituto Brasileiro de Geografia e Estatística, 2006) and Brazil is one of the few countries that has a national policy for organic and agroecological production (Brasil, Ministério do Desenvolvimento Agrário, 2013), most of the financial and political support is directed to large agribusinesses. Soils in these systems are only considered for their productive function and often viewed as substrates that are poor in nutrients and need to be amended by fertilizers. They are rarely viewed as a resource that needs soil specialists to care for it.

This brief context helps to explain why there are no bachelor's degrees in soil science and no soil careers at public and private institutions. Rather, the training and careers exist in agronomy and forestry. It also explains why soil science is almost exclusively housed in agrarian science units at the universities. Even so, graduate programs in soil science receive students from highly diversified undergraduate majors, such as geography, geology, civil, sanitary and environmental engineering, biology, etc., in addition to agronomy and forestry.

On the other hand, soil concepts have been present in basic education (elementary and secondary) since 1997's launch of the national curricula parameters where environment was defined as a transversal theme to spread through all basic education contents. This means that soils are mentioned in at least two of the nine years of basic education that are mandatory in Brazil. Still, there are some considerable difficulties with the soil content in basic education, mostly related not only to teachers' lack of training in the subject but also to limitations and lack of didactic materials and appropriate classroom space for hands-on activities. Those limitations have been tackled in various ways by different states and municipalities (because basic education is their responsibility). Currently, more than 30 soil education initiatives exist around the country with the goal to improve soil education in primary and secondary schools. We did not observe a lack of interest in soil science and agriculture by students, as has been reported in other countries (McKenna & Brann, 1992; Baveye et al., 2006; Havlin et al., 2010), and there has not been a decrease in enrollment in agronomy and agriculture bachelor's programs. On the contrary, there is still a lack of openings at the universities to admit all students with interest in soils.

Soil content is part of the large and transversal environment theme in basic education, but not in higher education, showing an undesirable split between basic and university education approaches to soils. Soil science at the universities has been fragmented and specialized, in part to train professionals and develop research and technologies for the dominant market needs. The focus in Brazil has been more on soil properties rather than taking the broader soil functions approach. As a result, for instance, soil biology is not taught in soil science departments. Despite the fact that the environmental functioning of soils has been promoted by many soil scientists, it is still not seen as an issue that needs to be investigated by soil specialists. When natural disasters happen, like water shortages, floods, and landslides and other mass movements, geologists and civil engineers are the ones called in for media interviews and to provide information and technical support. This has concerned the soil science community, and the Brazilian Soil Science Society (SBCS) has been working to change perceptions and promote specialized training in soils. After years of discussion within the soil science community, an ambitious program on soil training and mapping of the country (Programa Nacional de Solos do Brasil, PronaSolos, https://www.embrapa.br/pronasolos) was conceived with the ministry of education in 2016. It was launched in 2020, after continuous technical and political pressure, while its implementation remains unsure due to economic difficulties in the country. The program is expected to provide specific training in soils and soil mapping as a specialized post-bachelor degree to achieve the main goal of mapping 8.2 million km^2 of soils at 1:25,000 to 1:100,000 scales in the next 30 years.

24.4.3 Canada

Soil science courses at Canadian postsecondary institutions are most commonly offered by geosciences and environmental science units (59% of total units), while agricultural science units represent just 7% of the total units that offer soil science courses. This reflects a changing trend. In the early days of soil science postsecondary education in Canada, in the late nineteenth to early twentieth century, soil science units were housed within Faculties of Agronomy and were aligned with the initial soil survey work, financed by federal and provincial governments, with a strong agricultural focus (Anderson & Smith, 2011; Lavkulich, 2013). A notable decline in enrollment in agronomy and crop science programs in Canada occurred during the late 1990s and early 2000s (Baveye et al., 2006) with corresponding declines in the enrollment of students in soil science courses. These decreases were triggered by reductions in university budgets and were reflective of several interrelated issues. Many faculties of Agricultural Sciences responded to the budget cuts by rebranding themselves as Environmental Sciences, Renewable Land Resources, Land and Food Systems, etc., and this also involved relabeling of some of the soil science courses (Bill McGill, personal communication). In addition, budget cuts also forced university administrations to focus on the majors of greatest popularity (i.e., enrollment), and Soil Science majors traditionally had a low enrollment. Consequently, many soil science programs and courses ceased to exist. All of these issues were enhanced by the fact that the majority of current students came from urban centers, not from farming communities as was previously the case, and had a very limited understanding of roles played by soil (Brown, 2009).

Global change has driven some resurgence in soil science, since the surface of planet earth is recognized as a major venue for biological and (bio)chemical reactions and physical exchanges between the atmosphere, hydrosphere, and the lithosphere that are important to climate (Intergovernmental Technical Panel on Soils, 2015). Such realization sparked an interest in soil science in the environmental and geoscience communities leading to high enrollment of students majoring in geosciences and environmental sciences in soil science courses. In Canada, their enrollment is almost double that of majors from natural resource management and agronomy (Diochon et al., 2017). The fact that there are only three Soil Science departments in Canada today is indicative of a shift in overall teaching of soil science from the deeper disciplinary focus to a more general approach to address the needs of related disciplines (e.g., geosciences, environmental science, forestry, agriculture, natural resources). This is accompanied with the challenges of balancing teaching in-depth soil science concepts with creating a sense of wonder about roles of soils in various global issues (Krzic et al., 2018). Another challenge created by this shift is poor or missing field skills obtained by recent graduates. A survey of Canadian soil science professionals who are in a position to hire recent graduates revealed that 94% of respondents stated that graduates with whom they worked require additional training, either through a refresher course (34%) or more intensive training in soil science concepts and/or field skills (60%) (Masse et al., 2019).

There are also notable variations in the distribution of postsecondary units offering soil science courses across Canada, where the majority of these courses were offered in central (i.e., Ontario and Quebec) and western (i.e., British Columbia, Alberta, Manitoba, and Saskatchewan) provinces (Diochon et al., 2017). Coincidently, these are areas with the largest populations and the best integration of soil science into the high school curriculum (Krzic et al., 2014). In northern Canada, which has a very low population density and is facing unique and significant soil challenges related to a changing climate, there is only one soil science course. However, several universities located in other parts of the country (e.g., McGill, Carleton, Manitoba) have active field programs that give students research experience with northern soils, including areas dominated by permafrost. Lower than average percentages of institutions offering soils courses in eastern Canada, namely, in New Brunswick, Prince Edward Island, Nova Scotia, and Newfoundland and Labrador, is surprising since all of these provinces have a natural resource based economy and sustainable management is a key component for economic prosperity (Diochon et al., 2017).

There is no formal certification of professional soil scientists in Canada. The certifications exist at the provincial level but are mostly limited to agrologists or professional foresters, and those professional certifications are eclipsing a large portion of soil science professional practices. The Canadian Society of Soil Science (CSSS) is currently considering designating CSSS representatives who will help develop a set of rules similar to those currently used by soil science certification programs in other countries (e.g., the United States, Australia). Once a general framework is agreed upon, CSSS representatives could then work in collaboration with existing professional organizations, mainly the provincial agrology institutes, to establish soil scientist certification programs.

24.4.4 China

Thirty years ago, there were over 40 agricultural universities and over 300 agricultural professional schools in China, while today, there are about 30 agricultural universities and 150 agricultural professional technical institutions. There are an additional five institutes devoted to agricultural and/or soil science in the China

Academy of Sciences and 12 in the China Academy of Agricultural Sciences. Over 50,000 bachelor's students and more 5000 master's and doctoral students have graduated from Chinese programs in the last five years.

24.4.5 Germany

Soil science education in Germany is spread over a very broad spectrum of subjects, reflecting the diverse and important roles soil holds for humankind. At the beginning of the twentieth century, soil science in Germany was at a scientific crossroads, mainly between the sciences (geology and chemistry) and agricultural practice. Due to the pressing need to increase food production after World Wars I and II, soil science shifted toward agriculture, and most soil science studies took place in agriculture and contiguous programs. However, with the growing discussion of environmental issues in the 1970s, including topics like forest dieback, climate change, sustainable development, (bio-)diversity, and other subjects, the placement of soil science diversified. Most soil scientists in Germany today have a master's degree in fields such as geosciences, agriculture (including plant sciences and horticulture), environmental sciences (including biology), ecology and nature protection, and forestry.

24.4.6 Israel

Just one postsecondary institution offers a bachelor's program in soil science in Israel (Faculty of Agriculture, Food and Environment; The Hebrew University of Jerusalem (HUJI)). Except for occasional fluctuations, enrollment in soil and water sciences at HUJI has been quite steady over the last 20 years (Rony Wallach, personal communication). One other institution (Faculty of Civil & Environmental Engineering; Technion–Israel Institute of Technology) offers programs in water engineering and environmental engineering that cover soil-related topics needed for certification and a professional career in soil science. Following the merger of the Faculty of Agricultural Engineering with the Faculty of Civil Engineering in 2002 to form the Faculty of Civil & Environmental Engineering, the department has experienced a decline in enrollment, mostly with respect to Agricultural Engineering students. Luckily, the downward trend has ceased and stabilized, and there has been an increase in Environmental Engineering students (Alex Furman, personal communication).

24.4.7 Japan

Historically, the Japanese University Establishment Standards (JUES), made by the Ministry of Education, Culture, Sports, Science and Technology (MEXT), allowed only 29 names for bachelor's degrees. The names included Bachelor of Agriculture and Science. Soil science was taught as a subject within these degrees. In 1991, the JUES was modified, and universities are now allowed to name their degrees freely. As a result, the names of undergraduate degrees markedly increased from 29 to >700. However, as stated earlier, the word "soil" is not included in any of the degrees.

One of the reasons that Japanese universities do not provide any soil science degrees is because the term "soil science" is unlikely to be authorized by MEXT as a new degree name. The modification of the JUES in 1991 aimed to diversify and to internationalize the degrees offered by Japanese universities, reflecting the needs of the modern world. Thus, many of the newly established degrees contain words such as "information," "international," and "environment." The discipline "soil science" sounds somewhat stale and obsolete as a candidate for a new degree name.

24.4.8 Mexico

Agriculture in Mexico has been a strongly discouraged activity since the government and agricultural infrastructure were separated over the last 30 years (Reyes-Sánchez et al., 2006). As a result, of the 69 agricultural schools that existed in Mexico in the 1970s, only nine universities that offer the degrees Agricultural Engineer or Agronomist still exist (Warman, 2003). Soil science is consequently most commonly taught within biology, environmental sciences, and environmental engineering programs.

The study of soil science constituted the backbone of the Agricultural Engineering and Agronomy degrees with a labor force heavily financed through the National Development Plans that offered a wide and stable area of work (Greenberg et al., 2012). This is currently reflected through the selection that young people make when choosing the career that they will study, because in the absence of a wide range of job options in the agricultural field, soil science is not often selected as a field of study.

Despite the importance of the soil resource for the food supply, economy, health, security, and social stability of a country, it is not perceived as such by the young people when they choose a university degree. Soil is also not perceived as important by the general public, and there is a lack of interest in soil science by the Mexican government. Mexico and its education system need to understand that "the soil is a limited resource under increasing pressure, the fundamental basis for food security and the provision of important environmental services" (FAO, 2012). But it is unclear who should carry out the task of raising awareness regarding soil that would allow citizens to better understand the importance of soil for their daily lives. There are signs that interest in soils, and the study of soil science, is increasing due to obvious environmental problems such as climate change and especially consequences such as hunger and massive human migrations, as they affect the stability of governments and interests of transnational markets.

24.4.9 South Africa

In South Africa, soil science courses at postsecondary institutions are most commonly offered in agricultural programs. However, the increasing lack of interest in agriculture by the current young generation has led to low enrollment in agriculture programs and the wide spread trend of the term "soil" disappearing from many subjects and departments, together with the merging of soil science with other more general or more fashionable subjects (Díaz-Fierros Viqueira, 2015). This poses a potential shift of the current observed trends in soil science courses offered in South African universities. This shift is likely to place more emphasis on soils courses offered in environmental and geosciences. This has been observed in other countries, for example, in Canada where students majoring in environmental science are most frequently enrolling in soil science courses and their enrollment is almost double that of majors from natural resource management and agronomy (Diochon et al., 2017) or in the United States, where environmental science students now make up the single largest group of students enrolling in many soil science classes (Brevik et al., 2018).

24.4.10 The United Kingdom

There are a dwindling number of graduates with adequate skills in soil science for application in both academic and industrial sectors in the United Kingdom (Godwin et al., 2003; NERC, 2012). This indicates current undergraduate provision in soil science is lacking; only one degree program in the United Kingdom offered a degree in "Plant and Soil Science," and one had significant soil science content (four modules in soil science for one Environmental Science degree program). However, the majority (90%) of degree programs identified in this study offered only one module in soil science, and many were elective (optional), so a graduate could be awarded a degree in environmental science, geography, or agriculture without receiving any soil science training. The skills shortage at postgraduate level (NERC, 2012) recognized by the Natural Environment Research Council (NERC, the United Kingdom's main agency for funding research, postgraduate training and knowledge exchange in the environmental sciences) resulted in funding for advanced training programs for postgraduates in soil science (NERC, 2019; STARS, 2019) between 2014 and 2020.

There are no guidelines for minimum requirements to qualify as a soil scientist in the professional sector in the United Kingdom . To address this, the British Society of Soil Science (BSSS) developed a professional competency scheme ("Working with Soil") that outlined the minimum qualifications, knowledge, and skills required of a professional soil scientist to conduct a range of contract work. The BSSS offers a training program for foundation (describing and exposing a soil profile, soil classification, and soil-landscape relationships) and advanced (e.g., agricultural land classification) skills for professional soil scientists working in industry. For university graduates going into the professional sector, some of the soil science modules (within the degree programs identified in this study) offered similar content to the basic foundation skills in the "Working with Soil" program, although as previously discussed, many of these modules were optional to the student. In the agricultural sector, BASIS (BASIS, 2019) provides agronomists with recognized continuing professional development (CPD) for professional registration. Courses are dominated by agronomic topics, but certificates are available in "Soil and Water Management" and "Quality of Soils" that form part of a diploma in Agronomy. A BASIS soils foundation course was recently launched for agronomists and farmers (BASIS, 2019), presumably due to demand in the sector from farmers focusing on soil quality and health.

24.4.11 The United States

Only three "soil science" degree programs were identified in the United States at the bachelor's level, with six more programs in soil and water science. The most common degree programs that prepare students to work as soil scientists are associated with agricultural programs. This was shown by the frequent link to degrees in agronomy, agriculture, and crop and plant sciences. The links to agriculture are long-standing in the United States (Brevik et al., 2014), with the US Department of Agriculture (USDA) assisting with the establishment of the first soil science program at a university in the United States in 1903 (Lapham, 1949). However, the role of and needed training for soil scientists was changing as we entered the twenty-first century. As enrollment in agricultural programs, including soil science, declined from the 1980s into the early 2000s (McKenna & Brann, 1992; McCallister et al., 2005; Baveye et al., 2006), there was a corresponding increased interest in environmental science that altered soil science education and job opportunities (Pepper, 2000; Havlin et al., 2010). That shift was documented by Brevik et al. (2018), who found that the enrollment of agronomy, crop science, and horticulture students was on the decline in soil science classes in the United States in the early 2010s, while enrollment of environmental science students was increasing. The first environmental science curricula were not developed in the United States until the 1970s (Fortuin, 2015), making these relatively young academic programs that have room for growth. Approximately 67% of the environmental science bachelor's programs in the United States include soil science in their curricula (Brevik, 2009).

24.5 Comparisons

There was no single category of degrees that had high numbers for every country, indicating that there is not universal agreement on where soil science belongs as an academic subject. Categories that were highest for individual countries include "Agriculture and Agronomy" (highest for Brazil (undergraduate) and USA), "Biological and Ecological Sciences, Nature Protection" (highest for Mexico), "Environmental Sciences" (highest for Australia), "Geosciences" (highest for Canada, Germany, and the United Kingdom), and "Soil Science and Soil and Water Science" (highest for Brazil (graduate), China, Israel (tied with "miscellaneous"), and South Africa). Numbers of degree programs were fairly low for all countries for the categories "Crop and Soil Sciences" and "Natural Resources, Water Sci. and Resource Management." The "Miscellaneous" category was high for some countries (Brazil (undergraduate), Germany, Mexico), in the case of Brazil and Germany being related to forestry and for Mexico being related to environmental and forestry engineering. The "Miscellaneous" category also represented half the degree offerings in Israel, in which case the formal degree program was in civil and environmental engineering.

In countries such as Canada, Germany, and the United Kingdom, the results stand in contrast with the United States and Australian universities where soil science has traditionally been taught in agricultural programs (Baveye et al., 2006; Sharik et al., 2012; Brevik et al., 2020), followed by environmental science programs. The United Kingdom has the most similar distribution of type of units offering soil science courses to those in Canada, since in both countries geosciences and environmental sciences are the most dominant categories, followed by agricultural sciences in the United Kingdom and forestry and sciences in Canada. Germany is also similar to the United Kingdom, in particular, with geosciences being the most common program of study and agriculture and agronomy programs also being highly important in each of the two countries. It is interesting to note that among the countries included in this study that had more than two soil science programs, in Canada, agricultural programs represent the smallest proportion (7%) of all programs that offer soil science courses.

Of the countries in this study, the US results were most similar to those from Australia. In both countries, agriculture and agronomy and environmental sciences were the two most common programs pursued to obtain training in soil science, followed by plant sciences and soil science/soil and water science. The US results were also relatively similar to those from the United Kingdom but with fewer programs offered through the geosciences and more through the soil science/soil and water science route in the United States. Interestingly, there was an effort within the United States in the late 1800s to integrate national soil survey activities within the US Geological Survey (Amundson & Yaalon, 1995). If this effort had succeeded, the US results might have looked quite different!

China and South Africa were unique in that the soil science/soil and water science category was the largest category in each case. One of Israel's two degree programs that prepared students to work as soil scientists was also in soil science, with the other in engineering as previously mentioned. Agriculture and agronomy and environmental sciences were also both in the top three categories for China and South Africa. Every country except for Brazil, Israel, and Japan had at least one entry in at least six of the nine categories investigated, showing the breadth of programs that offer soil science training, even within many individual countries.

When comparing the status of soil education between the countries discussed herein, it is noteworthy that many of the discussions mention either a decline in basic soil sciences or a major shift in the fields that soil science is associated with. Several studies have reported on a global decline in soil science (Basher, 1997; Baveye et al., 2006; Terribile et al., 2013; Díaz-Fierros Viqueira, 2015), particularly from the 1990s into the early 2000s. Others have noted changing trends in the students taking soil science classes (Brevik et al., 2018; Diochon et al., 2017) or in the career ambitions of students in soil science programs (Havlin et al., 2010); in all of these cases, the student shift has been from an agronomic focus to an environmental focus. One striking example that might illustrate the state of pedology (and awareness of soils) is soil mapping in Israel—a fundamental tool that has so many applications. The most updated publicly available soil maps of Israel were prepared in the mid-1970s at a scale of 1:500,000 (Dan et al., 1976). These maps are composed of polygons of nongenetically merged soil taxa and are based on an archaic (local) classification system (Dan & Koyumdjisky, 1979). Newer soil mapping data is non-public (or not publicly functional) and disregards the broadly accepted terminology of Soil Taxonomy and/or WRB; therefore, it is very difficult to communicate information at the international level. The age of these maps and lack of ability to communicate internationally reflect the declining status of pedology in Israel.

24.6 Concluding Statements

No strong universal trends were detected among the world's academics regarding where soil science belongs in the academic curricula, rather, soil science training was scattered across a multitude of related fields. It can broadly be stated that most experts seem to view soil science as being some part of the agronomic, environmental, or geosciences but exactly which of these areas and the relative proportions of soil science training opportunities found within each varies considerably from one country to the next. Training specifically in soil science as a field of study does occur in several countries, but with the exception of China, it was not the dominant field of study for such training in any of the countries investigated, as both South Africa and Israel had other fields that equaled or nearly equaled soil science in terms of undergraduate training opportunities. Therefore, it can be concluded that each country's unique culture plays a part in determining where academic soil training is found.

Changes in cultural perceptions of the most important topics that need to be addressed seem to be driving many of the changes we are seeing regarding where soil science is taught in the countries where such changes are occurring. Many countries reported a shift from an agricultural emphasis to an environmental or geoscience emphasis (e.g., Australia, Canada, Germany, Mexico, South Africa, the United States). It remains to be seen what the long-term consequences of this shift will be in terms of the knowledge base of soil science graduates. However, there are already some indications that there is a shift from an in-depth disciplinary focus on soil science to covering soil-related topics in broader, interdisciplinary courses that produce students that have a superficial understanding of soil science concepts and lack important hands-on field skills. Such trends have been documented in places like Australia, Canada, the United Kingdom, and the United States. While it is exciting that closely related fields have "rediscovered" soils and seek to use soil information in their own work, it is important that we also educate and maintain a group of soil science specialists who understand soils in great depth.

Acknowledgments Danny Itkin thanks Onn Crouvi, Alex Furman, David (Dudu) Magal, and Rony Wallach for their remarks and suggestions. Yoshitaka Uchida thanks Drs. Toshiya Saigusa and Akinori Yamamoto for their opinions and suggestions. Thomas Scholten and Rainer Horn thank Dr. Christian Ahl for his help collecting the data and with suggestions. Jack Hannam thanks Prof. Paul Hallett and Dr. Emilia Urbanek for providing information on undergraduate degree programs involving soil science at the University of Aberdeen and Swansea University, respectively. The authors thank Katsutoshi (Toshi) Mizuta for his comments and suggestions to improve the manuscript. E.C. Brevik was partially supported by the National Science Foundation, Established Program to Stimulate Competitive Research (EPSCoR), under Grant Number IIA-1355466 during this study.

References

Amundson, R., & Yaalon, D. H. (1995). E. W. Hilgard and John Wesley Powell: Efforts for a joint agricultural and geological survey. *Soil Science Society of America Journal, 59*, 4–13. https://doi.org/10.2136/sssaj1995.03615995005900010001x

Anderson, D. W., & Smith, C. A. S. (2011). A history of soil classification and soil survey: Personal perspectives. *Canadian Journal of Soil Science, 91*, 675–694. https://doi.org/10.4141/cjss10063

Basher, L. R. (1997). Is pedology dead and buried? *Australian Journal of Soil Research, 35*, 979–994. https://doi.org/10.1071/S96110

BASIS (2019). *BASIS Information Booklet 2019*. http://www.basis-reg.co.uk/training#. Accessed 9 Sept 2019.

Baveye, P., Jacobson, A. R., Allaire, S. E., Tandarich, J. P., & Bryant, R. B. (2006). Whither goes soil science in the United States and Canada? *Soil Science, 171*, 501–518. https://doi.org/10.1097/01.ss.0000228032.26905.a9

Bouma, J. (2015). Engaging soil science in transdisciplinary research facing "wicked" problems in the information society. *Soil Science Society of America Journal, 79*, 454–458. https://doi.org/10.2136/sssaj2014.11.0470

Brasil, Ministério do Desenvolvimento Agrário (2013). *Brasil agroecológico: Plano Nacional de Agroecologia e Produção Orgânica (Planapo)*. MDA, Brasília, Brazil.

Brevik, E. C. (2009). The teaching of soil science in geology, geography, environmental science, and agricultural programs. *Soil Survey Horizons, 50*, 120–123. https://doi.org/10.2136/sh2009.4.0120

Brevik, E. C. (2019). Bachelors level soil science training at land grant institutions in the USA and its territories. *Natural Sciences Education, 48*, 180021. https://doi.org/10.4195/nse2018.12.0021

Brevik, E. C., Abit, S., Brown, D., Dolliver, H., Hopkins, D., Lindbo, D., Manu, A., Mbila, M., Parikh, S. J., Schulze, D., Shaw, J., Weil, R., & Weindorf, D. (2014). Soil science education in the United States: History and current enrollment trends. *Journal of the Indian Society of Soil Science, 62*(4), 299–306.

Brevik, E. C., Dolliver, H., Edinger-Marshall, S., Itkin, D., Johnson-Maynard, J., Liles, G., Mbila, M., Moorberg, C., Sanchez-de Leon, Y., Steffan, J. J., Ulery, A., & Vaughan, K. (2020). Undergraduate degrees that train students for soil science careers at universities in the USA and its territories. *Soil Science Society of America Journal*. https://doi.org/10.1002/saj2.20140

Brevik, E. C., & Sauer, T. J. (2015). The past, present, and future of soils and human health studies. *Soil, 1*, 35–46. https://doi.org/10.5194/soil-1-35-2015

Brevik, E. C., Vaughan, K. L., Parikh, S. J., Dolliver, H., Lindbo, D., Steffan, J. J., Weindorf, D. C., McDaniel, P., Mbila, M., & Edinger-Marshall, S. (2018). Trends in undergraduate soil science education at selected universities in the USA from 2009-2013. *Soil Science Society of America Journal, 82*, 295–306. https://doi.org/10.2136/sssaj2017.10.0346

Brown, L. R. (2009). *Plan B 4.0: Mobilizing to Save Civilization*. W.W. Norton & Co. Inc. https://doi.org/10.1017/S174217051000027X

Critchley, W. R. S., Reij, C., & Willcocks, T. J. (1994). Indigenous soil and water conservation: A review of the state of knowledge and prospects for building on traditions. *Land Degradation and Development, 5*, 293–314. https://doi.org/10.1002/ldr.3400050406

Dan, J., & Koyumdjisky, H. (1979). *The classification of Israel soils* (Agricultural research organization, Institute of Soils and Water, soil conservation and drainage department, special publication No. 137). Division of Scientific Publications, The Volcani Center.

Dan, J., Yaalon, D. H., Koyumdjisky, H., & Raz, Z. (1976). *The soils of Israel (with Map 1:500,000)* (Ministry of Agriculture, Agricultural Research Organization, Institute of Soils and Water, Soil Conservation and Drainage Department 159). Division of Scientific Publications, The Volcani Center.

Díaz-Fierros Viqueira, F. (2015). What does the future hold for soil science? *Spanish Journal of Soil Science, 5*(1). https://doi.org/10.3232/SJSS.2015.V5.N1.05

Diochon, A., Basiliko, N., Krzic, M., Yates, T., Olson, E., Masse, J., Amiro, B., & Kumaragamage, D. (2017). Profiling undergraduate soil science education in Canada: Status and projected trends. *Canadian Journal of Soil Science, 97*, 122–132. https://doi.org/10.1139/CJSS-2016-0058

FAO (2012). *Global Soil Partnership*. http://www.fao.org/global-soil-partnership/overview/why-the-partnership/en/. Accessed 24 Aug 2019.

FCEE (2018). *The Faculty of Civil & Environmental Engineering Course Catalogue (in Hebrew)*. http://cee.faculty-ms.technion.ac.il/wp-content/uploads/sites/34/2016/11/01-Civil-2016.pdf. Accessed 15 Nov 2018.

Field, D. J., Yates, D., Koppi, A. J., McBratney, A. B., & Jarett, J. (2017). Framing a modern context of soil science learning and teaching. *Geoderma, 289*, 117–123. https://doi.org/10.1016/j.geoderma.2016.11.034

Fortuin, K. P. J. (2015). *Heuristic principles to teach and learn boundary crossing skills in environmental science education*. Ph.D. diss. Wageningen Univ., Wageningen.

Godwin, R., Spoor, G., Finney, B., Hann, M., & Davies, B. (2003). *The current status of soil and water management in England*. Royal Agricultural Society of England.

Greenberg, J., Weaver, T., Browning-Aiken, A., & Alexander, W. (2012). The neoliberal transformation of Mexico. In T. Weaver, J. Greenberg, W. Alexander, & A. Browning-Aiken (Eds.), *Neoliberalism and commodity production in Mexico* (pp. 1–31). University Press of Colorado.

Hartemink, A. E., Balks, M. R., Chen, Z. S., Drohan, P., Field, D. J., Krasilnikov, P., Lowe, D. J., Rabenhorst, M., Rees, K., Schad, P., Schipper, L. A., Sonneveld, M., & Walter, C. (2014). The joy of teaching soil science. *Geoderma, 217-218*, 1–9. https://doi.org/10.1016/j.geoderma.2013.10.016

Hartemink, A. E., McBratney, A., & Minasny, B. (2008). Trends in soil science education: Looking beyond the number of students. *Journal of Soil and Water Conservation, 63*(3), 76A–83A. https://doi.org/10.2489/jswc.63.3.76A

Havlin, J., Balster, N., Chapman, S., Ferris, D., Thompson, T., & Smith, T. (2010). Trends in soil science education and employment. *Soil Science Society of America Journal, 74*, 1429–1432. https://doi.org/10.2136/sssaj2010.0143

Huckleberry, G. (2015). Defining the environmental context of indigenous agriculture in the Southwest: What we don't know about middle to late Holocene climate change and floodplain dynamics. In S. E. Ingram & R. C. Hunt (Eds.), *Arid lands agriculture: Understanding the past for the future* (pp. 89–130). University of Arizona Press.

Instituto Brasileiro de Geografia e Estatística. Censo Agropecuário (2006). *Agricultura Familiar – Brasil, Grandes Regiões e Unidades da Federação*. Primeiros Resultados. Disponível em. https://www.ibge.gov.br/. Accessed 21 June 2020.

Intergovernmental Technical Panel on Soils (2015). *Status of the world's soil resources*. Food and Agriculture Organization. http://www.fao.org/3/a-i5199e.pdf. Accessed 20 Feb 2019.

Krzic, M., Wilson, J., Basiliko, N., Bedard-Haughn, A., Hymphreys, E., Dyanatkar, S., Hazlett, P., Strivelli, R., Crowley, C., & Dampier, L. (2014). Soil 4 youth: Charting new territory in Canadian high school soil science education. *Natural Science Education, 43*, 73–80. https://doi.org/10.4195/nse2013.11.0034

Krzic, M., Yates, T., Diochon, A., Swalow, M., Basiliko, N., & Pare, M. (2018). Introductory soils courses: A frontier of soil science education in Canada. *Canadian Journal of Soil Science, 98*, 343–356. https://doi.org/10.1139/cjss-2018-0006

Landa, E. R., & Feller, C. (2010). *Soil and culture*. Springer.

Lapham, M.H. (1949): Crisscross trails: Narrative of a soil surveyor. - Willis E. Berg Publishers.

Lavkulich, L. M. (2013). Soil science. *The Canadian Encyclopedia*. https://www.thecanadianencyclopedia.ca/en/article/soil-science. Accessed 12 February 2019.

Lobry de Bruyn, L., Jenkins, A., & Samson-Liebig, S. (2017). Lessons learnt: Sharing soil knowledge to improve land management and sustainable soil use. *Soil Science Society of America Journal, 81*, 427–438. https://doi.org/10.2136/sssaj2016.12.0403

Masse, J., Yates, T. T., Krzic, M., Unc, A., Chen, C., Hodgson, K., Quideau, S., & Warren, J. (2019). Identifying learning outcomes for a Canadian pedology field school: Addressing the gap between new graduates' skills and the needs of the current job market. *Canadian Journal of Soil Science, 99*(4), 458–471. https://doi.org/10.1139/CJSS-2019-0040

McBratney, A., Field, D. J., & Koch, A. (2014). The dimensions of soil security. *Geoderma, 213*, 203–213. https://doi.org/10.1016/j.geoderma.2013.08.013

McCallister, D. L., Lee, D. J., & Mason, S. C. (2005). Student numbers in agronomy and crop science in the United States: History, current status, and possible actions. *NACTA Journal, 49*, 24–29.

McKenna, J. R., & Brann, D. E. (1992). Enhancement of recruiting activities to attract rural youth to careers in agronomy. *Journal of Natural Resources and Life Science Education, 21*, 84–86. https://doi.org/10.2134/jnrlse.1992.0084

NERC (2019). *Advanced training short courses 2014–2019: dirt science: An introduction to soil systems*. https://nerc.ukri.org/funding/available/postgrad/advanced/atsc/. Accessed 9 Sept 2019.

NERC (2012). *Most wanted II. Postgraduate and professional skills needs in the environment sector*. LWEC NERC Report. https://nerc.ukri.org/skills/postgrad/policy/skillsreview/2012/most-wanted2-report/. Accessed 9 Sept 2019.

OPM (2018). *Classification & qualifications, general schedule qualification standards, soil science series, 0470*. https://www.opm.gov/policy-data-oversight/classification-qualifications/general-schedule-qualification-standards/0400/soil-science-series-0470/. Accessed 28 Oct 2018.

Pepper, I. L. (2000). Environmental science: A new opportunity for soil science. *Soil Science, 165*, 41–46.

Reyes-Sánchez, L. B., Irazoque-Reyes, L. A., Barrón, J. M., Espinoza, R., & Ruvalcaba, R. M. (2006). *The Mexican Bajío, yesterday the grain growing country, today degraded soils, polluted waters and human poverty*. Pedoconflicts Workshop TORBA Sols & Sociétés IUSS, in the human development Report-2006. http://hdr.undp.org/sites/default/files/reports/267/hdr06-complete.pdf. Accessed 24 Aug 2019.

RHSFAFE (2018). *The Robert H. Smith Faculty of Agriculture, food and environment course Catalog (in Hebrew)*. http://departments.agri.huji.ac.il/shnaton/searth.html. Accessed 15 Nov 2018.

Sharik, T. L., Lilieholm, R. J., & Richardson, W. W. (2012). *Factors influencing undergraduate enrollment trends in natural resources*. Presentation at 9th Biennial Conference on University Education in Natural Resources, Fort Collins, CO. https://digitalcommons.usu.edu/cuenr/9thBiennial/Keynote/1/. Accessed 7 Nov 2018.

STARS (2019). *Soils training and research studentships*. http://wp.lancs.ac.uk/stars/. Accessed 9 Sept 2019.

Terribile, F., Basile, A., Bonfante, A., Carbone, A., Colombo, C., Langella, G., Iamarino, M., Manna, P., Minieri, L., & Vingiani, S. (2013). Future soil issues. In E. Costantini & C. Dazzi (Eds.), *The soils of Italy* (pp. 303–348). Springer. https://doi.org/10.1007/978-94-007-5642-7_11

Warman, A. (2003). La reforma agraria mexicana, una visión de largo plazo. *Journal of Land Reform, Land Settlement and Cooperatives, 39*, 85–94.

Webster's (1996). *Webster's new universal unabridged dictionary*. Barnes & Noble Books.

Chapter 25
Soil Proverbs: Cultural Language of the Soil

J. E. Yang, H. S. Kim, P. Borrelli, and M. B. Kirkham

25.1 Introduction

Soil is the basis of human life and necessary for the production of food and the maintenance of ecosystems. Soil and people are elaborately interwoven. Thus, soil has been closely connected to the culture and civilization of people living in a given place or country, including their religion, philosophy, life style, and health (Minami, 2009).

Proverbs are concise phrases that give advice or state a truism to people and that are handed over from generation to generation. Proverbs describe the accepted truths or traditions about how people act and live (Hou, 2013; Mieder, 2004; White, 1987). There are numerous proverbs available globally that are deep rooted in the soil. The scope of proverbs includes knowledge, social norms, religion, family values, and human life and welfare (Yang et al., 2018). Proverbs are used to transmit knowledge over long periods of time.

Schuster (1998) mentioned that studying proverbs can help people understand the similarities and differences of other cultures compared to our own. The proverbs not only have a didactic function to support one's arguments, educate younger generations, or enrich daily conversations, but they also contribute to the cultures of all nations despite the regional diversities.

Because soil has played a key role in the birth of culture globally, many soil scientists and stakeholders deal with the links between soil and culture. This approach develops soil awareness programs that are intended to enhance the importance of

J. E. Yang (✉) · H. S. Kim · P. Borrelli
Department of Biological Environment, Kangwon National University, Chuncheon, Republic of Korea
e-mail: yangjay@kangwon.ac.kr

M. B. Kirkham
Department of Agronomy, Kansas State University, Manhattan, KS, USA

N. Patzel et al. (eds.), *Cultural Understanding of Soils*,
https://doi.org/10.1007/978-3-031-13169-1_25

soil for the general public. Soil science is an undervalued science, and soil is an undervalued resource (Landa & Feller, 2009). Various activities, such as exhibitions, paintings, field trips, camps, and video clips, have been developed to link soil with literature, paintings, songs, and rituals. Many global initiatives, such as the United Nations (UN) World Soil Day, the UN International Year of Soil 2015, and the International Union of Soil Sciences (IUSS) International Decade of Soil, are typical examples of soil awareness activities. Professional societies, for example, the Soil Science Society of America with its program "soils4teachers" (https://www.soils4teachers.org/soils-and-culture), provide educational material to enhance the importance of soil by linking it with culture.

We published a book entitled "Global Soil Proverbs: Cultural Language of the Soil" (2018). This book lists over 600 soil-related proverbs presented in 32 chapters from 29 countries on six continents. The book was intended to collect soil proverbs with pertinent illustrations and, through them, share insights about the philosophy, culture, and life in each country, as they relate to soils.

The objective of this chapter is to introduce soil proverbs that are relevant to culture by categorizing them into scientific principles, social values and statements, human well-being and ecosystem, and religion and spirituality, which we did similarly in "Global Soil Proverbs" (Yang et al., 2018). The soil proverbs cited come from that book. Readers may refer to it in order to understand specifically the meaning of each soil proverb and to see it in its original language. The selected proverbs are mostly intercultural. A person reading them in any country can understand their meaning. Many of the proverbs cited in this chapter express the same idea yet in a country's own unique way of stating it.

25.2 Soil Proverbs Relating to the Common Elements of Culture

A. **Scientific Principles of Soil**

Proverbs have been employed as tools for education. Soil proverbs provide us with scientific principles of soil, and they correspond to different disciplines of soil science such as physics, chemistry, biology, fertility, management, conservation, and land use. There are more proverbs related to soil fertility and soil management than those in other sub-disciplines indicating that the importance of fertile soils and well managed soils for productivity has been understood for a long time in many countries. Below are lists of soil proverbs related to soil science disciplines.

Soil becomes hard after the rains (Korea)
There isn't a cultivator/soil-worker at all as good as the ant (Ireland)
Adding sand treats cracks of land (Iran)
Without manure the soil very soon will lose its fertility (Moldova)
Who plows in wet soil, plows only half (France)
Break the clods (to prepare a fine seed-bed), so that you get bread (Pakistan)

Sand over clay, money thrown away (USA)
Manure is not holy, but where it falls it makes miracles (Spain)
If you treat soil bad, soil will treat you bad (China)
We've golden soil and wealth for toil (Australia)
When you have to hoe on hard soil it is a bad day (Italy)
Whoever wants to know about the soil should ask rodents (Nigeria)
Crop without manure is as worthless, as is a flower without scent (India)
Marl makes rich fathers and poor sons (Germany)

B. **Social Values and Statements Related to Soil**

Agricultural societies have created cultures that are deeply built on social values and relations, such as family issues, marriage, being a neighbor, friendships, and traditions. Soil proverbs encompass such values in their succinct and pithy expressions. Below is a list of soil proverbs that are closely related with the social values and statements related to soil. Through these proverbs, we can see that some proverbs from other countries may sound familiar to another, that is, soil proverbs are intercultural. It is common for countries to have their own versions of a proverb that express a given social relationship. Proverbs from remote antiquity, when no communication tools were available, share common truths and advice that we recognize today. Through these proverbs, our ancestors shared ideal values that people are expected to follow.

Do not leave the land and do not sell it (Egypt)
Difficult to leave the land of hometown (China)
Sad as if one has sold one's land (Lithuania)
The causes of war are land and women (New Zealand)
"Soily" hands beget oily mouth (Nigeria)
Who licks the plate, has to lick the ground (Italy)
In this land, the person who does not work, does not eat (Spain)
You reap what you sow (Turkey)
What we do to the land, we do to ourselves (USA)
Live by the law, die by the soil (Malaysia)
Embrace soil to have gold (Turkey)
The earth feeds only those who love her, who respect her, who maintain her power, who are faithful husbands their whole lives (Moldova)

C. **Importance of Soil to Human Well-being and the Ecosystem**

Life cycles, which include the processes of birth, living, health, and wealth, have been the central theme of cultures. Soil proverbs from around the globe express such processes figuratively and deliver insights about how soils are the basis for life, health, wealth, and well-being. Ecosystem services and sustainability have been a major issue in the formation of culture, because they are the basis of human life and health. The benefits that soil can provide to people are known as soil ecosystem services. They consist of supporting, provisioning, regulating, and cultural services of the soil (Millennium Ecosystem Assessment, 2005). These services influence human security, which is based on food, life, health, and social relations, and these, in turn, are constituents of well-being. People in ancient times in different countries had the same philosophy and knowledge about soils' ecosystem services and sustainability that we do today. Proverbs that relay positive valuations of soil and health information to the modern era are found below.

Healthy soils healthy nation (India)
The land owns everybody (Nigeria)
The properties of the soil you are born on decide your fate, riches, honor, and wealth (Taiwan)
The land belongs to whomever labors upon it (Mexico)
Our farms should function as ecosystems. We should farm closer to nature's design (Australia)
Land is wealth (Uganda)
One seed in the soil turns out thousands and thousands of grains in the barn (China)
The soil is generous, not stingy (Iran)
The value of a man is equal to the value of his land; a stingy farmer is never rich (Italy)
Better have a handful of soil than a handful of gold (Turkey)
So is the man, so is the land (France)
What the earth gives, the earth also takes (Lithuania)
Take only what you need and leave the land as you found it (USA)
Though there be an abundance of treasures, they will vanish, but the land is immovable (New Zealand)

D. **Soil in Religion and Spirituality**

Every society has its own "spirits of the land." Although largely diverse spirits are global in nature, they commonly connect people's thinking and beliefs to the power of soil (land) to produce food, which, in turn, necessitate management and conservation of the soil. This consciousness of soil and spirits, itself, has been merged into concepts that are expressed in mythology, religion, and philosophy, and it continues its existence in our daily lives (Minami, 2009). Szulczewski (2015) stated that the importance of agriculture and fertile soil is evident starting with ancient religions, as revealed in the scriptures of many religions such as Buddhism, Hinduism, Judaism, Christianity, and Islam. Also, these concepts are evidenced in the birth of civilizations, such as the Egyptian, Greek, and Roman civilizations. Every society has recognized the crucial role of soil for survival, from birth, through life, and even in death. People have expressed this recognition in proverbs for a long time, as shown in the following examples.

As we walk upon the earth, let us be mindful...She is a sacred living being (USA)
All were borne from soil and when dead will be part of soil (India)
You are soil, from soil you eat, in soil you will come back; The earth is the mother of all men and even its burial (Italy)
Be friends with the earth (soil): from soil we came; the soil feeds us; to the soil we will return (Moldova)

I am the land, the land is me (New Zealand)
From dust we came and to dust we shall return (Nigeria)
The lighthouse of Alexandria has fallen down. He said, God delivers us from its dust (Egypt)
Each soil has its own fortune (Pakistan)
The soil is the great connector of lives, the source and destination of all (USA)
We are made from Mother Earth and we go back to Mother Earth (USA)
Body and land are one, not two (Korea)

25.3 Linkage of Soil Proverbs to Culture: Intercultural or Specific

Soil proverbs and culture have common characteristics that bridge them, proving proverbs to be part of culture. They are learned, shared, transmitted, and changeable. Soil proverbs are culture-loaded being expressed with concise language and are regarded as one of the most ancient and valuable manifestations of human culture. They express commonly held ideas and beliefs, or provide us with interesting glimpses at a people's geography, religion, history, social views, and attitudes: (Fig. 25.1) (Mieder, 2004; Schuster, 1998; Yang et al., 2018). This determines whether proverbs are part of one culture only, or intercultural.

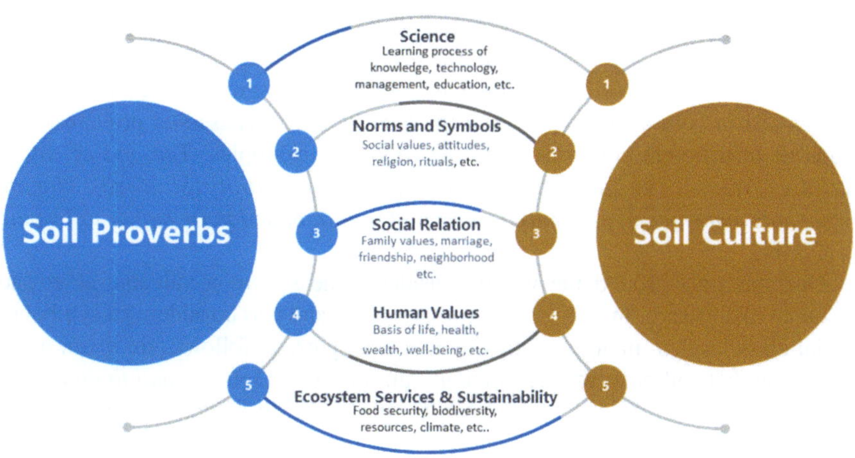

Fig. 25.1 Common characteristics bridging soil proverbs and soil culture

Lessons from the global soil proverb archives (Yang et al., 2018) revealed that most of the soil proverbs are more intercultural than specific. A person reading them in any country can understand their meaning. As noted, many of the proverbs cited in this chapter express the same idea yet in a country's own unique way of stating it.

Soil proverbs are intercultural, and in addition, they can help people understand a culture and cultural diversity. They are just like a mirror that will reflect certain properties or diversities of cultures (Hou, 2013; Schuster, 1998). In ancient times, agriculture was a main driving force to provide goods to people and sustain people's lives. Wherever people were, they shared a situation of living conditions, and thus, the respective cultures were born accordingly, even though they were subject to different locations and environment. Since the word "culture" was etymologically and figuratively used in the "agricultural" sense, the soil was improved by good husbandry, and the way of thinking and attitude toward soils was improved by soil proverbs, which were tools for educating people.

There are many ways to demonstrate that soil proverbs are multicultural, even though they originated from different countries and continents. The most abundant soil proverbs are related to soil fertility, fertilizers, and management, which are directly connected to food production. A soil proverb from Korea, *"A good farmer manages the soil, a mediocre farmer grows plants, and a poor farmer grows grasses,"* emphasizes the importance of soil fertility and management. This importance is stated in the same manner in different countries, as follows: *Managed well, soil circulates chemical elements, water, and energy for great human benefit. Managed poorly, it is impossible to imagine an optimistic future (USA); Who hoe, produce wine (France); Dig deep while sluggards sleep and you will have corn to sell and to keep (Ireland); The land is defended by its owners (Egypt); Mixing sand soil with clayey soil makes profits (China); The more you plough the land, the more you get out of it and the more you eat (Pakistan); Plough deep and reap rich (Ireland);* and *Plow very deeply, you will gather abundantly (France).*

The importance of soil organic matter cannot be overemphasized and has been recognized in both ancient times and the modern era, because soil proverbs have stated it in different countries as in the following examples: *The god of soil is manure (France); The darker the soil, the higher is soil fertility (Moldova); Manure is not holy, but where it falls makes miracles (Spain);* and *Manure makes soil young (Pakistan).*

Our ancestors around the world considered the soil as a valuable asset and resource to conserve. In some cases, owning good land was considered a symbol of social-class level in many societies. A few examples, as follows, are given as an indication that soil proverbs are intercultural: *An inch of soil is equal to an inch of gold (Taiwan); An inch of land is worth its weight in gold (Vietnam); Better have a handful of soil than a handful of gold (Turkey); Sand over clay, money thrown away (USA); "Soily" hands beget oily mouth (Nigeria); A shō (1.8 L) of soil is the same as a shō of gold (Japan); Body and land are one, not* two (Korea); and *The more you give to the land, the more it gives back (France).*

In nearly every culture, to speak in a religious way, people have considered that life originated from soil and returned to soil after the end of life. Thus, death has been a key, common component of the cultures. The following soil proverbs describe interculturally the death of people but in different ways of expressions: *From dust we came and to dust we shall return (Nigeria); One's remains decompose in the soil, but one's good deeds are remembered (Malaysia); The man rests in the soil (Italy); Let man die and become like soil (New Zealand);* and *Be friends with the earth (soil): from soil we came; the soil feeds us; to the soil we will return (Moldova).*

Soil was regarded as the power of a country to secure food for its people. For this reason, war was often inevitable to extend lands. Soil proverbs describe war, as given in the following examples: *Who buys land, very often buys war (Italy); Who has land, has war (Italy); To gain ground, to lose ground (Austria); He who has no root (land) will die in conquest of it (Egypt);* and *The causes of war are land and women (New Zealand).*

Many soil proverbs are intercultural, but at the same time, many are specific to a certain country. Proverbs of a particular culture reflect to some extent the values of that specific culture (White, 1987). Many of Asia's countries under the monsoon climatic condition have developed their own cultures related to rice, which is different from other countries where rice is not the staple crop. Such examples are demonstrated in the following soil proverbs: *Rotation from paddy soil into upland soil will yield out more for three years (China); Ten rice bags of husk rice can be produced by supplying fertilizer, but producing more than 15 rice bags requires improving soil fertility (Japan); A bowl of rice can be provided free of charge to others, but not a basket of compost (Korea);* These proverbs can be contrasted with those from countries where rice is not the staple crop: *Of wheat abundant harvest, wants in field manure that smell (France); The water that goes to the jujubier is more profitable for the olive tree (Tunisia); Sandy soil, a lot of grapes (Italy); White bread grows in black earth (Lithuania); After a warm winter, never expect a good yield of winter wheat (Moldova); The black soil good bread carries, the white soil as it reaches (Spain); You shall not put corn in a basket with holes (Mexico);* and *A rocky vineyard does not need a prayer, but a pick ax (USA).*

The following examples encompass specific soil proverbs. These proverbs seem to be inherent from different geographies, climates, religions, economies, histories, life styles, and customs that are important components of the specific culture. Readers may refer to the book "Global Soil Proverbs" (Yang et al., 2018) for their specific meanings.

If the water is available for ablution, the use of sand is discontinued (Egypt)
A farmer becomes rich if a shō (1.8 L) of soil accumulates in his bed (Japan)
Each type of soil has its own special natural characteristics (Taiwan)
Marl makes rich fathers and poor sons (Germany)
Marry a mountainy woman and you marry the whole mountain (Ireland)
The vineyard in harsh and arid soil will produce a lot of grapes (Italy)
Strong frost and deep snow lead to a productive year (Moldova)
Sit on the sand (The Netherlands)

When the fields do not produce, the saints do not have (Spain)
My reliable comrade is the soil (Turkey)
When our Land is truly free, the Quetzal will sing once again (Mexico)
Plowing the Back Forty; They were dirt poor; When you throw dirt, you lose ground; He bet the farm (USA)
We've Golden Soil and Wealth for Toil (Australia)
A warrior who works hard at growing food will not fail (New Zealand)
If you want to store wheat and barley, plow fertile lands and lowland areas (Tunisia)
Soil types determine the human health and their characteristics; One cannot adapt to a certain local climate condition or eating habits (China)

25.4 Summary

The objective of this chapter was to introduce soil proverbs that are relevant to culture by categorizing them into four topics: scientific principles of soil, social values and statements, human well-being and the ecosystem, and religion and spirituality. The proverbs cited come from the book "Global Soil Proverbs" (Yang et al., 2018). This book compiles more than 600 soil proverbs from 29 different countries and shares insights about the philosophy, culture, and life in each country, as they pertain to soils. The proverbs that relate to scientific principles are mainly concerned with soil fertility and management. It has been known from antiquity that productivity depends upon fertilization of the soil and good management of the soil. The proverbs that relay social values express the need for hard work to maintain the soil, which can be epitomized by the proverb, "You reap what you sow." The proverbs that consider social relations show the importance of soil for one's identity and for marrying into a family with good soils. The soil proverbs that cover human health and well-being confirm that "Land is wealth." Finally, the soil proverbs that cover ecosystem services and sustainability can be summarized by the North American proverb, "All the capital in all the banks cannot substitute for the soil of the land." The soil proverbs quoted in this chapter encapsulate millennia of cultural understanding of the soil.

Acknowledgments This work was supported by Korea Ministry of Environment, with the strategic EcoSSSoil Project, KEITI (Korea Environmental Industry and Technology Institute), Korea (Grant No. 2019002820004).

References

Hou, R. (2013). Proverbs reveal culture diversity. *CrossCultural Communication, 9*(2), 31–35.
Landa, E. R., & Feller, C. (2009). *Soil and culture*. Springer GmbH. https://doi.org/10.1007/978-90-481-2960-7
Mieder, W. (2004). *Proverbs: A handbook*. Greenwood Press.

Millennium Ecosystem Assessment (2005). *Ecosystems and human well-being. A framework for assessment*. www.millenniumassessment.org

Minami, K. (2009). Soil and humanity: Culture, civilization, livelihood and health. *Soil Science and Plant Nutrition, 55*, 603–615. https://doi.org/10.1111/j.1747-0765.2009.00401.x

Schuster, E. (1998). Proverbs: A path to understanding different cultures. *Journal of Extension, 36*(1), Tools of the Trade, 1TOT2.

Szulczewski, M. (2015). Soil's social and cultural connections. *Soil Horizons*, 1–3. https://doi.org/10.2136/sh2015-56-6-gc

White, G. (1987). Proverbs and cultural models. In D. Holland & N. Quinn (Eds.), *Cultural models in language and thought* (pp. 151–172). Cambridge University Press. https://doi.org/10.1017/CBO9780511607660.007

Yang, J. E., Kirkham, M. B., Lal, R., & Huber, S. (Eds.). (2018). *Global soil proverbs: Cultural language of the soil*. Catena Soil Science, an imprint of Schweizerbart Science Publishers.

Part VII
Concluding Remarks and Outlook

Chapter 26
Summary and Conclusions: Understanding Soil-Cultural Relations

Nikola Patzel, Sabine Grunwald, Eric C. Brevik, and Christian Feller

26.1 Diversity of Views

In the introduction of the book, we stated that both soil and culture are phenomena that can be viewed through many different lenses, and concepts are constructed using various perspectives. There are many different perceptions and representations of soil and earth, land, and nature in different societies. This was expanded in Part II, where 13 chapters explored how soil was culturally perceived and related to, from African, Asian, European, North American, and South American viewpoints. African, ancient European, Buddhist, Confucian, Hinduist, Islamic, Judaeo-Christian, and Shintoist religious perspectives were discussed, along with northern-hemispheric Shamanistic ones. Furthermore, Western urban perspectives were introduced, and a history of ideas of the now-globalized Western scientific views of soil was presented.

N. Patzel (✉)
Freelance Scientist, Office for Soil Communications, Überlingen, Germany
e-mail: nikola.patzel@soilcom.com

S. Grunwald
Soil, Water, and Ecosystem Sciences Department, University of Florida, Gainesville, FL, USA
e-mail: sabgru@ufl.edu

E. C. Brevik
College of Agricultural, Life, and Physical Sciences, Southern Illinois University, Carbondale, IL, USA
e-mail: eric.brevik@siu.edu

C. Feller
UMR Eco & Sols, Institut de Recherche pour le Développement, Montpellier, France
e-mail: christian.feller@ird.fr

N. Patzel et al. (eds.), *Cultural Understanding of Soils*,
https://doi.org/10.1007/978-3-031-13169-1_26

Part III explored some of the interplay between soil and sociopolitics. Part IV looked at two influential figures in the history of Western science, Palissy and Darwin, and explored their inner soil and how it may have influenced their illustrious careers. Part V investigated soil care concepts from multiple perspectives, proposed a foundation for soil ethics, and conceptually provided more explanation of the outer versus inner soil. Finally, Part VI focused on the place of soil in the educational systems of different countries and how societies around the world have used proverbs to communicate important ideas about soil.

The views, perspectives, personal and cultural backgrounds of the authors are extremely diverse and influenced their interpretations of the people and cultures they wrote about. This book project of the working group Cultural Patterns of Soil Understanding of the International Union of Soil Sciences (IUSS) did not seek to rank one view versus another, assigning levels of "right" and "wrong" or "better" and "worse" to them. Rather, it sought to celebrate the vast array of experiences and viewpoints that exist in our world, where each one has value and meaning to those who hold them and can provide insights to everybody who is receptive to exploring their inner depth and breath and the sociocultural-political implications that arise out of them.

Soil and culture look inseparable and deeply intertwined. Human origins are often traced back to the soil. There is a duality view of soils either as an object to be studied: measured and described (contemporary soil science education) or as a lived experienced subjective reality of humans. This distinction does not yet touch the question of whether the soil is an acting and relational subject on its own (as seen in Chap. 10). These distinctions are sometimes blurred or become one and the same (soil as subject-object). Soils are viewed through different individual and socially constructed lenses.

26.2 Soil and Earth: Fertile Grounds for the Creation of Humanity?

The cultural elements linked to soil that were reviewed in our case studies show the importance of soil in numerous creation myths of humanity in the major religious systems as well as in various indigenous and folk religions. Humans emerging from inside the earth or from deep waters of the ground or being formed out of animated soil material are the main features of many myths throughout the world.

Looking at ancient Europe (Chap. 2), one sees common structures in how soil and deep earth is part of the worldviews of polytheistic societies: Finno-Ugric, Balto-Slavic, Celtic and Germanic, Greco-Roman, and syncretic cultures, as well as in the somewhat animist and polytheist folk religious undercurrents in monotheistically dominated societies (Chap. 3). They all contain ideas of circularity in life that re-emerges from soil, with the essential help of female divine will. They show elements of the union of divine opposites to bring forth soil fertility. In places, and most prominently in the Balto-Slavic culture, there is even a symbolism of the female godhead to be the "gatekeeper" of all sources and pores of land and soil. In Greek mythology, humans descended from creatures of Gaia, the primordial Mother Earth. In one version of these myths of anthropogenesis, the mediating deity Prometheus, the one who later brought fire and craft skills to the human realm, formed humans (with Zeus' consent) from clay and water.

A birth-like emergence of land from the primordial ocean, beginning with the first raised mud then becoming a solid earth and soil, is described in the "earth-diver" pattern of cosmogony: "The earth-diver creation myth type can be found in many parts of the world, but is particularly important in Central Asia, India, and Native North America" (Leeming, 2010, p. 24). In Eurasian shamanism, such as among the Buryats (Chap. 5), the earth's origin is explained by an Earth diver myth. This type of myth is also common among Native American tribes (Chap. 9), although emergence myths are also noted. In these, humanity emerges, usually assisted by an Earth goddess, from their former dwelling place in the underworld, such as through a hole in the ground or a cave.

Chapter 8 is devoted to the Dogon in West Africa. They believe the god Amma, who had created Earth, copulated with it, resulting in the birth of a primordial pair of twins in human form. Then referring to a later state of human development, their myth describes humans becoming farmers like a "second creation," which at the same time is a revelation for great cultural change.

Chapter 10 is devoted to a Brazilian religious system originating from West Africa. This system, called Ifa, comes from the Yoruba myths of Benin and Nigeria. They believe that humanity is created from mud that the Queen of the dead, Nanã, lady of the wetland and stagnant waters, provided to Oxalá (the Creator of the World), and soil itself is the Goddess Onilé.

In ancient China (Chap. 13), the world creation myth is based on a cosmic egg. The goddess Nüwa is a creator figure in Chinese myths. She and her brother Fuxi are depicted with a human head on a snake body. Out of longing for company, Nüwa successively created humans in two episodes: first, by shaping yellow soil with her hands into creatures similar to herself, and then, by dipping a rope or cord in the mud and tossing it up to create human beings. They became the noble and the lower classes, respectively.

These few examples show the importance of earth/soil in the creation myths of humanity. It can also be noted that the examples in this book cover most of the global diversity of myths reported by Leeming (2010) in his Encyclopedia of Creation Myths of the World. Based on the work of Charles Long (1963), Leeming (2010) classified creation myths into five types: creation from nothing—*ex nihilo*—(e.g., in monotheism), from chaos from a primordial matter (including cosmic eggs, e.g., ancient China), from world parents (e.g., Dogon), from emergence, or from an earth diver (e.g., Native Americans). According to Leeming (2010), the *ex nihilo* type is mainly linked with either a figure of God the (heavenly) Father or just a non-defined ultimate divine, while emergence myths are mainly associated with Mother Earth.

These mythical images are often deeply buried in our psyche, and they can repeatedly emerge from the unconscious. It is then important to consider them for personal awareness and in communication between scientific and other perspectives and perceptions. Indeed, scientific perspectives themselves are partly inspired by mythical ideas and ideologemes. In Chap. 4, for example, it is shown how images associated with earth/land such as Mother Earth, the living soils with vital force and health, of circular renewal of organic material are linked in various ways to the founding myths of humanity discussed in this book. Rather philosophical (concealed mythical) ideas like the basic elements of the world (e.g., nutrient paradigm) or the natural body of soil being a kind of meta-organism have been proposed. Humans may be influenced by myths and their symbolic content in multifaceted and paradoxical ways. In stating this, it is important to note that while the term myth has often become associated with something that is not true in modern societies, any identification of "myth" with "lie" is just a reiteration of a classical rationalist impulse of "enlightenment," but this is not necessarily the case. Myths tell a basic story of humanity's and all the world's existence, mainly in a symbolic language forming an essential web of meaning. They are symbolic narratives that are part of a culture's mental or psycho-spiritual base and provide storylines about how we came into being and our belonging and deepest longings.

26.3 Psycho-Spiritual Motivations That Inform Relations with Soils

Soil is something that we can see, touch, smell, and feel using our senses. For scientists, soil is part of the physical world, having properties we can measure, quantify, and qualify. And yet, there is also a psycho-spiritual aspect to the human relationship with soil that comes out in many chapters within this book. Chapters 2, 3 and 4, which look at cultural patterns of soil cultivation in Europe, show how psycho-spiritual attitudes toward soil and earth changed over time in Europe, from a view where soil fertility was dependent on a feminine spiritual power (or a holy wedding in earth; Chap. 2) to a view of soils where humans are empowered by God to dominate the natural environment. This latter view has been the predominant Christian attitude that remained after the secular movement of rationalist "Enlightenment" (Chaps. 3 and 4).

Some societies continue to this day with a remarkably spiritual view of soil. Native Americans (Chap. 9) view themselves as being part of nature. Therefore, even when a Christian cares for God's creation, including soil, through environmentally responsible actions, it is not the same as a Native American following traditional belief caring for nature (and soil) because that nature (soil) is a part of them. It is the difference between caring for a highly valued possession versus caring for yourself.

Likewise, the Buryats and Soyots of Siberia (Chap. 16) consider digging into soil to be an intrusion into a deity, an action not to be taken lightly. This religious belief has influenced their relation with the land the Buryats and Soyots and their ancestors live on but again for reasons that are different from those of a Christian or Native American. Precolonial Indian-origin ideas of the sacredness and divinity of soil have influenced modern politics in West Bengal to the point that they have had a leading role in political change (Chap. 15).

In Buddhism, psychological and spiritual views differ among traditions and schools (Chaps. 11, 12 and 16 from Southern Siberia). For example, the *Theravāda* Buddhist school focuses on a deep meditative absorption (*samadhi*) practice that is considered to liberate the individual self, while concerns for nature, soil, and the environment are of lesser priority. In contrast, in *Mahāyāna* Buddhism, the focus is on the *bodhisattva* ideal embodying compassion as well as other prosocial emotions (perfections) and *bodhicitta* (awakened mind and heart) to liberate both oneself and all sentient beings. The *bodhisattva*-inspired ecosattva embodies both—meditative inner knowing of ultimate truth (immersion) and outer directed compassion for soil, the Earth, and sentience.

Through examples such as these, the chapters in this book clearly demonstrate that soils have multiple meanings other than just their physical manifestations and the ecosystem services they provide to us. For some cultures, and people all over the world, soils truly have a spiritual existence and relevance.

26.4 The Influence of Sociopolitical Structures on Soil-Human Relations

Notions of soils have been constructed or adopted in various cultures and times that define ownership, property, usage, or exploitation rights, which influence and express how people relate to soils. For example, Chap. 16 explains how the Russian federal government ordered the privatization of all Soviet collective farm assets and the division of agricultural and pastoral lands into individual parcels in the 1990s as part of an agrarian reform. In Buryat and Soyot cultures in Russia, such a shift toward private property ownership challenged the spiritual relations of local people that were grounded in Shamanistic (and Buddhist) religious beliefs that stress oneness and wholeness with land and a general taboo on tilling or other disturbance of the skin of the earth (soil). Land and social conflicts resulted. Animistic, pantheistic, or panentheistic views stress spirit and the divine rather than people's material possession of the land. Examples are found in Shamanistic mythology in Northern Eurasia (Chap. 5 on "Shamanism Mythology"), emptiness and nondual views in Buddhism and Shintoism (Chaps. 11 and 21), Chinese Daoism of the five elements and in the complementarity of Yin and Yang (Chap. 13), and spiritual indigenous views in North America (Chap. 9).

Historically, spiritual-religious motivated views relating to soil/land/nature as sacred were pronounced in various cultures, while some of them have been marginalized and replaced by the ideological legitimation of exploitative soil use and management (e.g., certain types of agricultural food production; Chap. 4). Agricultural stewardship to crop and farm soils may inherently imply ownership of soils to serve as "good stewards" of the planet. In Chap. 21, Grunwald stressed the dialectic opposition between stewardship as care for nature (care ethics) and the "myth of the garden" where nature is dominated and exploited by man, the human master "on behalf of the One God," prominently found in Abrahamic traditions. Not only in Islam (Chap. 6; see also Chap. 7 on Rumi) but also in Judaism and Christianity (Chap. 3), there is the explicit and theologically important distinction between God being the "only land owner" and the human landlords, stewards, or "Caliphs" to exert usage rights, being responsible lieges of the supreme Lord.

According to Grunwald and Kastner-Wilcox (Chap. 20), cultural identities are associated with specific moral viewpoints toward soil/land/nature—certain environmental ethics. These ecological or eco-identities were associated by Esbjörn-Hargens and Zimmerman (2009) to mirror developmental levels of human consciousness. The authors argued that the commodification of land and soil through property rights and ownership is associated with objectivism and rationalism. Both are prominently found at eco-identity levels that can be identified with the terms "Eco-Warrior, Eco-Manager, and Eco-Strategist" representing soils usage and exploitation for human profit, soil resource management, and conservation management of soils and crops, respectively. These kinds of soil notions are pronounced in both capitalist and communist economies (Chap. 4) that differ profoundly from soil conceptions that are rooted in animistic-spiritual-mythical or holistic-spiritual-wisdom views that go beyond or are different than materialistic conceptions of soils. These nonmaterialistic notions point to spiritual, subtle energetic, sacred, or divine realms in which conceptions of ownership and property that are social constructs lose meaning (even and perhaps especially when they are presented as divine allocations). Interestingly, on the one hand, connecting to nonmaterialistic aspects of soils may evoke deep belonging and care for soils and nature, but on the other hand, spiritual and religious notions of soils may provoke escapism into an idealized "non-soil" or "beyond soil and matter" realm (Chap. 12). For example, Buddhist notions of form-formless dichotomies invite escapism into an inconceivable ultimate reality that promises individual liberation, while *bodhisattva* notions of deep care for other beings combined with *ecosattva* notions emanate deep care for the environment. (Chapter cited above). Historically, Christianity (more than in Islam) has shown strong elements of escapism from and even a turn against nature, where the soil was seen as "mere dust" and even the visible life of the soil was sometimes associated with demons rather than the power of life (Chap. 3 on Monotheism).

In Chap. 15, Banerjee analyzed the history of ideas and political conflicts about soil and the land in West Bengal, India. The author argued that human labor on soil is central to the generation of worldviews, which are inextricably political, socio-economic, and theological mixtures. The notion that labor is the main act of cultivation (and appropriation) and the root of wealth and property is considered transhistorical and transcultural; this is not just an Indian phenomenon. Labor has been strongly genderized, with male labor being valued far more than female labor in Indian society. However, it was the Goddess Shakti who invited the God Shiva to become a farmer with her, thereby domesticating wild masculinity into a relationship with her and with a plot of land, providing a role model for humans. In West Bengal, human labor on soil has not only produced notions of rights but also of social interdependence. Agrarian cooperation has helped generate political cooperation. Banerjee argued that it is indisputable that forms of divinity matter in how people view and relate to soils as shown by the peasant farmer who labors on soil mirroring the role model of gods as peasants from the Hindu perspective.

In Chap. 19, Krzywoszynska emphasized that how soils are framed within knowledge cultures, political discourses, and other arenas of life has consequences for how soil needs are experienced, conceptualized, and responded to. From a social science perspective, soil care means social soil governance, away from individual agents toward broader socio-material relations. Krzywoszynska emphasized that collective framings link attention to soils with societally valued ways of living off and with the land. The author argues that appropriate collective notions of soils imply that they are cared for and managed by society. This general concept of care grew out of feminist literature and broadly refers to looking after the interests or showing concern for the needs of someone or, in the case of soil, something. Concepts of soil care are also communicated in Chaps. 20 and 21, with each chapter calling on humanity to rethink its relationship with soils. These chapters argue that the commodification of soils has overshadowed soil care, which threatens the soil resource making it a spiritless material object that is used for the benefit of people (e.g., food production in the solely material sense).

Within Europe, urban soils and gardens have emphasized the notion of individualized garden plots such as *Kleingärten* (tiny gardens), while the institutionalization of urban gardens views their soil as an asset that provides green space increasing the sustainability and vitality of urban domains (Chap. 14).

26.5 The Feminine and the Masculine with Respect to Soils

The multicellular life of the earth is practically always a two-in-one of female and male expressions, which also exists in soil in a constant interplay of differentiation and unification. Humans biologically consist of males and females. Their symbolic worlds and cultural narratives show many anthropomorphic or zoomorphic images of God, with female or male characteristics and socioculturally or spiritually constructed gendered features.

Different dimensions of human soil relations and ways of viewing soil show that ideas or practices can have a more feminine or masculine hue: What are the characteristics of soil itself that are seen as more feminine or masculine? Some sexual and motherly associations with soil are culturally expressed, for example, by the

Slavonic "Wet One" Goddess, by the phalloi of the Greek God Dionysus rising from soil like plants, or by the symbol of the cornucopia for fruit abundance (Chap. 2).

Did soil originate from a feminine matrix, as implied by the German and French terms "Muttergestein" and "roche mère" ("mother rock") in soil science? Are soil layers the skin of Mother Earth, as believed by the Buryats (Chaps. 5 and 16), or was soil created by a masculine spirit that blew life into matter and vivified it as indicated in Judaism, Christianity, and Islam (Chaps. 3, 6 and 7)? How culturally effective are the sexual analogies in planting, hole digging, or plowing? Who identifies these interventions directly or indirectly with, for example, a "divine plowman" or with fertilizing human actors, and what is the understanding of their respective roles? Are composting activities linked with circular (nonlinear) processes in life, and how are these ideas culturally embedded (Chap. 4)? Sexual symbols in arable farming and those of a "holy wedding" in the fertile ground was and is historically and cross-culturally widespread. These symbols have been described, for example, by Frazer (1951), Mannhardt (1875), and Dieterich (1925). It is therefore not surprising that social roles and power relations between the sexes also play a very important role in soil understanding, in shaping of agriculture, and in dealing with the soil. Grunwald (Chap. 20) touched on the marginalization and oppression of women in patriarchal cultures that gender relations toward soils and nature expressed through care, or lack thereof. The case studies in this book suggest that the practice of soil relations, the cultural relations toward soil and land in general, and the social status and role models of the women and men who cultivate, work, and perhaps own the land are strongly related to ideas of masculine and feminine principles. One common intercultural image of a Goddess or feminine principle present and acting in soil is often called Mother Earth.

In Europe, the term Mother Earth was used, for example, in a Homeric hymn some 2600 years ago: "*I will sing of the earth, the mother of all, the firmly founded, the most ancient of all beings. She nourishes all creatures, ... From her abundance they all live. ... Hail, Mother Earth.*" All polytheistic and monotheistic religions that were or are practiced in Europe cherish a feminine principle, which may or may not be divine, inside soil, or soil being part of it (Chaps. 2, 3, 4, and 5). Many old religions around the globe viewed the feminine divine to be a quite dominant factor of soil fertility, a hub of the wheel of life, a force of death and resurgence. Worldwide, one may recognize similarities specifically considering mythical religious ideas of divine Mother Earth with Siberian Shamanic religions (Chaps. 5 and 16) or those of the Americas (Chap. 9). In the same vein, the soil Goddess of the West African Yoruba people incorporates feminine aspects of soils (Chap. 10). *Mahāyāna* Buddhism suggests that the outer Mother Earth is mirrored within oneself as the inner Mother in humans. Thich Nhat Hanh pointed out that the natural Earth Mother analogy is experienced by the biological, maternal, and compassionate relation between mother and child (for refs. see Chap. 12). Polarized, including genderized, images are also prominently found in Daoism where the whole is conceptualized as Yin-Yang expressing the dichotomous and complementary categories of female/male or heaven (sky)/earth (soil) (Chap. 13). Patterns such as Mother Earth can also reproduce themselves under rather adverse conditions. In Christianity, the feminine part of the God image partly returned with devotion to Mary, "Mother of God" (compensating somewhat for "God the Father"; Chap. 3). In some contemporary agricultural movements such as regenerative agriculture and organic farming, the motive of the great Mother Earth is significant in many different cultural contexts and under different names. In Islam, creation is often attributed a rather "feminine" role (Chap. 3 on Monotheism). When developing an Islamic eco-theology (Chap. 6) or a new Christian ethics of care, for example, it will be necessary to address anew questions of the place of the often rejected or suppressed feminine and female in human societies and nature relations. This might also change the organization of societies, their power structures, land use rights, and so on (see Sect. 26.4 above).

The masculine in soil appeared as a spirit of the deep below, like the fertile god of death Hades among the ancient Greeks or similar Balto-Slavic images. The masculine could also be seen in plants rising up out of the soil, which were compared, for example, with the phallusses of the Greek God Dionysos. The masculine was perceived in plowing, sowing, and cultural heroes of agriculture such as Triptolemus, a messenger of cultural knowledge sent by Demeter (Chap. 2 on European polytheism and soil). Banerjee (Chap. 15) elaborated on the strong symbolism of plowing in West Bengal. In India, plowing a cross-like grid pattern generally symbolizes a marriage between heaven and earth, male and female Gods, and a vivifying of the Naga-snake. The motive of the snake occurs in many of our case studies: being an attribute of the goddess or symbolizing the chthonic male godhead (Chap. 8). The snake image is found all over the world linked with the fertility of soil and nature, with death and renewal, and even in the semantic field of the earthworm who pleasantly intrigued Charles Darwin on his way to the "Origin of Species" (Chap. 18). In Christianity and its popular tradition, a masculine image of God is linked to the soil in narratives of Christ's birth, state of death and resurrection from below.

Whether the appearance of God images with feminine or masculine attributes in myths reflects the spiritual or the spiritual reflects the social does not need to be decided here. Importantly, soil science and the human-soil relationship are part of inner and outer structures and forces in which the respective manifestations of the feminine and the masculine are at work. The masculine dominance in the religious wordviews of the Abrahamic religions of Judaism, Christianity, and Islam is partly compensated by, for example, the image of God-bearer Mary as the fertile soil. But huge questions are left with these traditions about the status of the feminine in respect to its divinity, relation with nature and status in societies. The conceptualized ideas of balance and harmony between feminine and masculine spiritual/divine principles in nature has had tantalizing overtones from a cultural perspective. The human struggle to strive for ideals of cultural and societal balance is expressed in concepts such as equality, equity, diversity, inclusivity, human and civil rights, and overcoming power structures that harm soils and people.

In contrast, the personification of the feminine in heaven, on earth, and in soil is splendidly differentiated in India (Chap. 15). According to ancient Chinese beliefs, the Yin-Yang principle focuses on fundamental qualities just inherent everywhere, and the symbolism of earth and soil including its fertility can be found strongly in the Chinese Jijing (Chap. 13).

The ethics of human nature relations and soil ethics are soaked with masculine and feminine features. The current concept of care ethics is closely tied to feminine nurturing conceptions of relating with nature. This ethics of care and reciprocity contrasts with a classical patriarchal ethics of responsibility in the Abrahamic religions, a stewardship ethics that is oriented on feudal structures that views the rulers of the land as the ruling landlords of the Creator God (Chap. 21).

Idealized concepts of feminine expressions in a society may be arguably at odds with patriarchal, hierarchical, fundamentalist, and colonial cultures and societal power structures that have dominated around the globe. Such systemic cultural strands are characterized by inequalities, suppression, oppression, and marginalization of feminine expressions, and female gender that are found cross-culturally.

26.6 Concepts for Protection and Preservation of Soils for Future Generations

There are many different kinds of human relations with soils described in the chapters of this book, and they directly and indirectly influence the way the soil is approached and managed. On one side of the spectrum, soil is viewed as sacred. For example, in the shamanistic communities of Siberia, soil is considered sacred, and one does not disturb it lightly (Chaps. 5 and 16). The view of soil as a deity contrasts sharply with modern Western scientific views of soil and growth economies as something to be manipulated and engineered, a view that has often led to soil degradation (Chap. 19).

Soil quality and soil health are concepts that are often associated with agriculture but which in practice extend beyond the agricultural context. In particular, these are concepts that go beyond the historical concept of soil fertility, which is mainly focused on crop nutrition and productivity. Although there is no clear consensus in regard to the definition of soil health, the analogy between soil health, human health, and environmental health is evident. The health concept stresses qualities that encompass multidimensional health perspectives that go beyond mere chemical, physical, and biological properties. From a broader perspective, soil health depends on healthy and mutually supportive relations between soil-individuals and soil-cultures to attain global food security and environmental health. Perspectives on soil quality and/or health are discussed in the Chap. 3 on Monotheism, Chap. 13 on China, and Chaps. 19 and 20 on soil care. Soil care shares some broad similarities in particular with soil health, with a strong focus on showing empathic concern for the needs of the soil. In various cultural settings, the commodification of soils and intensification of agriculture in industrialized contemporary culture in North America have replaced indigenous ways of caring and valuing soils with detrimental effects to soil and human health (Chap. 9). Soil health cannot be sustained into the future without an explicit soil ethical foundation that addresses moral issues related to how people use, and sometimes exploit and degrade, soils (Chap. 21).

Soil security is a newer concept that seeks to take advantage of the wide recognition of other security movements such as energy security, food security, and water security. Soil security explicitly brings cultural and social dimensions into the discussion (Chaps. 4, 21, and 22). Soil security and human security are closely interconnected. Humanity has become increasingly urbanized, raising concerns about the disconnect between people and soils, which extends to food production and logistics. Chapter 23 by Muggler et al. touched on the different relations of urban and rural cultural communities in Brazil that influence how people connect to soils and value them. A distant and anonymous relation to soil may pose a major barrier to achieving soil health, quality, or security. However, there have been many urban movements globally in recent decades, such as urban gardening or trying to achieve circular economies, that help reconnect urban people to their soils. Some of these are discussed in Chap. 14.

26.7 Tensions and Commonalities Among Cultural Perspectives Regarding Soil

26.7.1 Inter- and Intracultural Contrasts

The reason we speak of different cultures and cultural areas is that there are considerable differences in concepts, ideas, and values and how they are expressed and become alive for people. These are figuratively like the genotype (concepts, ideas, and values) and phenotype (expressions) of different realms, in other words, the basic patterns of ideas and values, as well as their visible characteristics. The latter is shown, for example, in the structural visual elements of a landscape that is cultivated and transformed by humans. In addition, intra-cultural differences are also particularly interesting for any approach to the soil. One can speak of subcultures or even cultural wars in regard to the valuation of soils and nature within a cultural area. For example, in North American indigenous creation myths, connections to nature versus contemporary mainstream American culture provide a stark contrast about views of land and soils (Chap. 9). The cultural ruptures between traditionally religious and social views of land (earth-centered Shamanistic culture mixed with Buddhism), the communist paradigm postulating communal land ownership, and capitalist ideas regarding private land ownership as part of socioeconomic transformation in southern Siberia (Russia) are discussed in Chap. 16. In West Bengal (India), colonial foreign governance was also reflected in the country's agrarian structure and its values, then succeeded by the values and structures of the globalized economy, before experiencing powerful religious and socially inspired reform movements based on the country's own cultural traditions and values (Chap. 15). Many of the case studies showed that intra-cultural conflicts often resulted in profound cultural change that also shifted the place of soils in people's worldviews.

Cultural conflicts and changes are sometimes amplified by tension between urban and rural communities. Are soils perceived as part of the everyday lived environment or separate from urban living spaces (Chap. 14 on urban soils and Chap. 23 on the rural-urban split in soil perception)? Tension arises when people's intimate deep relations and felt proximity with soil collide with an indifferent and distancing view toward soil. In such cases, the valuation of soils undergoes a cultural shift. In Chap. 20, Grunwald discusses how a distancing from soil in modernism is moving toward empathic concern for soil in metamodernism.

Social fragmentation, cultural dissociation, and marginalization of cultural understandings of relations to nature and the role of humans in it are important issues to be explored. All of these were or are the case with organic farming or agroecological movements in Europe (Chap. 4) and in some traditional aspects of Chinese (Chap. 13) and Japanese culture (Chap. 11). The shifts or conflicts between traditional spiritual/religious cultural views and modern secularized culture have profoundly impacted agricultural and soil management. Sooner or later, such tensions could lead to movements of cultural renewal considering cultural knowledge from a community's own history or enculturated from elsewhere. There is great potential for cultural change, as many of our chapters show! Skillfully navigating adaptations in soil-cultural relations invites dialogue on the soil ethics and morals that undergird relations to soils (Chap. 21). Soil care ethics that aim to replace the exploitative usage of soils that have caused soil degradation hold potential for cultural transformation (Chaps. 19 and 20).

26.7.2 *Historic Continuity and Recurrence of Cultural Patterns*

History repeats its basic human patterns. This is not only most impressively shown by ever-recurring wars, but it may also be important for the history of ideas. Agricultural reform movements, such as ecological farming or agroecology, have shown ancient cultural patterns in relation to the use of land, as do urban gardening movements and community agriculture of various kinds in Europe and other continents (Chaps. 4, 14, and 15).

Cultural movements may pendulate across the spectrum from "conservative" to "progressive" attitudes. This leads, for example, to cultural revival or fundamentalist movements on one side, or to ways of liberalism and progressivism that include many sociocultural and technical changes or even cultural ruptures. Disagreements, polarizations, and conflicts may intensify into culture wars that give birth to the dissociation of nations, even civil wars and secessions, as well as new cultural views and memes that change people's value of soils and nature.

There is also a notable tension between traditional/conservative/orthodox cultural views and contemporary globalization trends (Chap. 20). Whether cultural heritage or traditions are interpreted as cultural baggage and an arrest in cultural traditional views, an anchor or inspiration for sociocultural change influences how people relate to soils.

However, given the many conflicts on sociopolitical, science and technology, and other cultural issues, it is important not to lose sight of how many ideological cross-overs of the many transitions and mixtures of worldviews take place between and within cultural areas, as shown in almost all the book's chapters. There are examples of the acculturation and inculturation of new religions and worldviews from the beginning of known human history until today. The Persian poet Mowlana Rumi (Chap. 7), for example, may have integrated elements of pre-Islamic culture into Islamic culture; centuries later, Rumi and other Persian mystic poets like Haafez and Saadi were able to influence European poetry and philosophy (Goethe's West-Eastern Divan of 1819 is just one example). Today's globalization with the multiplied possibilities of intercultural personal travel and migration as well as the accessibility of foreign cultural material via the Internet mean that large parts of the world's population (if they are not informationally locked in by their government or other circumstances) can now also be inspired by other cultures in terms of nature and soil.

Cultural views may also be informed by the soul, the spirit, the psyche. These mysterious and mainly unconscious dimensions, too, have their structures and impart to us psycho-spiritual worlds of experience, touching and stimulating our "inner soil" (Chap. 22).

26.7.3 Social Construction Versus Revelation and Exclusion Versus Inclusion of Cultural Views

Seen from the lens of the perspective of myths and other stories that are of fundamental relevance for people, a worldview may be accepted to be true in its statement about how things really are. This truth may be revealed to humans by vision, prophecy, or other divine sources (see how land and soil is addressed in holy scriptures, in the Chaps. 2, 3, 4 and 6). On the contrary, religious systems can also be interpreted as being the product of social constructive processes. For example, environmental ethics are quite obviously social constructions, as is the concept of ownership. In contrast, the mythologeme of the "promised land" or of "the land of our ancestors" claims to be a divine allotment or at least a spiritual foundation of property. Thus, in interpreting mythologems (meaning units of basic narrative) and religious worldviews regarding their relation to land and soil, it makes a big difference whether they are seen as a revelation or a social construction. What matters is that these diverse worldviews become ingrained in cultures and inform how we live life and relate to soil and nature. Religious organizations like the Catholic Church or an Islamic Fikh council, or individual religious leaders may tend to accept only those natural and spiritual phenomena and human opinions to be "true" that are or can be brought into alignment with their constitutive beliefs and traditions, refusing others as heresy or superstition. The religious authority then determines what scientific ideas can serve as arguments for their own purpose.

Similarly, scientific institutions, like academies of science or advisory boards, or single scientific leaders may tend to accept only those phenomena and opinions to be "true" that are or can be brought into alignment with their scientific paradigm, theories, and methods, viewing other ideas as being false or at least nonscientific. Here, the scientific or other secular authority may use appropriate statements from religion that can serve as an ethical underpinning for their own values, beliefs, and identity. In modernity, specific cultural views formed exclusive identities that compete for power claiming to be the best and true. However, postmodern thinking has opened toward a more pluralistic view of soils and cultures in which different religious, spiritual, social, and ecological views coexist and are valued (Chap. 20).

The chapters in this book convey multiple cultural perspectives that have co-created a variety of understanding of soil-cultural relations. Such pluralistic postmodern embrace acknowledges that there is not a single religious, scientific, or other cultural view that is better than all others honoring cultural diversity and fullness of people's experiences. But cultural patterns that have tended to degrade soil in the ways they were realized face major cultural and practical challenges to co-create sustainable futures that value both people and soils. Soil degradation drivers are multiple: They can be overfertilization or underfertilizing, driven by industry or overpopulation, etc.

26.7.4 Intercultural Congruities

Chapter 25 succinctly shows that there are commonalities among cultures and across times regarding how humans relate to soils. For example, the biblical quote "for dust you are and to dust you will return" occurs in a slightly different formulation as the Nigerian proverb "From dust we came and to dust we shall return" or the Indian proverb "All were borne from soil and when dead will be part of soil". Images of humans being made from earth, soil, or emerging from underground are important mythologems as discussed above. Thus, interreligious symbols and ethics become recognizable, being connected to common patterns of imagination or mystic experience. This leaves the question open of where these religious or spiritual experiences and ideas and their effective powers came from. What do you believe to be the cause forming one's relation to soils: does faith move mountains or who is believed in? Does spirituality take us away from earthly reality or bring us into it? Do ethics make us act differently or are they just a veiling mask of other motivations?

26.7.5 What Can Be Learned from a Deeper Cultural Understanding of Soils?

Earth sustains all of humanity. But does the full human being also engage in a relationship with the earth? Do people view nature as a self-regulating system, or as a fulfilling creature and instrument of a God from heaven (as seen in the Abrahamic religions, and several other creation myths)? For humans whose worldview is shaped by their science or religion and for people who struggle with both orientations and their specific perspectives, basic questions arise. For example, the question whether there is a spirit believed to dwell in or to be nonexistent in the depths of the earth makes a big difference in an individual or whole culture's worldview. Just celebrating the diversity and inclusivity of cultural views would be too simple. The plurality of these cultural views need to be interpreted from the perspective of their outcomes and how they promote sustainability of soils and human life. Different cultural patterns can lead to different outcomes, including the preservation of naturally vital and fertile soils or their degradation and destruction. There may be inherently harmful approaches such as striving for nothing but short-term (financial) yield. But it is also important to note the ambivalent possibilities within a cultural pattern of relating to nature. When, for example, does a Christian stewardship mandate toward the soil tend to lead to a harmful usurpation of dominion, and when to sustainable care? Or when do symbolically charged soil cultivation techniques such as farming in small traditional squares in arable soil in Dogon society as well as tilling in cross-directions in India lead to protection against soil erosion, and when do they accelerate it? These questions invite us to recognize and assess cultural and pedological aspects of actions not separately, but simultaneously.

Cultural patterns may be ambiguous and are often unconsciously followed. They can have positive effects or pose great risks to soil sustainability and ultimately human health. Hence, becoming consciously aware of cultural patterns and how they inform our relationships to soil and nature provide more choices to make decisions. Options emerge to live more harmoniously with soil and nature that might otherwise not be recognized, ideally leading to more sustainable future life on Earth. Or, this must unfortunately also be said, people can use apparently positive cultural ideas or patterns in very manipulative and possibly destructive ways.

Striving for long-term unity in diversity among culture-soil relations, such as Western science recognizing the utility of indigenous soil knowledge, opens up possibilities that might help preserve both soils and human health and can improve intracultural and intercultural communication. Some cultural views have been pernicious and sometimes vicious, leading to extensive soil degradation because of their ambivalent or even paradoxical nature. For example, the introduction of free market principles in Siberia were in conflict with traditional Buryat beliefs and led to land fragmentation and related problems (Chap. 16). It is important to note that how people advocate for and use or manage soils depends on their worldviews. Thus, cultural conflicts and paradoxical views in regard to soil management may be resolved through negotiation, dialogue, learning from each other, education, and development of shared goals for common goods like soil, water, and climate. Even if cultural relations with soils are rooted in opposing psycho-spiritual or science-based motivations, as long as shared ideals, for example, justice, care, or sustainability for soils, can be identified, they can be pursued and achieved (Chap. 20).

In taking cultural understanding of soils seriously, a reader of this book may be tempted to use values or cultural ideas to enforce something that contradicts them, to increase the acceptance of one's own goals. Instead, ethical coherence of one's own behavior would be required to minimize cognitive dissonance between different areas of life. Awareness regarding the personal and collective power drive that is so ubiquitous and potentially harmful is very important.

The inner soil can inform land use, making various mythical ideas still very effective today, regardless of whether they are recognizably religious or have another, even secular, ideological appearance: For example, the myth of Prometheus (which reinforces the technical mastery of nature) and the myth of Demeter (which favors a relationship-oriented approach with religious aspects toward soils/earth). Both archetypal patterns may be unconsciously present in every human being (Chaps. 2 and 3). It may be critically important to develop some relationship with nature out of the dead ends of materialism and rationalism on the one hand and romanticism or religious ideologies or diverse one-sided fundamentalisms on the other. In this way, an ever-changing interplay of inner and outer images of the soil and its web of interrelatedness and meaning may emerge among us, borne by respect *and* compassionate care for nature, taking into account scientific knowledge, and being consciously in touch with the personal sense of meaning. The emerging metamodern sensibilities combining care, emotions, and knowing provide hope for balanced soil-cultural relations in the foreseeable future. The cultural ideas and structures that we described here are potent (although polyvalent) in their possibilities to provide orientation and the psychic energy to follow it. If we do not want to risk being naively biased and one-sided in how we interact with soil, this cannot be neglected in the present and future.

26.8 What Needs More Attention and Action?

26.8.1 Religion and Spirituality

Our goal with this book has been to provide a broad range of perspectives on links between culture and soil. We sought to juxtapose similarities and dissimilarities among cultural perceptions of soils. The chapters of the book speak to an interdisciplinary and transdisciplinary audience, and in approaching such people, the editors asked an interdisciplinary and transdisciplinary group of authors to work on this project. There were a large number of non-(or not only) soil scientists involved in this project. Two editors and four other authors have training as psychologists (Laurence Feller Girod, Sabine Grunwald, Nikola Patzel, Nilton Sousa da Silva, Sylvia Xiaorui Wen, Eva Wertenschlag-Birkhäuser). Five authors are an anthropologist or another type of social scientist (Milinda Banerjee, Jeff Homburg, Tatiana

Intigrinova, Anna Krzywoszynska, Germain Meulemans). In sum, 70% of our chapters were authored or co-authored by people with backgrounds that extend beyond soil science or come from other disciplines. However, we are missing some academic disciplines that have their own, different perspectives on soil and culture. Colleagues from those disciplines may feel inspired by this work to complement it with their own experiences and perspectives.

Like all such projects, this one is incomplete. We were not able to find anyone to provide a perspective from Oceania, and representation from Africa and South America was sparse considering the diversity of cultures across these continents. Some perspectives (Asian Shamanistic, Brazilian Yoruba, Christian, Dogon, Native American) were provided by authors who are not practitioners of those groups; there is a need for presentation of such viewpoints from people who are part of these cultural groups. Others (Buddhist, Hindu, Islamic, and Shintoist) were provided by authors who are practitioners of those religions, which provides both strengths and weaknesses regarding the perspectives they provided.

26.8.2 Cultural Dimensions of Soil Communication

The hard fact is that humans have degraded and destroyed life and soil worldwide and at an unprecedented rate. No religious orientations, practiced ethics, or scientific knowledge and approaches have been effective in large-scale preservation of soils in all their splendor and natural fertility. How do we need to communicate to face these problems and involve decision-makers and all stakeholders? How do we leverage professional societies, including IUSS, to help engage stakeholders? How do we engage the broader public? Accounting for cultural perspectives does not mean that scientifically obtained data are only one opinion among many. We need to identify the fundamental facts underlying our discussions (whether or not climate change or soil degradation are happening are scientific questions, not opinions); how we address them are political questions. At the same time, it is important to acknowledge that nonscientific wisdom and culturally oriented knowledge matter. Doing so is advantageous to finding acceptable solutions to major local or global soil issues.

The current main scientific paradigm, scientific peer groups, and individual scientists have also played a part in causing soil degradation. For example, the extensive use of monoculture cropping, pesticides, and overfertilization have contributed to many environmental problems. At the same time, science offers many possible solutions if recognized and adopted. Therefore, we conclude that explicit integration of cultural themes into soil use and management is needed, for example, through citizen science and participatory action research and beyond.

There lies incredible strength in exploitation to serve the myth of growth without decay. But decay comes after sprouting, growing, flourishing, and fruiting. The soils know why and how, bearing the memory of thousands and thousands of years of life's expansion to more complex forms, and a self-regulating tendency (ecosystem resilience etc.). Religions and spiritual webs of meaning know how and why. We may listen to soil again; we may listen to the voices and visions from within and maybe even talk about it. This has been largely lost in many, though not all, societies. We may acknowledge the patterns of soil formation since life came from water and entered land some 400 million years ago, and we may acknowledge the patterns of "knowing imagination" that formed in the human psycho-spiritual realm and may be rooted in all our phylogenetic ancestor's minds.

Questions about why communication has not elicited changes can be seen in view of the painful "behavioral gap": the incongruence between talking and doing, between knowing and acting, between believing and wanting, and so on. This is particularly evident with soil, where so many wonderful values and insights come from cultural areas and social communities whose loss of soil is striking. But not every failure is the fault of the one who fails. Not every religious, spiritual, and ethical orientation belongs on the dung heap of history just because its soil protection effect is not currently recognized or present. In addition to questions about the inadequacies of these religious, spiritual, and ethical orientations to protect soil, one can also ask what are the forces that lead people to degrade their own livelihoods? A critical look at the dark fiery side of natural science and its technologies as well as social and especially power relations and asymmetries in present economies that are socially constructed could be considered.

How do we overcome the behavioral gap between cultural orientation toward sustainability, and soil degradation and destruction? Soil communication may benefit in many regards from being in touch with metaphorical and symbolic dimensions of soil relations and management. Are soil communicators themselves rooted in or detached from their subject? We support the newly emerging trend in science to use metaphors for soil and earth in its concepts and communication (see Janzen et al., 2021). We do not ask the question of truth in this book. Of course, authors do not just report on the opinions of others, but also shine light on the assumptions or "verities" they personally believe in, and on some current socially accepted trends. We urgently invite more people to engage with those issues.

While global organizations such as the UN, FAO, and IUSS provide important knowledge and advocacy on soil sustainability, there are limitations to their effectiveness and accountability at national and regional levels. Talking about soil should not only be a one-sided affair, for example, hierarchically "superior" scientific institutions or authorities to "lower" (closer to the soil) practitioners. Continuing in this way only can sometimes be ineffective at reaching intended positive aims or may even promulgate soil degradation and harm to nature and people. Soil communication needs to be an integrated effort that includes bidirectional and multifaceted perspectives. When this occurs, more happens to emotionally, cognitively, and effectively sustain communities and soil.

What if soil education included more intercultural competence, more empathy to and care for soils and humans? It is important that the IUSS and various other communities of soil related scientists take the cultural dimension of soil seriously. The chapters of this book may open new pathways for a future-oriented reconnection to what is more than "mere matter" when dealing with nature and particularly soils and earth.

There are additional contemporary issues regarding soil and culture that could benefit from being addressed, or more completely addressed, in future research. These include the cultural patterns linked to artificial intelligence and agro-technologies and their impact on soil management, including the chasm that pedometrics created between virtual simulation of soils and natural soils. Another example is that soil is seen and modeled as part of the Earth's climate system. This can lead to extremely narrow functionalist views that reduce the modelers' view of soil to a very simplistic representation that also does not account for the cultural dimension. Here, too, the cultural dimension taken into account may partly determine which paths are taken, including whether we create sustainable futures. We want to emphasize that the broadness of feminine spiritualities and spiritual femininities needs much more attention.

26.9 Concluding Statements

It is not possible to largely understand soils from the perspective of other cultures if we do not acknowledge and maybe even try to connect to the web of meaning created with their myths, spiritual beliefs, rituals, and symbolic actions and representations, and core values. An important difference for human soil perception, understanding, and relation is whether the world (earth/land/soil) is considered to be spiritually animated, that is, full of soul, spirit, divinities, God, or not. One may say the question is whether a religious or spiritual worldview is dominant versus a rather secular (rationalist) scientific or materialist worldview. Approaches as different as those of spirituality and natural science, even if one accepts both in principle, often lead to thinking, feeling, and acting in compartments and lacking integration, while others embrace diverse perspectives.

This book invites readers to follow their personal quest to defragment knowledge systems and explore different dimensions of human life. The chapter authors strove to integrate different types of understanding while seeking inter- and transdisciplinarity and to stimulate all of us to become a bit more "whole and full" in our process of individuation through life. It is also about awareness and mindfulness in soil science and overall nature relations and its use. It is about respecting and maybe accepting apparently incompatible viewpoints on soil and people. It is about the awareness that everybody partakes in the history of ideas, their origin and social context, with all the ambiguities, perils and opportunities these ideas provide.

References

Dieterich, A. (1925). *(Mother earth. An essay on folk religion, 3rd expanded ed.)* Mutter Erde. Ein Versuch über Volksreligion. Dritte, erweiterte Auflage, besorgt von Eugen Fehrle. Leipzig: Teubner.

Esbjörn-Hargens, S., & Zimmerman, M. (2009). *Integral ecology: Uniting multiple perspectives on the natural world*. Boston: Shambhala Publications.

Frazer, J. (1951). *The golden bough. A study in magic and religion* (3rd ed., 8 parts in 12 vols + 1 suppl). (There exist many editions of that opus that are different in size, the first edition appeared in 1890.)

Goethe, J. W. von (1819). West-Östlicher Divan. Stuttgart: Cotta. (Engl. transl. By E. Dowden 1914: West-eastern divan, London: Dent).

Janzen, H. H., Janzen, D. W., & Gregorich, E. G. (2021). The “soil health” metaphor: Illuminating or illusory? *Soil Biology and Biochemistry, 159*, 108167.

Leeming, D. A. (2010). *Creation myths of the world: An encyclopedia* (2nd ed.). Santa Barbara: ABC-CLIO.

Long, C. (1963). *Alpha: The myths of creation*. Chico: Scholar’s Press.

Mannhardt, W. (1875). (*Forest and field cults. Vol. 1: The tree cult of the Germanic people and their neighbouring tribes. Mythological studies.*) Wald- und Feldkulte. Erster Band: Der Baumkultus der Germanen und ihrer Nachbarstämme. Mythologische Untersuchungen. Berlin: Borntraeger. Esp. pp. 480–488.

Index

N. Patzel et al. (eds.), *Cultural Understanding of Soils*,
https://doi.org/10.1007/978-3-031-13169-1

P

Q

R

MIX
Papier aus verantwortungsvollen Quellen
Paper from responsible sources
FSC® C105338

If you have any concerns about our products,
you can contact us on
ProductSafety@springernature.com

In case Publisher is established outside the EU,
the EU authorized representative is:
Springer Nature Customer Service Center GmbH
Europaplatz 3, 69115 Heidelberg, Germany

Printed by Libri Plureos GmbH
in Hamburg, Germany